Microbial Growth in Drinking-Water Supplies

Microbial Growth in Drinking-Water Supplies

Problems, Causes, Control and Research Needs

Edited by

Dirk van der Kooij and
Paul W. J. J. van der Wielen

IWA Publishing
London • New York

Published by **IWA Publishing**
Alliance House
12 Caxton Street
London SW1H 0QS, UK
Telephone: +44 (0)20 7654 5500
Fax: +44 (0)20 7654 5555
Email: publications@iwap.co.uk
Web: www.iwapublishing.com

First published 2014

British Library Cataloguing in Publication Data
A CIP catalogue record for this book is available from the British Library

ISBN: 9781780400402 (Paperback)
ISBN: 9781780400419 (eBook)

Cover images:

Row 1, picture 1: A slime-producing bacterium isolated from drinking water.
Row 1, picture 2: *Legionella pneumophila* in *Hartmannella vermiformis* (*source*: Kuiper M. W., Wullings B. A., Akkermans A. D. L., Beumer R. and van der Kooij D. (2004) Intracellular proliferation of *Legionella pneumophila* in *Hartmannella vermiformis* in aquatic biofilms grown on plasticized polyvinyl chloride. *Appl. Environ. Microbiol.,* **70**(11), 6826–6833.
Row 1, picture 3: Colonies of *L. pneumophila* grown on Buffered Charcoal Yeast Extract agar.
Row 2, picture 1: *Asellus aquaticus* originating from a drinking water distribution system (courtesy: J. Hein M. van Lieverloo).
Row 2, picture 2: Bacteria on granular activated carbon collected from a water treatment pilot plant.
Row 2, picture 3: 'Tap water, to your health!' (courtesy: J. Hein M. van Lieverloo).

Contents

Authors and co-authors

Dr. Anne Camper is Professor of Civil Engineering, Adjunct Professor of Microbiology and Associate Dean for Research and Graduate Studies, College of Engineering at Montana State University. She has been involved with the Center for Biofilm Engineering since its inception. Her B.S. and M.S. degrees are in environmental microbiology, and her Ph.D. is in civil/environmental engineering. Research interests are in biofilm growth and control in drinking and industrial water systems; the fate, transport and survival of pathogens in biofilm systems; and the physiology of biofilm bacteria. She has been the principal investigator or co-principal investigator on over 50 grants related to drinking-water treatment or distribution. She also directs community based participatory research funded by the National Institute of Health with the Crow Tribe in Montana and as the principal investigator on projects supporting Native American graduate students engaged in health related degree programmes.

Dr. Hans-Curt Flemming is Professor at the University of Duisburg-Essen and one of the Scientific Directors of the IWW Water Centre, Mülheim/Ruhr. After studying chemistry, he received his Ph.D. at the Max-Planck Institute for Immunology in Freiburg. From 1978 to 1994 he worked on biofouling of ion exchangers and reverse osmosis membranes at the Institute for Civil Engineering of the University of Stuttgart. There, he founded a research group on biofouling and biofilms. In 1993, he achieved his Habilitation degree in Biofilms at the University of Stuttgart. From 1994 to 1996 he established the Department of Biotechnology at the Institute for Civil Engineering at the Technical University Munich. Since 1996 he has held the Chair for Aquatic Microbiology, University of Duisburg-Essen and joined the board of directors at IWW. In 2001, he founded the Biofilm Centre and has been its Managing Director ever since. Dr. Flemming has (co)authored more than 250 publications in refereed journals and 10 books.

Dr. Beate Hambsch has a Masters degree in Food Technology at the University of Munich and received her Ph.D. in Chemical Engineering at the University of Karlsruhe in 1992. She joined the DVGW-Technologiezentrum Wasser Karlsruhe (TZW) (Centre of applied research of the German Gas and Waterworks Association) in 1985 and since 1995 she has been head of the Microbiology Department. Besides joint research projects with utilities in the field of drinking-water microbiology and bacteriological monitoring programmes she is or was responsible for various national research projects especially in the field of regrowth, coliform bacteria and other hygienically relevant bacteria (e.g. *Pseudomonas aeruginosa*) as well as in material research and testing. At TZW she was also leading the EU-funded projects on materials in contact with drinking water regarding the enhancement of microbial growth.

Dr. Peter Huck is Professor and NSERC (Natural Sciences & Engineering Research Council of Canada) Chairholder in Water Treatment in the Department of Civil & Environmental Engineering at the University of Waterloo. Dr. Huck has over 30 years' experience in drinking-water treatment and over 130 refereed publications.

Dr. Mark LeChevallier is the Director of Innovation & Environmental Stewardship for American Water; a water utility operating in more than 30 states and Canada; serving over 15 million people. He received his Bachelor of Science and Masters degrees in Microbiology from Oregon State University, and his Ph.D. in Microbiology from Montana State University. Dr. LeChevallier has authored over 100 research papers and has received awards from the American Water Works Association for outstanding contributions to the science of water treatment. He was the recipient of the George Warren Fuller award and the Able Wolman Award from the American Water Works Association. He is a fellow of the American Academy of Microbiology.

Hein van Lieverloo is an independent scientist and consultant at Viaeterna, Rosmalen, the Netherlands. He received his Masters degree in Biology at the University of Nijmegen in 1988 and joined the Microbiology group at Kiwa Water Research in 1989. He was in charge of the invertebrate research project for 10 years. Since 1995, he was also involved in research on the microbial contamination of distributed drinking water. Quantitative microbial risk assessment became the focus of his work in 1999. In 2006, he switched to NIZO Food Research (Ede, the Netherlands) where he developed the quantitative microbial risk assessment approach for the Netherlands dairy industry. Since 2008 he has been an independent scientist and consultant at Viaeterna specialising in risk assessment for food and drinking-water companies.

Prof. Dr. Gertjan Medema is chief science officer and principal microbiologist at KWR Watercycle Research Institute. He is chair on Water & Health at Delft University of Technology. He has a Masters degree in Microbiology from Leiden University and a Ph.D. degree on *Cryptosporidium* and *Giardia* in water supply from Utrecht University. The focus of his research in the past 25 years (70 publications) is the transmission of pathogenic micro-organisms via the environment and the effectiveness of barriers against pathogens. The objective is to provide a scientific basis for effective management of pathogen risks by the water sector. He coordinates the Joint Research Programme of the Dutch water utilities (Bedrijfstakonderzoek [BTO]) and the research strategy at KWR.

Dr Michèle Prévost is professor at the NSERC Industrial Chair on Drinking Water, Ėcole Polytechnique Montréal. She became involved with biostability issues in 1987 while conducting applied research on biological filtration for her doctoral work. She introduced several biodegradable organic carbon methods and biological filtration to Canada. She then joined and coordinated the Biodegradable Organic Matter Work Group to foster technical exchanges concerning biodegradable organic matter measurement methods. The group held a series of scientific meetings in Nancy (France), Nieuwegein (The Netherlands), Karlsruhe (Germany), Barcelona (Spain) and Maison-Laffitte (France), as well as several international conferences. This involvement resulted in the publication of an American Water Works Association (AWWA) book on Biodegradable Organic Matter.

Biostability is an excellent example of an international collaboration that led to the development of long-term scientific collaborations, fruitful exchanges and long-lasting friendships. New challenges in biostability have emerged with the control of opportunistic pathogens. It is no surprise as Louis Pasteur once said: '*Gentlemen, it is the microbes which will have the last word*'.

Dr. Dirk (Dick) van der Kooij was Principal Microbiologist at KWR Watercycle and Research Institute, Nieuwegein, the Netherlands, and part-time professor in Environmental Microbiology at the University of Wageningen. He retired in July 2011. He received his Masters degree in Environmental Engineering and Waste Water Treatment at the Agriculture University of Wageningen (presently Wageningen University) in 1972 and in 1984 his Ph.D. in Microbiology at the same university. In 1972, he was the first microbiologist at Kiwa and established a research group on Microbiology, mainly involved with the Joint Research Programme (BTO) of the Water Supply Companies in the Netherlands. He is a recipient of the Van Marle Award of the Royal Society of Water Supply Engineers. He has (co)authored over 75 peer-reviewed publications and about 50 publications in Dutch professional journals.

Dr. Paul van der Wielen is Principal Microbiologist at KWR Watercycle Research Institute, Nieuwegein, the Netherlands, where he is head of the 'Biological Activity' group. In 1996 he received his Masters degree in Microbial Ecology at the University of Groningen and in 2002 his Ph.D. degree at the Utrecht University. Before starting his work at KWR Watercycle Research Institute in 2004, he worked for 2.5 years on the microbial ecology of hypersaline anoxic basins in the Mediterranean Sea, which resulted in publications in Nature and Science. He is a member of the editorial board of Applied and Environmental Microbiology. His research interest is in the microbial ecology of the drinking-water ecosystem, in particular growth of opportunistic pathogens in drinking-water supplies, reducing biological activity in drinking water and improving biological active filters in drinking-water treatment.

Dr. Jan Vreeburg is Principal Researcher in the field of drinking-water distribution and infrastructure at KWR Watercycle and Research Institute, Nieuwegein, the Netherlands. He received his Masters degree in Civil Engineering at the University of Technology in Delft (TUD). His expertise covers all aspects of the transportation and distribution of drinking water, with emphasis on water quality. In 2007 he received his Ph. D. degree in drinking-water discolouration at TUD, where he held a position of part-time assistant professor till October 2012. From November 2012 onwards he has been part-time associate professor at Wageningen University within the field of Management of Urban Infrastructures. He also is a consultant for water companies making ties between practice and research. He has participated in several EU-funded projects including TECHNEAU and PREPARED and is (co) author of many research papers.

Dr. Yoshimasa Watanabe is Professor Emeritus and Distinguished Professor at Hokkaido University, Sapporo, Japan. He received his Bachelors degree in Sanitary Engineering from Hokkaido University in 1967 and his Masters degree in 1969. In 1972, he received his Ph.D. in Sanitary Engineering from Hokkaido University, and was employed as Associate Professor at Miyazaki University, Japan, being promoted to Professor in 1988. In 1975–1976 he was a Visiting Associate at the California Institute of Technology, US. He joined Hokkaido University in 1993 as Professor of Sanitary Engineering. His research interests are in hybrid water and wastewater treatment technology using membrane, phosphorous recovery from sludge and urban water metabolism systems. He received the IWA Award 2008 for Outstanding Contribution to Water Management and Science, given in honour of Dr Karl Imhoff and Dr Pierre Koch.

Co-authors

John Ashworth, Water Regulations Advisory Scheme, 30 Fern Close, Pen-y-Fan Industrial Estate, Oakdale, Gwent, UK.

Bernd Bendinger Ph.D., DVGW Forschungsstelle TUHH, Hamburg University of Technology, Germany.

Marie Claude Besner Ph.D., Polytechnique de Montréal, Montréal and City of Laval, Canada.

E.J. Mirjam Blokker Ph.D., KWR Watercycle Research Institute, Nieuwegein, the Netherlands.

Prof. Martin Exner Ph.D., Institute for Hygiene and Public Health, University of Bonn, Germany.

Jürgen Gebel Ph.D., Institute for Hygiene and Public Health, University of Bonn, Germany.

Leo Heijnen, Ing., KWR Watercycle Research Institute, Nieuwegein, the Netherlands.

Wim Hoogenboezem Ph.D., Het Waterlaboratorium HWL, Haarlem, the Netherlands.

Ronald Italiaander, KWR Watercycle Research Institute, Nieuwegein, the Netherlands.

Prof. Thomas Kistemann Ph.D., Institute for Hygiene and Public Health, University of Bonn, Germany.

Patrick Laurent Ph.D., UNICEF, Somalia Support Centre (USSC), Nairobi, Kenya.

Gabriela Schaule Ph.D., IWW Water Centre, Mülheim an der Ruhr, Germany

Prof. Pierre Servais Ph.D., Université Libre de Bruxelles, Belgium.

Barbara K. Siembida-Lösch, Ph.D., Postdoctoral Fellow, Department of Civil and Environmental Engineering, University of Waterloo, Waterloo, Ontario, Canada.

Patrick W.M.H. Smeets Ph.D., KWR Watercycle Research Institute, Nieuwegein, the Netherlands.

Prof. Marek M. Sozański, Ph.D., Institute of Environmental Engineering, Poznań University of Technology, Poznań, Poland.

Prof. Ulrich Szewzyk Ph.D., TU Berlin, Department of Microbial Ecology, Germany.

Gerrit Veenendaal M.Sc., Water Supply Company Drenthe NV, Assen, the Netherlands.

Harm R. Veenendaal, KWR Watercycle Research Institute, Nieuwegein, the Netherlands.

Jost Wingender Ph.D., Biofilm Centre, Faculty of Chemistry, University of Duisburg-Essen, Germany.

Acknowledgements

We wish to express our appreciation to all those who have contributed to this book, in particular the authors of the various chapters. These chapters are based on the presentations given at the International BTO-Conference 'Microbial Growth in Drinking Water Distribution Systems and Tap Water Installations' on 11 and 12 May 2011 at Santpoort in the Netherlands. Preparing a presentation for a conference is time consuming, but writing a book chapter on the same issue requires much dedication and additional time to put all details down in black and white. Unfortunately, a few presenters could not contribute to this book due to a change in job position. *Legionella* is a highly important issue in relation to water supply and, therefore, has been described by one of us, based on a review of the literature. Furthermore, a General Introduction has been added to provide information about the historical context and the relevance of microbial growth in water supplies. All chapters have been reviewed by us and chapters written by (one of) us have been peer reviewed by others. However, the authors remain responsible for the contents of their chapter.

We hope that the presented information will stimulate further research aiming at resolving the complex issue of microbial growth in drinking-water supplies.

The editors
Dick van der Kooij
Paul W.J.J. van der Wielen

Foreword

On 11 and 12 May 2011, an international conference was held in the Netherlands, focusing on microbial growth in drinking-water distribution systems and potable-water installations. Invited speakers from Canada, European countries, Japan and the US, all with many years of experience in applied research and water supply, presented their latest observations and vision on this issue to about 100 participants, and discussed research needs with them.

The conference was organized as part of the Joint Research Programme (Bedrijfstakonderzoek [BTO]) of the water-supply companies in the Netherlands. The BTO Programme started in 1972, in a period with an increasing pollution of the freshwater environment and a rapidly growing drinking-water demand. Its principle is quite simple. The participating companies put money in the research programme and get knowledge in return, knowledge they can use because problem articulation and implementation of the results in their drinking water operations are the two main pillars of the BTO Programme. A continuous exchange of information between researchers and specialists of the water supply companies is facilitated by a framework of committees and working groups.

Over the last 40 years, the BTO Programme has proven to be very successful. The Dutch contributions to this conference would not have been possible without it. Within the BTO Programme, the concept of production and distribution of hygienically safe and biologically stable drinking water without the use of a persistent disinfectant was developed, based on fundamental research in which Dick van der Kooij played a major role.

The conference showed that sharing knowledge also pays at an international level. After sharing visions, knowledge and experiences during the first day of the conference, knowledge gaps and research needs for the future were identified on the second day. This book contains the results of this two-day conference.

The selected issues and the experience of the authors have resulted in a compilation of the current knowledge on microbial growth in drinking-water distribution systems and potable-water installations that goes beyond the level of conference proceedings. The presented information shows that microbial growth in drinking water has been a research issue for more than a century. The knowledge collected during this period resulted in an increasing awareness of the complexity of the microbial processes in water supply systems and the necessity to elucidate these processes to ensure water quality, water safety and customers' confidence.

We hope that the information provided in this book will be useful to all those who are directly involved in ensuring the quality of drinking water, and also to researchers, teachers and students, interested in water-supply microbiology.

Prof. Jan Peter van der Hoek Ph.D., MBA
Chair BTO Programme Committee

Trudy M.H. Suylen Ph.D.
Chair BTO Committee on Microbiology

Chapter 1

General introduction

D. van der Kooij and P. W. J. J. van der Wielen

1.1 WATER-SUPPLY MICROBIOLOGY

1.1.1 Discoveries and impact

The discovery of microorganisms (in water) by Antonie van Leeuwenhoek in 1676 and the publication by Louis Pasteur in 1861 on his observations showing that the so-called spontaneous generation in aqueous solutions was caused by microorganisms, are well-known milestones in microbiology (Brock 1975). In 1876 Robert Koch was the first to report that microorganisms can cause disease. A few years later, it became clear that the outbreaks of cholera, which had already been associated with the consumption of contaminated water (Snow 1855), were caused by a specific microorganism of faecal origin (Koch 1884). In 1886 also the organism responsible for typhoid fever was detected in contaminated water. Furthermore, injection of samples of untreated water and sediments into rodents led to the discovery of a number of other, mostly non-human, pathogens (Frankland and Frankland 1894). The observations on the presence of disease-causing microorganisms in the freshwater environment stimulated the establishment of water-supply companies in many countries, with the objective and the obligation to provide the consumers with drinking water, complying with water quality standards incorporated in national regulations.

Further progress in science and engineering in the past century has resulted in extended descriptions of pathogenic microorganisms and indicator bacteria, methods for their detection and identification, health-based targets and also water treatment processes (World Health Organization [WHO] 2004). Nevertheless, microbiological water quality is still a major issue in drinking-water supply and ensuring drinking-water safety demands continuing vigilance and improvement of methods and measures. The main reasons for concern are: (i) the large impact of microbiologically-contaminated drinking water on public health, (ii) the ongoing contamination of the environment with microorganisms of faecal origin, (iii) the increasing proportion of the population that is highly susceptible to infectious diseases, (iv) the discovery of not-yet identified pathogenic microorganisms, and last but not least (v) maintaining customers' confidence. Further scientific developments are needed to cope with these concerns for example, improvement of methods for identification and selective detection of microorganisms involved in disease and refinement of quantitative microbial risk assessment in relation to drinking-water safety.

Furthermore, microorganisms indigenous to the freshwater environment play an essential role in water treatment by removing biodegradable organic and inorganic compounds from the water. Hence, to achieve compliance with regulatory standards for microbiological, chemical and physical water quality characteristics, water supply companies are challenged to combine microbiological processes with physico-chemical processes ('barriers') in water treatment. Maintaining the achieved water quality during storage, transportation and distribution, and in potable-water installations, with or without the use of a disinfectant, is another challenge.

Overall, prevention of microbial contamination and control of microbial growth are required to provide the consumer with safe tap water of the desired quality.

1.1.2 A century of progress

In the early days of drinking-water supply, microscopy was used to investigate (micro)biological problems (de Vries 1890; see also Olson and Nagy 1984). In 1881 Koch developed a solid medium for the cultivation of bacteria and a few years later he introduced the plate count method for determining the number of cultivable heterotrophic bacteria in water (Koch 1883). Numerous applications of this method in freshwater environments and its use in water supplies showed that elevated heterotrophic plate counts (HPCs) were associated with a possible microbial contamination of the water (Frankland and Frankland 1894). Subsequently, a plate count of 100 CFU/ml was introduced as the first microbiological quality standard for treated water (Koch 1893; Payment *et al.* 2003). Consequently, increases of HPCs observed in samples of stored drinking water caused much confusion. These observations revealed that treated water has a potential for microbial growth, which was enhanced by the use of cotton-wool stoppers enabling access of air to the sample (Frankland and Frankland 1894).

By the end of the 19th century and at the beginning of the 20th century, many studies were conducted aiming at the characterization of bacteria contributing to the HPCs of drinking water on nutrient gelatine plates. These studies (e.g. Zimmerman 1890; Kayser 1900; Haenle 1903), which included injection of the isolated bacteria into animals, did not lead to the detection of pathogens.

However, exposure of experimentally-injured frogs to drinking water accidentally led to the discovery of the animal pathogen *Bacillus hydrophilus fuscus* (Sanarelli 1891). This organism was later renamed *Aeromonas hydrophila* (Schubert 1967). The developments related to the use of HPCs in drinking water quality monitoring in the 20th century have been described in detail elsewhere (Payment *et al.* 2003).

The detection of *Bacillus coli-communis* (later renamed *Escherichia coli*) by Escherich in 1885 as a predominant microorganism in human faeces was the start of the faecal indicator concept, primarily based on *E. coli* and thermotolerant coliforms (Eijkmann 1904). This development was a major step forward in detecting microbial (faecal) contamination of water. The observations of coliforms multiplying in distribution systems (e.g. Baylis 1930) initiated much debate about the significance of using coliforms in water supplies as an indicator of faecal contamination (Howard 1940). This debate continued throughout the 20th century in Europe and the US. Growth of coliforms in distribution systems in the US attracted much attention the 1980s and 1990s and it was shown that attachment onto water-exposed surfaces protected these organisms against the effects of disinfection (Herson *et al.* 1987; LeChevallier *et al.* 1988b). To date, total coliforms are maintained as indicator organisms in drinking-water regulations in the US (Environmental Protection Agency [EPA] 1999) and in Europe (European Union [EU] 1998), with stringent standards for *E. coli* as the real indicator of faecal contamination.

By the end of the 19th century, so-called iron bacteria, in particular *Crenothrix polyspora,* and certain invertebrates were microscopically identified as causes of considerable nuisance in water supplies (de Vries 1890). In the 20th century, iron-precipitating bacteria (Heymann 1928; Buswell 1938; Lueschow and Mackenthun 1962), invertebrates (Smalls and Greaves 1968), and also microorganisms involved in taste and odour problems (Sigworth 1957) in water supplies, continued to attract the attention of (micro)biologists, because of the negative impact on customers' satisfaction.

The increasing contamination of the aquatic environment with organic compounds of anthropogenic origin in the second half of the 20th century forced the water-supply companies in Europe and the US to extend surface-water treatment with additional processes, for example ozonation and granular activated carbon (GAC) filtration (Sontheimer *et al.* 1978). The discovery of organic halogens, including trihalomethanes (THM), as by-products of chlorination (Rook 1974) promoted further changes in water treatment, aiming at limiting THM formation, because of its possible role in cancer. In the Netherlands, a number of water-supply companies terminated post chlorination (Schellart 1986). The introduction of ozonation and GAC filtration yielded elevated HPCs in drinking water in a number of supplies and intensified the discussions about the significance of microbial growth in distribution systems and the need for control measures. Rittmann and Snoeyink (1984) introduced "biological stability" to characterize drinking water that "does not support the growth of microorganisms to a significant extent in distribution systems". Limiting microbial growth emerged as a challenge to microbiologists and engineers, and stimulated the development of methods to assess the microbial-growth potential of treated water (see Chapters 2, 10 and 11).

In the last quarter of the 20th century, an increasing variety of microorganisms with pathogenic properties were detected in water from distribution systems and/or premise plumbing systems, for example *Mycobacterium kansasii* (McSwiggan and Collins 1974), *Pseudomonas aeruginosa* (Hoadley 1977), *Legionella pneumophila* (Tobin *et al.* 1980) *Aeromonas* spp. (LeChevallier *et al.* 1980; Burke *et al.* 1984) and *Mycobacterium avium* (du Moulin and Stottmeier 1986). These observations strengthened the notion that regrowth poses a potential health threat, in particular to immunocompromised persons.

Outbreaks of disease ('diarrhoea') in the US and UK that were associated with drinking water and caused by the persistent pathogenic protozoa *Giardia lamblia* and *Cryptosporidium parvum,* raised concerns about the safety of chlorinated drinking water supplies that comply with the standards for coliforms (Rose 1990). These observations and the large outbreak of cryptosporidiosis in Milwaukee in 1993 had a large impact on the definition of microbiologically-safe drinking water. The Safe Drinking Water Act (1974) was the first step in a process of adapting regulations on drinking-water quality and water treatment in the US. Adaptations included the removal capacity for persistent protozoa and viruses, the indicator position of coliforms and the limitation of disinfection by-products concentrations (EPA 1999; Chapter 2). In European drinking-water quality regulations (EU 1998) *E. coli* and enterococci were included as the primary bacterial indicators for microbial safety in relation to faecal contamination. Monitoring of *Clostridium perfringens* is only obligatory for surface-water supplies and total coliforms are considered an operational indicator. In the Netherlands, also the evaluation of the removal capacity of surface-water treatment for persistent pathogenic protozoa and viruses of faecal origin, to show compliance with a maximum acceptable infection risk of 10^{-4} per person per year, was introduced in drinking-water quality regulations (Ministerie van Verkeer, Ruimtelijke Ordening and Milieubeheer [VROM] 2001; Ministerie van Infrastructuur en Milieu [IM] 2011). To date Water Safety Plans, sometimes in combination with Quantitative Microbial Risk Assessment (QMRA), are common practice in a number of countries (Chapter 5).

The development and introduction of polymerase chain reaction (PCR)-based methods by the end of the 20th century facilitated the detection and identification of (opportunistic) pathogens in source water, in treated water and in drinking-water distribution systems with a short time between sampling and analysis result (e.g. Bej *et al.* 1990). These methods can also be used for the elucidation of microbiological processes in water treatment and distribution (Chapter 11).

Overall, within about one century, water-supply microbiology has developed from using HPCs for monitoring of the microbiological drinking-water quality to the application of techniques for detection and identification of a wide range of microorganisms, the quantitative assessment of the microbial safety of drinking water, evaluation of water treatment, including microbiological processes, and extended knowledge about the microbiology of distributed drinking water. This book focuses on microbial growth and activity in distribution systems and potable-water installations: (i) the impact on water quality, (ii) the nature of these processes, (iii) the environmental conditions in these systems affecting microbial growth and activity, and (iv) the effects of water treatment and (v) distribution on the microbial-growth potential.

1.2 REGROWTH: PROBLEMS AND ASSESSMENT

Microbial growth in drinking-water distribution systems (regrowth) and in potable-water installations in houses and buildings can have a negative impact on water quality. The main regrowth problems as well as methods used for assessing the extent of regrowth are summarized below.

1.2.1 Problems

Heterotrophic plate counts (HPCs) exceeding the standard defined in drinking-water quality regulations following the introduction of ozonation and/or GAC filtration in water treatment, attracted much attention (Dietlicher *et al.* 1970; Stalder and Klosterkötter 1976; LeChevallier *et al.* 1980; Maki *et al.* 1986; Reasoner *et al.* 1989). Also, the potential health risk of elevated HPCs became an issue of much debate (Geldreich *et al.* 1972; Müller 1972; LeChevallier *et al.* 1980). Although bacteria with virulence factors (e.g. cytotoxicity, invasiveness, haemolytic activity) contributed to HPCs of drinking water, these bacteria were not considered a threat to human health (Payment *et al.* 1994; Edberg *et al.* 1997). Furthermore, HPCs do not correlate with the presence of coliforms in drinking water (Edberg *et al.* 1989) and the colony counts of aeromonads (Havelaar *et al.* 1990), nor with opportunistic pathogens, for example *Legionella, Mycobacterium avium* in potable-water installations (Liu *et al.* 1993; Habicht and Müller 1988; Aronson *et al.* 1999). The significance of HPCs in drinking-water quality management is described in detail in an International Water Association (IWA)-WHO publication (Bartram *et al.* 2003). The opinion that (bacteria contributing to) these plate counts do not have a direct public-health significance has become more generally accepted (Payment *et al.* 2003). Elevated HPCs in distributed drinking water are, however, considered as a signal for failure of maintaining a disinfectant residual, a possible contamination of the system, and fouling of the distribution system, for example sediment accumulation and/or biofilm formation.

Coliforms have been observed in many water supplies as a result of regrowth (Wieringa 1985; LeChevallier 1990; Smith *et al.* 1990; Chapter 2). Sediments accumulating in the mains can promote growth of these organisms (Baylis 1930). More recent studies revealed that multiplication of coliforms is associated with corroding cast-iron pipes and elevated temperatures (>15°C) (Camper *et al.* 1996; LeChevallier *et al.* 1996; Besner *et al.* 2002; Chapters 2 and 4). Certain natural or synthetic construction materials in contact with drinking water during storage and distribution can also enhance multiplication of coliforms, for example redwood in reservoirs (Seidler *et al.* 1977), coatings and rubber materials (Ellgas and Lee 1980; Colbourne 1985) and rubber-coated valves (Kilb *et al.* 2003) (Chapter 12). Concern remains about the public-health significance of multiresistant coliforms, including *Klebsiella,* in potable-water systems in hospitals (Talbot *et al.* 1980; Jarvis *et al.* 1985; Geldreich and Rice 1987; Ainassie *et al.* 2002; Verweij 2011).

Opportunistic pathogenic microorganisms, for example *L. pneumophila*, *P. aeruginosa* and certain nontuberculous mycobacteria (NTM), are a health concern associated with regrowth. Both in the US and Europe, the incidence of reported cases of Legionnaires' disease (LD) has increased since the turn of the century (Chapter 6). Ongoing outbreaks of LD, for example at a flower show in the Netherlands in 1999, many community-acquired cases, and the presence of *Legionella* in potable-water installations in hospitals and hotels, have resulted in the introduction of regulatory standards for *Legionella* in potable water in a number of European countries (Chapters 6 and 11). Chapter 7 addresses the nature and presence of other opportunistic pathogens in drinking-water distribution systems. Effects of environmental conditions in potable-water installations, for example temperature, nature of materials, and stagnation, on growth and survival of *L. pneumophila*, *P. aeruginosa* and NTM are described in Chapter 8.

Free-living protozoa are a common component of the microbial community in aquatic environments and are also present in water-supply systems (Michel *et al.* 1995; Thomas *et al.* 2008; Valster *et al.* 2009; Thomas and Ashbolt 2011; Valster *et al.* 2011). Most of these organisms feed on bacteria present in water, biofilms and sediments. Certain amoebae can serve as hosts for *L. pneumophila* (Rowbotham 1980; Wadowsky *et al.* 1988; Chapter 6) or harbour other opportunistic pathogenic bacteria (Thomas *et al.* 2006; Thomas *et al.* 2008). Some free-living amoebae have pathogenic properties, for example *Acanthamoeba* spp. causing keratitis and *Naegleria fowleri* causing meningitis (Chapter 7). Investigations in water supplies in the Netherlands and in the Caribbean region showed that the diversity and abundance of free-living protozoa is affected by temperature and water composition (Chapter 11).

Fungi have been observed with culture methods in many surface-water supplies and groundwater supplies, as well as in potable-water systems (Burman 1965; Nagy and Olson 1982; Niemi *et al.* 1982; Zacheus *et al.* 1995; Doggett 2000; Hageskal *et al.* 2006). Their ubiquitous presence has been confirmed with PCR-based techniques (Valster *et al.* 2009; Valster *et al.* 2011; van der Wielen and van der Kooij 2013; Chapter 7). Fungi have been related to taste and odour problems (see below) and some of these organisms have been implicated in disease (Chapter 7).

Discoloured water, mainly related to the presence of iron(hydr)oxides, (still) is a major cause of customer complaints, both in Europe (Vreeburg and Boxall 2007) and in the US (Schock and Lytle 2011). The presence of elevated concentrations of iron in drinking water originating from corroding pipes has been attributed to the activity of iron-oxidizing bacteria (Heymann 1928, Lee *et al.* 1980; Victoreen 1984; Clancy 1995). *Gallionella*, a characteristic iron-oxidizing bacterium, was observed on pipe surfaces in a distribution system by using scanning electron microscopy (Ridgway and Olson 1981). Recently, PCR-based methods have been used for the detection of this organism (de Vet *et al.* 2011). Many types of heterotrophic bacteria can adsorb iron present in finished water and/or originating from corrosion processes, resulting in the accumulation of iron-containing sediments (Clark *et al.* 1967; Macrae and Edwards 1972). Interactions between cast iron and microorganisms and elevated concentrations of iron in the distribution system, originating from finished water, are addressed in Chapters 4 and 14.

Taste and odour problems can be caused by microbial activity in distribution systems. Actinomycetes that produce the earthy/musty smelling geosmin and 2-methylisoborneol have been cultivated from drinking-water distribution systems, but an association of these organisms with taste and odour problems remains uncertain (Silvey *et al.* 1953; Burman 1965; Dott und Waschko-Dransmann 1981; Zaitlin and Watson 2006). Many cyanobacteria, present in surface water are also able to produce geosmin (Ketelaars 1994). Algae- and cyanobacteria-associated compounds have been implicated with off-flavours of drinking water due to an insufficient removal in water treatment (Krasner *et al.* 1989). These compounds can also accumulate in biofilms in the distribution system (Skjevrak *et al.* 2004). Microbial transformation of chlorinated phenols, which are by-products of chlorination, into trichloroanisoles with an odour threshold value at the ng/litre level, was identified as a main cause of off-flavour complaints in chlorinated supplies (Burtschell *et al.* 1959; Nyström *et al.* 1992; Montiel *et al.* 1999; Piriou *et al.* 2001). Furthermore, microbial activity under anaerobic conditions in deposits and biofilms in mains and reservoirs can lead to the formation of the highly odorous dimethyl-trisulphide (Wajon *et al.* 1986; Krasner *et al.* 1989; Franzmann *et al.* 2001).

Invertebrates visible to the naked eye, for example the arthropod *Asellus aquaticus*, may cause consumers' complaints about the quality of drinking water and block water meters (Smalls and Greaves 1968; Levy *et al.* 1986; Christensen *et al.* 2011; van Lieverloo *et al.* 2012). A large variety of invertebrates can proliferate in distribution systems by grazing on biofilms and sediments. Their abundance and diversity depends on water composition (natural organic matter

(NOM) concentration), pipe materials and retention time (Christensen *et al.* 2011; van Lieverloo *et al.* 2012; Chapter 9).

Operational problems caused by microbial growth include (i) accelerated decomposition of monochloramine by nitrifying bacteria, resulting in a decrease of disinfection capacity, the formation of nitrite and elevated HPCs (Skadsen 1993; Lipponen *et al.* 2002), (ii) elevated concentrations of iron or copper in the water and/or leakage caused by microbiologically-influenced corrosion (MIC) of metals (Lee *et al.* 1980; Bremer *et al.* 2001), (iii) leakage due to biodeterioration of rubber sealings in pipe joints by actinomycetes (Leeflang 1968) and (iv) pressure loss in water mains (Characklis 1973b; Picologlou *et al.* 1980).

Overall, microbial growth in distribution systems and in potable-water installations may impair the safety of drinking water usage, can negatively affect consumers' appreciation of and confidence in water quality, or hamper water distribution. Therefore, understanding of the nature and the causes of these problems is needed for defining adequate preventive and curative measures to control regrowth.

1.2.2 Regrowth assessment

Culture techniques based on the plate-count method are still widely used to detect and quantify bacteria in drinking water. The classic HPC methods that are used for monitoring rely on pour plates, high substrate concentrations and short incubation times. Media with relatively low concentrations of substrates, in combination with the spread-plate method and prolonged incubation time yield clearly higher HPCs and have been introduced primarily as a research tool (Reasoner and Geldreich 1985). However, these HPCs and those on media used in routine monitoring generally represent only a small (<1%) and variable fraction of the total concentration of bacteria present in drinking water (McCoy and Olson 1987; Servais *et al.* 1992; Block *et al.* 1993; Chapter 11). Hence, HPCs do not represent the level of microbial growth in water supplies.

Selective media, often in combination with membrane filtration to collect the organisms, are well suited for the enumeration of certain types of bacteria that can multiply in distribution systems and/or potable-water installations, for example coliforms, *Aeromonas*, *P. aeruginosa* and *L. pneumophila*. Also culture-independent methods have been developed for the detection of specific microorganisms, many of which are not (yet) cultivable. The high specificity makes PCR-based methods attractive for the quantitative detection of selected microorganisms in water (Chapters 7 and 11), including bacteria, that play a role in the (trans)formation of specific compounds, for example ammonia (van der Wielen and van der Kooij 2009), ferrous iron by *Gallionella* (de Vet *et al.* 2012), methane (Erwin *et al.* 2005) and sulphate (van der Wielen and Heijs 2007).

A variety of methods are available for the assessment of microbial biomass in water, biofilms and sediments. A common approach in aquatic microbiology is microscopic counting of the suspended bacteria collected by membrane filtration (Hobbie *et al.* 1977), followed by measuring the volume of the cells and conversion of the volume to biomass (dry weight) (e.g. Tranvik and Höfle 1987; Eiler *et al.* 2003). The use of total cell counts as a biomass parameter in water supply has a number of limitations, including: (i) requirement of advanced equipment and skills, (ii) differentiation between living and dead bacteria is difficult, (iii) cell size measurements are needed for determining cell volume and (iv) discrepancies between values of the conversion factor used for expressing volume in biomass-C (Watson *et al.* 1977; Bratbak 1985). Total cell numbers can also be obtained by using flow cytometry (Vives-Rego *et al.* 2000; Hammes and Egli 2010). However, this method does not seem to be suited for cell volume measurements and the estimation of the total cell count of biofilms and sediments is hampered by clusters of cells and attachment of cells to inorganic compounds, for example iron.

Analysis of specific biomass components is also used for measuring the concentration of microbial biomass, for example adenosinetriphosphate (ATP) (Hamilton and Holm-Hansen 1967; Karl 1980); DNA (McCoy and Olson 1985), membrane phospholipids (Balkwill *et al.* 1988), muramic acid (King and White 1977) and lipopolysaccharides (Korsholm and Søgaard 1987). For ATP a conversion factor (0.4% of cell carbon) is used (Karl 1980), but the ATP content of cells depends on the growth conditions (Karu and Vilu 1983; Kjelleberg and Dahlbäck 1984) and dead cells do not contain ATP. The amount of attached biomass has also been determined by assessing enzyme activity (Laurent and Servais 1995; Butterfield *et al.* 2002), the uptake or transformation of compounds labelled with radioactivity for example thymidine, glucose or leucine (Servais *et al.* 1991; Boe-Hansen *et al.* 2002a) or measuring the concentration of total carbohydrates or proteins (Percival *et al.* 1998; Hijnen *et al.* 2011). Estimation of the amount of microbial biomass in drinking water by measuring the concentration of polysaccharides or proteins requires a concentration step, because these concentrations are below the detection limits of the analytical methods. The biomass of invertebrates can be estimated by isolation, concentration and counting of the number of each type of organisms by using microscopy in combination with conversion factors for the biomass of the different types of organisms (van Lieverloo *et al.* 2012; Chapter 9). The significance of several methods for determining the amount of bacterial biomass is demonstrated with observations in the Netherlands (Chapter 11).

1.3 CAUSES OF REGROWTH

The microbial-growth potential of drinking water is defined by the concentration and nature of biodegradable compounds serving as a source of energy for microorganisms. The (rate and extent of the) utilization of these compounds in distribution systems and in potable-water installations depends on the growth kinetics of the microorganisms for these compounds and a number of environmental conditions. These conditions include: (i) water temperature, (ii) residence time, (iii) disinfectant residual and (iv) materials in contact with drinking water, (v) presence of inorganic nutrients (in particular N and P) and (vi) presence of oxygen or another electron acceptor. Drinking water normally contains oxygen and inorganic nutrients, but phosphate limitation has been reported (Lehtola *et al.* 1999; Lehtola *et al.* 2002). The kinetics of bacterial growth at low concentrations of biodegradable organic compounds, as well the effects of water temperature and distribution system conditions on regrowth and biofilm formation are summarized below. The effects of construction materials in contact with drinking water on microbial growth are described in Chapter 4 (corroding iron) and Chapter 12 (synthetic materials). Chapter 13 addresses the effects of hydraulic conditions on sediment accumulation and microbial activity in the deposits.

1.3.1 Growth kinetics and growth potential assessment

1.3.1.1 Growth kinetics

Multiplication of bacteria in drinking water implies that these organisms are able to utilize low concentrations of organic and/or inorganic compounds as a source of energy and/or carbon. The utilization of and growth at low concentrations of biodegradable compounds in the aquatic environment became an issue of research in the second half of the 20th century. Zobell and Grant (1943) reported that bacterial growth in seawater occurred at concentrations of organic substrates as low as 0.25 mg/litre. A few decades later, utilization of organic compounds at the μg/litre level was demonstrated by using compounds labelled with radioactivity (Parsons and Strickland 1962; Hamilton *et al.* 1966; Wright and Hobbie 1966; Geesey and Morita 1979). Quantification of the enhancement of bacterial growth at the μg/litre level was also achieved in simple batch tests with pure cultures of bacteria by using colony counts as the biomass parameter (van der Kooij and Hijnen 1980). These tests demonstrated that many low molecular weight compounds and also certain high molecular weight compounds promoted bacterial growth when added to drinking water at a concentration of a few μg/litre.

The relationship between the concentration of a growth-limiting compound and the growth rate, as defined by Monod (1949), is given by:

$$V = V_{max} \times S/(S + K_s) \qquad (1)$$

where: V is growth rate (doublings h^{-1}); V_{max} is maximum growth rate, S is substrate concentration (e.g. μg C/litre) and K_s is S for $V = 0.5\ V_{max}$. From eq. (1) it can be derived that $S \ll K_s$ gives:

$$V = S \times V_{max}/K_s \qquad (2)$$

which implies that the growth rate is proportional to the substrate concentration and depends on V_{mas}/K_s. This quotient is defined as the specific substrate affinity

($\mu g\ C \cdot litre^{-1} \cdot h^{-1}$). The numerical value of the specific substrate affinity represents V at S = 1 μg C/litre, unless $K_s \ll 1$ μg C/litre with V approaching V_{max}. Table 1.1 shows the growth kinetics of 10 substrate-bacterium combinations, derived from batch tests and selected on the basis of a high specific substrate affinity. The maximum growth rate derived for S = 1 μg C/litre was about 0.09 h^{-1}, which corresponds with a doubling time of about 11 h. For *Polaromonas*, an organism predominating in GAC filters (Magic-Knezev *et al.* 2009), the growth rate at 1 μg C/litre corresponded with V_{max} (Table 1.1). The highest V_{max} observed for a bacterium-substrate combination at 15°C was 0.5 h^{-1} (doubling time of 2.0 h) (van der Kooij and Hijnen 1985).

Tests with individual compounds added to tap water and with mixtures of selected compounds, revealed that many substrates are utilized simultaneously when present at a concentration of 1 μg C/litre each. Also amino acids which did not support growth when tested individually were utilized when present in a mixture of substrates (van der Kooij *et al.* 1982a, b; van der Kooij and Hijnen 1988a; Sack *et al.* 2010). Even addition of a mixture of amino acids at an individual compound concentration of 0.05 μg C/litre promoted growth of *Aeromonas hydrophila* in tap water (van der Kooij and Hijnen 1988). Certain polysaccharides and proteins can promote growth as well at the μg/litre level (van der Kooij and Hijnen 1981; van der Kooij and Hijnen 1985; Sack *et al.* 2010; Sack *et al.* 2011). The yield values (CFU/μg C) of the different combinations of substrate and organism, shown in Table 1.1, range from 4.2×10^6 to 2.3×10^7 CFU/μg C. The lowest yield value (2.9×10^6 CFU/μg C) was observed for growth of *Spirillum* sp. strain NOX on oxalate (van der Kooij and Hijnen 1984). The yield value is affected by cell size, the physiological properties of the organism and the Gibbs energy dissipation associated with the production of biomass from the involved substrate (Heijnen and van Dijken 1992). *P. fluorescens* strain P17, a nutritionally highly versatile bacterium, and *Spirillum* sp. strain NOX, specialized in the utilization of carboxylic acids, have been selected for determining the concentration of easily assimilable organic carbon (AOC) in drinking water (Chapter 11).

Estimations of growth kinetics for coliforms and *P. aeruginosa* gave 10 to 100 times higher values for K_s than those presented in Table 1.1 (van der Kooij *et al.* 1982b; van der Kooij 1997; van der Kooij 2002). Camper *et al.* (1991) showed that yeast extract (0.1 mg/litre) promoted growth of a variety of coliforms in distilled water. However, tests with mixtures of about 20 low-molecular-weight compounds, each at a concentration of 1 μg C/litre revealed that carbohydrates are the favourite growth substrate of certain coliform bacteria, for example *Klebsiella* spp., *Enterobacter* sp., *Citrobacter* spp., isolated from drinking water (van der Kooij 1997). However, growth was slow at a carbohydrate concentration of 10 μg C/litre and the organisms did not multiply in slow sand filtrate. The coliform growth response test with a pure culture of *Enterobacter cloacea* yielded less growth in treated water than the organisms used in the AOC test

Table 1.1 Growth kinetics of 10 bacterium-substrate combinations with an elevated growth rate at a concentration of 1 μg C/litre*. Incubation temperature: 15°C.

Substrate	Organism	V_{max} (h^{-1})	K_s (μg C/litre)	Growth rate (h^{-1}) at 1 μg C/litre	Growth yield (CFU/μg C)	Ref.
Arginine	*Aeromonas hydrophila*	0.09	0.32	0.09**	5.6×10^6	1
Acetate	*Polaromonas* sp. strain P-315	0.09	0.36	0.09**	1.4×10^7	2
Laminarin	*Flavobacterium johnsoniae*	0.29	3.4	0.085	1.3×10^7	3
Maltotetraose	*Flavobacterium* sp. strain S12	0.45	5.4	0.083	2.3×10^7	4
Maltopentaose	*Flavobacterium* sp. strain S12	0.44	5.7	0.077	2.3×10^7	4
Maltotriose	*Flavobacterium* sp. strain S12	0.41	5.7	0.072	2.0×10^7	4
Starch	*Flavobacterium* sp. strain 166	0.27	3.9	0.069	2.0×10^7	5
Glucose	*Flavobacterium* sp. strain 166	0.21	3.3	0.064	1.6×10^7	5
Acetate	*Pseudomonas fluorescens* P17	0.18	4.0	0.045	4.2×10^6	6
Oleate	*Aeromonas hydrophila*	0.21	2.1	0.04	1.0×10^7	1

*growth rate at 1 μg C/litre is calculated from kinetics; **observed maximum growth rate (V_{max}; h^{-1}); Ref.: 1, van der Kooij and Hijnen (1988); 2, Magic-Knezev and van der Kooij (2006); 3, Sack *et al.* (2011); 4, van der Kooij and Hijnen (1985); 5, van der Kooij and Hijnen (1981); 6, van der Kooij *et al.* (1982a).

(Rice *et al.* 1991). Growth of coliforms in distribution systems therefore is an indication of an elevated substrate concentration, for example carbohydrates present in sediments or released from redwood in storage reservoirs (Talbot and Seidler 1979; LeChevallier *et al.* 1996; Chapter 2).

Natural organic matter (NOM) in freshwater environments is mainly composed of refractory humic and fulvic acids. Despite the high resistance to microbial decay, humic and fulvic acids can promote slow microbial growth with a low (0.4–20%) efficiency (Tranvik and Höfle 1987; Eiler *et al.* 2003). The growth-promoting properties of elevated concentrations of humic and fulvic acids have also been demonstrated on surfaces of corroding iron (Camper 2004; Chapter 4). In surface water certain NOM components are subject to photochemical breakdown by sunlight (UV), leading to the formation of low molecular weight compounds, such as aldehydes and carboxylic acids, which are rapidly utilized (Kieber *et al.* 1989). Similar compounds are produced by oxidation processes in water treatment. The effects of water treatment processes, the conditions in the distribution system, for example the type of construction materials in contact with water, and the presence of sediments, on microbial growth (potential) are discussed in more detail in the following chapters.

1.3.1.2 Assessment of the microbial-growth potential of drinking water

Analytical chemistry for drinking-water quality assessment has expanded tremendously since the discovery of disinfection by-products and observations of pesticides in the aquatic environment and is further stimulated by the continuing contamination of the aquatic environment with "new" chemicals (Mons *et al.* 2013). NOM became also a subject for detailed characterization procedures, using a variety of chemical techniques (Liao *et al.* 1982) and physical methods, for example ultraviolet absorbance, assessment of the molecular-weight distribution (Thurman *et al.* 1982; Huber and Frimmel 1994) and fluorescence spectroscopy (Bagoth *et al.* 2009). Chemical methods are available for the quantification of low concentrations of a variety of low-molecular weight compounds, that is carboxylic acids, aldehydes (Peldzus *et al.* 1996), carbohydrates (Cheng and Kaplan 2001) and amino acids (Lindroth and Mopper 1979). However, assessment of the total concentration of compounds contributing to the microbial-growth potential is still not possible with chemical analysis because of (i) the large number of different compounds present at a low concentration, (ii) inadequacy of isolation, separation and concentration techniques, (iii) inability to identify compounds, and (iv) lack of information about the role of identified compounds to serve as a source of energy and/or carbon. Therefore, bioassays and tests combining a microbiological procedure with chemical analysis have been developed to assess the microbial-growth potential of (drinking) water or the concentration of biodegradable compounds present in the water. A number of these methods and their applications are described in Chapters 2, 10 and 11. An extended review on methods for the characterization of NOM and the assessment of the concentration of biodegradable organic matter in drinking water is given elsewhere (Prévost *et al.* 2005).

1.3.2 Temperature

The temperature of surface water depends on latitude and attains a maximum level of 28 to 30°C in tropical regions (Lewis 1987). In temperate regions, for example the Netherlands, drinking-water temperature in the distribution system of surface-water supplies ranges from a few degrees Celsius in winter to about 20°C in summer. Water temperatures over 30°C are commonly attained in warm tap water installations (Chapter 6). The growth rate of bacteria is low at temperatures <5°C, even in the presence of growth substrates, and generally increases about twofold with a 10°C increase of temperature within certain temperature ranges. Donlan *et al.* (1994) observed much higher biofilm formation rates on cast iron, based on HPC measurements, in water pipes at 15–25°C than at temperatures below 10°C. Consequently, elevated HPC values are more frequently observed in summer than in winter. The preference of bacteria and other microorganisms for a specific temperature range has a profound impact on microbiological processes in water treatment and distribution. For example, nitrification in chloraminated supplies occurs more frequently at water temperatures >15°C (Wilczak *et al.* 1996). Also multiplication of coliforms, resulting in noncompliance with water quality standards, has been observed more frequently in drinking-water distribution systems at water temperatures exceeding 15°C, than at lower temperatures (LeChevallier *et al.* 1996; Chapter 2).

P. aeruginosa is observed more frequently in drinking water in tropical areas and in water installations in buildings at elevated temperatures than in distribution systems in temperate zones (Favero *et al.* 1971; Schubert *et al.* 1975; Völker *et al.* 2010; Wingender *et al.* 2010). Fluorescent pseudomonads were found in all samples collected from water supplies in the Netherlands, but *P. aeruginosa,* was only observed once (van der Kooij *et al.* 1977). At 15°C, *P. aeruginosa* was unable to compete effectively with the related species *P. fluorescens*, which may utilize the same compounds in tap water (van der Kooij *et al.* 1982b). *Aeromonas* spp. occur in larger numbers in drinking water in the summer than in the winter (LeChevallier *et al.* 1982; Burke *et al.* 1984; Havelaar *et al.* 1990).

A striking example of the effect of temperature on microbial growth in water supplies is the proliferation of *L. pneumophila* in warm tap water systems in temperate climates and in drinking-water distribution systems in (sub)tropical regions (see Chapters 6 and 8). No legionellae were detected by a culture method in drinking water supplies at temperatures below 15°C, but by using selective quantitative PCR methods, a wide range of uncultured, mostly not-yet described *Legionella* spp. were observed (Wullings *et al.* 2006; Wullings *et al.* 2011).

These observations indicate that *L. pneumophila* is unable to compete with other species of this genus at temperatures below 25°C. Growth of this organism is also limited at temperatures above 42°C (Yee *et al.* 1982; Wadowsky *et al.* 1985). In a warm-water installation at a water temperature of 55 to 60°C, HPCs on plates incubated at 55°C were higher than HPCs incubated at 25°C, indicating that thermophilic heterotrophic bacteria are present at elevated temperatures (Bagh *et al.* 2004).

Water temperature also affects eukaryote communities, including free-living protozoa, fungi and metazoa in water supplies. A large variety of free-living protozoa have been observed in water supplies at temperatures below 20°C (Michel *et al.* 1995; Valster *et al.* 2009). *Hartmannella vermiformis*, a protozoan host for *L. pneumophila,* was also observed in these supplies, but was present in higher numbers in supplies in the Caribbean at temperatures close to 30°C (Valster *et al.* 2011). For the pathogenic protozoan *N. fowleri* an optimum temperature range of 27 to 35°C has been reported (Sykora *et al.* 1983) and *N. australiensis* was observed at temperatures ranging from 25 to 43°C (Scaglia *et al.* 1983). *N. fowleri* was also detected in drinking water supply wells in Arizona, at temperatures ranging from 18 to 42°C (avg. 29°C) (Bright *et al.* 2009). In Australia, a drinking water temperature above 25°C for more than 4 months per year is considered a risk factor for growth of *N. fowleri* (Trolio *et al.* 2008). However, non pathogenic *Naegleria* spp., which are not thermotolerant, have been observed frequently at temperatures <20°C (Michel *et al.* 1995; Thomas *et al.* 2008). Metazoan activity is also affected by water temperature, but no seasonal effects were observed on the occurrence of *Asellus aquaticus* in groundwater supplies in Denmark (Christensen *et al.* 2011).

In summary, temperature has a major influence on the (micro)biological activity and the composition of the microbial community in drinking-water supplies. However, publications about microorganisms in water supplies in many cases do not provide (accurate) data on water temperature.

1.3.3 Biofilms, sediments and hydraulics

1.3.3.1 Biofilms

Hydrodynamic conditions for example (variations in) flow rate, residence time and diameter of the pipes (surface to volume ratio) affect microbial growth in distribution systems. Microorganisms that are not attached to drinking-water exposed surfaces and sediments leave the system with the water after a residence time of a few hours to a few days. Hence "adhere to day; here tomorrow", as defined for oral microorganisms (Kolenbrander and London 1993), also applies to microorganisms in drinking water. Consequently, microorganisms that grow attached to surfaces or in sediments constitute the main proportion of biomass in the distribution system and potable-water installations (LeChevallier *et al.* 1987; Van der Wende *et al.* 1989; Block *et al.* 1993).

Generally, the following steps are distinguished in the process of biofilm formation: (i) adsorption of microorganisms to the surface starting within a few minutes to one hour of exposure, (ii) formation of extracellular polymeric substances (EPS) by the cells, resulting in irreversible attachment ("polymer bridging") (Vandevivere and Kirchman 1993), (iii) multiplication of the bacteria, forming microcolonies embedded in the EPS matrix, which also attracts inorganic compounds, and (vi) continuing release of bacteria from the biofilm due to swarming, changes in shear and sloughing (Marshal *et al.* 1971; Characklis 1973a). Attachment protects the bacteria against disinfectants (Herson *et al.* 1987; LeChevallier *et al.* 1988a) and EPS protects against protozoan grazing (Weitere *et al.* 2005) and facilitates substrate utilization (Flemming and Wingender 2010). The processes leading to the formation of biofilms, which initially were designated as biological films or slimes (Characklis 1973a), have been the subject of many studies and are described in Chapter 8.

Biofilm formation on surfaces exposed to drinking water has been studied by (i) using devices with sample pieces exposed to the water under investigation, (ii) in pilot plants, and (iii) by collecting segments of distribution system pipes. Pederson (1990) reported that the concentration of bacteria attached on steel and unplasticized polyvinyl chloride (PVC-U) exposed to drinking water increased exponentially (doubling time 11 days), reaching a maximum level of about 5×10^6 cells/cm^2 after about 150 days on both materials. Biofilm concentrations on PVC-U and cemented coupons ranged from 7.8×10^6 cells/cm^2 after 40 h of residence time to 4.4×10^6 cells/cm^2 after 240 h in a pilot plant operated continuously for one month with treated surface water (Block *et al.* 1993). HPCs in these biofilms declined to <1% of the total number of bacteria with increasing residence time and generation times decreased with increasing residence time from 17 days to 288 days. HPCs on cast iron exposed to different types of treated surface water containing a monochloramine residual reached steady state levels of 10^5 to 10^7 CFU/cm^2 after about 30 days of contact at temperatures between 20 and 25°C (Donlan *et al.* 1994). HPCs on stainless steel in a model distribution system supplied with drinking water also reached a maximum level (about 5×10^4 CFU/cm^2) after about 30 days and decreased thereafter (Boe-Hansen *et al.* 2002b). However, a level of 2.6×10^6 cells/cm^2 was attained after 200 days of exposure to drinking water (AOC: 6 μg C/litre), which was more than 3 log units above the HPCs. The doubling time of the biofilm in this system was estimated at 23 days and a 50% reduction time of 53 days was established by using ^{14}C-benzoate (Boe-Hansen *et al.* 2002a). In a so-called biofilm monitor fed with treated groundwater, the biofilm concentration on glass attained a maximum of about 10^7 cells/cm^2 and 10^5 CFU/cm^2 after an exposure time of 100 days (van der Kooij *et al.* 2003). These studies show that (i) the growth rate of bacteria in drinking-water biofilms is low, (ii) HPCs represent a small and variable fraction of the number of attached bacteria and (iii) surface coverage is limited, when assuming complete coverage at 10^8 cells/cm^2.

Measurement of exoproteolytic activity on coupons exposed to drinking water with a low or no chlorine residual yielded biomass concentrations ranging from less than 0.02 to 1 μg C/cm^2 (Ninquette *et al.* 2000). The lowest concentrations were observed on polyethylene (PE) and PVC-U and the highest on grey iron. Servais *et al.* (1992) reported biomass concentrations of 0.48 to 0.65 μg C/cm^2 on cast-iron coupons exposed to chlorine-free drinking water in Paris. The concentration of active biomass on the surfaces of PVC-U pipes (Ø 10 cm) in distribution systems in the Netherlands ranged from less than 100 pg ATP/cm^2 in drinking water prepared from aerobic groundwater (dissolved organic carbon [DOC] < 0.5 ppm) to about 6000 pg ATP/cm^2 in drinking water prepared from anaerobic groundwater (NPOC 8 ppm) (van der Kooij *et al.* 1995). Conversion by using a biomass-C/ATP ratio of 250 (Karl 1980) gives a range of 0.025 to 0.5 μg C/cm^2, which is similar to the range of biofilm concentrations on PVC-U exposed to drinking water at 15 locations in the distribution system of Brussels (Ninquette *et al.* 2000). Hence, the biofilm concentration on surfaces exposed to drinking water is affected by (i) the water composition, for example, concentration and nature of NOM, (ii) the type of material, and (iii) the presence of a disinfectant residual. Interactions between surfaces of corroding steel, water composition and disinfectant residual are described in Chapter 4 and the effect of material on biofilm growth in Chapter 12.

Hydrodynamic conditions affect the transfer of microorganisms and growth substrates to the surface as well as cell detachment (sloughing). The flow pattern, which depends on the combination of flow velocity and pipe diameter, is turbulent at a Reynolds number >2300. This (dimensionless) Reynolds number (Re) is defined by:

$$Re = D \times V/\nu \tag{3}$$

where: D = internal pipe diameter (cm), V = flow velocity (cm/s) and ν = kinematic viscosity of water, which is 0.01 cm^2/s at 20°C. Typical flow velocities in distribution systems range from <10 cm/s to 50 cm/s, corresponding with turbulent flow in most cases (Chapter 14). An increase in velocity increases the degree of turbulence and reduces the thickness of the laminar boundary layer at the exposed surface. According to Fick's first law, the flux (amount of substance per unit area per unit time) is proportional to the concentration gradient and the diffusion coefficient of the involved substance. Consequently, the supply of growth-promoting compounds and nutrients from the bulk water to the attached bacteria, is limited by diffusion through the laminar layer and increases with an increase of Re. The effect of biofilm thickness on these processes in distribution systems generally will be limited at a low percentage of surface coverage, but may play a role in microcolonies. Utilization of biodegradable compounds therefore mainly depends on growth kinetics, that is, rapid uptake of compounds for which microorganisms have a high substrate affinity (Table 1.1) vs. slow uptake of compounds that require binding to biomass and extracellular enzymatic degradation. Interactions between water quality and attached microorganisms also depend on the surface (S) to volume (V) ratio, which increases proportionally with the decrease of pipe diameter. The S/V ratio in plumbing systems is much higher than in distribution systems and total numbers of bacteria in water sampled from service lines were much higher than those in the distribution system (Prévost *et al.* 1997). Microbial-growth phenomena in plumbing systems are described in more detail in Chapter 8.

Attachment forces of up to 68 N/mm^2 have been reported for bacteria attached to surfaces (Fang *et al.* 2000; Tsang *et al.* 2006). Hence, very high shear forces are needed for detachment of bacteria from the surface, but forces of autoaggregation and coaggregation in multispecies biofilms are less strong and variable over time (Rickard *et al.* 2004). Certain processes in the biofilm, for example, development of motile organisms, limitation of substrate or oxygen, can also promote detachment (Hunt *et al.* 2004). Grazing by protozoa or certain metazoa leads to removal of attached biomass as well. The effects of hydrodynamic conditions also depend on substrate concentration, the presence of a disinfectant, temperature and the nature of the exposed material (Ollos *et al.* 2003).

1.3.3.2 Sediments

The presence of sediments in distribution systems has been demonstrated in many studies (e.g. McCoy and Olson 1987; Gauthier *et al.* 1999; Chapter 9). These sediments act as a niche for microbial growth (Tuovinen and Hsu 1982; Allen and Geldreich 1977; Zacheus *et al.* 2001; Lethola *et al.* 2004; Torvinen *et al.* 2004), are a food source for invertebrates (Christensen *et al.* 2011; Chapter 9) and provide protection against disinfectants (Gauthier *et al.* 1999). Resuspension of iron-containing sediments is a major cause of customer complaints about discoloured water (Vreeburg and Boxall 2007; Chapter 14). Accumulation of sediments depends on flocculation and sedimentation of colloids and particulate matter, including biomass (components) and inorganic constituents such as iron, manganese, calcium carbonate and silica (McCoy and Olson 1987; Vreeburg 2007) and on biomass detachment (sloughing). Thus, enhanced removal of particles, biomass (components) and growth-promoting compounds in water treatment reduces sediment accumulation during distribution. Also a daily maximum flow velocity of 0.4 m/s, in a self-cleaning distribution system, further reduces sediment accumulation (Vreeburg *et al.* 2009). Sediment accumulation in distribution systems and control measures are described in detail in Chapter 14.

1.3.4 Construction materials

The concentration of biofilms on corroding iron generally is higher than on noncorrosive materials such as PVC-U, PE, glass, and stainless steel (Ninquette *et al.* 2000; Ollos 2003). This aspect has been attributed to nutrient adsorption and the inactivation of the disinfectant by the corrosion products (LeChevallier *et al.* 1988a; Camper *et al.* 2004; Chapter 4). The observation of elevated numbers of *Asellus aquaticus* in cast-iron distribution pipes, as compared to PVC-U pipes, in a non-chlorinated supply (Christensen *et al.* 2011) is consistent with the effect of cast iron on nutrient accumulation. Synthetic materials releasing biodegradable compounds can enhance microbial growth as well (Chapter 12).

1.3.5 Disinfectant residual

In most countries, a residual concentration of a disinfectant (chlorine or monochloramine) is maintained in drinking-water distribution systems to ensure microbial safety and limit regrowth. In the Netherlands (van der Kooij *et al.* 1999) and in a number of other European countries, for example, Germany, Austria, Denmark and Switzerland, drinking water in many cases is distributed without a disinfectant residual. Maintaining microbial safety in distribution systems without a disinfectant residual is described in Chapter 5. A residual can effectively limit regrowth, provided that it reaches the tap. However, formation of biodegradable by-products, for example, chloro-acetic acids, as a result of interactions between chlorine and NOM, and the presence of ammonia due to decay of monochloramine (Wilczak *et al.* 1996; Lipponen *et al.* 2002) enhances regrowth. Interactions between disinfectant residual, NOM and corroding surfaces and their effects on regrowth phenomena are described in Chapter 4. Monochloramine is less reactive than free chlorine with NOM, corrosion products (Vikesland and Valentine 2000) and biomass (components) and more effective towards (microorganisms in) biofilms (LeChevallier *et al.* 1988a; LeChevallier *et al.* 1990). In a study of 31 water systems in North America, coliforms were more frequently observed and present in higher numbers in supplies with free chlorine as compared to supplies with monochloramine (LeChevallier *et al.* 1996). Monochloramine also more effectively prevents growth of *Legionella* within building-water systems (Kool *et al.* 1999; Chapter 6), but can enhance growth of nontuberculous mycobacteria (Pryor *et al.* 2004). Obviously, maintaining a disinfectant residual can have negative and positive effects on regrowth, the balance of which depends on local conditions, for example, concentration of NOM, temperature, pipe material and residence time.

1.4 SCOPE AND AIM

The information presented in this chapter shows that microbial growth in distribution systems and potable-water installations is a complex phenomenon, both microbiologically and regarding the effects on water quality. The high microbial diversity is associated with the large variety of environmental conditions, including water composition, material type, physico-chemical characteristics and microniches in biofilms and sediments within distribution systems and potable-water installations. The microorganisms present under these conditions include many different, mostly yet-unknown genera and species, with undefined physiological characteristics. For decades, microbiologists and engineers have tried to elucidate conditions affecting microbial growth with the aim to prevent water quality deterioration during distribution. The following chapters present the current knowledge about the significance and the causes of a number of regrowth phenomena.

Acknowledgements

This review has been made as part of the Joint Research Programme (Bedrijfstakonderzoek [BTO]). The authors acknowledge the comments made by Gertjan Medema on this chapter.

1.5 REFERENCES

Allen M. J. and Geldreich E. E. (1977). Distribution line sediments and bacterial regrowth. In: Proceedings American Water Works Association Water Quality Technology Conference, American Water Works Association, Denver, CO, pp. 1–6.

Anaissie (2002). The hospital water supply as a source of nosocomial infections; a plea for action. *Arch. Intern. Med.*, **162**, 1483–1492.

Aronson T. A., Holtzman A., Glover N., Boian M., Froman S., Berlin O. G., Hill H. and Stelma G. Jr. (1999). Comparison of large restriction fragments of *Mycobacterium avium* isolates recovered from AIDS and non-AIDS patients with those of isolates from potable water. *J. Clin. Microbiol.*, **37**, 1008–1012.

Bagh L. K., Albrechtsen H. J., Arvin E. and Ovesen K. (2004). Distribution of bacteria in a domestic hot water system in a Danish apartment building. *Water Res.*, **38**, 225–235.

Bagoth S. A., Dignum M., Grefte A., Kroesbergen J. and Amy G. L. (2009). Characterization of NOM in a drinking water treatment process with no disinfectant residual. *Water Sci. Technol. Wat. Supply*, **9**(4), 379–386.

Balkwill D. L., Leach F. R., Wilson J. T., McNabb J. F. and White D. C. (1988). Equivalence of microbial biomass measures based on membrane lipid and cell wall components, adenosine triphosphate, and direct counts in subsurface aquifer sediments. *Microb. Ecol.*, **16**, 73–84.

Bartram J., Cotruvo J., Exner M., Fricker C. and Glasmacher A. (2003). Heterotrophic plate counts and drinking-water safety: the significance of HPCs for water quality and public health. Emerging issues in water and infectious disease series. IWA Publishing, London, UK.

Baylis J. R. (1930). Bacterial aftergrowths in water distribution systems. *Wat. Works Sewerage*, **77**(10), 355–338.

Bej A. K. (1990). Multiplex PCR amplification and immobilized capture probes for detection of bacterial pathogens and indicators in water. *Mol. Cell. Probes*, **4**(5), 353–365.

Besner M. C., Gauthier V., Servais P. and Camper A. (2002). Explaining the occurrence of coliforms in distribution systems. *J. Am. Wat. Works Assoc.*, **94**, 95–109.

Block J. C., Haudidier K., Paquin J. L., Miazga J. and Levi Y. (1993). Biofilm accumulation in drinking water distribution systems. *Biofouling*, **6**, 333–343.

Boe-Hansen R., Albrechtsen H. J., Arvin E. and Jørgensen C. (2002a). Bulk water phase and biofilm growth in drinking water at low nutrient conditions. *Water Res.*, **36**, 4477–4486.

Boe-Hansen R., Albrechtsen H. J., Arvin E. and Jørgensen C. (2002b). Dynamics of biofilm formation in a model drinking water distribution system. *J. Wat. Supply Aqua*, **51**(7), 399–406.

Bratbak G. (1985). Bacterial biovolume and biomass estimations. *Appl. Environ. Microbiol.*, **49**(6), 1488–1494.

Bremer P. J., Webster B. J. and Wells D. B. (2001). Biocorrosion of copper. *J. Am. Wat. Works Assoc.*, **93**, 82–91.

Brigth K. R., Marciano-Cabrall F. and Gerba C. P. (2009). Occurrence of *Naegleria fowleri* in Arizona drinking water supply wells. *J. Am. Wat. Works Assoc.*, **101**(11), 43–50.

Brock T. D. (1975). Milestones in microbiology. American Society for Microbiology, Washington, DC.

Burke V., Robinson J., Gracey M., Peterson D. and Partridge K. (1984). Isolation of *Aeromonas hydrophila* from a metropolitan water supply: seasonal correlation with clinical isolates. *Appl. Environ. Microbiol.*, **48**, 361–366.

Burman N. P. (1965). Taste and odour due to stagnation and local warming in long lengths of piping. *Proc. Soc. Water Treat. Exam.*, **14**, 125–131.

Burtschell R. H., Rosen A. A., Middleton F. M. and Ettinger M. B. (1959). Chlorine derivatives of phenol causing taste and odor. *J. Am. Wat. Works Assoc.*, **51**, 205–214.

Buswell A. M. (1938). Microscopic growth in distribution systems and their food supply. *J. Am. Wat. Works Assoc.*, **30**(10), 1651–1654.

Butterfield P. W., Bargmeyer A. M., Camper A. C. and Biederman J. A. (2002). Modified enzyme activity assay to determine biofilm biomass. *J. Microbiol. Methods*, **50**, 23–31.

Camper A. K. (2004). Involvement of humic substances in regrowth. *Int. J. Food Microbiol.*, **92**, 355–364.

Camper A. K., McFeters G. A., Characklis W. G. and Jones W. L. (1991). Growth kinetics of coliform bacteria under conditions relevant to drinking water distribution systems. *Appl. Environ. Microbiol.*, **57**, 2233–2239.

Camper A. K., Jones W. L. and Hayes J. T. (1996). Effect of growth conditions and substratum composition on the persistence of coliforms in mixed-population biofilms. *Appl. Environ. Microbiol.*, **62**(11), 4014–4018.

Characklis W. G. (1973a). Attached microbial growths; I Attachment and growth. *Water Res.*, **7**, 1113–1127.

Characklis W. G. (1973b). Attached microbial growths; II Frictional resistance due to microbial slimes. *Water Res.*, **7**, 1249–1258.

Cheng X. and Kaplan L. A. (2001). Improved analysis of dissolved carbohydrates in stream water with HPLC-PAD. *Anal. Chem.*, **73**, 458–461.

Christensen S. C. B., Nissen E., Arvin E. and Albrechtsen H. J. (2011). Distribution of *Asellus aquaticus* and microinvertebrates in a non-chlorinated drinking water supply system. Effects of pipe material and sedimentation. *Water Res.*, **45**, 3215–3224.

Clancy J. L. (1995). Iron bacteria. In: Problem organisms in water: identification and treatment, Manual of Water supply Practices, AWWA M7, American Water Works Association, Denver, CO, pp. 7–18.

Clark F. M., Scott R. M. and Bone E. (1967). Heterotrophic iron-precipitating bacteria. *J. Am. Wat. Works Assoc.*, **59**, 1036–1042.

Colbourne J. S. (1985). Materials usage and their effects on the microbiological quality of water supplies. *J. Appl. Bacteriol. Symp. Suppl.*, **71**, 47S–59S.

de Vet W. W. J. M., Dinkla I. J. T., Abbas B. A., Rietveld L. C. and van Loosdrecht M. C. M. (2011). Biological iron oxidation by *Gallionella* spp. in drinking water production under fully aerated conditions. *Water Res.*, **45**, 5389–5398.

de Vries H. (1890). Die Pflanzen und Thiere in den dunklen Räumen der Rotterdamer Wasserleitung. *Bericht der biologischen Untersuchungen der Crenothrix-Commission zu Rotterdam vom Jahre 1887*. Gustav Fischer, Jena.

Dietlicher K. (1970). Wiederverkeimung ozonisierter Schnellfiltrate im Rohrnetz. *Schriftenr. Ver. Wasser-, Boden-, Lufthyg., Berlin-Dahlem*, **31**, 171–186.

Doggett M. S. (2000). Characterization of fungal biofilms within a municipal water distribution system. *Appl. Environ. Microbiol.*, **66**(3), 1249–1251.

Donlan R. M., Pipes W. O. and Yohe T. L. (1994). Biofilm formation on cast iron substrata in water distribution systems. *Water Res.*, **28**(6), 1497–1503.

Dott W. and Waschko-Dransmann D. (1981). Vorkommen und Bedeutung von Actinomycetales im Trinkwasser. *Zbl. Bakt. Hyg. I. Abt. Orig. B*, **173**, 217–232.

du Moulin G. C. and Stottmeier K. D. (1986). Waterborne mycobacteria: an increasing threat to health. *ASM News*, **52**, 525–529.

Edberg S. and Smith D. B. (1989). Absence of association between total heterotrophic and total coliform bacteria from a public water supply. *Appl. Environ. Microbiol.*, **55**(2), 380–384.

Edberg E. C., Kops S., Kontnick C. and Escarzaga M. (1997). Analysis of cytotoxicity and invasiveness of heterotrophic plate count bacteria (HPC) isolated from drinking water on blood agar. *J. Appl. Bacteriol.*, **82**, 455–481.

Eijkmann C. (1904). Die Gärungsprobe bei 46°C als Hilfsmittel bei der Trinkwasseruntersuchung. *Zentralbl. Bakteriol. I Abt. Orig.*, **37**(5), 742–752.

Eiler A., Langenheder S., Bertilsson S. and Tranvik L. J. (2003). Heterotrophic bacterial growth efficiency and community structure at different natural organic carbon concentrations. *Appl. Environ. Microbiol.*, **69**(7), 3701–3709.

Ellgas W. M. and Lee R. (1980). Reservoir coatings can support bacterial growth. *J. Am. Wat. Works Assoc.*, **72**(12), 693–695.

EPA (1999). 25 years of the Safe Drinking Water Act: History and trends. EPA 816-R-007. United States Environmental Protection Agency, Office of Water.

Erwin D. P., Erickson I. K., Delwiche M. E., Colwell F. S., Strap J. L. and Crawford R. L. (2005). Diversity of oxygenase genes from methane- and ammonia-oxidizing bacteria in the eastern snake river plain aquifer. *Appl. Environ. Microbiol.*, **71**(4), 2016–2025.

EU (1998). Council Directive 98/83/EC of 3 November 1998 on the quality of water intended for human consumption. *Off. J. Eur. Commun.*, **L330**, 32–53.

Fang H. H. P., Chan K. Y. and Xu L. C. (2000). Quantification of bacterial adhesion forces using atomic force microscopy (AFM). *J. Microbiol. Methods*, **40**, 89–97.

Favero M. S., Carson L. A., Bond W. W. and Petersen N. J. (1971). *Pseudomonas aeruginosa:* growth in distilled water from hospitals. *Science*, **177**, 836–838.

Flemming H. C. and Wingender J. (2010). The biofilm matrix. *Nature Rev. Microbiol.* 2 August 2010; doi:10.1038/nrmicro2415.

Frankland P. and Frankland P. (1894) Microorganisms in water. Their significance, identification and removal. Longmans, Green and Co. London and New York.

Franzmann P. D., Heitz A., Zappia L. R., Wajon J. E. and Xanthus K. (2001). The formation of malodorous dimethyl oligosulphides in treated groundwater: the role of biofilms and potential precursors. *Water Res.*, **35**(2), 1730–1738.

Gauthier V., Gérard B., Portal J. M., Block J. C. and Gatel D. (1999). Organic matter as loose deposits in a drinking water distribution system. *Water Res.*, **33**(4), 1014–1026.

Geesey G. G. and Morita R. Y. (1979). Capture of arginine at low concentrations by a marine psychrophilic bacterium. *Appl. Environ. Microbiol.*, **38**(6), 1092–1097.

Geldreich E. E. and Rice E. W. (1987). Occurrence, significance and detection of *Klebsiella* in water systems. *J Am. Wat. Works Assoc.* **79**, 74–80.

Geldreich E. E., Nash H. D., Reasoner D. J. and Taylor R. H. (1972). The necessity of controlling bacterial populations in potable waters: community water supply. *J. Am. Wat. Works Assoc.*, **64**, 596–602.

Habicht W. and Müller H. E. (1988). Occurrence and parameters of frequency of *Legionella* in warm water systems of hospitals and hotels in Lower Saxony. *Zbl. Bakt. Hyg. B*, **186**, 79–88.

Haenle O. (1903). Die Bacterien-Flora der Metzer Wasserleitung, Heitz und Mündel, Strassburg.

Hageskal G. H., Knutsen A. K., Gaustad P., Sybren de Hoog G. and Skaat I. (2006). Diversity and significance of mold species in Norwegian drinking water. *Appl. Environ. Microbiol.*, **72**(12), 7586–7593.

Hamilton R. D. and Holm-Hansen O. (1967). Adenosine triphosphate content of marine bacteria. *Am. Soc. Limnol. Oceanogr.*, **12**, 319–324.

Hamilton R. D., Morgan K. M. and Strickland J. D. H. (1966). The glucose uptake kinetics of some marine bacteria. *Can. J. Microbiol.*, **12**, 995–1003.

Hammes F. and Egli T. (2010). Cytometric methods for measuring bacteria in water: advantages, pitfalls and applications. *Anal. Bioanal. Chem.*, **397**, 1083–1095.

Havelaar A. H., Versteegh J. F. and During M. (1990). The presence of *Aeromonas* in drinking water supplies in The Netherlands. *Zentralbl. Hyg. Umweltmed.*, **190**, 236–256.

Heijnen J. J. and van Dijken J. P. (1992). In search of a thermodynamic description of biomass yields for the chemoheterotrophic growth of micro-organisms. *Biotechnol. Bioeng.*, **39**, 833–858.

Herson D. S., McGonigle B., Payer M. A. and Baker K. H. (1987). Attachment as a factor in the protection of *Enterobacter cloacae* from chlorination. *Appl. Environ. Microbiol.*, **53**(5), 1178–1180.

Heymann J. A. (1928). De organische stof in het waterleidingbedrijf. *Water Gas*, **12**, 61–65, 69–72.

Hijnen W. A. M., Cornelissen E. R. and van der Kooij D. (2011). Threshold concentrations of biomass and iron for pressure drop increase in spiral-wound membrane elements. *Water Res.*, **45**, 1607–1616.

Hoadley A. W. (1977). Potential health hazards associated with *Pseudomonas aeruginosa* in water. In: Bacterial Indicators/Health Hazards Associated with Water, A. W. Hoadley and B. J. Dutka (eds), American Society of Testing and Materials, Philadelphia, PA, pp. 80–114.

Hobbie J. E., Daley R. J. and Jasper S. (1977). Use of nucleopore filter for counting bacteria by fluorescence microscopy. *Appl. Environ. Microbiol.*, **33**(5), 1225–1228.

Howard N. J. (1940). Bacterial depreciation of water quality in distribution systems. *J. Am. Wat. Works Assoc.*, **32**(9), 1501–1506.

Huber S. A. and Frimmel F. H. (1994). Direct gel chromatographic characterization and quantification of marine dissolved organic carbon using high-sensitivity DOC detection. *Environ. Sci. Technol.*, **28**(6), 1194–1197.

Hunt S. M., Werner E. M., Huang B., Hamilton M. A. and Stewart P. 2004. Hypothesis for the role of nutrient starvation in biofilm detachment. *Appl. Environ. Microbiol.*, **70**(12), 7418–7425.

IM (2011). Drinking Water Decree (in Dutch). *Staatsblad*, **293**, 1–90.

Jarvis W. R., Munn V. P., Highsmith A. K., Culver D. H. and Hughes J. M. (1985). The epidemiology of nosocomial infections caused by *Klebsiella pneumoniae*. *Infection Control*, **6**(2), 68–74.

Kahru A. and Vilu R. (1983). On characterization of the growth of *Escherichia coli* in batch culture. *Arch. Microbiol.*, **135**, 12–15.

Karl D. M. (1980). Cellular nucleotide measurements and applications in microbial ecology. *Microbiol. Rev.*, **44**(4), 739–796.

Kayser H. (1900). Die Flora der Strassburger Wasserleitung. Dissertation, Kgl. Bayer. Hofbuchdruckerei H. Kayser, Kaiserslautern.

Ketelaars H. A. M. (1994). Ursachen von Geruchs- und Geschmacksproblemen in der Trinkwasserversorgung und ihre Lösung: eine Übersicht. In: Die Biologie der Trinkwasserversorgung aus Talsperren, H. A. M. Ketelaars, D. J. Nienhüser and R. C. Hoehn (eds), Academic Book Center, De Lier, The Netherlands, pp. 133–153.

Ketelaars D. J., McDaniel J. and Mopper K. (1989). Photochemical source of biological substrates in sea water: implications for carbon cycling. *Nature*, **341**, 637–639.

Kilb B., Lange B., Schaule G., Flemming H. C. and Wingender J. (2003). Contamination of drinking water by coliforms from biofilms grown on rubber-coated valves. *Int. J. Hyg. Environ. Health*, **206**, 563–573.

King J. D. and White D. C. (1977). Muramic acid as a measure of microbial biomass in estuarine and marine samples. *Appl. Environ. Microbiol.*, **33**(4), 777–783.

Kjelleberg S. and Dahlback B. (1984). ATP level of a starving surface-bound and free-living marine *Vibrio* sp. *FEMS Microbiol. Letters*, **24**, 93–96.

Koch R. (1883). Über die neuen Untersuchungsmethoden zum Nachweis der Mikrokosmen in Boden, Luft und Wasser. In: Gesammelte Werke, Bd 1, Georg Thieme, Leipzig 1912, pp. 274–284.

Koch R. (1884). Über die Cholerabakterien. In: Gesammelte Werke, Bd 2 Teil 1, Georg Thieme, Leipzig 1912, pp. 61–68.

Koch R. (1893). Wasserfiltration und Cholera. In: Gesammelte Werke, Bd. 2 Teil 1, Georg Thieme, Leipzig 1912, pp. 183–206.

Kolenbrander P. E. and London J. (1993). Adhere today, here tomorrow: oral bacterial adherence. *J. Bacteriol.*, **175**(11), 3247–3252.

Kool J. L., Carpenter J. C. and Fields B. S. (1999). Effect of monochloramine disinfection of municipal drinking water on the risk of nosocomial Legionnaires' disease. *Lancet*, **353**, 272–277

Korsholm E. and Søgaard H. (1987). Comparison of endotoxin (LPS) measurements and plate counts for bacteriological assessment of drinking water. *Zbl. Bakteriol. Microbiol. Hyg. B.*, **185**(1–2), 121–130.

Krasner S. W., Barrett S. E., Dale M. S. and Hwang C. J. (1989). Free chlorine versus monochloramine for controlling off-tastes and off-odors. *J. Am. Wat. Works Assoc.*, **81**, 86–93.

Laurent P. and Servais P. (1995). Fixed bacterial biomass estimated by potential exoproteolytic activity. *Can. J. Microbiol.*, **41**, 749–752.

LeChevallier M. W. (1990). Coliform regrowth in drinking water: a review. *J. Am. Wat. Works Assoc.*, **82**, 74–86.

LeChevallier M. W., Seidler R. J. and Evans T. M. (1980). Enumeration and characterization of standard plate counts in chlorinated and raw water supplies. *Appl. Environ. Microbiol.*, **40**(5), 922–930.

LeChevallier M. W., Evans T. M. and Seidler R. J. (1982). *Aeromonas sobria in chlorinated drinking water supplies. Microb. Ecol.*, **8**, 325–333.

LeChevallier M. W., Babcock T. M. and Lee R. G. (1987). Examination and characterization of distribution system biofilms. *Appl. Environ. Microbiol.*, **53**(12), 2714–2724.

LeChevallier M. W., Cawthon C. D. and Lee R. G. (1988a). Inactivation of biofilm bacteria. *Appl. Environ. Microbiol.*, **54**(10), 2492–2499.

LeChevallier M. W., Cawthon C. D. and Lee R. G. (1988b). Factors promoting survival of bacteria in chlorinated water supplies. *Appl. Environ. Microbiol.*, **54**(3), 649–654.

LeChevallier M. W., Lowry C. D. and Lee R. G. (1990). Disinfecting biofilms in a model distribution system. *J. Am. Wat. Works Assoc.*, **82**(7), 87–99.

LeChevallier M. W., Welch N. J. and Smith D. B. (1996). Full-scale studies of factors related to coliform regrowth in drinking water. *Appl. Environ. Microbiol.*, **62**(7), 2201–2211.

Lee S. H., O'Connor J. T. and Banerji S. K. (1980). Biologically mediated corrosion and its effects on water quality in distribution systems. *J. Am. Wat. Works Assoc.*, **72**(11), 636–645.

Leeflang K. W. H. (1968). Biologic degradation of rubber gaskets used for sealing pipe joints. *J. Am. Wat. Works Assoc.*, **60**, 1070–1076.

Lehtola M., Miettinen I. T., Vartiainen T. and Martikainen P. J. (1999). A new sensitive bioassay for determination of microbially available phosphorus in water. *Appl. Environ. Microbiol.*, **65**(5), 2032–2034.

Lehtola M., Miettinen I. T. and Martikainen P. (2002). Biofilm formation in drinking water affected by low concentrations of phosphorus. *Can. J. Microbiol.*, **48**, 494–499.

Lehtola M. J., Nissinen T. K., Miettinen I. T., Martikainen P. J. and Vartiainen T. (2004). Removal of soft deposits from the distribution system improves the drinking water quality. *Water Res.*, **38**, 601–610.

Levy R. V., Hart F. L. and Cheetham D. D. (1986). Occurrence and public health significance of invertebrates in drinking water systems. *J. Am. Wat. Works Assoc.*, **78**(9), 105–110.

Lewis W. M. (1987). Tropical limnology. *Ann. Rev. Ecol. Syst.*, **18**, 159–184.

Liao W., Christman R. F., Johnson J. D., Millington D. S. and Haus J. R. (1982). Structural characterization of aquatic humic material. *Environ. Sci. Technol.*, **16**, 403–410.

Lindroth P. and Mopper K. (1979). High performance liquid chromatographic determination of subpicomole amounts of amino acids by precolumn fluorescence derivatization with o-phtaldialdehyde. *Anal. Chem.*, **51**(11), 1667–1674.

Lipponen M. T. T., Suutar M. H. and Martikainen P. J. (2002). Occurrence of nitrifying bacteria and nitrification in Finnish drinking water distribution systems. *Water Res.*, **36**, 4319–4329.

Liu W. K., Healing D. E., Yeomans J. T. and Elliott T. S. J. (1993). Monitoring of hospital water supplies for *Legionella*. *J. Hosp. Infect.*, **24**, 1–9.

Lueschov L. A. and Mackenthun K. M. (1962). Detection and enumeration of iron bacteria in municipal water supplies. *J. Am. Wat. Works Assoc.*, **54**, 751–756.

Macrae I. C. and Edwards J. F. (1972). Adsorption of colloidal iron by bacteria. *Appl. Microbiol.*, **24**(5), 819–823.

Magic-Knezev A. and van der Kooij D. (2006). Nutritional versatility of two *Polaromonas* related-bacteria isolated from biological granular activated carbon filters. In: Recent progress in slow sand and alternative biofiltration processes, Gimbel Rolf, J. D., Graham Nigel and Collins M. Robin (eds), IWA Publishing, London, UK, pp. 303–311.

Magic-Knezev A., Wullings B. and van der Kooij D. (2009). *Polaromonas* and *Hydrogenophaga* species are the predominant bacteria cultured from granular activated carbon filters in water treatment. *J. Appl. Microbiol.*, **107**, 1457–1467.

Maki J. S., LaCroix S. J., Hopkins B. S. and Staley J. T. (1986). Recovery and diversity of heterotrophic bacteria from chlorinated drinking waters. *Appl. Environ. Microbiol.*, **51**, 1047–1055.

Marshall K. C., Stout R. and Mitchell B. (1971). Mechanism of the initial events in the sorption of marine bacteria to surfaces. *J. Gen. Microbiol.*, **68**, 337–348.

McCoy W. F. and Olson B. H. (1985). Fluorometric determination of the DNA concentration in municipal drinking water. *Appl. Environ. Microbiol.*, **49**(4), 811–817.

McCoy W. F. and Olson B. H. (1987). Analysis of the microbiological particulates in municipal drinking-water by scanning electron microscopy/X-ray energy spectroscopy. *Zbl. Bakt. Hyg. B*, **183**, 511–529.

McSwiggan D. A. and Collins C. H. (1974). The isolation of *M. kansasii* and *M. xenopi* from water systems. *Tubercle*, **55**, 291–297.

Michel R., Hoffmann R., Giese A. and Müller K.-D. (1995). Untersuchung von drei Grundwasserwerken auf Vorkommen von Acanthamoeben, Naeglerien and anderen freilebenden Amöben. *Acta Hydrochim. Hydrobiol.*, **23**(5), 202–211.

Monod J. (1949). The growth of bacterial cultures. *Ann. Rev. Microbiol.*, **29**, 265–268.

Mons M. N., Heringa M. B., van Genderen J., Puyker L. M., Brand W., van Leeuwen C. J., Stoks P., van der Hoek J. P. and van der Kooij D. (2013). Use of the threshold of toxicological concern (TTC) approach for deriving target values for drinking-water contaminants, *Water Res.*, **47**, 1666–1678.

Montiel A., Rigal S. and Welle B. (1999). Study of the origin of musty taste in the drinking water supply. *Wat. Sci. Technol.*, **40**(6), 171–177.

Müller G. (1972). Koloniezahlbestimmungen im Trinkwasser. *Gas-Wasserfach*, **113**, 92–100.

Nagy L. A. and Olson B. H. (1982). The occurrence of filamentous fungi in drinking water distribution systems. *Can. J. Microbiol.*, **28**(6), 667–671.

Niemi R. M., Knuth S. and Lundström K. (1982). Actinomycetes and fungi in surface water and in potable water. *Appl. Environ. Microbiol.*, **43**(2), 378–388.

Niquette P., Servais P. and Savoir R. (2000). Impacts of pipe materials on densities of fixed bacterial biomass in a drinking water distribution system. *Water Res.*, **34**(6), 1952–1956.

Nyström A., Grimvall A., Kranz-Rülcker C., Savenhed R. and Akerstrand K. (1992). Drinking water off-flavour caused by 2,4,6-trichloroanisales. *Wat. Sci. Technol.*, **25**(2), 241–249.

Ollos P. J., Huck P. M. and Slawson R. M. (2003). Factors affecting biofilm accumulation in model distribution systems. *J. Am. Wat. Works Assoc.*, **95**(1), 87–108.

Olson B. H. and Nagy L. A. (1984). Microbiology of potable water. *Adv. Appl. Microbiol.*, **30**, 73–132.

Parsons T. R. and Strickland J. D. H. (1962). On the production of particulate organic carbon by heterotrophic processes in sea water. *Deep-Sea Res.*, **8**, 211–222.

Payment P., Coffin E. and Paquette G. (1994). Blood agar to detect virulence factors in tap water heterotrophic bacteria. *Appl. Environ. Microbiol.*, **60**(4), 1179–1183.

Payment P., Sartory D. P. and Reasoner D. J. (2003). The history and use of HPC in drinking-water quality management. In: Heterotrophic plate counts and drinking-water safety, J. Bartram, J. Cotruvo, M. Exner, C. Fricker and A. Glasmacher (eds), Emerging Issues in Water and Infectious disease Series, WHO. IWA Publishing, London, UK, pp. 20–48.

Pederson K. (1990). Biofilm development on stainless steel and PVC surfaces in drinking water. *Water Res.*, **24**(2), 239–243.

Peldszus S., Huck P. M. and Andrews S. A. (1996). Determination of short-chain aliphatic, oxo- and hydroxy-acids in drinking water at low microgram per liter concentrations. *J. Chromatogr. A.*, **723**, 27–34.

Percival S. L., Knapp J. S., Edyvean R. and Wales D. S. (1998). Biofilm development on stainless steel in mains water. *Water Res.*, **32**(1), 243–253.

Picologlou B. F., Zelver N. and Characklis W. G. (1980). Biofilm growth and hydraulic performance. *J. Hydr. Div.*, **106**, 733–746.

Piriou P., Malleret L., Brucet A. and Kiéné L. (2001). Trichloroanisole kinetics and musty tastes in drinking water distribution systems. *Wat. Sci. Technol. Wat. Supply*, **1**(4), 11–18.

Prévost M., Laurent P., Servais P. and Joret J. C. (2005). Biodegradable organic matter in drinking water treatment and distribution. American Water Works Association, Denver, CO.

Prévost M., Rompre A., Baribeau H., Coallier J. and Lafrance P. (1997). Service lines: their effect on microbiological quality. *J. Am. Wat. Works Assoc.*, **89**(7), 78–91.

Pryor M., Springthorpe S., Riffard S. and Broks T. (2004). Investigations of opportunistic pathogens in municipal drinking water under different supply and treatment regimes. *Wat. Sci. Technol.*, **50**, 83–90.

Reasoner D. J. and Geldreich E. E. (1985). A new medium for the enumeration and subculture of bacteria from potable water. *Appl. Environ. Microbiol.*, **49**, 1–7.

Reasoner D. J., Blannon J. C., Geldreich E. E. and Barnick J. (1989). Nonphotosynthetic pigmented bacteria in a potable water treatment and distribution system. *Appl. Environ. Microbiol.*, **55**, 912–921.

Rice E. W., Scarpine P. V., Reasoner D. J., Logsdon G. S. and Wild D. K. (1991). Correlation of coliform growth response with other water quality parameters. *J. Am. Wat. Works Assoc.*, **83**(7), 98–102.

Rickard A. H., McBain A. J., Stead A. T. and Gilbert P. (2004). Shear rate moderates community in freshwater biofilms. *Appl. Environ. Microbiol.*, **70**(12), 7426–7435.

Ridgway H. F., Means E. G. and Olson B. H. (1981). Iron bacteria in drinking-water distribution systems: elemental analysis of *Gallionella* stalks, using x-ray energy-dispersive microanalysis. *Appl. Environ. Microbiol.*, **41**, 288–297.

Rittmann B. E. and Snoeyink V. L. (1984). Achieving biologically stable drinking water. *J. Am. Wat. Works Assoc.*, **76**(10), 106–114.

Rook J. (1974). The formation of haloforms during chlorination of natural waters. *Water Treat. Exam.*, **23**, 234–243.

Rose J. B. (1990). Emerging issues for the microbiology of drinking water. *Water Engineering and Management*, July, 23–29.

Rowbotham T. J. (1980). Preliminary report on the pathogenicity of *Legionella pneumophila* for freshwater and soil amoeba. *J. Clin. Pathol.*, **33**, 1179–1183.

Sack E. L. W., van der Wielen P. W. J. J. and van der Kooij D. (2010). Utilization of oligo- and polysaccharides at microgram-per-litre levels in freshwater by *Flavobacterium johnsoniae*. *J. Appl. Microbiol.*, **108**, 1430–1440.

Sack E. L. W., van der Wielen P. W. J. J. and van der Kooij D. (2011). *Flavobacterium johnsoniae* as a model organism for characterizing biopolymer utilization in oligotrophic freshwater environments. *Appl. Environ. Microbiol.*, **77**(19), 6931–6938.

Sanarelli G. (1891). Über einen Mikro-Organismus des Wassers, welcher für Thiere mit veränderlicher und constanter Temperatur pathogenist. *Zbl. Bakt.*, **9**, 193–199; 222–228.

Scaglia M., Strosselli M., Grazioli V., Gatti S., Bernuzzi A. M. and de Jonckheere J. F. (1983). Isolation and identification of pathogenic *Naegleria australiensis* (*Amoebida, Vahlkampfiidae*) from a spa in northern Italy. *App. Environ. Microbiol.*, **46**(3), 1282–1285.

Schellart J. A. (1986). Disinfection and bacterial regrowth: some experiences before and after stopping the safety chlorination by the Amsterdam Water Works. *Wat. Supply*, **4**(3), 217–225.

Schock M. R. and Lytle D. A. (2011). Internal corrosion and deposition control. In: Water quality & Treatment, J. K. Edzwald (ed.), American Water Works Association, McGraw–Hill Companies, New York, pp. 20.1–20.103.

Schubert R. H. W. (1967). The taxonomy and nomenclature of the genus *Aeromonas* Kluyver and Van Niel 1936. Part 1. Suggestions on the taxonomy and nomenclature of the aerogenic Aeromonas species. *Int. J. Syst. Bact.*, **17**, 23–27.

Schubert R. and Scheiber P. (1975). Das Vorkommen von *Pseudomonas aeruginosa* im Grundwasser, Oberflächenwasser und in Wasserversorgungsleitungen in tropischen Gebieten. *Gas Wasserfach*, **116**, 413–415.

Seidler R. J., Morrow J. E. and Bagley S. T. (1977). Klebsielleae in drinking water emanating from redwood tanks. *Appl. Environ. Microbiol.*, **33**, 893–900.

Servais P., Billen G., Ventresque C. and Bablon G. (1991). Microbial activity in GAC filters at the Choisy-le-Roi treatment plant. *J. Am. Wat. Works Assoc.*, **83**, 62–68.

Servais P., Laurent P., Billen G. and Levi Y. (1992). Étude de la colonisation bactérienne des réseaux de distribution. *T.S.M. –L'Eau*, **87**(6), 321–326.

Sigworth E. A. (1957). Control of odor and taste in water supplies. *J. Am. Wat. Works Assoc.*, **48**(12), 1507–1518.

Silvey J. K. G. and Roach W. W. (1953). Actinomycetes in the Oklahoma City water supply. *J. Am. Wat. Works Assoc.*, **45**, 409–416.

Skadsen J. (1993). Nitrification in a distribution system. *J. Am. Wat. Works Assoc.*, **85**, 95103.

Skjevrak I., Lund V., Ormerod K., Due A. and Herikstad H. (2004). Biofilm in water pipelines; a possible source for off-flavours in the drinking water. *Water Sci. Technol.*, **49**(9), 211–217.

Smalls I. C. and Greaves G. F. (1968). A survey of animals in distribution systems. *J. Soc. Wat. Treat. Exam.*, **17**, 150–187.

Smith D. B., Hess A. F. and Hubbs S. A. (1990). Survey of distribution system coliform occurrences in the United States. In: Proceedings AWWA WQTC Conf., San Diego, CA, pp. 1103–1115.

Snow J. (1855). On the mode of communication of cholera. 2nd edn. John Churchill, London.

Sontheimer H., Heilker E., Jekel M. R., Nolte H. and Vollmer F. H. (1987). The Mülheim process. *J. Am. Wat. Works Assoc.*, **70**, 393–396.

Stalder K. and Klosterkötter W. (1976). Untersuchungen zur Wiederverkeimung von Trinkwasser nach Ozonbehandlung. *Zentralbl. Bakteriol. Hyg.*, **161**, 474–481.

Sykora J. L., Keleti G. and Martinze A. J. (1983). Occurrence and pathogenicity of *Naegleria fowleri* in artificially heated waters. *Appl. Environ. Microbiol.*, **45**(1), 974–979.

Talbot H. W. and Seidler R. M. (1979). Cyclitol utilization associated with the presence of *Klebsiellae* in botanical environments. *Appl. Environ. Microbiol.*, **37**(5), 909–915.

Talbot H. W., Yamamoto D. K., Smith M. W. and Lee R. J. (1980). Antibiotic resistance and its transfer among clinical and nonclinical *Klebsiella* strains in botanical environments. *Appl. Environ. Microbiol.*, **39**(1), 97–104.

Thomas J. M. and Ashbolt N. J. (2011). Do free-living amoebae in treated drinking water systems present an emerging health risk? *Environ. Sci. Technol.*, **45**, 860–869.

Thomas V., Herrera-Rimann K., Blanc D. S. and Greub G. (2006). Biodiversity of amoebae and amoeba-resistant bacteria in a hospital water network. *Appl. Environ. Microbiol.*, **72**(4), 2428–2438.

Thomas V., Loret J. J., Jousset M. and Greub G. (2008). Biodiversity of amoebae ands amoebae-resisting bacteria in a drinking water treatment plant. *Environ. Microbiol.*, **10**(10), 2728–2745.

Thurman E. M., Wershaw R. L., Malcolm R. L. and Pinckney D. J. (1982). Molecular size of aquatic humic substances. *Org. Geochem.*, **4**, 27–35.

Tobin J. O., Beare J., Dunnill M. S., Fisher-Hoch S., French M., Mitchell R. G., Morris P. J. and Muers M. F. (1980). Legionnaires's disease in a transplant unit: isolation of the causative organism from shower baths. *Lancet*, **2**(8186), 118–121.

Tranvik L. J. and Höfle M. G. (1987). Bacterial growth in mixed cultures on dissolved organic carbon from humic and clear waters. *Appl. Environ. Microbiol.*, **53**(3), 482–488.

Trolio R., Bath A., Gordon C., Walker R. and Wyber A. (2008). Operational management of *Naegleria* spp. in drinking water supplies in Western Australia. *Wat. Sci. Technol. Wat. Supply.*, **8**(2), 207–215.

Tsang P. H., Li G., Braun Y. V., Freund L. B. and Tang J. X. (2006). Adhesion of single bacterial cells in the micronewton range. *PNAS*, **103**(15), 5764–5768.

Tuovinen O. H. and Hsu J. C. (1982). Aerobic and anaerobic microorganisms in tubercles of the Columbus, Ohio, water distribution system. *Appl. Environ. Microbiol.*, **44**(3), 761–764.

Valster R. M., Wullings B. A., Bakker G., Smidt H. and van der Kooij D. (2009). Free-living protozoa in two unchlorinated drinking water supplies, identified by phylogenic analysis of the 18S rRNA gene sequences. *Appl. Environ. Microbiol.*, **75**(14), 4736–4746.

Valster R. M., Wullings B. A., van den Berg R. and van der Kooij D. (2011). Free-living protozoa and *Legionella* in three Caribbean water supplies. *Appl. Environ. Microbiol.*, **77**(20), 7321–7328.

van der Kooij D. (1977). The occurrence of *Pseudomonas* spp. in surface water and in tap water as determined on citrate media. *Antonie van Leeuwenhoek J. Microbiol.*, **43**, 187–197.

van der Kooij D. (1997). Multiplication of coliforms at very low concentrations of substrates in tap water. In: Coliforms and E. coli: problem or solution? D. Kay and C. Fricker (eds), The Royal Society of Chemistry, Cambridge, UK, pp. 195–207.

van der Kooij D. (2002). Assimilable organic carbon (AOC) in drinking water: determination and significance. In: Encyclopedia of Environmental Microbiology, D. Bitton (ed.), John Wiley & Sons, New York, pp. 312–327.

van der Kooij D. and Hijnen W. A. M. (1980). Growth of *Aeromonas hydrophila* at low concentrations of substrates added to tap water. *Appl. Environ. Microbiol.*, **39**, 1198–1204.

van der Kooij D. and Hijnen W. A. M. (1981). Utilization of low concentrations of starch by a *Flavobacterium* species isolated from tap water. *Appl. Environ. Microbiol.*, **41**, 216–221.

van der Kooij D. and Hijnen W. A. M. (1984). Substrate utilization by an oxalate-consuming *Spirillum* species in relation to its growth in ozonated water. *Appl. Environ. Microbiol.*, **47**, 551–559.

van der Kooij D. and Hijnen W. A. M. (1985). Determination of the concentration of maltose and starch-like compounds in drinking water by growth measurements with a well-defined strain of a *Flavobacterium* species. *Appl. Environ. Microbiol.*, **49**(7), 65–71.

van der Kooij D. and Hijnen W. A. M. (1988). Nutritional versatility and growth kinetics of an *Aeromonas hydrophila* strain isolated from drinking water. *Appl. Environ. Microbiol.*, **54**, 2842–2851.

van der Kooij D., Visser A. and Oranje J. P. (1982a). Multiplication of fluorescent pseudomonads at low substrate concentrations in tap water. *Antonie van Leeuwenhoek J. Microbiol.*, **48**, 229–243.

van der Kooij D., Oranje J. P. and Hijnen W. A. M. (1982b). Growth of *Pseudomonas aeruginosa* in tap water in relation to utilization of substrates at concentrations of a few micrograms per liter. *Appl. Environ. Microbiol.*, **44**(5), 1086–1095.

van der Wende E., Characklis W. G. and Smith D. B. (1989). Biofilms and bacterial drinking water quality. *Water Res.*, **23**(10), 1313–1322.

van der Kooij D., Vrouwenvelder J. S. and Veenendaal H. R. (1995). Density and composition of biofilms in drinking water distribution systems in the Netherlands, In: Proceedings AWWA Water Quality Technology Conference, 12–16 November, American Water Works Association, Denver, CO, pp. 1055–1062.

van der Kooij D., van Lieverloo J. H. M., Schellart J. and Hiemstra P. (1999). Maintaining quality without a disinfectant residual. *J. Am. Wat. Works Assoc.*, **91**(1), 55–64.

van der Kooij D., Vrouwenvelder J. S. and Veenendaal H. R. (2003). Elucidation and control of biofilm formation processes in water treatment and distribution using the unified biofilm approach. *Wat. Sci. Technol.*, **47**(5), 83–90.

van der Wielen P. W. J. J. and Heijs S. K. (2007). Sulfate-reducing prokaryote communities in two deep hypersaline anoxic basins in the Eastern Mediterranean deep sea. *Environ. Microbiol.*, **9**(5), 1335–1340.

van der Wielen P. W. J. J. and van der Kooij D. (2009). Ammonia-oxidizing bacteria and archae in groundwater treatment and drinking water distribution systems. *Appl. Environ. Microbiol.*, **108**(14), 1430–1440.

van der Wielen P. W. J. J. and van der Kooij D. (2013). Nontuberculous mycobacteria, fungi and opportunistic pathogens in unchlorinated drinking water in the Netherlands. *Appl. Environ. Microbiol.*, **79**(3), 825–834.

vandevivere P. and Kirchman D. L. (1993). Attachment stimulates exopolysaccharide synthesis by a bacterium. *Appl. Environ. Microbiol.*, **59**(10), 3280–3286.

van Lieverloo J. H. M., Hoogenboezem G., Veenendaal W. and van der Kooij D. (2012). Variability of invertebrate abundance in drinking water distribution systems in the Netherlands in relation to biostability and sediment volumes. *Water Res.*, **46**, 4918–4932.

Verweij P. E. (2011). Personal communication.

Victoreen H. T. (1984). Controlling corrosion by controlling bacterial growth. *J. Am. Wat. Works Assoc.*, **76**, 87–89.

Vikesland P. J. and Valentine R. L. (2000). Reaction pathways involved in the reduction of monochloramine by ferrous iron. *Environ. Sci. Technol.*, **34**, 83–90.

Vives-Rego J., Lebaron P. and Nebe-von Caron G. (2000). Current and future applications of flow cytometry in aquatic microbiology. *FEMS Microbiol. Rev.*, **24**, 429–448.

Volker S., Schreiber C. and Kistemann T. (2010). Drinking water quality in household supply infrastructure; a survey of the current situation in Germany. *Int. J. Hyg. Env. Health*, **231**, 204–209.

Vreeburg J. H. G. (2007). Discolouration in Drinking Water Systems: A Particular Approach. PhD thesis, Delft University of Technology. Delft, the Netherlands.

Vreeburg J. H. G. and Boxall J. B. (2007). Discolouration in potable water distribution systems. *Water Res.*, **41**, 519–529.

Vreeburg J. H. G., Blokker E. J. M., Horst P. and van Dijk J. C. (2009). Velocity-based self-cleaning residential drinking water distribution systems. *Wat. Sci. Technol. Wat. Supply*, **9**(6), 635–641.

VROM (2001). Besluit van 9 januari 2001 tot wijziging van het Waterleidingbesluit in verband met de richtlijn betreffende de kwaliteit van voor menselijke consumptie bestemd water (Ministry of Housing, Spatial Planning and Environment). *Staatsblad*, **31**, 1–26.

Wadowsky R. M., Wolford R., McNamara A. M. and Yee R. B. (1985). Effect of temperature, pH and oxygen level on the multiplication of naturally occurring *Legionella pneumophila* in potable water. *Appl. Environ. Microbiol.*, **49**(5), 1197–1205.

Wadowsky R. M., Butler L. J., Cook M. K., Verma S. M., Paul M. A., Fields B. S., Keleti G., Sykora J. L. and Yee R. B. (1988). Growth-supporting activity for *Legionella pneumophila* in tap water cultures and implication of Hartmannellid amoeba as growth factors. *Appl. Environ. Microbiol.*, **54**(11), 2677–2682.

Wajon J. E., Kavanagh B. V., Kagi R. I., Rosich R. S., Alexander R. and Fleay B. J. (1986). Swampy odour in the drinking water of Perth, Western Australia. *Water*, September, 28–32.

Watson S. W., Novitsky T. J., Quinby H. L. and Valois F. W. (1977). Determination of bacterial number and biomass in the marine environment. *Appl. Environ. Microbiol.*, **33**, 940–946.

Weitere M., Bergfeld T., Rice S. A., Matz C. and Kjelleberg S. (2005). Grazing resistance of *Pseudomonas aeruginosa* biofilms depends on type of protective mechanism, developmental stage and protozoan feeding mode. *Environ. Microbiol.*, **7**(10), 1593–1601.

WHO (2004). Guidelines for drinking water quality. 3rd edn, Volume 1, Recommendations. World health Organization, Geneva.

Wierenga J. T. (1985). Recovery of coliforms in the presence of a free chlorine residual. *J. Am. Wat. Works Assoc.*, **77**, 83–88.

Wilczak A., Jackangelo J. G., Marcinko J. P., Odell L. H., Kimeyer G. J. and Wolfe R. L. (1996). Occurrence of nitrification in chloraminated distribution systems. *J. Am. Wat. Works Assoc.*, **88**, 74–85.

Wingerden J., Hambsch B. and Schneider S. (2009). Mikrobiologisch-hygienische Aspekte des Vorkommens von *Pseudomonas aeruginosa* im Trinkwasser. *DVGW Energie/Wasser Praxis*, **60**, 1–7.

Wright R. T. and Hobbie J. E. (1966). Use of glucose and acetate by bacteria and algae in aquatic ecosystems. *Ecology*, **47**(3), 447–464.

Wullings B. A. and van der Kooij D. (2006). Occurrence and genetic diversity of uncultured *Legionella* spp. in drinking water treated at temperatures below 15°C. *Appl. Environ. Microbiol.*, **72**(1), 157–166.

Wullings B. A., Bakker G. and van der Kooij D. (2011). Concentration and diversity of uncultured *Legionella* spp. in two unchlorinated drinking water supplies with different concentrations of natural organic matter. *Appl. Environ. Microbiol.*, **77**(3), 634–641.

Yee R. B. and Wadowsky R. M. (1982). Multiplication of *Legionella pneumophila* in unsterilized tap water. *Appl. Environ. Microbiol.*, **43**(6), 1330–1334.

Zaitlin B. and Watson S. B. (2006). Actinomycetes in relation to taste and odour in drinking water: myths, tenets and truths. *Water Res.*, **40**, 1741–1753.

Zhang S. and Huck P. M. (1996). Removal of AOC in biological water treatment processes: a kinetic model. *Water Res.*, **30**(5), 1195–1207.

Zobell C. E. and Grant C. W. (1943). Bacterial utilization of low concentrations of organic matter. *J. Bacteriol.*, **45**, 555–564.

Zacheus O. M. and Martikainen P. J. (1995). Occurrence of heterotrophic bacteria and fungi in cold and hot water distribution systems using water of different quality. *Can. J. Microbiol.*, **41**, 1088–1094.

Zacheus O. M., Lehtola M. J., Korhonen L. K. and Martikianen P. J. (2001). Soft deposits, the key site for microbial growth in water distribution networks. *Water Res.*, **35**(7), 1757–1765.

Zimmermann O. E. R. (1890). Die Bakterien unserer Trink- und Nutzwasser, insbesondere des Wassers der Chemnitzer Wasserleitung, Carl Brunner, Chemnitz.

Chapter 2

Measurement of biostability and impacts on water treatment in the US

M. W. LeChevallier

2.1 INTRODUCTION

To understand the role of biodegradable organic matter (BOM) and biological stability (e.g. biostability) in drinking water systems in the United States of America (US), it is necessary to first appreciate the regulatory context for provision of safe drinking water in the US. The US Environmental Protection Agency (USEPA) is required by the Safe Drinking Water Act (USEPA 1974) to establish regulations for contaminants that have a known or anticipated adverse human health effect. As a result of concerns over acute microbial health risks, and numerous waterborne outbreaks where disinfection was either not used or inappropriately applied, the USEPA has enacted several regulations where water supplies must be disinfected:

- The Surface Water Treatment Rule (SWTR; USEPA 1989a) requires disinfection of all surface water supplies and groundwater under the influence of surface water. Disinfection is required to achieve at least 0.5 log of *Giardia* inactivation and 2 logs of viruses (for an overall treatment of 3 logs for *Giardia* and 4 logs for viruses). The required maintenance of a continuous disinfectant of at least 0.2 mg/L in the treated plant effluent and a detectable disinfectant residual at 95% of the coliform monitoring locations in the distribution system.
- The Long Term II Enhanced Surface Water Treatment Rule (LT2; USEPA 2006a) requires surface water utilities (and groundwater under the influence) with poor source water quality (e.g. *Cryptosporidium* levels >1 oocysts/L) to provide additional treatment including disinfection. Unfiltered surface water systems must provide for 3 logs of *Cryptosporidium* inactivation.
- The Groundwater Rule (USEPA 2006b) requires 4 logs of virus inactivation for groundwaters contaminated by fecal contamination; however, there is no

requirement for maintenance of a disinfectant residual in the distribution system. Systems that rely of disinfection must continuously monitor the residual at the point of treatment.

In addition, the Safe Drinking Water Act specifies that state primacy agencies can enact regulations that are more stringent than the federal regulations. As a result, many states require disinfection for all public water supplies (surface and ground water). In many cases BOM can be thought of as a disinfectant by-product because chemical oxidation of drinking water supplies can convert refractory organic matter into BOM.

Concern about the biological stability of treated drinking water supplies in the US is generally triggered by the growth of total coliform bacteria in distribution system biofilms. Because disinfection is considered a "best available technology" for control of total coliform bacteria (USEPA 1989b), disinfection practices to try to control the growth of microorganisms can actually aggravate the problem. Therefore, the approach to achieving high quality, biologically-stable water in the US is a balance between maintaining the required levels of disinfectants and minimizing the levels of BOM.

2.2 MEASUREMENT OF BIODEGRADABLE ORGANIC MATTER IN WATER

Despite the plethora of published methods to measure the biological stability of water (Huck 1990; Volk 2002), most water systems and research studies in the US have used the assimilable organic carbon (AOC) or the biodegradable dissolved organic carbon (BDOC) assays. It is important to note that there have been many variations of these tests (Tables 2.1 and 2.2) so one has to be careful when reviewing the literature to understand the origins of the assay. Most studies have used the AOC assay by van der Kooij and Hijnen (1984) or the BDOC assay by Joret and Levy (1986) with some slight modifications. For example, Kaplan *et al.* (1993) showed that 40 mL vials could be used (instead of 1-L flasks) with minimal impact on the AOC test. Likewise, LeChevallier *et al.* 1993b) measured adenosine triphosphate (ATP) instead of plate counts to determine the bacterial levels in the AOC test.

Both the AOC and BDOC assays are difficult to perform and require a high level of proficiency. The need to prepare specially cleaned glassware, maintain stocks of the test organisms, perform extensive dilutions, count colonies, and calculate AOC equivalents exceed the abilities and resources of most water systems. A recent survey of North American water utilities (Evans *et al.* 2010), even those that were practicing biological filtration, found that few routinely measured AOC or BDOC levels – presumably due to the difficulty and expense of the assays. The commercial rate, for example, for an AOC assay at US national testing laboratories can exceed $400 per test.

Table 2.1 AOC methodologies.

Author	Water preparation	Inoculation (final inoculum concentration)	Incubation conditions	Measured parameter	Result expression
Van der Kooij *et al.* (1982); Van der Kooij and Hijnen (1984)	Pasteurization	Pure strains *P. fluorescens* P17 and *Spirillum* NOX (500 CFU/ml)	20 days 15°C	CFU/ml	AOC calibration in known solutions of sodium acetate (μg C eq. acetate/l)
Kaplan *et al.* (1993)	Pasteurization	*P. fluorescens* P17 + *Spirillum* NOX (500–1000 CFU/ml)	9 days 20°C	CFU/ml	AOC
LeChevallier *et al.* (1993b)	Pasteurization	*P. fluorescens* P17 + *Spirillum* NOX (10^4 CFU/ml)	5 days 22°C	ATP	AOC
Bradford *et al.* (1994)	Filtration	*P. fluorescens* P17	12 h 20°C	Cell elongation Epi. counts	AOC
Kemmy *et al.* (1989)	Filtration	Mixture of 4 strains: *Pseudomonas fl. Curtobacterium Corynebacterium* 1 species of coryneform type	6 days 20°C	CFU/ml	AOC Calibration in solution of mixed organic compounds (μg C/l)
Stanfield and Jago (1987)	Filtration	Bacteria from – Raw water – Sand filtered water	Until max. Growth 20°C	ATP	AOC – standard conversion factor (μg C/l)
Reasoner and Rice (1989)	Filtration	*Enterobacter cloacae, Escherichia coli, Klebsiella oxytoca*	5 days 20°C	CFU/ml	log N5/N0

Table 2.2 BDOC methodologies.

Author	Water preparation	Inoculation	Incubation conditions	Measured parameter	Result expression
Servais *et al.* (1987)	Filtration	Suspended Bacteria from: – river water	28 days 20°C	DOC	BDOC = DOCi-DOCf
Joret and Levi 1986); Joret *et al.* (1989)	–	Bacteria fixed on sand	1 week 20°1C	DOC	BDOC = DOCi-DOCmini
Mogren *et al.* (1989)	–	Bacteria fixed on sand	5 days 20°1C	DOC	BDOC = DOCi–DOC 5 days
Frias *et al.* (1992)	Filtration	Column with bacteria fixed on porous glass particles	2.5 h 20°1C	DOC	BDOC = DOCinflow-DOCoutflow

2.3 CONCENTRATIONS OF AOC AND BDOC IN US DRINKING WATER SUPPLIES

Volk and LeChevallier (2000) determined that AOC levels in 94 North American distribution systems ranged from 20 to 214 μg/L, with a median of 100 μg/L, and that levels of BDOC from 30 North American water systems ranged from 0 to 1.7 mg/L, with a median level of 0.38 mg/L (Fig. 2.1 and Fig. 2.2). These values are similar to those of other North American surveys (Kaplan and Bott 1989; Kaplan *et al.* 1993; Bradford *et al.* 1994; Najm *et al.* 2000).

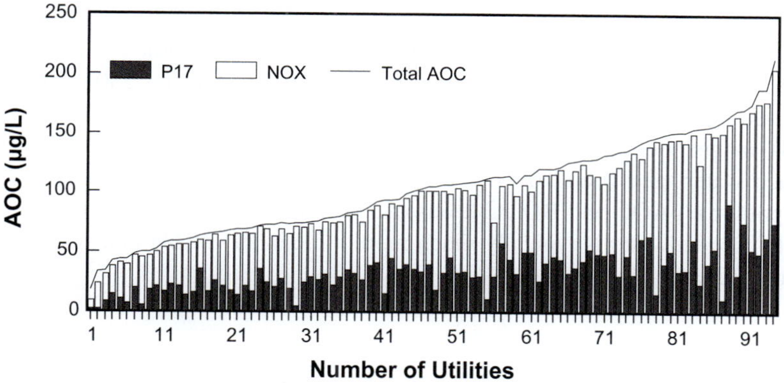

Figure 2.1 AOC levels in 94 North American water systems. Utilities are numbered 1 through 94 along x axis. (From: Volk and LeChevallier 2000. Copyright © Journal AWWA).

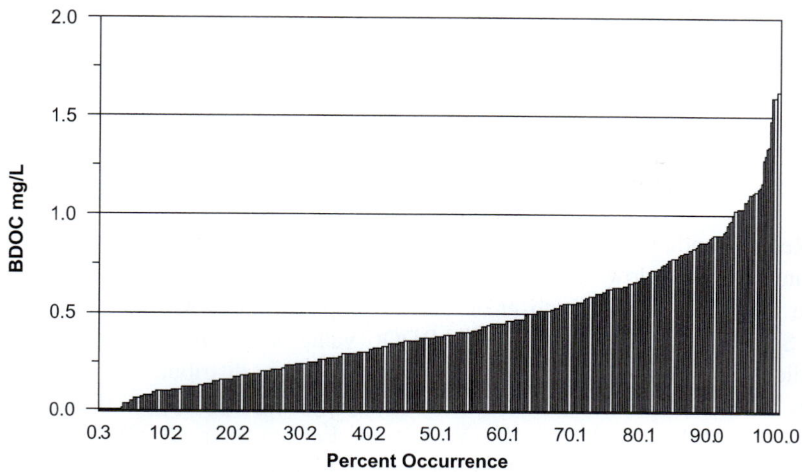

Figure 2.2 BDOC levels in 30 North American water systems. (From: Volk and LeChevallier 2000. Copyright © Journal AWWA).

Kaplan and Bott (1989) found raw water AOC levels ranged from 48 to 607 µg/L at 10 sites from two surface water supplies in Pennsylvania. Kaplan *et al.* (1992) examined 53 surface water sources in North America and found 62 percent of the raw water AOC values were above 100 µg/L. The mean and median AOC levels were 145 and 143 µg/L, respectively. Bradford *et al.* (1994) reported AOC levels ranging between 94 and 275 µg/L (mean 168 µg/L) for two surface water reservoir sites and 75 to 731 µg/L (mean 317 µg/L) for 16 locations along the Santa Ana River in California. Kaplan and LeChevallier (1993) examined 13 surface water treatment plants in 10 states and found plant effluent AOC levels ranging between 34 and 247 µg/L (median 146 µg/L). In detailed studies of four treatment plants, Najm *et al.* (2000) showed a range of plant effluent varied over the 12 months of monitoring. For example, in one system studied, raw water and plant effluent AOC levels ranged from around 50 µg/L to 250 µg/L, while groundwater concentrations varied between 10 and 79 µg/L. Similarly, BDOC levels ranged between 0.05 and 0.85 mg/L in the plant effluent and blended surface water, and 0 and 0.3 mg/L in the groundwater. These results illustrate two points: (i) AOC and BDOC levels can be variable within a system over time, and (ii) groundwater is typically much more biologically stable that surface water (Figure 2.3).

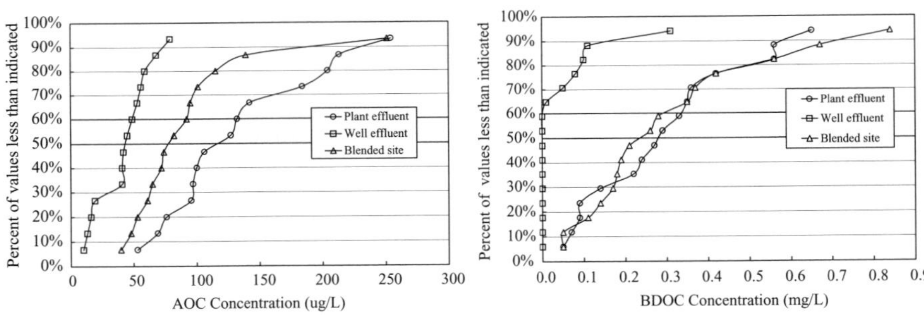

Figure 2.3 Cumulative distribution plots of AOC and BDOC concentrations in the surface water plant effluent, groundwater effluent and blended site. (From Najm *et al.* 2000. Copyright © 1996–2012 Water Research Foundation.)

Recommendations regarding the threshold concentration of biodegradable organic matter below which water can be considered biologically stable have been offered. Van der Kooij *et al.* (1989) proposed an AOC value of 10 µg/L, and Servais *et al.* (1995) suggested a BDOC value of 0.15 mg of carbon per liter. Studies of coliform occurrences in North American distribution systems have been associated with AOC concentrations greater than 50 µg/L (LeChevallier *et al.* 1991a) and 100 µg/L (Volk and LeChevallier 2000) and BDOC consumption in the French distribution system higher than 0.15 mg of carbon per liter (Volk and Joret 1994). High levels of AOC can stimulate coliform growth in distribution system biofilms (LeChevallier *et al.* 1996; Volk and LeChevallier

2000). On average, free chlorinated systems with AOC levels greater than 100 μg/L had 82% more coliform-positive samples, and the coliform densities were 19 times higher than in free chlorinated systems with average AOC levels less than 99 μg/L. However, levels of AOC alone do not dictate the occurrence of coliform bacteria in drinking water but are only one factor. High water temperature, high AOC/BDOC levels, and low disinfectant residuals especially may result in microbial growth problems (Volk and LeChevallier 2000). Figure 2.4 illustrates a decision tree that graphically depicts combinations of threshold values above which the probability of coliform occurrence is increased (Volk and LeChevallier 2000). As more of the threshold values are exceeded (temperature, AOC level, disinfectant residual), the probability of coliform occurrences is increased.

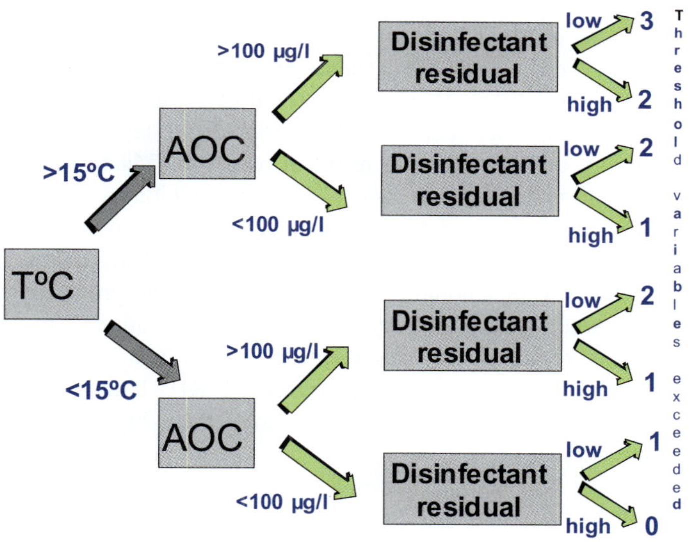

Figure 2.4 Decision tree for coliform occurrences in drinking water. (From: Volk and LeChevallier 2000. Copyright © Journal AWWA).

Data summarized in Table 2.3 show that the frequency of coliform occurrence was less than 2% when no threshold criteria were exceeded and increased to 16% when all three criteria were exceeded. The magnitude (the number of positive samples per event) also increased with a greater exceedance of threshold criteria. Similar models developed for specific systems have yielded higher predictive probabilities (Volk *et al.* 1992). Growth of the opportunistic pathogen, *Mycobacterium avium*, has been correlated with levels of AOC and BDOC in drinking water supplies (Falkinham *et al.* 2001), and biofilm densities of *M. avium* in controlled studies increased with increasing levels of AOC (Norton *et al.* 2004).

Table 2.3 Relationship between threshold criteria and coliform occurrence.

Number of positive criteria*	Total number of events	Coliform positive samples	Number of coliform episodes	Frequency of coliform observation (%)
0	160	3	3	1.9
1	292	18	15	5
2	191	24	16	8.4
3	62	26	10	16

*Criteria include temperature >15°C, AOC levels >100 μg/L, and disinfectant residual <0.5 mg/L for free chlorine or <1.0 mg/L for chloramines.

Organic carbon is often a growth-limiting nutrient; however, limitation of bacterial growth because of the availability of phosphorus also has been demonstrated in some distribution systems (Lehtola *et al.* 2002). In such cases, the use of phosphate-based corrosion inhibitors could lead to a decrease in biological stability of the water and result in regrowth (Miettinen *et al.* 1997). However, in chlorinated system, such corrosion inhibitors can reduce the disinfectant demand from oxidation of metallic surfaces and thus improve the inactivation of biofilm bacteria (LeChevallier *et al.* 1993a). Excess ammonia resulting from the use of chloramination may result in the growth of nitrifying bacteria and in water quality problems related to nitrification (e.g. accelerated loss of chlorine residuals, increased nitrite levels, and stimulated growth of HPC bacteria). Rittmann and Snoeyink (1984) found that ammonia concentrations in groundwater supplies frequently were high enough to cause biological instability.

2.4 IMPACT OF WATER TREATMENT ON BOM

The processing of water through a potable water treatment plant can result in profound changes in the nature and composition of biodegradable organic matter. The water treatment process itself can fractionate the organic matter into numerous subcategories making it a challenge to monitor and track the impact of treatment on the biological stability of the finished water. Summarized below are some of the key factors affecting BOM treatment.

2.4.1 Watersheds

The degree of watershed protection has been associated with the level of AOC in finished drinking water (Volk and LeChevallier 2002). Watersheds that were fully protected had AOC levels 42% lower than unprotected watersheds. This is not too surprising since wastewater or runoff from livestock or agricultural activities can be linked to molecular markers such as caffeine or faecal steroids in watersheds heavily impacted by anthropogenic activity (Kaplan *et al.* 2002).

2.4.2 Disinfection

Oxidation of NOM has been widely reported to increase the biodegradable fraction of organic carbon in water. For example, ozone applied alone or in conjunction with hydrogen peroxide or heterogeneous catalysts can oxidize a larger spectrum of organic compounds and increase the oxidation level of molecules (Paillard *et al.* 1991; Volk *et al.* 1997). Using the different biological tests for estimating BOM levels, researchers reported an increase in biodegradability of dissolved natural organic matter following ozonation (Van der Kooij and Hijnen 1984; Volk *et al.* 1993; Prévost *et al.* 2005). Studies have shown that chlorination can produce similar increases in AOC levels after treatment (van der Kooij 1987; LeChevallier *et al.* 1991a). Other oxidants like chlorine dioxide and potassium permanganate can also increase the level of AOC in treated supplies.

Of greater concern is that different measures of biodegradable organic matter may yield different conclusions depending on the treatment process. For example, application of pre-chlorination at one utility in New Jersey increased AOC levels by an average of 240% (Figure 2.5), whereas BDOC levels were reduced (8%) in the settled water (Volk and LeChevallier 2002). Therefore, measurement of BDOC at this plant would not have indicated the importance of the pre-disinfection process on the biostability of this water supply.

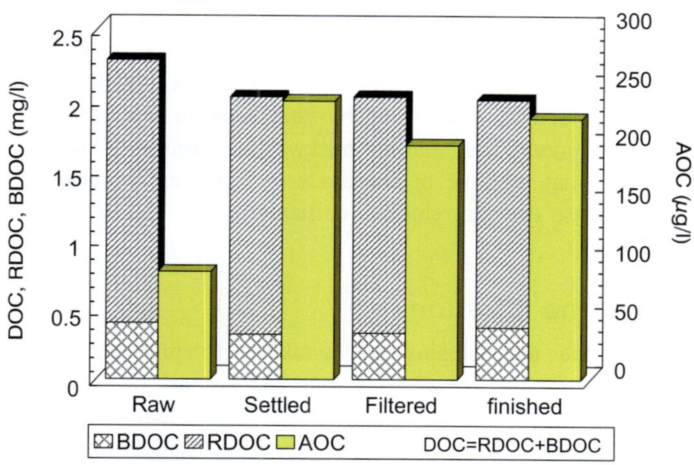

Figure 2.5 Changes in AOC and BDOC through treatment at one location (Site NJ102a) in New Jersey, USA. Note how AOC levels are increased after prechlorination in the settled water whereas BDOC levels are slightly reduced (From: Volk and LeChevallier 2002. Copyright © Journal AWWA).

2.4.3 Coagulation and sedimentation

The removal of organic carbon by coagulation/sedimentation depends on the water matrix characteristics and the settling conditions (coagulant type and doses, coagulation conditions, settling type, etc.). (Jacangelo *et al.* 1995). Volk *et al.* (2000) examined coagulation conditions for 10 systems and found that enhanced coagulation for removal of DOC showed concurrent improvements in removal of BDOC, but variable impacts on AOC (Table 2.4). Coagulation, especially at low pH levels, preferentially removes large molecular weight humic molecules (Edzwald 1994; Krasner and Amy 1995) and the low efficiency of AOC removal during coagulation would be expected if the AOC pool was composed of small, non-humic compounds. As indicated before, it is important to recognize that the apparent effectiveness of the treatment processes can depend on the choice of biostability measurement.

2.4.4 Granular media filtration

When there is no prechlorination prior to filtration, biomass can form on the filter media in a biofilm, leading to the assimilation of biodegradable materials from the water. Organic removal during filtration is impacted by the quality and quantity of organic matter, the media type, the contact time, and the backwashing strategy. Granular activated carbon is most commonly used as a biological filter because the large surface area of the media can accommodate a high level of biomass (LeChevallier *et al.* 1991b), although sand and anthracite media can also be effective biologically active filters. Macro-porous activated carbons provide bacteria with a large number of attachment sites protected from the abrasive action of backwash. The performance of biologically active filters varies with temperature and removal of biodegradable material also increases when contact times vary from 5 to 20 minutes (Huck *et al.* 1991; LeChevallier *et al.* 1991b). Finally, the effects of backwashing on biofilter bacteria can vary depending on the type of support media used and the backwashing strategy. Backwashing with chlorinated water may remove or inactivate a significant portion of the fixed biomass and decrease organic removal, particularly for sand or anthracite filters (Miltner *et al.* 1992).

2.4.5 Membrane filtration

Different membrane technologies are available to produce drinking water including: reverse osmosis (removal of ions and organic matter), nanofiltration membrane (removal of organic compounds larger than 400 Daltons and polyvalent ions), ultrafiltration (removal of colloids and molecules larger than 10,000 Daltons), and micro-filtration (for the elimination of turbidity particles, parasites, bacteria, and some viruses). Micro-filtration and ultrafiltration are not suitable for organic matter removal unless pre-treatments are applied. The addition of powdered activated carbon increased DOC removal efficiencies of

Table 2.4 Organic carbon removal during enhanced coagulation.

Site	Required removal	DOC	BDOC	AOC	DOC removal (%)		BDOC removal (%)		AOC removal (%)	
	(%)	(mg/l)	(mg/l)	(μg/l)	Baseline	Enhanced	Baseline	Enhanced	Baseline	Enhanced
JB	40	2.62	0.87	62	13	45	12	40	0	0
MC	50	10.74	2.44	112	56	56	40	40	89	89
SJ	30	6.08	1.75	482	28	51	54	74	0	0
HO	35	5.59	1.96	133	36	58	51	63	0	1
NO	30	2.75	0.71	152	12	18	0	1	12	19
EW	20	2.63	0.73	242	31	31	23	23	64	64
TA	25	15.09	2.14	120	70	76	43	55	80	Inhibition
PL	20	2.5	0.39	113	10	28	15	5	11	2
GR	40	3.48	0.70	138	17	32	23	21	0	Inhibition
AL	25	5.1	1.08	302	14	34	44	56	0	47
		AVG	**removal**	**(%)**	25	43	30	38	–	–

Source: From Volk *et al.* (2000).

ultrafiltration membranes up to 40–50% (Richard 1994). Various configurations of nanofiltration membranes can result in a high retention of DOC and BDOC (from 50 to 99%), but may not have similar efficiencies for removal of small molecular weight compounds that contribute to AOC.

Experiments to fractionate organic compounds on the basis of molecular weight showed that for river water most of the AOC was in the fraction greater than 1000 daltons (Table 2.5). However, for reservoir water most of the AOC was contained in the small (<1000 mw) molecular weight fraction. No AOC removal was observed when this reservoir water was tested for enhanced coagulation (site GR, Table 2.4).

Table 2.5 Fractionation of AOC and DOC by ultra-filtration.

Raw water	Water fraction	AOC concentration (μg/L)	DOC concentration (mg/L)
Mississippi River	Raw water	184 + 30	5.9
	Retentate	84 + 28	4.3
	Filtrate	43 + 9	1.9
Putnam Reservoir	Raw water	138 + 29	3.2
	Retentate	31 + 9	2.2
	Filtrate	150 + 11	1.2

Source: From: Volk *et al.* (2000).

Similar fractionation of the biostability of nanofiltered water has been observed for a water system in Florida (Escobar and Randall 1999). The nanofiltration membrane achieved complete removal of BDOC, but did not affect the level of AOC (Table 2.6). Although the impact of nanofiltration on AOC and BDOC levels will be dependant upon the specifics of the membrane, the manufacturer, its pore size and physical properties, the purpose of this example is to show that determination of only AOC or BDOC may lead to incomplete conclusions regarding the biostability of water.

2.4.6 Bank infiltration

Bank-filtered water is surface water seeping from the bank or bed of a river or lake to the production wells of a water treatment plant (LeChevallier and Au 2004). During the ground passage, water quality changes due to microbial, chemical and physical processes and by mixing with groundwater. The efficiency of the underground passage process depends on a number of factors: the quality of the surface water (turbidity, dissolved organic matter, oxygen, ammonia, and nutrients), the composition and porosity of the soil, the residence time of the water in the soil, and temperature. This efficiency can vary over time depending on the difference between the source water (e.g. river stage) and ground water levels. This difference can influence the degree of ground water mixing and residence time of the infiltrated surface water. Variations in the performance of bank infiltration for removal of AOC and BDOC have been reported by Weiss *et al.* (2004) for three different infiltration systems. For two of the infiltration systems AOC and BDOC removals ranged from 70 to 93%, and 90–95%, respectively. However the third bank infiltration system achieved AOC and BDOC removals of only 14 and 24% despite a set back distance of 36.6 m. It is not known why the third system performed so poorly for BOM removal, particularly when bacterial spores and bacteriophage were completely removed (Weiss *et al.* 2003).

Table 2.6 Impact of Nanofiltration on Removal of AOC and BDOC.

Site	University of Central Florida		American Water		Lyonnaise des Eaux
	AOC (μg/L)	BDOC (mg/L)	AOC (μg/L)	BDOC (mg/L)	BDOC (mg/L)
Raw Water	62	1.96	57	1.77	–
Nanofilter Effluent	98	<0	74	0	0
Plant effluent	62	0	70	0	–
Distribution System	100	0	74	0	0

Source: Adapted from: Najm *et al.* (2000).

2.4.7 Recycling of backwash water

For many water treatment plants, particularly in arid or water-scarce areas with limited raw water resources, it is often necessary to reuse backwash water. When the water is recycled, recycling of microbial contaminants is a concern and the USEPA has enacted restrictions on the handling of filter backwash water (USEPA 2001). Cornwall *et al.* (2001) however, reported that the spent filter backwash can contain high levels of AOC, ranging from 240 to 740 μg/L. When this water is periodically returned to the head of the treatment process it could result in spikes in the levels of BOM.

2.5 MATERIALS IN CONTACT WITH WATER

Some pipe materials may leach constituents that support bacterial growth. For example, pipe gaskets and elastic sealants (containing polyamide and silicone) can be a source of nutrients for bacterial proliferation. Organisms associated with joint-packing materials include *Pseudomonas aeruginosa, Chromobacter* spp., *Enterobacter aerogenes,* and *Klebsiella pneumoniae* (Schoenen 1986). Pump lubricants should be non-nutritive to avoid bacterial growth in treated water (White and LeChevallier 1993). Coating compounds for storage reservoirs and standpipes can contribute to organic polymers and solvents that may support regrowth of heterotrophic bacteria (Schoenen 1986; Thofern *et al.* 1987). Liner materials may contain bitumen, chlorinated rubber, epoxy resin, or tar-epoxy resin combinations that can support bacterial regrowth (Schoenen 1986). PVC pipes and coating materials may leach stabilizers that can result in bacterial growth. Studies performed in the United Kingdom showed coliform isolations were four times higher when samples were collected from plastic taps than from metallic faucets (cited in Geldreich and LeChevallier 1999). Although procedures are available to evaluate the growth stimulation of different materials (Bellen *et al.* 1993); these tests are not typically applied in the US.

2.6 DEVELOPMENT OF A BIOLUMINESCENCE AOC METHOD

2.6.1 Bioluminescence AOC assay

The prior section illustrates the need for water utilities to monitor changes in AOC and BDOC levels at all stages of the treatment process, however, as demonstrated by Evans *et al.* (2010), few systems perform measurements of AOC or BDOC – presumably because these methods are very complicated. Although both tests are conceptually easy to perform (e.g. plate counts or DOC measurements), in practice they require a certain "art" and require careful attention by the analyst to the details of the methodology. The development of a bioluminescence AOC method overcomes some of the complexity of the conventional AOC test.

Haddix *et al.* (2004) developed bioluminescent strains of the AOC test bacteria (*Pseudomonas fluorescens* P-17 and *Spirillum* NOX) with a *luxCDABE* operon fusion and inducible transposons. The resulting strains showed a linear relationship between plate count densities and luminescence in water with different levels of AOC, and the growth of the luminescent strains compared well to the parent strains ($r^2 = 0.97$ and 0.99 for P17 and NOX, respectively).

The growth of the mutagenized strains can easily be monitored using a programmable luminometer to determine the maximum (N_{max}) cell yield (Weinrich *et al.* 2009). Standard curves using acetate carbon (ranging from 0 to 1000 µg/L) resulted in coefficients of determination (r^2) between luminescence units and acetate carbon of 0.95 for P-17 and 0.89 for NOX (Figure 2.6).

The advantage of using a programmable luminometer is that measurements can be taken at frequent intervals during the growth cycle. All this data can be fitted into the classical Monod growth curve to determine the bacterial growth kinetics and the maximum growth rate (μ_{max}) (Figure 2.7). One source of error in the conventional AOC is the possibility that the analyst could miss the peak in the growth curve by conducting only daily measurements; which would result in an underestimation of the true AOC level.

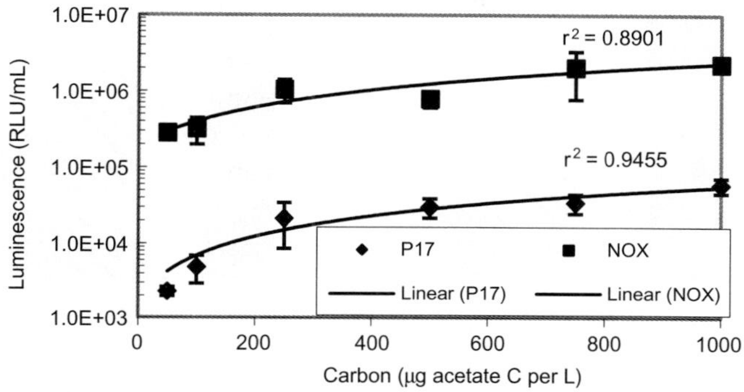

Figure 2.6 Standard curve results of AOC luminescence.

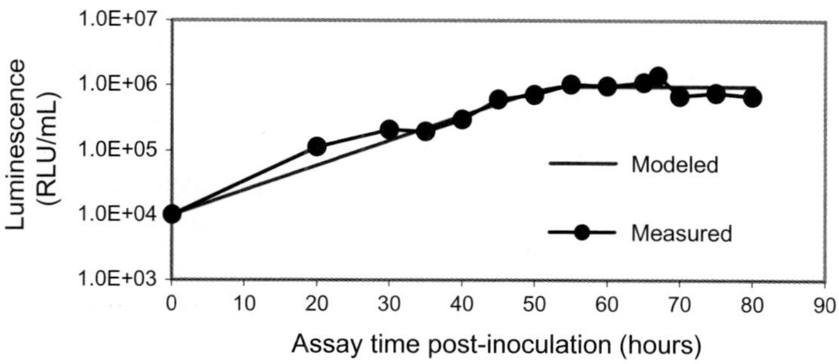

Figure 2.7 Example of a bioluminescence growth curve (with strain NOX).

In natural water samples containing a mixture of biodegradable organic compounds, a multi-substrate model was found to fit the data better (Weinrich *et al.* 2009). The modelling approach also determines the rate of substrate utilization (μ_{max}), which provides additional information regarding the biodegradability of the nutrients in the water sample. Figure 2.8 shows results from the plant effluent, storage tank and three points within a reclaimed water distribution system and demonstrates that the AOC from the plant effluent was rapidly utilized, but that AOC at the end of the distribution system (DS3) was more slowly biodegraded in the bioluminescence AOC test.

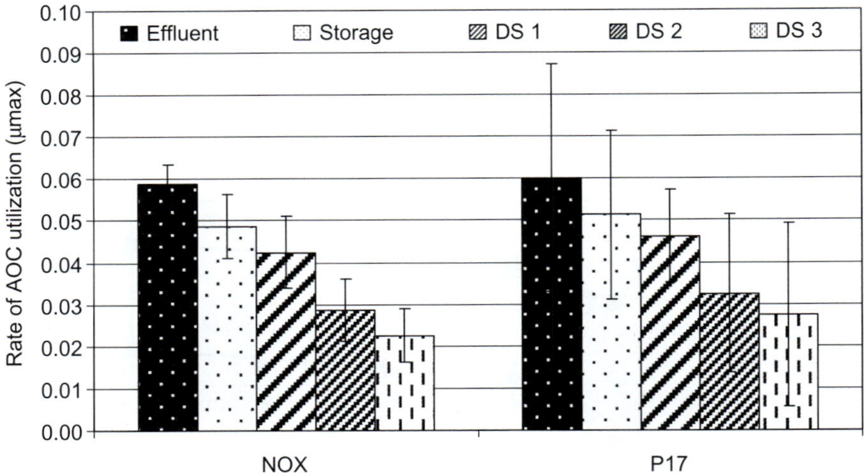

Figure 2.8 Mean rates of AOC utilization (µmax) in a reclaimed water distribution system. Error bars are ± SD. (From: Weinrich *et al.* 2009. Copyright © American Society for Microbiology).

2.6.2 Application of the bioluminescence AOC assay

The ease of the bioluminescence AOC test has allowed its application for a range of water types including raw water, processed water, finished drinking water, and reclaimed waste water.

Limitations on the availability of new sources of water, scarcity of existing supplies, and concerns about the impact of climate change on the future sustainability of water supplies are all factors increasing the interest in the use of reclaimed wastewater. One of the concerns about use of highly treated reclaimed water is the potential for bacterial growth in reclaimed distribution systems. The bioluminescence test was used to monitor reclaimed water systems and found that average AOC levels range between 150 and 1400 µg/L acetate carbon equivalents (Weinrich *et al.* 2010a). Jjemba *et al.* (2010) showed that the high levels of AOC in reclaimed waters could result in the occurrence of opportunistic pathogens, notably *Aeromonas*, *Legionella*, *Mycobacterium*, and *Pseudomonas*, more frequently than indicator bacteria (enterococci, coliforms, and *E. coli*). Weinrich *et al.* (2010a) reported that membrane bioreactors produced effluent waters with AOC levels averaging 147 µg/L; six times lower than the average of 889 µg/L for the other wastewater technologies examined. The authors emphasized that reclaimed water with low levels of AOC is feasible depending on the choice of treatment technology, specific operating parameters, and the final disinfectant. All membrane bioreactor systems used UV for final disinfection since chemical disinfection (chlorine or ozone) could increase effluent AOC levels by as much as 14-fold (Weinrich *et al.* 2010a).

Water systems that do provide biologically active filtration to control BOM levels in treated water supplies typically don't monitor or manage the filters to achieve a particular level of treatment – it's usually a passive process. Brown and Huck (2009), however, showed that biological filters can be operated in a manner (with phosphate addition) to improve the removal of dissolved organic carbon, manganese, 2-methylisoborneol (MIB) and certain pharmaceuticals. An on-going Water Research Foundation project (#4231) is using the bioluminescence AOC test with other measures of on-line performance to monitor the biological activity of the filters. Promising technologies include on-line monitors of dissolved oxygen, TOC, and UV_{254} absorbance. Monitoring of the filters showed 95% removal of AOC in September when water temperatures were 25°C and only 50% removal in January when temperatures were 3°C (Figure 2.9). Analysis of the substrate utilization rate (μ_{max}) showed that the remaining BOM in September was more slowly utilized than organic compounds in January.

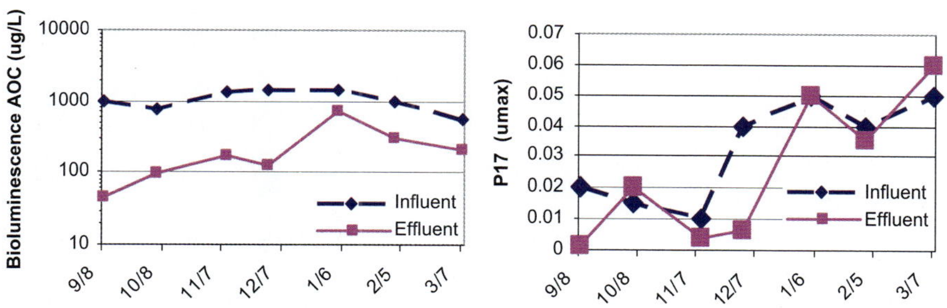

Figure 2.9 Removal of AOC across a biologically active GAC filter (left). Substrate utilization rate for P17 (right).

Another Water Research Foundation project (#4312) is evaluating "An Operational Definition of Biostability in Drinking Water" for North American water systems. The objective of the project is to develop an integrated decision support system that embodies the factors affecting biostability and practical indicators of biostability including: 1) change in biodegradable organic matter in the distribution system, 2) disinfectant demand and biodegradation of haloacetic acids; 3) biofilm formation and corrosion rates, and 4) biological consumption of ammonia, nitrate, and phosphate. When completed, along with the results of the biological filtration project (#4231), the research will equip North American water utilities with the tools to monitor and control biological filters and goals for maintaining biological stability in the distribution system.

2.6.3 Development of a salt water bioluminescence AOC test

Biofilm development on reverse osmosis membranes is a major problem for seawater desalination systems. The effectiveness of pre-treatment processes for BOM removal has not well described, in part because salt water methods for AOC measurement had not been available until recently. The traditional AOC strains, P17 and NOX, will not grow in saline solutions. Building upon the previous bioluminescence assay, Weinrich *et al.* (2010b) developed a salt water AOC assay using *Vibrio harveyi*, a marine organism that exhibits constitutive luminescence and exhibited a broad nutritional diversity; capable of utilizing amino acids, carboxylic acids, carbohydrates, aldehydes and alcohols, but not ascorbic acid (Figure 2.10).

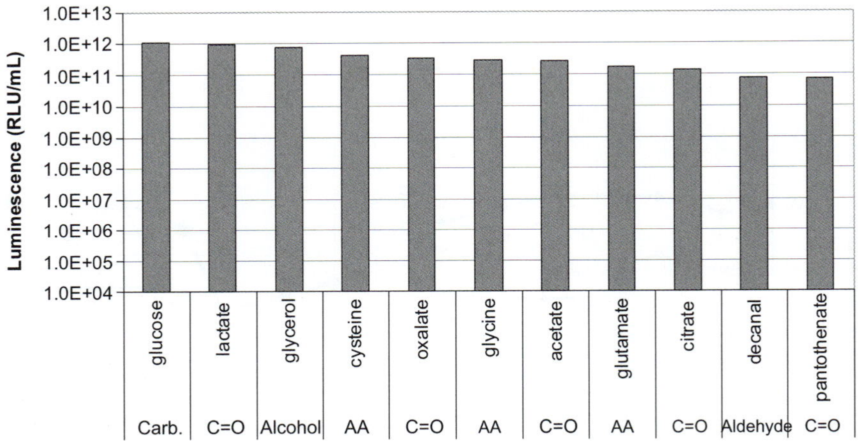

Figure 2.10 Nutritional characterization of V. harveyi (Adapted from: Weinrich *et al.* 2010b).

Application of the salt water assay for monitoring the pre-treatment of a desalination facility in Florida showed high levels of AOC in the source water that were further increased by chlorine dioxide prechlorination (Figure 2.11). Nearly 80% of AOC was removed by biological activity in the sand filters, and up to 32% of AOC removal occurred within the diatomaceous earth filters, possibly as a result of biological activity or adsorption within the diatomaceous earth filter cake. In contrast TOC removal was only about 10% through the entire process. It is interesting to note that AOC levels increased in samples collected in the feed water to the reverse osmosis membranes, which was most likely due to the addition of antiscalant chemicals, some of which have been shown to promote biofouling (Vrouwenvelder *et al.* 2000).

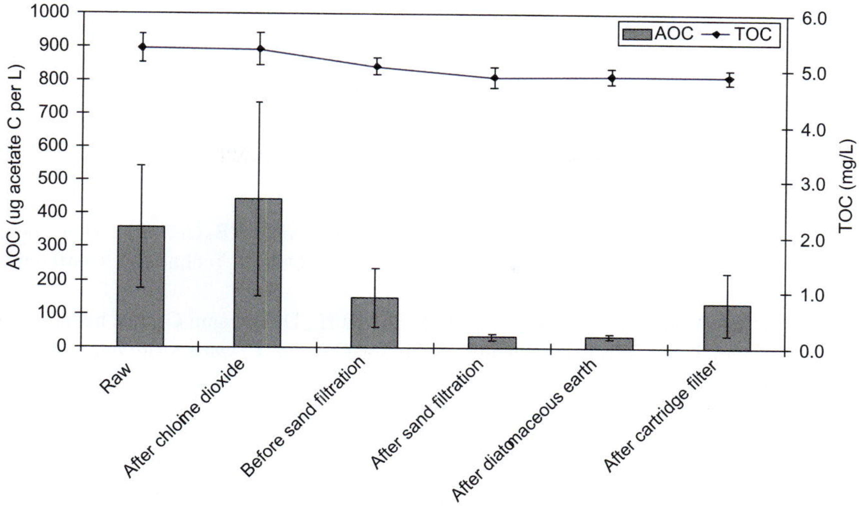

Figure 2.11 Changes in AOC and TOC levels during desalination pre-treatment.

2.7 CONCLUSIONS

It is imperative that North American water utilities address issues related to the biological stability of drinking water supplies. Challenges related to the potential health effects of disinfection by-products and chlorinous tastes and odors signal that reliance on disinfectant residual alone is not feasible to control changes in water quality. Moreover, biological filtration can have many benefits including more stable water quality; improve regulatory compliance, reduced corrosion of infrastructure, and lower customer complaints. Operators need to recognize that changes in source water and treatment, including disinfection, coagulation, filtration, treatment chemicals, and operations can all have profound changes on the concentration and nature of BOM in treated supplies. Because treatment processes fractionate the types of organic carbon present in water, it is important to use multiple methods (e.g. AOC and BDOC) to characterize the biodegradable fraction. Despite the availability of methods to measure BOM, most utilities do not use them, presumably because they are complicated and the resulting data is not useful in making operational changes. The bioluminescence AOC test gives equivalent results to the conventional assay, but is more rapid and gives additional data on the rate of substrate utilization. Finally, with growing challenges on the future sustainability of water supplies; water recycling, reuse, and desalination are growing trends – all of which are impacted by the presence of BOM. The water industry in North America (and throughout the rest of the world) has made substantial investments in granular media filters for water treatment. The trend in this next decade will be to employ biologically active filtration to leverage additional water quality benefits from these water treatments.

2.8 REFERENCES

Bellen G. E., Abrishami S. H., Colucci P. M. and Tremel C. (1993). Methods for Assessing the Biological Growth Support Potential of Water Contact Materials. Denver, CO, Water Research Foundation.

Bradford S. M., Palmer C. J. and Olson B. H. (1994). Assimilable Organic Carbon Concentrations in Southern California Surface and Groundwater. *Water Res.*, **28**, 427–435.

Brown J. and Huck P. (2009). Biofiltration for Organic Carbon, MIB, Geosmin, Manganese, EDCs and Membrane Fouling. Proc. AWWA Water Quality Technology Conference, AWWA, Denver, Colo.

Cornwell D. A., MacPhee M. J., McTigue N. E., Arora H., DiGiovanni G., LeChevallier M. and Taylor J. S. (2001). Treatment Options for *Giardia, Cryptosporidium*, and Other Contaminants in Recycled Backwash Water. Awwa Research Foundation and American Water Works Association. Denver, CO.

Edzwald J. K. (1994). Coagulation Concepts for Removal of TOC. In: Proc. AWWA Water Quality Technology Conference. San Francisco, Calif.

Escobar I. C. and Randall A. A. (1999). Influence of NF on Distribution System Biostability. *Journal AWWA*, **91**(6), 76–89.

Evans P. J., Opitz E. M., Daniel P. A. and Schulz C. R. (2010). Biological Drinking Water Treatment Perceptions and Actual Experiences in North America. Water Research Foundation, Denver, Colo.

Falkinham J. O., III, Norton C. D. and LeChevallier M. W. (2001). Factors influencing numbers of *Mycobacterium avium, Mycobacterium intracellulare*, and other mycobacteria in drinking water distribution systems. *Appl. Environ. Microb.*, **67**(3), 1225–1231.

Frias J., Ribas F. and Lucena F. (1992). A method for the measurement of biodegradable organic carbon in waters. *Water Res.*, **26**, 255–258.

Geldreich E. E. and LeChevallier M. W. (1999). Microbial water quality in distribution systems. In: Water Quality and Treatment, R. D. Letterman (ed.), 5th edn, New York, McGraw–Hill. Denver, CO, American Water Works Association, pp. 18.1–18.49.

Haddix P. L., Shaw N. J. and LeChevallier M. W. (2004). Characterization of bioluminescent derivatives of assimilable organic carbon test bacteria. *Appl. Environ. Microb.*, **70**(2), 850–854.

Huck P. M. (1990). Measurement of biodegradable organic matter and bacterial growth potential in drinking water. *Journal AWWA*, **82**(7), 78–86.

Huck P. M., Fedorack P. M. and Anderson W. B. (1991). Formation and removal of assimilable organic carbon during biological treatment. *Journal AWWA*, **83**, 69–90.

Jacangelo J. G., DeMarco J., Owen D. M. and Randtke S. J. (1995). Selected processes for removing NOM, an overview. *J. Amer. Water Works Assoc.*, **87**, 64–77.

Jjemba P. K., Weinrich L. A., Cheng W., Giraldo E. and LeChevallier M. W. (2010). Regrowth of potential opportunistic pathogens and algae in reclaimed-water distribution systems. *Appl. Environ. Microb.*, **76**(13), 4169–4178.

Joret J. C. and Levi Y. (1986). Méthode Rapide D'évaluation du Carbone Éliminable Des Eaux Par Voie Biologique. *Trib. Cebedeau*, **510**(39), 3–9.

Joret J. C., Levi Y. and Gibert M. (1989). The measurement of biodegradable organic carbon (BDOC), a tool in water treatment. *Water Supply*, **7**, 41–45.

Kaplan L. A. and Bott T. L. (1989). *Nutrients for Bacterial Growth in Drinking Water: Bioassay Evaluation*. U.S. Environmental Protection Agency, EPA/600/S2-89/030, Cincinnati, Ohio.

Kaplan L. A. and LeChevallier M. W. (1993). Assimilable Organic Carbon Measurement Techniques, American Water Works Association Research Foundation, Denver, CO.

Kaplan L. A., Reasoner D. J., Rice E. W. and Bott T. L. (1992). A survey of assimilable organic carbon, biodegradable organic carbon and coliform growth response in US drinking waters. *Rev. Sci. Eau* **5**(n special), 207–224.

Kaplan L. A., Bott T. L. and Reasoner D. J. (1993). Evaluation and simplification of the assimilable organic carbon nutrient bioassay for bacterial growth in drinking water. *Appl. Environ. Microb.*, **59**, 1532–1539.

Kaplan L. A., Standley L. J., Newbold J. D., Standridge J. H., Mager A. L., Kluender S. M., Peterson L. L., Smith D. B., Hession W. C. and Luitweiler P. (2002). Evaluation of Sources of Pathogens and NOM in Watersheds [Project #251]. Awwa Research Foundation, Denver, CO.

Kemmy F. A., Fry J. C. and Breach R. A. (1989). Development and operational implementation of a modified and simplified method for determination of assimilable organic carbon (AOC) in drinking water. *Water Science and Technology*, **21**, 155–159.

Krasner S. W. and Amy G. L. (1995). Jar-test evaluation of enhanced coagulation water. *Journal AWWA*, **87**(10), 93–107.

LeChevallier M. W. and Au K. K. (2004). Water Treatment and Pathogen Control. ISBN 92 4 156255 2, WHO, Geneva, Switzerland.

LeChevallier M. W., Schulz W. and Lee R. G. (1991a). Bacterial nutrients in drinking water. *Appl. Environ. Microb.*, **57**(3), 857–862.

LeChevallier M. W., Becker W. C., Schorr P. and Lee R. G. (1991b). Evaluating the performance of biologically active rapid filters. *Journal AWWA*, **84**(4), 136–146.

LeChevallier M. W., Lowry C. D., Lee R. G. and Gibbon D. L. (1993a). Examining the relationship between iron corrosion and the disinfection of biofilm bacteria. *Journal AWWA*, **85**(7), 111–123.

LeChevallier M. W., Shaw N. E., Kaplan L. A. and Bott T. L. (1993b). Development of a rapid assimilable organic carbon method for water. *Appl. Environ. Microb.*, **59**, 1526–1531.

LeChevallier M. W., Welch N. J. and Smith D. B. (1996). Full-scale studies of factors related to coliform regrowth in drinking water. *Appl. Environ. Microb.*, **62**(7), 2201–2211.

Lehtola M. J., Miettinen I. T., Vartiainen T. and Martikainen P. T. (2002). Changes in content of microbially available phosphorus, assimilable organic carbon and microbial growth potential during drinking water treatment processes. *Water Res.*, **36**(15), 3681–3690.

Miettinen I. T., Vartiainen T. and Martikainen P. J. (1997). Phosphorus and bacterial growth in drinking water. *Appl. Environ. Microb.*, **63**(8), 3242–3245.

Miltner R. J., Summers R. S., Wang J., Swertfeger J. and Rice E. W. (1992). The response of biological filters to backwashing. In: Proceedings AWWA Water Quality Technology Conference. Toronto, ONT, Canada.

Mogren E. M., Scarpino P. and Summers R. S. (1989). Measurement of biodegradable dissolved organic carbon in drinking water. Proc. Annual AWWA conf. Cincinnati, OH.

Najm I. N., Boulos L., LeChevallier M., Norton C., Volk C., Randall A., Escobar E., Kiene L. and Campos C. (2000). Case Studies of the Impacts of Treatment Changes on Biostability in Full Scale Distribution Systems. Awwa Research Foundation and American Water Works Association. Denver, CO.

Norton C. D., LeChevallier M. W. and Falkinham J. O. III. (2004). Survival of *Mycobacterium avium* in a model distribution system. *Water Res.*, **38**, 1457–1466.

Paillard H., Dore M. and Bourbigot M. M. (1991). Les perspectives d'application de l'ozone catalytique en traitement des eaux a potabiliser. 10th Ozone World Congress, IOA Monaco, France, March 19–21.

Prévost M., Laurent P., Servais P. and Joret J. C. (2005). Biodegradable Organic Matter in Drinking Water Treatment and Distribution. Denver, CO, American Water Works Association.

Reasoner D. J. and Rice E. W. (1989). US EPA Experiences with AOC and Coliform Growth Response Assay. Workshop on Measurement of AOC in the Field of Drinking Water Treatment, Karlsruhe, RFA (Nov. 1989).

Richard Y. (1994). *Rehabilitation of Water Treatment Plant Impact of Membrane Filtration.* Proc. 1994 International Water Supply Association Conference, Zurich.

Rittmann B. E. and Snoeyink V. L. (1984). Achieving biologically stable drinking water. *Journal AWWA*, **76**(10), 1076–1084.

Schoenen D. (1986). Microbial growth due to materials used in drinking water systems. In: Biotechnology, H. J. Rehm and G. Reed (eds), Vol. 8, Weinheim, Germany, VCH Verlagsgesellschaft, pp. 628–647.

Servais P., Billen G. and Hascoet M. C. (1987). Determination of the biodegradable fraction of dissolved organic matter in waters. *Water Res.*, **21**, 445–450.

Servais P., Laurent P. and Gatel D. (1995). Characterization of dissolved organic matter biodegradability in waters. Impact of water treatment and bacterial regrowth in distribution systems. In: Proc. Of the Annual AWWA Water Quality Technology Conference. New Orleans, LA, AWWA.

Stanfield G. and Jago P. H. (1987). The Development and Use of a Method for Measuring the Concentration of Assimilable Organic Carbon in Water. WRC Environment Report, PRU 1628-M. Manheim, UK.

Thofern E., Schoenen D. and Tuschewitzki G. J. (1987). Microbial surface colonization and disinfection problems. *Offentl. Gesundh.*, **49**, 14–20.

USEPA (1974). Safe Drinking Water Act. http://www.epa.gov/safewater/sdwa/index.html (accessed 28 September 2011).

USEPA (1989a). The Surface Water Treatment Rule. http://water.epa.gov/lawsregs/rulesregs/sdwa/swtr/index.cfm (accessed 28 September 2011)

USEPA (1989b). The Total Coliform Rule. http://water.epa.gov/lawsregs/rulesregs/sdwa/tcr/regulation.cfm (accessed 28 September 2011)

USEPA (2001). National Primary Drinking Water Regulations, Filter Backwash Recycling Rule. *Federal Register*, **66**(111), 31085–31105.

USEPA (2006a). The Long Term II Enhanced Surface Water Treatment Rule. http://www.epa.gov/safewater/disinfection/lt2/index.html (accessed 28 September 2011)

USEPA (2006b). The Groundwater Rule. http://water.epa.gov/lawsregs/rulesregs/sdwa/gwr/index.cfm (accessed 28 September 2011).

van der Kooij D. (1987). The effect of treatment on assimilable organic carbon in drinking water. In: Treatment of drinking water for organic contaminants, P. M. Huck and P. Toft (eds), Pergamon Press, New York, pp. 317–328.

van der Kooij D. and Hijnen W. A. M. (1984). Substrate utilization by an oxalate consuming *Spirillum* species in relation to its growth in ozonated water. *Appl. Environ. Microb.*, **47**, 551–559.

van der Kooij D., Visser A. and Hijnen W. A. M. (1982). Determining the concentration of easily assimilable organic carbon in drinking water. *Journal AWWA*, **74**, 540–545.

van der Kooij D., Hijnen W. A. M. and Kruithof J. C. (1989). The effects of ozonation, biological filtration and distribution on the concentration of easily assimilable organic carbon (AOC) in drinking water. *Ozone: Sci. Eng.*, **11**, 297–311.

Volk C. (2002). Biodegradable dissolved organic carbon in drinking water. In: Encyclopedia of Environmental Microbiology, G. Bitton (ed.), Volume 4, John Wiley & Sons, Inc., New York, pp. 484–507.

Volk C. and Joret J. C. (1994). Paramètres prédictifs de l'apparition des coliformes dans les réseaux de distribution d'eau d'alimentation. *Revue des Sciences de l'Eau*, **7**, 131–152.

Volk C. J. and LeChevallier M. W. (2000). Assessing biodegradable organic matter. *Journal AWWA*, **92**(5), 64–76.

Volk C. J. and LeChevallier M. W. (2002). Effects of conventional treatment on AOC and BDOC levels. *Journal AWWA*, **94**(6), 112–123.

Volk C., Renner C. and Joret J. C. (1992). La mesure du CODB, un index du potentiel de reviviscence bactérienne des eaux. *Sciences de l'Eau*, **5**, 189–205.

Volk C., Renner C., Roche P., Paillard H. and Joret J. C. (1993). Effect of ozone on the production of biodegradable dissolved organic carbon (BDOC) during water treatment. *Ozone: Sci. Eng.*, **15**, 389–404.

Volk C., Roche P., Joret J. C. and Paillard H. (1997). Comparison of the effect of ozone, ozone hydrogen peroxide system and catalytic ozone on the biodegradable organic matter of a fulvic acid solution. *Water Res.*, **31**, 650–656.

Volk C., Bell K., Ibrahim E., Verges D., Amy G. and LeChevallier M. (2000). Impact of enhanced and optimized coagulation on removal of organic matter and it's biodegradable fraction in drinking water. *Water Res.*, **34**(12), 3247–3257.

Vrouwenvelder J. S., Manolarakis S. A., Veenendaal H. R. and van der Kooij D. (2000). Biofouling potential of chemicals used for scale control in RO and NF membranes. *Desalination*, **132**, 1–10.

Weiss W. J., Bouwer E. J., Ball W. P., O'Melia C. R., LeChevallier M. W., Arora H. and Speth T. F. (2003). Riverbank filtration – fate of DBP precursors and selected microorganisms. *Journal AWWA*, **95**(10), 68–81.

Weiss W. J., Bouwer E. J., Ball W. P., O'Melia C. R., Aboytes R. and Speth T. F. (2004). Riverbank filtration, effect of ground passage on NOM Character. *Aqua*, **53**(2), 61–83.

White D. R. and LeChevallier M. W. (1993). Oil lubricated wells and AOC in groundwater. *Journal AWWA*, **85**(8), 112–114.

Weinrich L. A., Giraldo E. and LeChevallier M. W. (2009). Development and application of a bioluminescence-based test for assimilable organic carbon in reclaimed waters. *Appl. Environ. Microb.*, **75**(12), 7385–7390.

Weinrich L. A., Jjemba P. K., Giraldo E. and LeChevallier M. W. (2010a). Implications of organic carbon in the deterioration of water quality in reclaimed water distribution systems. *Water Res.*, **44**, 5367–5375.

Weinrich L. A., Schneider O. D. and LeChevallier M. W. (2010b). Bioluminescence-based method for measuring assimilable organic carbon in pretreatment water for reverse osmosis membrane desalination. *Appl. Environ. Microb.*, **77**(3), 1148–1150.

Chapter 3

Removal of organic matter in water treatment systems – Case studies in Japan

Y. Watanabe

3.1 INTRODUCTION

In Japan, about 75% of the drinking water originates from surface water including rivers and lakes. Therefore, surface-water pollution has influenced the drinking-water treatment systems. After the Second World War, the most traditional system in Japan has been the conventional rapid sand filtration system consisting of coagulation, flocculation, sedimentation, sand filtration and post-chlorination. However, since surface-water pollution, caused by the discharge of municipal wastewater, became serious, many water works introduced pre-chlorination or midpoint-chlorination to oxidize organic matter and ammonia nitrogen. The introduction of multiple chlorination resulted in trihalomethane production and taste and odour problems. To solve these problems ozonation and activated carbon adsorption were introduced into the conventional rapid sand filtration system, which subsequently is called an advanced water purification system. This paper describes: (i) a case study in the advanced water purification system in Osaka Water Works and the effect of biological oxidation of purified water by this system on the biological regrowth, and (ii) application of membrane filtration coupled with ozonation, activated carbon adsorption/biological oxidation and biofilm oxidation for the removal of organic matter from surface water.

3.2 ADVANCED WATER PURIFICATION SYSTEM IN OSAKA WATER WORKS

3.2.1 Reduction of chlorine dosage

Osaka Water Works had used the conventional rapid sand filtration system before 1999. In that period, tap water in Osaka city had an unpleasant taste and odour caused by the high chlorine dosage. Osaka Water Works use surface water from downstream (Kunijima point) of Yodo River. This river comes from Lake Biwa, the largest lake in Japan, with a water volume of 27.5 billion m^3. The length of the river is 75 km and 16 million people in the Kansai area use this surface water as the drinking water resource. Many municipalities use Yodo River water and treated municipal wastewaters from those municipalities, including Kyoto city, are discharged into the Yodo River. Therefore, the Yodo River is part of an unintentional water recycling system. The annual average biological oxygen demand (BOD) at Kunijima point was between 2 and 3 mg/litre. In 1999, Osaka Water Works introduced the use of ozonation and granular activated carbon (GAC) filtration into the conventional rapid sand filtration system (advanced water purification system) to reduce the chlorine dosage (Fuchigami and Terashima 2003). Figure 3.1 shows the water treatment system before and after 1999.

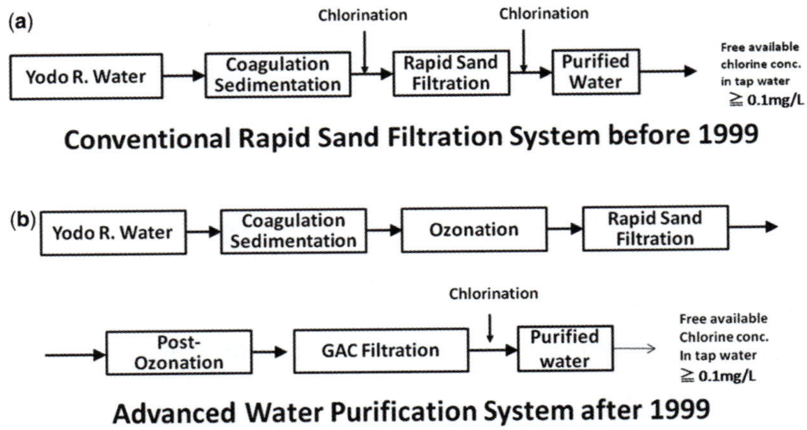

Figure 3.1 Water purification system in Osaka Water Works before and after 1999 (Courtesy of Fuchigami and Terashima).

Figure 3.2 compares the chlorine consumption rate constant of the water treated by conventional treatment and by the advanced water purification system. The first order rate constant was calculated using the residual chlorine concentration measured at various points of the water distribution pipes in Osaka City. According to the Japanese drinking-water standard, free available chlorine concentration in tap water should be higher than 0.1 mg/litre. The advanced water purification system produces drinking water with a much lower chlorine demand.

3.2.2 Bacterial-regrowth control by AOC reduction and less chlorine dosage

Kasahara *et al.* (2002) measured the removal efficiency of dissolved organic carbon (DOC) and assimilable organic carbon (AOC) in the Yodo River water at

Kunijima point. Polyaluminium-chloride was added at 3 mg Al/litre at pH 7. The experimental results clearly demonstrate that the AOC concentration of the water treated in the conventional system used to be high (Table 3.1). The average AOC of the water treated in the advanced system was about 40 µg/litre. As a result, Osaka Water Works succeeded in reducing the chlorine dosage to produce drinking water of better taste using the advanced water purification system.

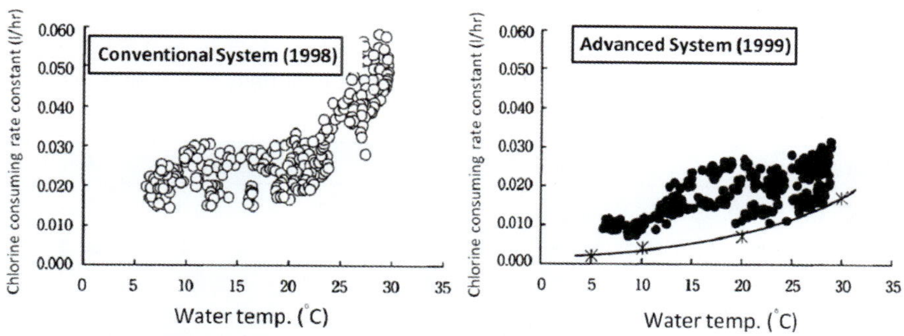

Figure 3.2 Chlorine consuming rate constant based on monitored residual chlorine (Courtesy of Fuchigami and Terashima).

Table 3.1 Effect of coagulation on AOC removal in Yodo River water (Courtesy of Kasahara and Ishikawa).

Sampling date		AOC (µg/litre)			DOC (mg/litre)		
		Raw water	Coagulated water	Removal (%)	Raw water	Coagulated water	Removal (%)
99	6–16	86			3.0		
	7–29	81			1.6		
	10–6	113	93	18	1.8	1.0	44
	11–4	243			1.8		
	12–8	186			2.0		
00	1–7	221	206	7	2.0	0.9	55
	7–28	288	105	64	4.2	2.4	43
	9–1	606	146	76	4.1		
	9–20	533	179	66	4.8	3.1	35
	10–26	406	98	76	3.2	2.4	25
	11–19	230	149	35	3.7	2.3	38

Kasahara *et al.* (2009) conducted an experimental study on the AOC removal using annular reactors (AR) under various hydraulic retention times. The influent water was the treated water of the advanced system. Table 3.2 shows the experimental results. The AOC concentration of the treated water was 40 μg/litre and decreased to 10–15 μg/litre in the annular reactor effluent. The yield of bacteria in the hydraulic retention time of 0.1 and 4 hours was almost the same as that of heterotrophs and P17 strains, respectively. The effluent from the annular reactor was chlorinated at two ranges of free available chlorine and the time variations of HPC and total bacteria counts (DTMC) were measured.

Table 3.2 Experimental results in 4 annular reactors (Courtesy of Kasahara and Ishikawa).

Parameter	Influent	AR 1	AR 2	AR 3	AR 4
HRT (h)*	–	0.1	0.1	0.1	4
Surface velocity (m/sec)	–	1.32	0.43	0.07	0.43
AOC (μg/litre)	40	13	14	10	15
HPC (CFU/mL)**	N.D.	2.4×10^5	7.1×10^5	1.6×10^6	1.5×10^5
Yield (CFU/(μg)	–	1.6×10^5	1.2×10^5	1.6×10^5	3.2×10^6

*, hydraulic retention time; **, heterotrophic plate count.
Source: Ksahara *et al.* (2009).

Figure 3.3 shows the variation of HPC and DTMC with time when the residual free available chlorine concentration was 0.1–0.4 mg/litre and 0.4–0.8 mg/litre. HPC in the attached biomass decreased by 4 logs compared with the case of no residual chlorine.

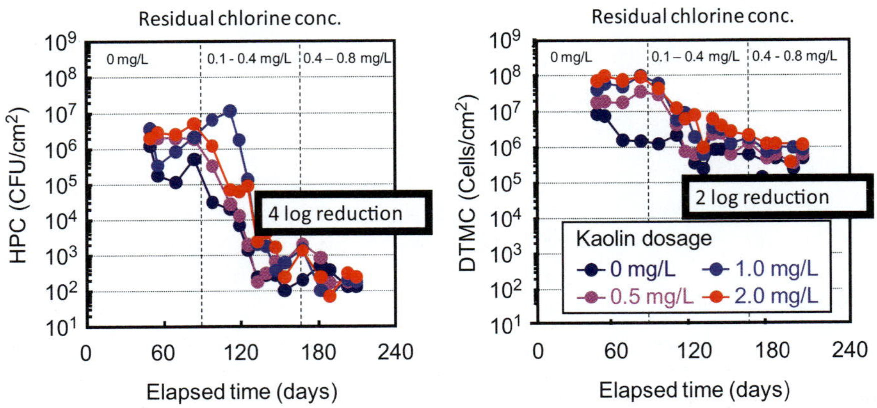

Figure 3.3 Variation of HPC and total bacteria counts (DTMC) in attached biomass at various residual chlorine concentrations (Courtesy of Kasahara and Ishikawa).

3.3 ORGANIC REMOVALS IN A HYBRID MEMBRANE FILTRATION SYSTEM

3.3.1 PVDF MF membrane filtration coupled with pre-ozonation

Japanese membrane manufacturing companies have developed an ozone-resisting membrane (MF) module made of PVDF (polyvinylidene fluoride) with potting material, which has a high resistance to ozone (Mori *et al.* 1998). In the developed membrane module, water containing residual ozone can be directly filtered. This system can provide a consistently high permeate flux for various raw waters, especially high turbidity water and secondary treated municipal wastewater. Jang *et al.* (2002) studied the effect of residual ozone on fouling reduction using the ozone-resisting PVDF membrane with a nominal pore size of 0.1 μm. Figure 3.4 shows a schematic diagram of the experimental system.

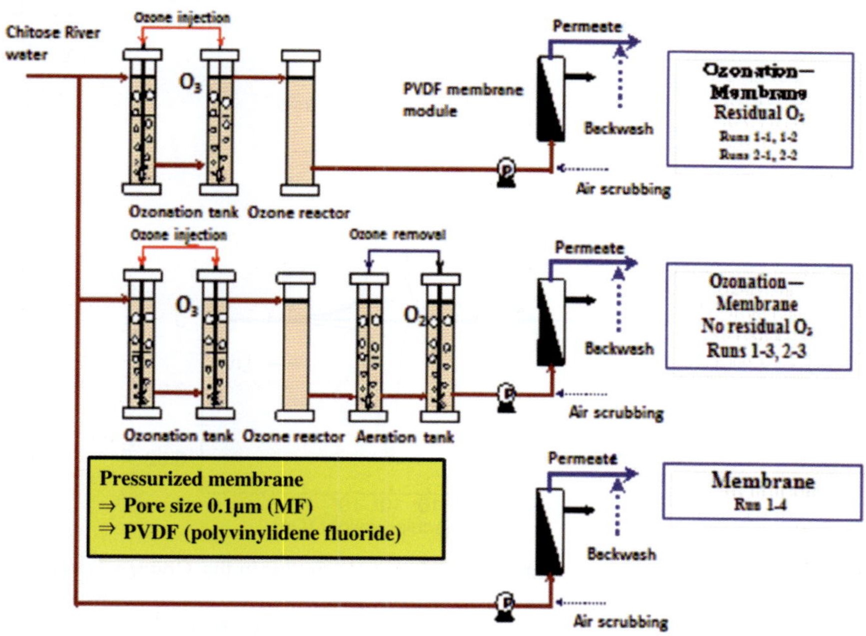

Figure 3.4 Experimental system of pre-ozonation and membrane filtration.

Chitose River water was fed into the membrane system. The average water quality is shown in Table 3.3 and Figure 3.5 shows the molecular weight distribution of organic matter contained in the raw water used in this study. This organic matter could be roughly divided into two fractions: large molecules with a hydrophilic nature (little UV absorbance at 260 nm, E260) and small molecules with a hydrophobic nature (high E260).

Table 3.3 Quality of Chitose River water.

Characteristic	Value (average)
Temperature (°C)	11.5
pH	7.1
Turbidity (NTU)*	16.5
UV absorbance at 220 nm (cm^{-1})	0.41
UV absorbance at 260 nm (cm^{-1})	0.099
TOC (mg/litre)*	2.43
DOC (mg/litre)	2.29
THMFP (mg/litre)*	0.086
Manganese (mg/litre)	0.10
Soluble manganese (mg/litre)	0.074
Ammonia-N (mg/litre)	0.22

*NTU, nephelometric turbidiy units; TOC, total organic carbon; THMFP, trihalomethane formation potential.

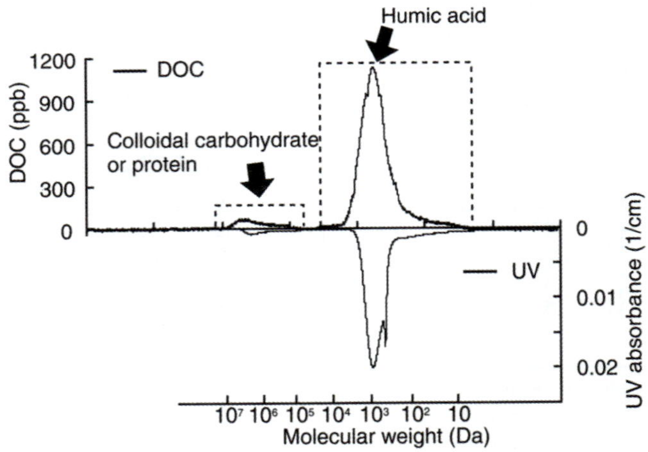

Figure 3.5 Molecular weight distribution of organic matter in the Chitose River.

In experimental Runs 1-1 and 1-2, the ozone dosage was 2.0 and 4.2 mg O_3/litre, whereas the residual ozone concentration was 0.73 and 1.13 mg O_3/litre, respectively. The ozone contact time was 7.8 min in all experimental runs. In Runs 2-1 and 2-2, ozone dosage was 1.4 and 1.9 mg O_3/litre and the residual ozone concentration was 0.41 and 0.61 mg O_3/litre, respectively. Figure 3.6 shows the change in permeate flux with increasing operation time in Runs 1 and 2 where the constant permeate pressure mode of operation was 50 kPa and physical cleaning methods of backwashing and air scrubbing were carried out. It clearly demonstrates that the residual ozone reduced the membrane fouling.

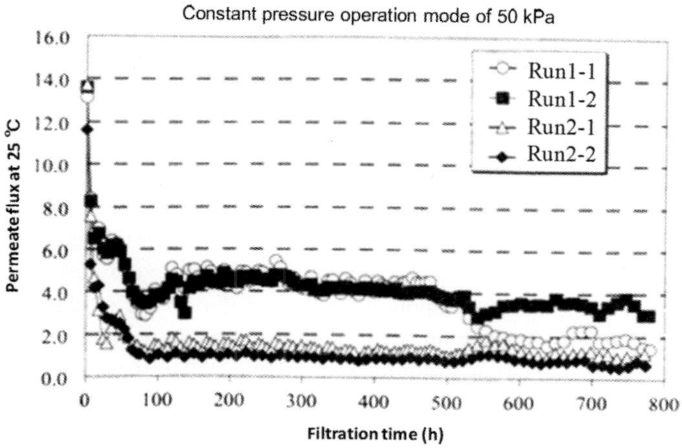

Figure 3.6 Effect of pre–ozonation and residual ozone on membrane permeate flux. See Figure 3.4 and text for the conditions in Runs 1-1, 1-2, 2-1 and 2-2.

After continuous operation for about 800 hours, chemical cleaning of the fouled membrane was conducted. The following three chemical solutions were used: NaOH solution of 1%, NaClO solution of 5 mg/litre, and oxalic and nitric acid of 2 and 5%, respectively. Figure 3.7 shows the extracted TOC in each chemical solution. As seen in Figure 3.7, pre-ozonation with residual ozone significantly decreased the amount of organic substances attached to the membrane causing the physically-irreversible fouling. This may come from the following two ozone induced reactions: (i) degradation of organic substances and (ii) destabilization of particles on the membrane surface. The ozone induced particle destabilization reaction has been reported by many researchers (Chandrankanth and Amy 1998; Grasson and Weber 1988; Jekel 1994).

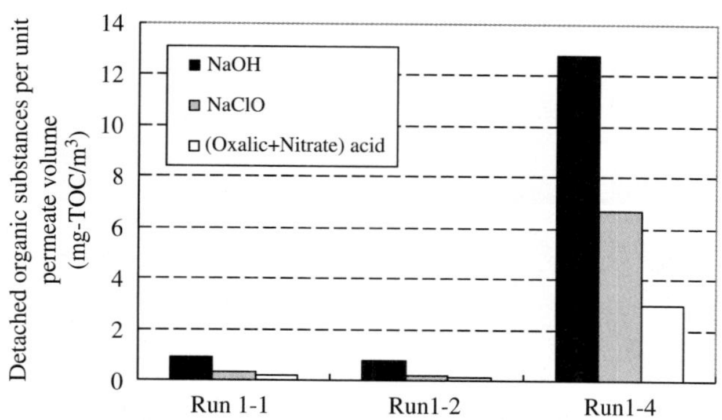

Figure 3.7 Effect of residual ozone on amount of detached organic substances. See Figure 3.4 and text for conditions in Runs 1-1, 1-2 and 1-4.

In another research (Lee *et al.* 2004), the size distribution of the fouling particles in the backwash water was measured. It was found that the average size of the particles was about 20 and 50 μm without and with ozonation, respectively. The increasing particle size increased the rate of back transport of organic particles, leading to a decrease in the accumulation of organic particles on the membrane surface. In this experiment the permeate TOC (i.e. DOC) concentration in the membrane filtration system without and with pre-ozonation was the same (2.4 mg/litre) but E260 and E260/DOC were 0.062 and 0.034 cm^{-1}, and 0.026 and 0.013 cm^{-1}/mg/litre, respectively. These results demonstrate that the biodegradability of organic compounds increased due to the oxidation by ozone. Figure 3.8 shows the change in AOC concentration in various water samples. AOC produced by ozonation is not retained by the PVDF membrane, but without pre-ozonation, the PVDF membrane removed AOC, especially the fraction that is degraded by P17, to some extent. Considering the molecular weight distribution in the raw water, shown in Figure 3.5, a larger part of the colloidal proteins and carbohydrates may be removed by the membrane separation.

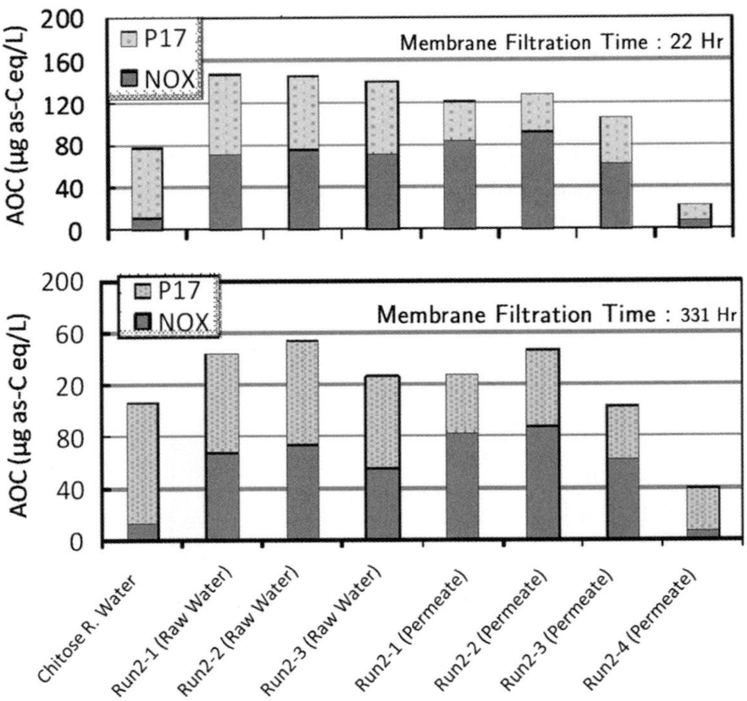

Figure 3.8 Effect of ozonation on AOC concentration at various points. Raw water is treated (ozonated) water prior to membrane filtration. See Figure 3.4 and text for the conditions in Runs 1-1, 1-2, 2-1, 2-2, 1-3 and 1-4.

3.3.2 PTFE MF membrane filtration coupled with powdered activated carbon adsorption and biological/chemical oxidation

Suzuki *et al.* (2001) developed a submerged membrane filtration system coupled with powdered activated carbon (PAC) adsorption and chemical/biological oxidation. This membrane system has been developed to treat raw waters containing high levels of soluble matter such as biodegradable organics, humic substances, manganese and ammonia nitrogen. In this system, soluble less-biodegradable organics are adsorbed to PAC, and suspended particles including PAC are separated by membrane filtration. The soluble biodegradable organics, manganese and ammonia nitrogen are biologically or chemically oxidized. In the case of chemical oxidation (with pre-chlorination), soluble manganese is oxidized with chlorine and the catalytic reaction of powdered activated carbon, and the oxidized manganese is removed by membrane separation. Ammonia nitrogen is also oxidized by chlorine in a pre-chlorination tank. In the case of biological oxidation (without pre-chlorination), the iron-oxidizing bacteria and ammonia-oxidizing bacteria, which are accumulated in a submerged membrane tank, oxidize the soluble manganese and ammonia, respectively. A schematic diagram of the pilot plant is shown in Figure 3.9.

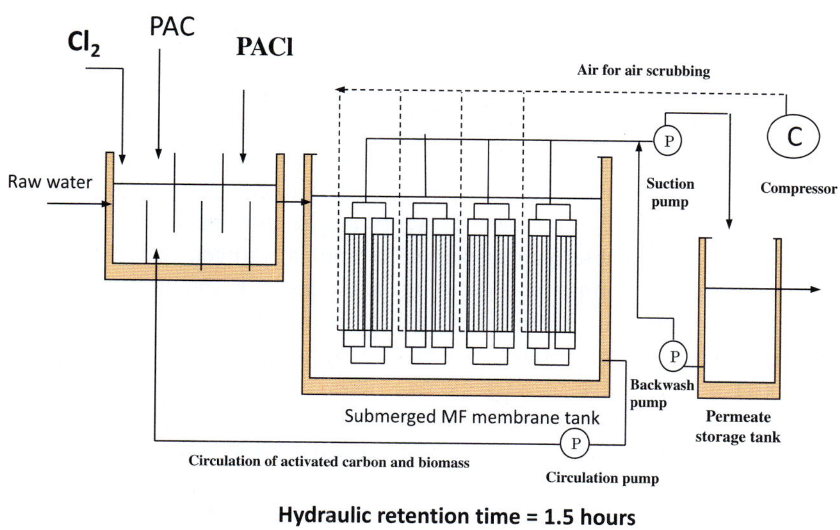

Figure 3.9 Pilot plant of membrane filtration system with powdered activated carbon adsorption and biological/chemical oxidation (Suzuki *et al.* 2001).

The volume of the membrane-submerged tank was 4 m^3 and the surface area of the submerged membranes was 86–120 m^2. The raw water was fed into the mixing tank (detention time 10–15 min). Four types of Poly-Tetra-Fluoro-ethylene (PTFE) membranes were used. When the first, second and third type of membrane were used, the raw water to the pilot plant was taken from the existing water purification plant, which already contained PAC at a concentration of 5–30 mg/litre. In the mixing tank, hypochlorite was added for chemical oxidation, and the sludge containing biomass and PAC were returned from the membrane submerged tank and mixed with the raw water in the mixing tank. PAC (average diameter of 10 μm) was also dosed into the mixing tank at a constant concentration of 13 mg/litre when the fourth type of membrane was used, where the river water was directly fed into the plant. Polyaluminium-chloride was added to coagulate the small suspended particles in the mixing tank. The pre-treated water was fed into the next tank where the hollow fibre PTFE membranes were submerged and intermittent aeration was performed to supply oxygen to the microorganisms. Air was supplied during 1 min at an intensity of 0.64 Nm^3/min every 4 min. The raw water in the pilot plant study was surface water from Chitose River with high concentrations of humic substances, soluble manganese and ammonia nitrogen (Table 3.3). The dosage of hypochlorite in the mixing tank was 4–6 mg Cl_2/litre. The coagulant dosage was 2–3 mg Al/litre.

In the pilot plant experiments, symmetrical (first type) or composite PTFE membranes with a pore size of 0.1 μm were used. The thickness of the top layer in the composite membrane varied at 60, 30 and 15 μm (second, third and fourth type). The pore density also changed. The top layer thickness and pore density in the fourth type of membrane are about 15 μm and 80%, respectively. This hybrid membrane system is able to efficiently remove dissolved compounds such as organics, manganese, and ammonia nitrogen. The soluble manganese and ammonia nitrogen were oxidized biologically or chemically and small humic substances were adsorbed to the powdered activated carbon. The removal efficiencies for TOC, E260 and THMFP are shown in Figure 3.10. The comparison of the soluble manganese and ammonia nitrogen between chemical oxidation with pre-chlorination and biological oxidation is shown in Figure 3.11. As previously reported, chemical oxidation is necessary to oxidize soluble manganese when the raw water temperature become less than 10°C (Kimura *et al.* 2001).

Figure 3.12 shows the change of the permeate flux, trans-membrane pressure (TMP) and raw water temperature during the experiment with the symmetrical membrane. In this experiment, the hybrid MF membrane system was operated without pre-chlorination. The average flux was relatively low at less than 0.3 m/day and the TMP increased to 70 kPa after 5 months of operation. To improve the permeability of the PTFE membrane, the membrane structure was changed from symmetrical to composite. The permeability of the composite membranes with a different top layer thickness was compared in the pilot plant experiment. The thinner the top layer thickness, the better the permeability. Figure 3.13 shows the change of the membrane flux, TMP and raw water temperature with increasing operation time when the fourth type of composite PTFE membrane was used. The permeate flux was four times higher (1.2 m/day) than the flux in the symmetrical membrane (0.3 m/day) (Figures 3.12 and 3.13).

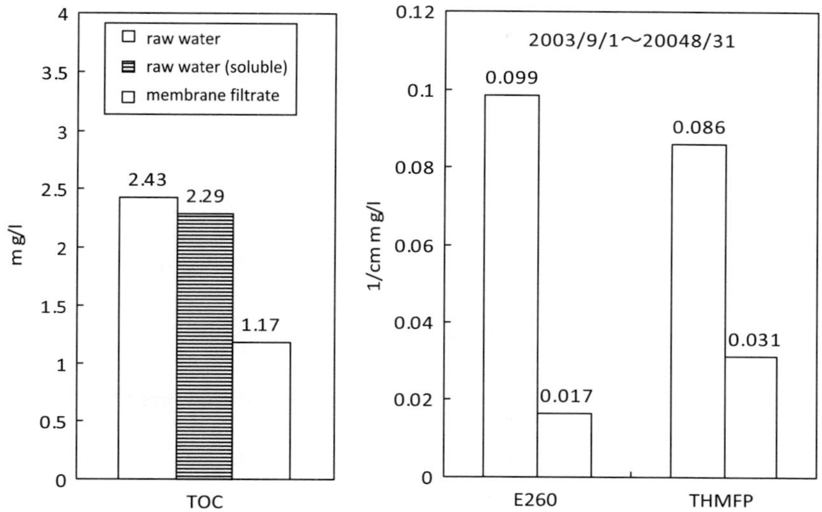

Figure 3.10 Quality of raw water and permeate (organic matter).

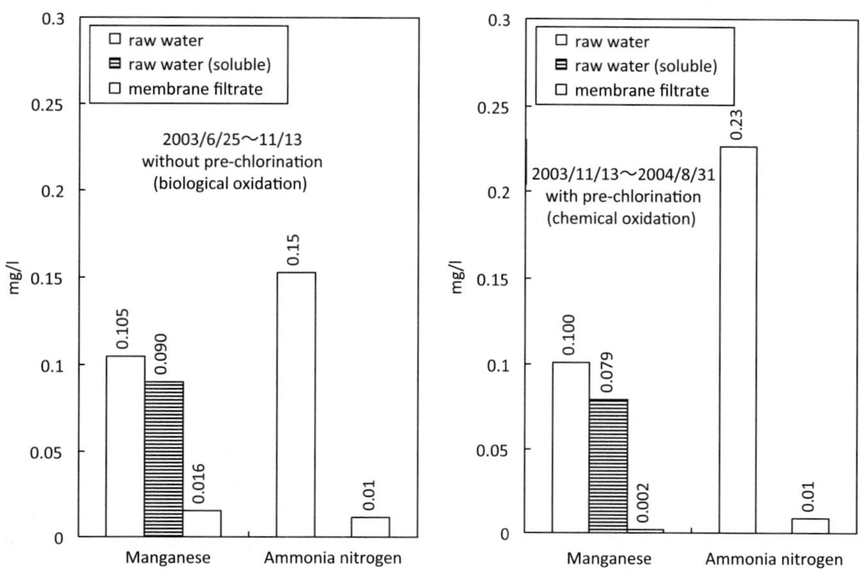

Figure 3.11 Quality of raw water and permeate (inorganic matter).

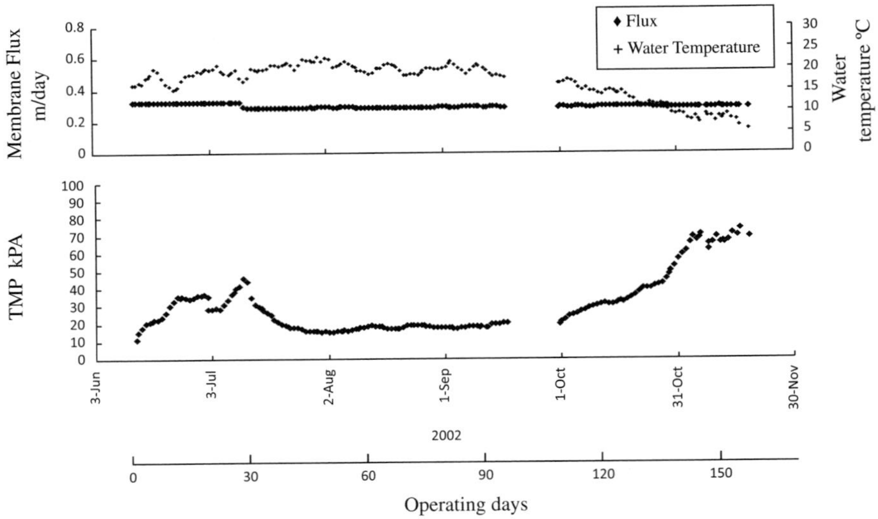

Figure 3.12 TMP change and permeate flux (symmetrical membrane).

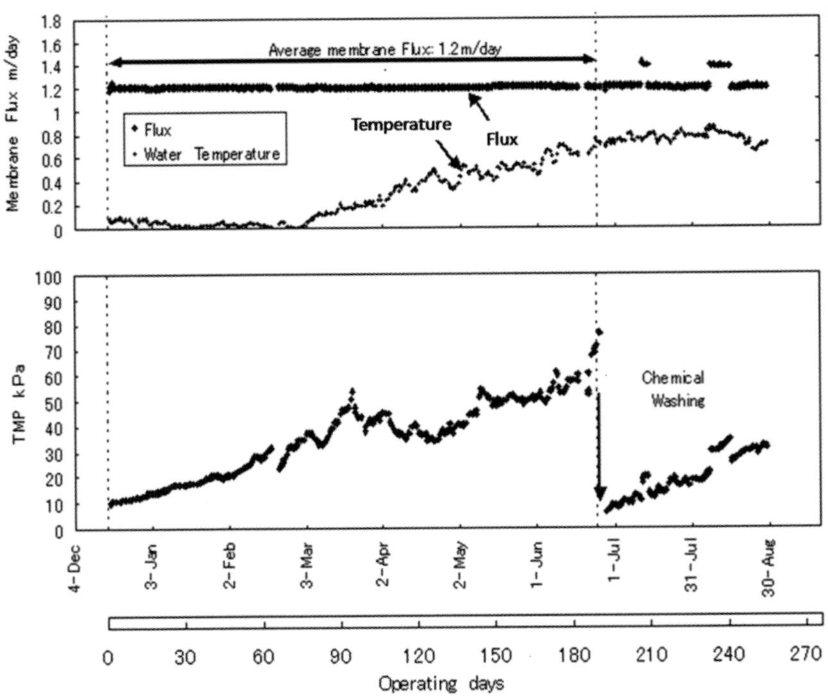

Figure 3.13 TMP change and permeate flux (fourth type of PTFE membrane).

3.3.3 Biofilm-membrane reactor for advanced drinking water treatment

Kimura *et al.* (2001) developed a novel biofilm-membrane reactor (BMR) in which a biofilm, dominated by nitrifying microorganisms, was fixed on the surface of a rotating ultrafiltration (UF) membrane. With this reactor (Figure 3.14), strict solid-liquid separation, oxidation of ammonia nitrogen and AOC removal can be performed simultaneously.

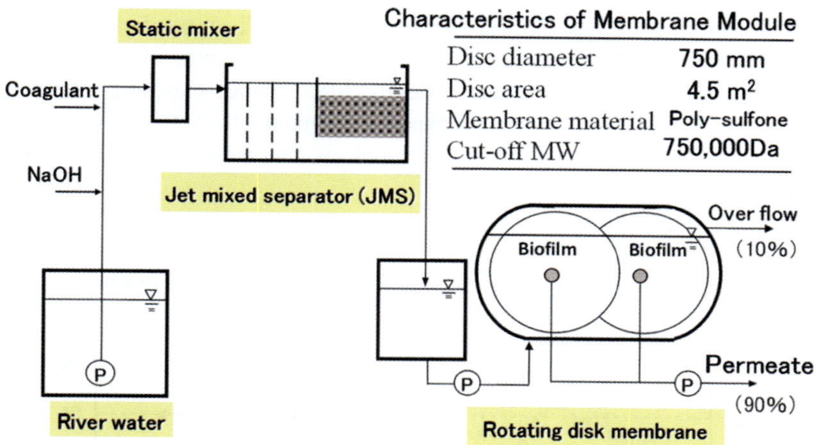

Figure 3.14 Biofilm-membrane system with pre-coagulation/sedimentation Cut-off MW: molecular weight cut-off.

Coagulation with polyaluminium-chloride and flocculation/sedimentation in a jet mixed separator (JMS) with inclined tube settler (Watanabe *et al.* 1990, 1998) were implemented as pre-treatment because throughout the year, the raw water contained a considerable amount of organic substances (up to 10 mg DOC/litre) (Table 3.3). The fluctuation of turbidity in the raw water from Chitose River is substantial, from 5 to 300 NTU. The water temperature decreases to below 5°C during winter, resulting in relatively high concentration of ammonia nitrogen and manganese. Polyaluminium-chloride was dosed at a concentration of 5–10 mg Al/litre. NaOH was dosed to control the pH of coagulated water in the range of 6.5–7.5. The hydraulic detention time in the JMS was fixed at 1 hour. The BMR was always operated with overflow from the membrane chamber, resulting in a water recovery ratio of 0.9 to 0.95. Intermittent operation, 30 minutes filtration and 2 minutes pause, was implemented. The membrane flux and disk rotating speed were fixed at 0.5 m/d and 10 rpm, respectively. Prior to the continuous operation, microorganisms (dominated by nitrifying organisms) were seeded in the membrane chamber. Dead-end filtration was carried out in order to fix the microorganisms on the membrane surface. Sponge cleaning was implemented to detach the excess biofilm when the trans-membrane pressure (TMP) increased to about 70 kPa; for details about sponge cleaning, refer to Kimura *et al.* (2000). Cleaning can decrease the TMP to about 15 kPa. When the turbidity of the effluent from the JMS was constantly less than 2 NTU, sponge cleaning was conducted once in a half year. The pre-treatment also removed humic substances and large organic molecules, resulting in the control of membrane fouling.

Figure 3.15 shows the change of the ammonia nitrogen concentration and raw water temperature during the experimental period. When nitrification became inactive at low water temperatures, phosphate was intermittently added to the BMR influent and nitrification was restored. After sponge cleaning, nitrification also became inactive due to the biomass detachment; the activity was restored after some time, depending on the amount of the detached biomass during cleaning.

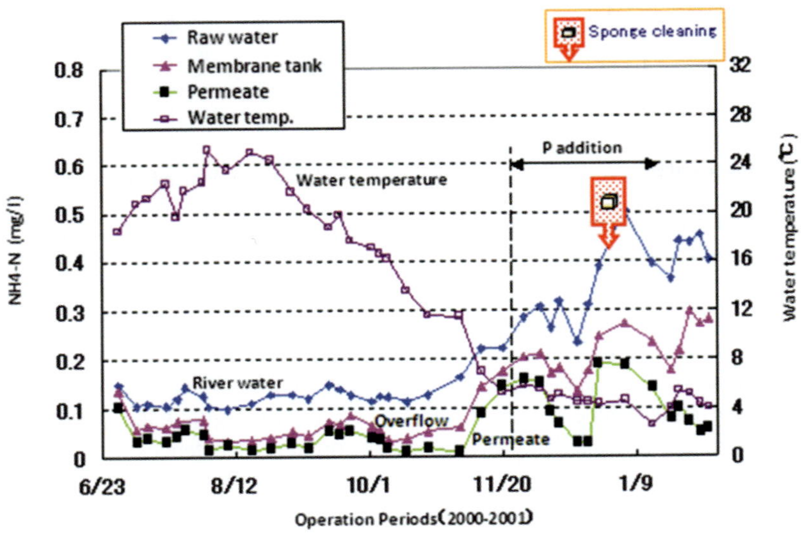

Figure 3.15 Variation of water temperature and NH_4-N concentrations in raw water and permeate.

Figure 3.16 shows the results of the AOC assessment. The AOC concentration decreased to below 10 µg/litre when a biofilm on the membrane had developed to a thickness of about 80 µm. AOC removal efficiency was high, even in a low water temperature such as 4.9°C. However, it gradually decreased when a low temperature period continued.

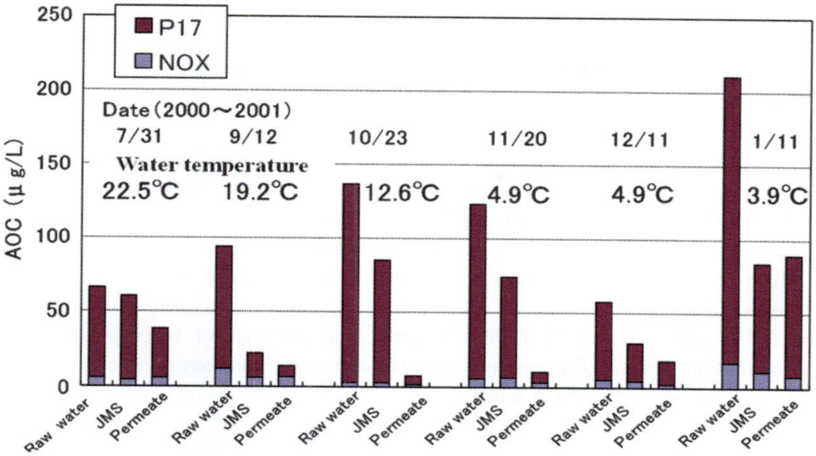

Figure 3.16 Variation of AOC concentrations in raw water and permeate.

3.4 CONCLUSION

The first part of this paper described the case study at Osaka Water Works. After the introduction of the advanced water purification system, Osaka Water Works succeeded in reducing the chlorine dosage because of the higher removal efficiency for organic matter and ammonia nitrogen. The additional biological oxidation also reduced AOC further. Hence, AOC reduction combined with a lower chlorine dosage will be the most promising way to economically produce drinking water of a better taste and to control bacterial regrowth in the distribution system.

The second part of this paper includes the effect of membrane filtration coupled with pre-ozonation or powdered activated carbon adsorption and biological/chemical oxidation. In the membrane filtration with pre-ozonation, the residual ozone effectively controls membrane fouling, but ozonation may increase AOC in the case of the raw water with high humic substances. Submerged membrane filtration with powdered activated carbon adsorption and biological oxidation is a suitable water purification system when the concentration of organic matter and ammonia nitrogen is high. The biofilm-membrane reactor with pre-coagulation/sedimentation removed AOC, but AOC removal efficiency decreased when a low water temperature period continued.

3.5 REFERENCES

Chandrankanth M. S. and Amy G. (1998). Effect of NOM source variation and calcium complexation capacity on ozone-induced particle destabilization. *Water Res.*, **32**, 115–124.

Fuchigami T. and Terashima K. (2003). Residual chlorine management of purified water by advanced water purification system in Osaka Water Works. *Journal JWWA*, **72**(7), 12–24 (in Japanese).

Grasso D. and Weber W. J. Jr. (1988). Ozone-induced particle destabilization. *J. AWWA*, **83**, 73–81.

Jang N. Y., Watanabe Y. and Ozawa G. (2002). A study on the MF membrane filtration of river water with pre-ozonation. *Journal JWWA*, **71**(2), 2–14 (in Japanese).

Jekel M. R. (1994). Flocculation effects of ozone. *Ozone Science & Engineering*, **16**, 55–66.

Kasahara S. and Ishikawa M. (2002). Evaluation of assimilable organic carbon in the Yodo River and effect of coagulation on its control. *Water Sci. Tech. Water Supply*, **12**(5–6), 465–472.

Kasahara S. and Ishikawa M. (2009). Growth characteristics of heterotrophs in water distribution system. *Proceedings of Environmental Engineering Forum, Jap. Soc. Civil Eng.*, **46**, 176–178 (in Japanese).

Kimura K., Watanabe Y. and Ohkuma N. (2000). Filtration resistance and efficient cleaning methods of the membrane with fixed nitrifiers. *Water Res.*, **34**(11), 2895–2904.

Kimura K., Watanabe Y. and Ohkuma N. (2001). A novel biofilm-membrane reactor for advanced drinking water treatment – pilot scale study. *Water Sci. Tech. Water Supply*, **1**(5–6), 157–168.

Lee S. Y., Jang N. Y. and Watanabe Y. (2004). Effect of residual ozone on membrane fouling reduction in ozone resisting micro filtration (MF) membrane system. *Water Sci. Tech.*, **50**(12), 287–292.

Mori Y., Oota T., Hashino M., Takamura M. and Fuyii Y. (1998). Ozone-microfiltration system. *Desalination*, **117**(10), 211–218.

Suzuki T., Watanabe Y. and Ozawa G. (2001). Performance of hybrid MF membrane system combining activated carbon adsorption and biological oxidation. *Water Science and Technology: Water Supply*, **1**(5/6), 253–259.

Watanabe Y., Fukui M. and Miyanoshita T. (1990). Theory and performance of jet mixed separator. *Water SRT, Aqua*, **39**(6), 387–395.

Watanabe Y., Kasahara S. and Iwasaki Y. (1998). Enhanced flocculation/sedimentation process by a jet-mixed separator. *Wat. Sci. Technol.*, **37**(10), 55–67.

Chapter 4

Organic matter, pipe materials, disinfectants and biofilms in distribution systems

A. K. Camper

4.1 INTRODUCTION

Bacteria in drinking water distribution systems can originate from several sources, including breakthrough from treatment, introduction via contamination events, or from growth. In well run facilities, treatment breakthrough should be minimal and contamination should be negligible and/or transient. In most instances, the presence of bacteria in distribution systems is likely to result from growth, particularly in biofilms, on pipe walls and sediments. The detachment and entrainment of organisms that proliferate in the biofilm leads to increased microbial counts in the water. These factors were first illustrated by Characklis (1988) in a systematic study conducted to explain coliform regrowth events in drinking water distribution systems.

Individual factors such as organic matter and disinfectants impact the growth of biofilms in distribution systems. However, clear correlations between single factors and biofilms/microbial growth are typically not found, and experience has shown that combinations of factors can be more important. This is particularly true where the organisms are impacted by interactions between water quality (organic carbon, disinfectant type and concentration) and the physical attributes (pipe materials) of the distribution system. These aspects are specific for each distribution system and are also governed by regulations that are attributes of individual countries. For example, some countries will distribute water with little to no disinfectant residual, while others mandate measureable disinfectant levels. In some situations, disinfectant by-product formation is limited by the use of chloramines, while in others removal of organic precursors is the desirable technology. Similarly, microbial growth in biofilms may be targeted for control through optimized disinfection strategies or the removal of organic carbon used by heterotrophic bacteria.

This chapter will summarize observations at the laboratory, pilot scale, and full scale on the interactions between the type and concentration of organic matter, chlorine and chloramine, and pipe materials on the growth of biofilms. The intent is to demonstrate that biofilm growth in distribution systems is complex and measurement of single factors may not provide adequate insight on its causes. This is related to the concept of biostability of the water: what constituents are responsible for controlling microbial growth in finished drinking water?

4.1.1 Organic matter and heterotrophic bacterial growth

Since heterotrophic bacteria require organic carbon for growth, organic matter is one aspect of water quality that can be responsible for biofilm proliferation in distribution systems. From an engineering perspective, it would be beneficial to have a test or measurement of the organic carbon that predicts microbial growth. The drinking water industry has focused on two separate measurements. One is biodegradable organic carbon (BDOC) that is the portion of the organic carbon mineralized by heterotrophic organisms. It is determined by measuring the difference in the initial and final concentration of dissolved organic carbon after incubation with a mixed microbial population. Another approach is assimilable organic carbon (AOC), representing the portion of the biodegradable organic carbon that can be converted to cell numbers by either a single organism or a defined suite of bacteria. AOC values are expressed as a carbon concentration by means of a conversion factor or calibration.

Past work focused on developing threshold levels for AOC and/or BDOC where microbial growth is limited. In systems with minimal secondary disinfectant (the Netherlands), van der Kooij *et al.* (1989) and van der Kooij and Hijnen (1990) showed a correlation between AOC (P17 + NOX) and regrowth with a suggested threshold level of 10 μg acetate C eq/L (van der Kooij 1992). In disinfected systems, LeChevallier *et al.* (1991) suggested a threshold of 50 μg acetate C eq/L (P17 + NOX) for controlling coliform regrowth. In a subsequent field study of 31 utilities, distribution systems receiving water with lower AOC concentrations (<100 μg/L) tended to have fewer regrowth problems (LeChevallier *et al.* 1996). Similarly, there have been numerical values associated with BDOC levels and microbial growth. Servais *et al.* (1991) set a level of 0.2 mg/L. This was modified according to temperature by Joret *et al.* (1994), with levels of 0.15 mg/L at 20°C and 0.30 mg/L at 15°C. In other situations there has been a lack of correlation between AOC/BDOC levels and microbial growth; these mainly negative results are often not reported in the literature.

There are several reasons why AOC and/or BDOC alone may not successfully predict biological stability. An obvious reason is if the amount of organic carbon and its growth potential is masked by the presence of a disinfectant residual. Another possibility is that the organic carbon level is variable, and there are insufficient measurements taken to give a realistic dataset for comparison.

Another aspect is that organic carbon may not be the limiting factor, as has been shown in Finnish waters where phosphorus tends to be at low levels and limiting regrowth of bacteria (Lehtola *et al.* 2002). Another reason is that there are constituents of natural organic matter (NOM) contributing to microbial growth that are not measured by the AOC or BDOC methods. This includes humic and fulvic acids found in surface and some ground waters.

Humic and fulvic acids are complex organic molecules composed of an aromatic backbone with attached functional groups, including carboxyls, enols, alcohols, ethers, ketones, aldehydes, esters, amines and amidines along with amino acids, amino sugars, peptides and alipahtics (Stevenson 1982, 1985). The molecules tend to be hydrophobic and when in solution are coiled with an exterior hydrophilic and the interior hydrophobic region. Amon and Benner (1996) noted that in both freshwater and marine environments, high molecular weight NOM supported more bacterial growth than low molecular weight compounds. An excellent book on the involvement of dissolved organic matter (DOM) in aquatic ecosystems has been edited by Findlay and Sinsabaugh (2003). In that book, a chapter by Foreman and Covert (2003) describes how microbial diversity can be linked to classes of DOM, and cites further work where the production of exoenzymes by bacteria leads to the metabolism of DOM. Another chapter of interest by Benner (2003) discusses the bioavailability of DOM in aquatic systems with reference to BDOC.

It is also known that NOM will adsorb to surfaces. For example, it is common knowledge that iron-based coagulation chemicals are effective at reducing levels of NOM in water (Edwards and Amirtharaja 1985). The affinity of NOM for iron oxides has been demonstrated in soil and other systems (Parfitt *et al.* 1977; Tipping 1981; Tipping *et al.* 1981; Davis 1982; Gu *et al.* 1996). When humic substances adsorb at interfaces at low ionic strength, it is assumed that they form flat configurations (Stumm and Morgan 1996; Wershaw 1999). The interaction of humic substances on iron oxides is assumed to be through ligand exchange or H-binding mechanisms (Stumm and Morgan 1996). From a microbial perspective, the attached humic molecule may uncoil, expand, and expose the functional groups (Gu *et al.* 1995) that may make them available for microbial degradation. Volk *et al.* (1997) suggested that biofilm bacteria may be capable of using humic materials, as was also shown by Gagnon *et al.* (2004). The potential involvement of humic and fulvic acids in biofilm growth was summarized by Camper (1997, 2004) and by Fisher (2003). All of these sources suggest that sorption of humic substances to biofilms permits their utilization and supports the growth of heterotrophic bacteria.

4.1.2 Disinfectans, NOM and microbial growth

4.1.2.1 Primary disinfection

Primary disinfectants (typically ozone, UV, or chlorine), along with inactivating pathogenic organisms, may have the effect of changing the nature of the organic matter in water. This is particularly true with ozone, where it has been shown that elevated levels of lower molecular weight compounds are produced and that these compounds can be associated with increased biofilm levels in distribution systems (van der Kooij *et al.* 1989; Price 1994; LeChevallier *et al.* 1996). Goel *et al.* (1995) reported that the fraction of recalcitrant natural organic matter in water made available for microbial growth was increased after ozonation, but the numerical value varied from site to site. This has also been substantiated by van der Kooij *et al.* (1982); Werner and Hambsch (1986); Servais *et al.* (1987); and Speitel *et al.* (1993). Buls (2000) also showed that ozone applied to soil-derived humic substances increased the bioavailable fraction. In pilot studies, it was shown that ozonated water supported biofilm growth, but if the same non-ozonated source water was chlorinated at a low level (0.05 mg/L) no biofilm was formed on plastic pipes (Lund and Ormerod 1995).

It has been hypothesized that UV treatment may increase the bioavailability of natural organic matter in water. Kieber *et al.* (1990) demonstrated that natural UV irradiation of humic substances at long exposure times resulted in the production of low molecular weight carbonyl groups. Although the exposure times were in excess of what would be encountered in the disinfection of drinking water, it is possible that UV could result in the same potential for regrowth that often accompanies ozonation. One investigation with well established biofilms that were subjected to UV treated water showed increased biofilm growth (Lund and Hongve 1994). In another case, a one log increase in biofilm numbers was seen post-UV treatment (Snicer *et al.* 1996). However, this phenomenon is not universal. In a study specifically designed to look at biofilm growth in a model distribution system (Shaw *et al.* 2000), four source waters were treated with UV light with a low pressure mercury vapor source at 254 nm and at a dose of 130 mW-s/cm^2. There was no statistical evidence for increased assimilable organic carbon or biodegradable organic carbon after UV treatment, although there were slight upward trends in these parameters for a few samples. These authors suggest that utilities should test their sources before adopting UV disinfection. Buls (2000) showed that reasonable doses of UV did not increase the bioavailability of humic substances, and that biofilm growth was not enhanced by UV treatment of the organic carbon. In a batch study, Lehtola *et al.* (2003) showed that UV treatment actually decreased AOC and small molecular weight organic molecules, and that the release of microbially available phosphorus was not enhanced at reasonable UV doses.

Chlorine, as a widely used disinfectant, is known to react with organics and create disinfection by-products. Although it is perceived by the industry that chlorination can increase the biodegradable fraction of organics, there are few publications that document this phenomenon, and the data are mostly anecdotal. Reckhow *et al.* (1990) demonstrated that chlorinated water reacted with total organic carbon (TOC) and suggested that the reaction was with the aromatic fraction of the TOC. If true, then there is the possibility that the aromatic characteristics of the TOC would be modified and hence increase biodegradability. In another study, LeChevallier *et al.* (1992) applied chlorinated/dechlorinated TOC to a biofilter, and noted increased TOC removal when compared to the performance when the TOC was not prechlorinated. Hanna *et al.* (1991) did a more thorough investigation that showed an increase of lower molecular weight species after chlorination, including compounds (mostly carbohydrates) more susceptible to microbial utilization. When specifically investigating the chlorination of humic substances, Buls (2000) showed that at high enough doses (4 mg/L) of chlorine, biodegradability was positively impacted, although the change in BDOC after chlorination averaged only 5%.

4.1.2.2 Secondary disinfection

The most common secondary disinfectants for distributed water are free chlorine or combined chlorine. The latter is most usually in the form of monochloramine. The section below is a comparison between these two disinfectants, with a focus on the impact these disinfectants have on biofilms in the distribution systems.

The increased use of monochloramine as a secondary disinfectant is primarily due to its lower potential to form regulated disinfection by-products. In addition, there have also been observations that in certain circumstances monochloramine may be more effective at controlling heterotroph and coliform regrowth than free chlorine. The increased effect is likely due to the slower reaction rate of monochloramine, which permits greater penetration into the biofilm than free chlorine. This hypothesis is substantiated by observations in the field where particle-associated coliforms were more protected from chlorination if the particles were greater than 7 micrometers, while the performance of monochloramine was unaffected (Berman *et al.* 1988). More detailed laboratory experiments with defined populations of bacteria have also demonstrated that there is a reaction-diffusion limitation for free chlorine that may not exist for monochloramine in biofilms (Xu *et al.* 1996; Chen and Stewart 1996; Stewart and Raquepas 1995). There are also studies of full-scale distribution systems where monochloramine and free chlorine have been compared. In one study, chloramines were more effective at reducing the number of total coliforms and HPC in biofilms than chlorine (Neden *et al.* 1992). In a field survey of 31 utilities, 0.51% of 35,159 samples were positive for coliforms at those utilities that used chloramines, whereas in chlorinated systems 0.97% of 33,196 samples were positive for coliforms. More importantly, the average density of coliforms in the water of the chlorinated systems was 35 times higher than that in the chloraminated systems (LeChevallier *et al.* 1996). In this case, however, only the bulk water and not the biofilms were analyzed. In a case study, where a distribution system that was in violation of the Total Coliform Rule switched to monochloramine and corrosion control, the coliform problems were eliminated (Schreppel *et al.* 1997). These studies suggest that the use of monochloramine as a secondary disinfectant is at least as effective as free chlorine and in general tends to provide better microbiological control for HPC and coliforms.

However, there are also practical issues that must be considered if monochloramine is to be used as a secondary disinfectant to improve microbiological quality of the water. The decay of monochloramine to produce free ammonia or the over-addition of ammonia during monochloramine formation can lead to nitrification problems. Nitrification can be difficult to control (see overview in Rahman *et al.* 2011) and may lead to an increase in the number of heterotrophic bacteria found in the water (Wolfe *et al.* 1990).

4.1.3 Pipe materials

For the purposes of this chapter, the focus is on distribution system mains, rather than premises plumbing which use another suite of materials. Distribution system mains are composed of a limited set of materials, primarily iron/steel, pre-stressed cement and plastics. Older distribution systems may be composed of cast iron, but more recently, ductile iron has been the material of choice. Ductile iron can be lined with cement, or in the past, with asbestos cement or asphaltic materials. Epoxy lining have also been used. In some locations, newer installations have primarily been with polyvinyl chloride (PVC) pipes. An overview of the types of pipes and the times they have been installed in the United States is available in a document by the American Water Works Service Company (2002). Each of these materials has its inherent advantages and disadvantages for cost, longevity, ease of installation, and impact on water quality.

Compared to disinfectants and organics in water, the ability to mitigate effects on microbiological water quality resulting from pipe materials is limited because of the inaccessibility and cost of replacing and relining pipes. The microbiological problems tend to be more serious in systems with corroding iron or steel surfaces. Several researchers have shown that biofilms on iron surfaces are more problematic than those on other types of surfaces (LeChevallier 1997; Delanou *et al.* 1997; Neden *et al.* 1992; Niquette *et al.* 2000; Kerr *et al.* 1998). In a field survey, LeChevallier *et al.* (1996) showed that distribution systems with more miles of unlined cast iron pipe had higher coliform occurrences than systems with less unlined cast iron pipe. Another study showed that the biofilm densities on iron were significantly higher than those on PVC with densities on cement intermediate to plastics and iron (Niquette *et al.* 2000). Neden *et al.* (1992) reported that cast iron pipes had the highest HPC, whereas PVC had the lowest. Laboratory studies conducted by Camper (1996) showed that mild steel surfaces in reactors had 10-fold higher HPC and coliforms than polycarbonate surfaces. In another laboratory study, Kerr *et al.* (1998) found highest counts and species diversity on cast iron when compared to medium-density polyethylene and uplasticized PVC. These studies illustrate that pipe materials appear to impact the amount of biofilm in distribution systems, with particular emphasis on the negative association with corroding iron.

Part of the association of corroding iron pipes with increased biofilm growth may be that the reactive metal surfaces act as a sink for disinfectants (van der Wende and Characklis 1990, LeChevallier *et al.* 1990; LeChevallier *et al.* 1996) in addition to adsorbing the humics. It then stands that perhaps the application of a corrosion control strategy may reduce the impact of iron on biofilm growth. Corrosion control through the addition of phosphate may result in the formation of iron-phosphate compounds that are less reactive than the oxides. pH control could change the electrostatic surface characteristics, which could then regulate the adsorption of organics to the pipe surface (Stumm and Morgan 1996; Chang *et al.* 1997; Beckett 1990).

4.2 INTERACTIONS OF FACTORS AND BIOFILM GROWTH

4.2.1 Importance of organic carbon and chlorine on biofilms

The ability of heterotrophic bacteria to proliferate in the presence of humic substances as the only carbon and energy source in the presence and absence of disinfection was shown in a series of experiments using laboratory annular reactors containing inert polycarbonate coupons (see Ellis *et al.* (2000), Camper *et al.* (2000a) and Butterfield *et al.* (2002b). The reactors were fed with Bozeman tap water treated with a granular activated carbon (GAC) column to remove background chlorine and then a biologically activated carbon (BAC) column to reduce the amount of biodegradable organic matter in the water. The BAC treated water also provided an undefined, mixed population inoculum to the reactors. Direct comparisons were made between three carbon sources; amino acids, carbohydrates, and humic substances extracted from Elliot Silt Loam soil from the International Humic Substances Society. Influent organic carbon concentrations were 0.5, 1.0 or 2.0 mg/L with sufficient nitrogen and phosphorus added to ensure that the reactors were carbon limited. Paired reactors were operated, with one of each pair receiving sufficient free chlorine to create a 0.2 mg/L residual in the reactor effluent after a 2 hr residence time. Over time periods of up to one year, the biofilm and effluent of the reactors were monitored for total (fluorescent stain and microscopy) and culturable cells (by R2A plates incubated at room temperature for 7 days). Analyses were also made of the influent and effluent dissolved organic carbon.

A surprising result was that in the absence of chlorine the humic substances supported biofilm heterotrophic numbers comparable to those of the amino acids and carbohydrates. At a level of 0.5 mg/L, humic substances produced ca. 10^6 total cells and 10^5 culturable cells/cm^2. Amino acids produced 8×10^6 total cells and 10^6 culturable cells/cm^2, and with carbohydrates (culturable) cell numbers were slightly lower. As the substrate levels increased to 2 mg/L, there was a general trend towards more biofilm cells, and the humics produced cell numbers closer to those of the amino acids of near 5×10^7 per cm^2 total cells and about a half log lower culturable cells. When substrate (organic carbon) utilization was followed, almost all of the amino acids and carbohydrates were removed, and humic removal approached 78%. Interestingly, the coupons from the humics reactors became brown over time, suggesting that the humic substances were being retained in the biofilm (Figure 4.1). Note that no iron was present in the reactors nor was any added to the influent feed water. Analysis showed this to be in the range of 8.3–11 μg C/cm^2. In the chlorinated reactors there was little effect of disinfection if amino acids were the carbon source, but total cell numbers declined by at least one log in total counts and three logs in culturable cells in the carbohydrate and humic systems. These data suggested that the effect of chlorination was more towards disinfection than increasing bioavailability of the humic substances for microbial growth.

Figure 4.1 Humic substances in a biofilm grown on a polycarbonate annular reactor slide (bottom) compared to a clean slide (top).

As a point of comparison, specific growth rates and yields for the biofilms were calculated for each of the carbon sources (see Butterfield *et al.* 2002b for calculations). These yields were: amino acids, 0.08–0.12 μg cell C/μg C; carbohydrates, 0.04–0.09 μg cell C/μg C; and humics, 0.03–0.05 μg cell C/μg C. The highest yield values were at the 0.5 mg/L carbon concentration. Specific growth rates were also highest at the lowest carbon concentration. For humic substances, the rate was 0.04/hr at 0.5 mg/L, and decreased by half for each subsequent doubling of the carbon concentration. Interestingly, the growth rates were all higher when the system was chlorinated. At 0.5 mg/L, the chlorinated growth rate was 0.06/hr with the same decrease as carbon concentrations were doubled. These results suggest that the organisms in the reactors were exhibiting zero order kinetics. Practically, the data indicate that drinking water biofilm bacteria are very well adapted to grow at low carbon concentrations. In situations typical of US water treatment where organic carbon is not removed to levels less than 0.5 mg/L, control of biofilms and detached cells may need to come from disinfection.

4.2.2 Importance of iron-corrosion products

One of the original observations that suggested the importance of corroding iron in supporting the growth of biofilms was reported in 1996 (Camper 1996). The experimental setup consisted of annual reactors that initially contained all mild steel coupons. This system was fed with easily degraded organics (ethanol, proprionaldehyde, parahydroxybenzoic acid, sodium benzoate and sodium acetate in equimolar concentrations by carbon) that have minimal reaction with chlroine. As the experiment continued, the mild steel coupons were systematically removed for sampling and replaced with polycarbonate surfaces. At the end of the experiment, the polycarbonate surfaces from the original mild steel reactor were also sampled. A parallel system contained only polycarbonate. When fed the same influent of biologically treated tap water with indigenous organisms, added coliforms, and an easily degradable organic carbon described above source, the polycarbonate-only reactors accumulated approximately one log fewer HPC and coliforms than what was found on the mild steel coupons in the second reactor. Interestingly, the polycarbonate coupons in the otherwise mild steel reactor also had elevated bacterial levels along with visible amounts of iron corrosion products. The results are indicative of an impact of these corrosion products on biofilm accumulation.

The above study demonstrated that AOC utilization was enhanced in the presence of iron. This observation was followed by laboratory tests to determine if corrosion products from an iron pipe could adsorb humic substances and support microbial growth (Abernathy and Camper 1998). Crushed corrosion products were placed in a column and a humic solution of approximately 3 mg/L was passed through the column for a week. On day 7, approximately 60% of the carbon was being removed by the corrosion products. The exposed corrosion products were then dried, exposed to UV light disinfection, and added to flasks containing AOC free water and 10^4 indigenous drinking water organisms. Within 2 days the organism numbers had increased to 10^8 CFU/ml, indicating that the sorbed organics were biologically available.

4.2.3 Iron, organics and disinfectants

Initial work investigating interactions between the concentration of organic carbon, pipe material and disinfectants was done using an easily degradable carbon source and pilot scale pipe loops of mild steel. Biologically treated drinking water with an average AOC of 20 μg/L was added either with or without 0.5 mg/L organic carbon and with or without 1 mg/L free chlorine. As with other experiments using corroding iron, there are limited analyses available to assess microbial growth; the corrosion products interfere with total counts, ATP and DNA extraction. Therefore, HPC analyses were used. The data showed that the biofilm HPC increased when the organic carbon was added. The effect was mitigated by the addition of free chlorine (Camper 1996). Follow-on studies in the same mild steel pipe loops and annular reactors with mild steel or polycarbonate coupons investigated the importance of transient increases in easily degraded organic carbon and the presence/absence of chlorine or monochloramine. Organic carbon was increased stepwise (0, 0.10, 0.25, 0.50 mg/L). In most cases the higher organic carbon concentrations resulted in more biofilm HPC in the annular reactors, but this was not consistent. Pipe loops showed no change in HPC with increased carbon loading. When chlorine or monochloramine was added to the annular reactors, there was little change in culturable HPC numbers on the mild steel surfaces and decreases on polycarbonate. In the mild steel pipe loop, disinfection also had little effect on the biofilm HPC (Camper and Jones 2000). Although these results were not as illuminating as had been expected, the beneficial outcome was the observation that the interaction of factors contributing to regrowth is not always systematic. The work also showed that in these simulated distribution systems very low levels of organic carbon was required to support biofilm HPC levels between 10^6 and 10^7 per cm^2.

Other studies were done to specifically investigate the reaction rates of corrosion products with adsorbed humics (Abernathy and Camper 1998). Corrosion products were exposed to 5 mg/L free chlorine or monochloramine and the first order reaction rates determined. The rate for free chlorine was 0.055 hr^{-1} mg^{-1} L, and the value for chloramines was 0.0012 hr^{-1} mg^{-1} L. These rates show the difference in behaviour between free chlorine and monochloramine in that monochloramine is less reactive and may therefore penetrate more deeply into corrosion products and biofilms.

Based on the knowledge that iron oxides tend to accumulate humic substances, the concept was used to design a membrane pretreatment system to remove the organics to reduce organic matter and biofouling. Laboratory scale porous media filters were constructed with biologically active carbon from Laval, Canada or new iron oxide coated sand. These filters were fed humic amended tap water (2 mg/L) with or without prechlorination (1 mg/L for 30 min resulting in no measurable residual downstream). The water was then passed by an RO membrane in tangential flow or through another bead column to determine biofouling potential. Biologically activated carbon and microbially colonized iron oxide coated sand exhibited similar efficacy as biological pretreatment options by removing approximately 20% of the TOC and decreasing downstream fouling on membranes and the glass bead column. There was no advantage from prechlorination, suggesting little change in the nature of the organics used in these experiments (Wend *et al.* 2003).

An extensive study was conducted to determine the impact of four different distribution systems surfaces, three levels of organics and two disinfectants on biofilm formation (Camper *et al.* 2003). The four materials were ductile iron,

PVC, epoxy coating, and cement lining. Coupons of these materials were placed in duplicate annular reactors. In phase 1, there were no added humic substances (background AOC of 20 µg/L) and no disinfectant. In phase 2, chlorine was added to one reactor and monochloramine to the other at a residual of 0.2 mg/L. In phase 3, humic substances were introduced at 0.5 mg/L and the influent disinfectants adjusted to yield effluent residuals of 0.2 mg/L. Phase 4 had a higher humics level of 2 mg/L and disinfectant residuals of 0.2 mg/L. Each phase lasted three to four months.

In phase one (no disinfectant, no added carbon) the biofilm numbers were no different on the four materials ($\log_{10}$ means of $5.5/cm^2$ to $6/cm^2$). Phase 2 (no added carbon, presence of disinfectants) data showed that the iron surfaces had a mean of one log more HPC than the other surfaces regardless of the type of disinfectant. When humic substances were present in phase 3, the numbers on the other surfaces increased, but were still higher by at least a half log on the iron surfaces. In phase 4, the other surfaces had comparable numbers of HPC ($10^5/cm^2$) with both disinfectants while the iron coupons were 1.5 logs higher. When comparing the iron surfaces across the phases, the biofilms in phase four (2.0 mg/L humics, added disinfectant) were significantly greater than the biofilm in the control phase. This indicates that the residual disinfectant level of 0.2 mg/L was insufficient to control the increased growth at the highest carbon level in the iron reactors when compared to the control phase. However, for the biofilms on the remaining materials, the biofilms in phase four did not significantly differ from the biofilms on the same materials in phase one where no carbon or disinfectant was added. This indicates that the disinfectant level of 0.2 mg/L was sufficient to control the increased growth at the highest carbon level in epoxy, cement, and PVC reactors when compared to the control phase. This effect was seen for both monochloramine and free chlorine.

The project also had an 18 months field study where annular reactors with three different materials (one per reactor, ductile iron, cement, epoxy) were installed at four utilities, either at the treatment plant or at locations in the distribution system. In these systems, total organic carbon concentrations ranged from 4 mg/L (a high TOC groundwater) to less than 1 mg/L. Two systems used free chlorine and two used chloramines. The high TOC groundwater systems and low TOC surface water utility had no difference in biofilm numbers on the three materials. The surface water systems with the intermediate TOC values used chloramines, and they had higher HPC on iron compared to the other surfaces. Overall, the field testing results indicate that either iron had the highest regrowth, or the type of material had no influence on the number of bacteria present.

An outcome of the study was the ability to provide practical insights to water utilities. In general, and as shown by others as noted in the introductory material to this chapter, higher levels of carbon tend to support greater biofilm growth on all pipe materials, but the effect is most pronounced on iron. Therefore, reducing organic carbon either as AOC, BDOC or humic substances should reduce biofilm growth in distribution systems. However, these levels may need to be very low and in some cases not feasible. Biofilms can still flourish even with very low carbon concentrations, especially on corroding iron surfaces. The findings suggest that utilities with significant amounts of iron pipe may have the greatest regrowth potential and this can be compounded if the distributed water is high in organics. For some systems the greatest gain in water quality may be from replacement or relining of corroding distribution system piping rather than by more extensive water treatment.

4.2.4 Iron, organics, disinfectants and corrosion control

4.2.4.1 Corrosion products and iron oxide coated beads

To further investigate the interactions between organic carbon, iron oxides, and disinfectants, studies were designed to utilize iron oxide coated glass beads or crushed corrosion products from distribution system pipes (Butterfield *et al.* 2002a). Since utilities may be able to employ corrosion control strategies to mitigate the importance of iron corrosion in this process, it was included as part of the experimental design.

Small columns were filled with either iron oxide coated glass beads or corrosion products that had approximately the same specific surface area as the glass beads. Controls with uncoated glass beads were also used. The columns were initially exposed in batch mode to water with 50 mg/L humics, flushed, and then continually fed with biologically treated tap water amended with nitrogen, phosphorus, and 3 mg/L C as humics. Some columns were chlorinated to provide a residual of 0.2 mg/L. For corrosion control, the pH was either 7.3, 8, or 9 and a set of pH 7.3 columns also received 3 mg/L phosphate. Columns were destructively sampled after at least two months of operation and the amount of biofilm determined by heterotrophic plate counts, total direct counts, and potential exoproteolytic activity (PEPA; Butterfield *et al.* 2002c). The treatment options were analyzed with respect to minimizing the amount of biofilm on the media surfaces.

The iron oxide coated beads adsorbed approximately 10 μg C/g beads, and the corrosion products had at least five times more adsorbed carbon than the iron oxide coated beads. The corrosion products also had higher chlorine demand. The range of analyses indicated that chlorine application was the most effective treatment to reduce biofilm on both the beads and corrosion products. This was particularly true in the columns that also had phosphate as a corrosion control. Interestingly, phosphate alone was not particularly effective. The second most important treatment was using water without humics; biofilms in the control reactors were lower than those in systems that had been fed humic substances. pH modification showed no significant effect on the amount of biofilm. This is not surprising, since in this case corrosion had already occurred and pH modification likely had no impact on the adsorption of the humics to the iron surfaces. Overall, the results showed that adsorbed humic substances support biofilm growth and mitigation could be attained by removing TOC from the water or countering growth through disinfection. The benefit of corrosion control was less conclusive.

4.2.4.2 Laboratory and pilot distribution system studies

Recognizing that the interaction of factors was complicated and that the study of individual parameters may not provide realistic results, a comprehensive laboratory and pilot scale study was designed and conducted (Abernathy and Camper 1998). The project was focused on humic substances as the organic source, chlorine and monochloramine, and corroding iron surfaces and was guided by the potential interactions shown in Figure 4.2. The laboratory studies used annular reactors composed of ductile iron pipe and glass/polycarbonate reactors with mild steel slides and the pilot system was 4″ mild steel.

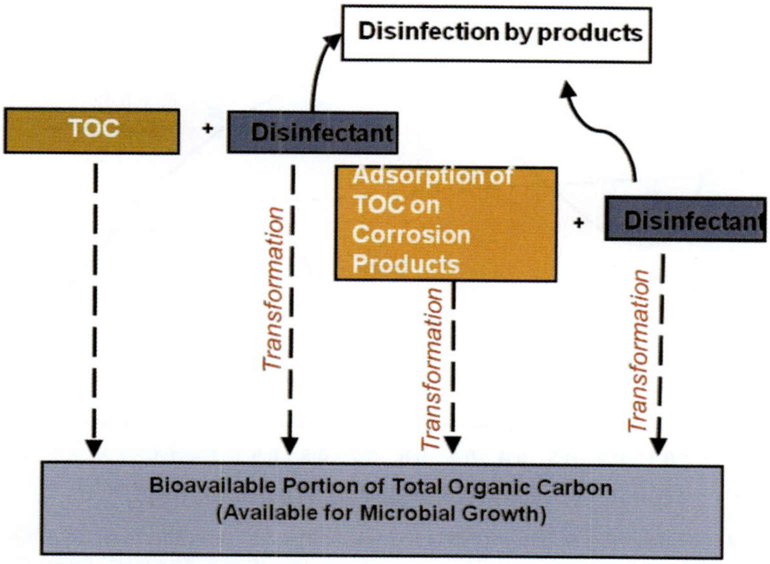

Figure 4.2 Considerations of the fate of organics in water distribution systems and their potential biodegradability. (Adapted from Abernathy and Camper, 1998).

The first experiments investigated the importance of increasing chlorine doses on the biofilm and mass of accumulated corrosion products on mild steel coupons in annular reactors. The reactors were fed with 0.25 mg/L easily degradable carbon (ethanol, proprionaldehyde, parahydroxybenzoic acid, sodium benzoate and sodium acetate), water from the biologically active carbon column, and balanced with nitrogen and phosphorus. The reactors were colonized for 76 days prior to the introduction of the disinfectant. Chlorine was then added to provide effluent residuals at increasing intervals from 0 to 1.25 mg/L. Biofilm data are shown in Figure 4.3. As chlorine dose increased, the HPC numbers also increased as compared to the unchlorinated control until a threshold of approximately 0.75 mg/L was reached. Simultaneously, corrosion product mass increased from approximately 2 mg/cm^2 to 23 mg/cm^2. At higher chlorine doses the mass of corrosion products also increased. However, at chlorine residuals above 0.75 mg/L there appeared to be sufficient disinfection "power" to reduce the bacterial numbers in spite of the greater mass of corrosion products. These data suggest that each distribution system may have a target disinfectant concentration over which biofilm control is seen, and lower levels where it may be enhanced due to either the conversion of the humic substances to more biodegradable fractions or to the increased surface area and reaction/depletion of the disinfectant at a corroding surface that allows for microbial growth.

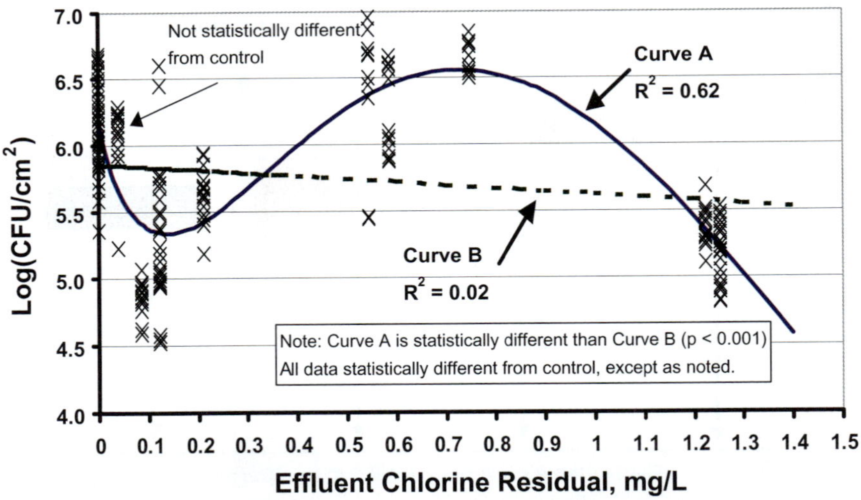

Figure 4.3 Response of biofilm HPC to increasing concentrations of chlorine when grown on mild steel coupons in an annular reactor. (From Abernathy and Camper, 1998).

A logical follow-up question was whether there was an interaction between the efficacy of chlorine at a constant dose of 2 mg/L, increasing concentrations of corrosion inhibitor (polyphosphate) from 2.4–8 mg/L, and biofilm numbers in annular reactors with mild steel coupons. Organic carbon was once again held constant at 0.25 mg/L. As before, the reactor that received chlorine alone had the highest biofilm numbers and amount of corrosion products even though the residual was still 1 mg/L. The chlorinated reactor had 7×10^6 CFU/cm^2 and a corrosion product density of 17 mg/cm^2. At an 8 mg/L polyphosphate dose, the HPC in the biofilm were reduced by one log, the effluent chlorine residual increased to 1.2 mg/L, and the corrosion product mass decreased to 9.9 mg/cm^2. There was a 6 mg/L loss of phosphate across the reactor, suggesting significant deposition on the surface of the coupons. Intermediate doses of polyphosphate did not decrease corrosion product mass and decreased the chlorine residual over that of the chlorine-only treatment. The manufacturer of the polyphosphate had suggested using a dose of 5 mg/L polyphosphate, which was not sufficient to reduce corrosion in this system.

The remainder of the bench-scale work used annular reactors composed of ductile iron outer cylinders with removable coupons. Comparisons were made between a control reactor and disinfected reactors (chlorine, and monochloramine, both at 3.25 mg/L) that ran for two months. The chlorinated reactor consistently had HPC biofilm numbers (ca. 10^8/cm^2) approximately one log higher than the control and the monochloramine treated system numbers decreased over time to less than 10^5/cm^2. As seen in Figure 4.4, the monochloramine treated reactor also had less obvious accumulation of corrosion products, which could be related to the decreased number of attached organisms.

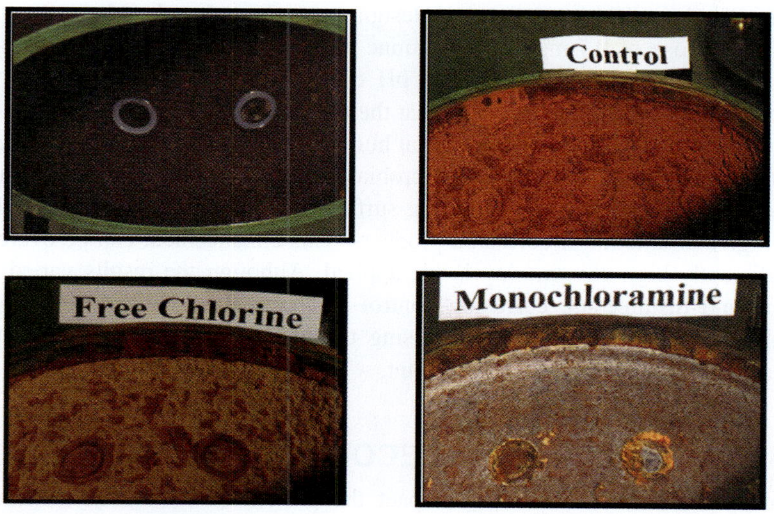

Figure 4.4 Ductile iron outer cylinders of annular reactors exposed to chlorine and monochloramine for two months. The upper left is the surface at the beginning of the experiment. Note the increased amount of corrosion on the control and chlorinated surfaces. (From Abernathy and Camper, 1998).

These experiments were expanded to include the impact of zinc orthophosphate and polyphosphate on biofilm accumulation. Again, ductile iron reactors were used as was chlorine and monochloramine and a control with no disinfectant. In the chlorinated systems, chlorine alone resulted again in higher cell numbers in the biofilm by an average of 1 log. Orthophosphate reduced the cell numbers by 1.5 logs in the chlorinated system; whereas polyphosphate was slightly less effective (cell numbers were reduced with 1.1 logs). Visual inspection also showed that the reactor with orthophosphate had the lowest accumulation of corrosion products. When monochloramine was added, the biofilm numbers increased over the control by 0.5 logs, which is in contrast to the test where monochloramine was compared to chlorine. Polyphosphate did not impact the biofilm abundance, but the zinc orthophosphate system was noticeably affected with a reduction in cell numbers of 1.8 logs. As with the prior experiments, the visually clean surface (zinc orthophosphate + monochloramine) had the lowest amount of biofilm.

The above laboratory results guided the design of pilot tests in a mild steel pipe loop reactor. Five parallel loops were operated with 0.25 mg/L added carbon (see information above). One was a control with no chlorine, one was with chlorine alone. The others had corrosion control strategies of pH control, another received zinc orthophosphate, and a third received polyphosphate. The pH adjusted control was maintained automatically to provide a slightly positive Langelier Saturation Index. In this system, the biofilm numbers were on average within a half log of each other, with the chlorine + polyphosphate loop having the highest HPC biofilm numbers. This experimental design was repeated with chloramines with an influent dose of 0.7 mg/L. Chloramine alone was effective at decreasing the amount of HPC in the biofilm. The pH control and polyphosphate loops had significantly lower biofilm numbers that the chloraminated loop.

Overall, the results demonstrated that humics adsorb to corrosion products, and that the adsorbed carbon supports microbial growth. Chlorine alone can actually increase biofilm numbers on corroding surfaces, and this increase was correlated with a greater amount of corrosion products. This enhancement can be overcome if sufficiently high chlorine levels are applied. Although the results varied with system and disinfectant, corrosion control strategies can also reduce microbial populations, hypothetically by decreasing the amount of corrosion products or increasing the efficacy of the disinfectant.

4.3 CONCLUSIONS AND RECOMMENDATIONS

There are numerous factors that impact the growth of biofilm in distribution systems. Several of these such as temperature or seasonal changes in water use patterns are beyond the control of water utilities. Factors that may be subject to manipulation as discussed in this chapter are organic carbon, disinfectant type and concentration, pipe materials, and corrosion control.

As shown using AOC and BDOC assays, biofilms in drinking water systems are adapted to grow under very low nutrient conditions. What these tests may not measure is the additional carbon that could be available to biofilm organisms in distribution systems in the form of more recalcitrant humic substances. As described in this chapter, humic substances do support the growth of heterotrophic bacteria in biofilms and should be considered as a parameter when biofilm control is necessary. Unfortunately, there are no direct measures of the bioavailability of humic substances in the environment, although this topic is of interest in microbial ecology/carbon cycling as well as for the drinking water industry. Water source is important, since it can be assumed that humics from soil and algal production found in surface waters will be more degradable than those found in groundwater.

Humic substances and natural organic matter in water may be rendered more biodegradable by the application of oxidizing agents used for disinfection. For example it is known that ozone as a primary disinfectant is capable of increasing biodegradability. The changes in biodegradability of natural organic matter treated with chlorine are not as consistent; in some cases there is increased use after treatment and in others there is no change. In contrast, monochloramine and UV treatment are less likely to increase the biodegradability of organic carbon in distribution systems.

The concept of biostability is relevant to this discussion. Biostability is a balance between the growth potential of the organic matter in the water and inactivation by disinfection. In systems with little or no disinfection, biostability is governed by the amount of organic matter. Biostability can also be attained in higher organic waters when there is sufficient inactivation of the bacteria via secondary disinfection. The strategy that most US utilities use to mitigate biofilms is to increase the disinfectant residual or convert from free chlorine to chloramine, which is considered to be more effective for controlling regrowth. Using disinfectants to control biostability may not be desirable from a customer acceptance, disinfection by-product formation potential, nitrification (in the case of chloramines) or regulatory perspective. These drivers, along with cost and ease of implementation, will determine if the best mechanism to achieve biostability is through removal of organics or changing disinfection strategies.

Related to discussions on biostability is the importance of pipe materials and their influence on biofilm growth. Natural organic matter in water can be adsorbed to the pipe surface, and this is most problematic if the distribution system has a significant amount of corroding iron. Corroding pipe materials can also cause the decay of oxidizing disinfectants. The combination of increased organic matter immobilized in biofilms and the decreased efficacy of disinfection may increase regrowth of HPC and coliforms.

Mitigation of the influence of corroding pipe materials can be either through the replacement/relining of pipes or corrosion control. In the first case, replacement and relining strategies can prioritize sections of the distribution system where older cast iron pipes exist. This approach could minimize breaks and improve water delivery by removing tuberculated and occluded pipes. Water quality could be enhanced by reducing the potential for red and yellow water, increased ability to maintain disinfectant residuals, and decreased microbial counts. If pipe replacement and lining is not realistic, corrosion control implemented to reduce lead and copper corrosion can mitigate some of the deleterious effects of corroding iron on microbial water quality. It is not certain if this is because corrosion control causes of decreased adsorption of natural organic matter or increased efficacy of disinfectants.

The contents of this chapter have been focused on distribution system mains. Similar factors are important in premises plumbing, and these issues will become more important as water conservation efforts become more prevalent. For example, lower water use will create greater water age in plumbing systems, and if recycled/reclaimed water is used, there will be elevated organic carbon loads. Another concern in premises plumbing is the presence of opportunistic pathogens. Although the majority of biofilm bacteria are environmental organisms, a few can be opportunistic pathogens. Factors that affect their growth (carbon concentration, type and concentration of disinfectant, and potentially plumbing materials) are important in protection public health.

Lab, pilot, and field work has shown that there can be a combination of factors related to the growth of bacteria in water distribution and plumbing systems. The research and practical experience has also illuminated the fact that although generalizations can be made, each system is unique. Using research and relevant field results, utilities should consider site specific testing to ensure that the choices they make for improving water quality have the desired effect. This will avoid unintended consequences that have been seen in systems where changes in treatment processes have been made.

4.4 REFERENCES

Abernathy C. and Camper A. K. (1998). Interactions between Pipe Materials, Organics, Corrosion Inhibitors, and Disinfectants on Distribution Biofilms. Volume 1. Bench and Pilot-scale Studies. National Water Research Institute, Fountain Valley, CA.

American Water Works Service Co., Inc. (AWWSC) (2002). Deteriorating buried infrastructure management challenges and strategies. Available on line at http://www.epa.gov/safewater/tcr/pdf/infrastructure.pdf (accessed June 15, 2011).

Amon R. M. W. and Benner R. (1996). Bacterial utilization of different size classes of dissolved organic matter. *Limnology and Oceanography*, **41**, 41–51.

Beckett R. (1990). The Surface Chemistry of Humic Substances in Aquatic Systems. In: Surface and Colloid Chemistry in Natural Waters and Water Treatment, R. Beckett (ed.), Plenum Press. New York, NY, pp. 3–16.

Benner R. (2003). Chapter 5: Molecular indicators of the bioavailability of dissolved organic matter. In: Aquatic Ecosystems: Interactivity of Dissolved Organic Matter, S. A. Findlay and R. L. Sinsabaugh (eds), Academic Press, pp. 121–137.

Berman D., Rice E. W. and Hoff J. C. (1988). Inactivation of particle-associated coliforms by chlorine and monochloramine. *Appl. Environ. Microbiol.*, **54**, 507–512.

Buls J. (2000). Making Humic Substances More Biodegradable. MS thesis, Civil/Environmental Engineering, Montana State University, Bozeman, MT, USA.

Butterfield P. W., Camper A. K., Biederman J. A. and Bargmeyer A. M. (2002a). Minimizing biofilm in the presence of iron oxides and humic substances. *Water Res.*, **36**, 3893–3910.

Butterfield P. W., Camper A. K., Ellis B. D. and Jones W. L. (2002b). Chlorination of model distribution system biofilm: Implications for growth and organic carbon removal. *Water Res.*, **36**, 4391–4405.

Butterfield P. W., Bargemeyer A. M., Camper A. K. and Biederman J. A. (2002c). Modified enzyme activity assay to determine biofilm biomass. *J. Microbiol. Meth.*, **50**, 23–31.

Camper A. K. (1996). Factors Limiting Microbial Growth in the Distribution System: Pilot and Laboratory Studies. American Water Works Association Research Foundation, Denver, CO

Camper A. K. (2004). Involvement of humic substances in regrowth. Special edition: HPC bacteria in drinking water: Public health implications. *Int. J. Food Micro.*, **92**(3), 355–364.

Camper A. K. and Jones W. L. (2000). Factors Affecting Microbial Growth in Model Distribution Systems. American Water Works Association Research Foundation, Denver, CO

Camper A. K., Ellis B., Butterfield P., Anderson B., Huck P., Volk C. and LeChevallier M. W. (2000). Biological Stability of Water in Treatment Plants and Distribution Systems. American Water Works Association Research Foundation, Denver, CO.

Camper A. K., Brastrup K., Sandvig A., Clement J., Spencer C. and Capuzzi A. J. (2003). Impact of distribution system materials on bacterial regrowth. *J. Amer. Water Works Assoc.*, **95**(7), 107–121.

Chang Y., Li C. W. and Benjamin M. M. (1997). Iron oxide coated media for NOM sorption and particulate filtration. *J. Amer. Water Works Assoc.*, **89**, 100–113.

Characklis W. G. (1988). Bacterial Regrowth in Distribution Systems. American Water Works Association Research Foundation, Denver, CO

Chen X. and Stewart P. S. (1996). Chlorine penetration into artificial biofilm is limited by a reaction-diffusion interaction. *Environ. Sci. Technol.*, **30**(6), 2078–2083.

Davis J. A. (1982). Adsorption of natural dissolved organic matter at the oxide/water interface. *Geochimica et Cosmochimica Acta*, **46**, 2381–2393.

Delanou A. D., Holt J. C., Woodward S., McMath M. and Smith S. E. (1997). Effect of pipe materials on biofilm growth and deposit formation in water distribution systems. In: Proc. AWWA Water Quality Technology Conference, Denver, CO. http://www.sciencedirect.com/science/article/pii/S004313540500391X-bbib9 (accessed June 15, 2011).

Edwards G. A. and Amirtharajah A. (1985). Removing color caused by humic acids. *J. Am. Water Works Assoc.*, **77**, 50–57.

Ellis B., Butterfield P., Jones W. L., McFeters G. A. and Camper A. K. (2000). Effects of carbon source, carbon concentration, and chlorination on growth related parameters of heterotrophic biofilm bacteria. Microbial Ecology, **38**, 330–347.

Findlay S. A. and Sinsabaugh R. L. (eds) (2003). Aquatic Ecosystems: Interactivity of Dissolved Organic Matter. Academic Press, San Diego, CA.

Fischer H. (2003). The role of biofilms in the uptake and transformation of dissolved organic matter. In: Aquatic Ecosystems: Interactivity of Dissolved Organic Matter, S. A. Findlay and R. L. Sinsabaugh (eds), Academic Press, San Diego, CA, pp. 285–313.

Foreman C. M. and Covert J. S. (2003). Linkages between dissolved organic matter composition and bacterial community structure. In: Aquatic Ecosystems: Interactivity of Dissolved Organic Matter, S. A. Findlay and R. L. Sinsabaugh (eds), Academic Press, San Diego, CA , pp. 343–362.

Gagnon G. A., Dykstra T. S., O'Leary K. C., Andrews R. C., Chauret C. and Volk C. (2004). Impact of UV Disinfection on Biological Stability American Water Works Research Foundation, Denver, CO.

Goel S., Hozalski R. and Bouwer E. J. (1995). Biodegradation of NOM: Effect of NOM source and ozone dose. *J. Amer. Water Works Assoc.*, **87**(1), 90–105.

Gu B., Schmitt J., Chen Z., Liang L. and McCarthy J. F. (1995). Adsorption and desorption of different organic matter fractions on iron oxide. *Geochimica et Cosmochimica Acta*, **59**, 219–229.

Gu B., Mehlhorn T. L., Liang L. and McCarthy J. F. (1996). Competitive adsorption, displacement, and transport of organic matter on iron oxide: I. Competitive adsorption. *Geochimica et Cosmochimica Acta*, **60**, 1943–1950.

Hanna J. V., Johnson W. D., Quezada R. A., Wilson M. A. and Lu X. L. (1991). Characterization of aqueous humic substances before and after chlorination *Environ. Sci. Technol.*, **25**, 1160–1164.

Joret J. C., Volk C., Randon G. and Cote P. (1994). Control of biodegradable organic matter during drinking water treatment. Proc. International Seminar on Biodegradable Organic Matter. Montreal, Quebec, 23–24.

Kerr C. J., Osborn K. S., Robson G. D. and Handley P. S. (1998). The relationship between pipe material and biofilm formation in a laboratory model system. *J. Appl. Microbiol.*, **85**, 29S–38S.

LeChevallier M. W. (1997). Disinfection of biofilms in drinking water systems. In: Proc. Water Quality Technol. Conf. Amer. Water Works Assoc., Denver, CO.

LeChevallier M. W., Lowry C. D. and Lee R. G. (1990). Disinfecting biofilms in a model distribution system. *J. Amer. Water Works Assoc.*, **82**, 87–92.

LeChevallier M. W., Schulz W. and Lee R. G. (1991). Bacterial nutrients in drinking water. *Appl. Environ. Microbiol.*, **57**, 857–862.

LeChevallier M. W., Becker W. C. and Schorr P. (1992). Evaluating the performance of biologically active rapid filters. *J. Amer. Water Works Assoc.*, **84**, 136–142.

LeChevallier M. W., Welch N. J. and Smith D. B. (1996). Full-scale studies of factors related to coliform regrowth in drinking water. *Appl. Environ. Microbiol.*, **62**, 2201–2211.

Lehtola M. J., Miettinen I. T. and Martikainen P. J. (2002). Biofilm formation in drinking water affected by low concentrations of phosphorus. *Can. J. Microbiol.*, **48**, 494–499.

Lehtola M. J., Miettinen I. T., Vartiainen T., Rantakokko P., Hirvonen A. and Martikainen P. J. (2003). Impact of UV disinfection on microbially available phosphorus, organic carbon, and microbial growth in drinking water. *Water Res.*, **37**, 1064–1070.

Lund V. and Hongve D. (1994). Ultraviolet irradiated water containing humic substrates inhibits bacterial metabolism. *Water Res.*, **28**, 1111–1116.

Lund V. and Omerod K. (1995). The influence of disinfection processes on biofilm formation in water distribution systems. *Water Res.*, **29**, 1013–1021.

Niquette P., Servais P. and Savoir R. (2000). Impacts of pipe materials on densities of fixed bacterial biomass in a drinking water distribution system. *Water Res.*, **34**, 1952–1958.

Neden D. G., Jones R. J., Smith J. R., Kirmeyer G. J. and Foust G. W. (1992). Comparing chlorination and chloramination for controlling bacterial regrowth. *J. Amer. Water Works Assoc.*, **84**, 80–89.

Parfitt R. L., Fraser A. R. and Farmer V. C. (1977). Adsorption on hydrous oxides III. Fulvic acid and humic acid on goethite, gibbsite and imogolite. *J. Soil Science*, **28**, 289–296.

Price M. L. (1994). Ozone and Biological Treatment for DBP Control and Biological Stability. Amer. Water Works Assoc. Denver, CO.

Rahman M. S., Encarnacion G. and Camper A. K. (2011). Nitrification and potential control mechanisms in simulated premises plumbing. *Water Res.*, **45**, 5511–5522.

Reckhow D. A., Singer P. C. and Malcolm R. L. (1990). Chlorination of humic materials: Byproduct formation and chemical interpretations. *Environ. Sci. Technol.*, **24**, 1655–1664.

Schreppel C. K., Fredericksen D. W. and Geiss A. A. (1997). The positive effects of corrosion control on lead levels and biofilms. Proc. Water Qual. Technol. Conf. Amer. Water Works Assoc., Denver, CO

Servais P., Billen G. and Hascoet M. C. (1987). Determination of the biodegradable fraction of dissolved organic matter in water. *Water Res.*, **21**, 445–450.

Servais P., Billen G., Ventresque C. and Bablon G. P. (1991). Microbial activity in GAC filters at the Choisy-le-Roi treatment plant. *J. Amer. Water Works Assoc.*, **83**, 62–68.

Shaw J. P., Malley J. P. Jr. and Willoughby S. A. (2000). Effects of UV-irradiation on organic matter. *J. Amer.Water Works Assoc.*, **92**, 157–167.

Snicer G. A., Malley J. P., Magolin A. B. and Hogan S. P. (1996). Evaluation of ultraviolet technology in drinking water treatment. Proc. Water Qual. Technol. Conf. Amer. Water Works Assoc., Denver, CO

Speitel G. E. Jr., Symons J. M., Diehl A. C., Sorensen H. W. and Cipparone L. A. (1993). Effect of ozone dosage and subsequent biodegradation of DBP precursors. *J. Amer. Water Works Assoc.*, **85**(5), 86–95.

Stevenson F. J. (1982). Humus Chemistry, John Wiley & Sons, NY.

Stevenson F. J. (1985). Geochemistry of Soil Humic Substances. In: Humic Substances in Soil, Sediment, and Water, G. R. Aiken, D. M. McKnight, R. L. Wershaw and P. MacCarthy (eds), John Wiley & Sons, New York, NY.

Stewart P. S. and Raquepas J. B. (1995). Implications of reaction-diffusion theory for the disinfection of microbial biofilms by reactive antimicrobial agents. *Chem. Eng. Sci.*, **50**(19), 3099–3104.

Stumm W. and Morgan J. J. (1993). Aquatic Chemistry. 3rd edn, John Wiley & Sons, Inc., New York, NY.

Tipping E. (1981). The adsorption of aquatic humic substances by iron oxides. *Geochimica et Cosmochimica Acta*, **45**, 191–199.

Tipping E., Woof C. and Cooke D. (1981). Iron oxide from a seasonally anoxic lake. *Geochimica et Cosmochimica Acta*, **45**, 1411–1419.

van der Kooij D. (1992). Assimilable organic carbon as an indicator of bacterial regrowth. *J. Amer. Water Works Assoc.*, **84**, 57–63.

van der Kooij D. and Hijnen W. A. M. (1990). Criteria for defining the biological stability of drinking water as determined with AOC measurements. Proc. Water Qual. Technol. Conf. Amer. Water Works Assoc. Denver, CO van der Kooij D., Visser A. and Hijnen W. A. M. (1982). Determining the concentration of easily assimilable organic carbon in drinking water. *J. Amer. Water Works Assoc.*, **74**(10), 540–545.

van der Kooij D., Hijnen W. A. M. and Kruithof J. C. (1989). The effects of ozonation, biological filtration, and distribution on the concentration of easily assimilable organic carbon in drinking water. *Ozone Sci. Engrg.*, **11**, 297–303.

van der Wende E. and Characklis W. G. (1990). Biofilms in potable water distribution systems. In: Drinking Water Microbiology, G. A. McFeters (ed.), Springer-Verlag.

Volk C. A., Volk C. B. and Kaplan L. A. (1997). Chemical composition of biodegradable dissolved organic matter in streamwater. *Limnol. Oceanogr.*, **42**, 39–44.

Wend C., Stewart P., Jones W. and Camper A. K. (2003). Biological treatment for membrane water treatment systems. *Water Res.*, **37**(14), 3367–3378.

Wershaw R. L. (1999). Molecular aggregation of humic substances. *Soil Science*, **164**, 803–813.

Werner P. and Hambsch B. (1986). Investigation of the growth of bacteria in drinking water. *Water Supply*, **4**(3), 227–234.

Wolfe R. L., Lieu N. I., Izaguirre G. and Means E. G. (1990). Ammonia-oxidizing bacteria in chloraminated distribution systems: Seasonal occurrence, distribution and disinfection resistance. *Appl. Environ. Microbiol.*, **56**(2), 451–462.

Xu X., Stewart P. S. and Chen X. (1996). Transport limitation of chlorine disinfection of *Pseudomonas aeruginosa* entrapped in alginate beads. *Biotech. Bioeng.*, **49**, 93–100.

Chapter 5

Safe distribution without a disinfectant residual

G. J. Medema, P. W. M. H. Smeets, E. J. M. Blokker and J. H. M. van Lieverloo

5.1 INTRODUCTION

5.1.1 Safe distribution of water: to disinfect or not to disinfect?

Drinking water is transported from the treatment works to the consumers' tap in a distribution network of pipes and reservoirs. In situations with high raw water quality and/or extensive multiple barrier treatment to produce high quality drinking water, this water should maintain its high quality during transport.

Microorganisms and microorganism-associated processes in the network can affect water quality:

- Ingress of pathogenic microbes due to loss of physical and hydraulic integrity of the network and the presence of a nearby contamination source (such as sewer lines);
- Growth of microorganisms in the distribution network, some of which may be opportunistic pathogens, such as *Legionella pneumophila,* non-tuberculous mycobacteria, *Pseudomonas aeruginosa, Acanthamoeba, Naegleria fowleri* and others.

These issues not only affect water quality but may also pose a health risk to the consumer. In addition, growth of non-pathogenic microorganisms can lead to (see Chapter 1):

- non-compliance with water quality standards for total coliforms, heterotrophic plate counts;
- aesthetical issues such as presence of invertebrates and taste and odour problems;
- corrosion of pipe materials.

In affluent nations, there are two principal approaches to control the microbial water quality in the distribution network. The first approach is to maintain a disinfectant residual in the water during distribution to provide a barrier against ingress of microbial contaminants and limit microbial growth. The second approach is to control the risk of ingress through strict maintenance of the physical and hydraulic integrity of the network and to control the growth of microbes by distributing biologically stable water and using materials that do not leach nutrients. In the latter approach, drinking water is distributed without the presence of a disinfectant residual. This approach is found in some European countries, the former in most of the other affluent nations.

The different approaches have been the subject of debate that amalgamated in the late nineties (Haas 1999; LeChevallier 1999; Trussel 1999; van der Kooij *et al.* 1999a, 1999b). In these debates, arguments for and against distribution with a disinfectant residual are highlighted. The most important arguments for the use of a disinfectant residual are:

- The presence of a residual disinfectant reduces the risk of microbial contaminants that may enter the distribution network. With the increasing complexity of the distribution network, the costs of the network and the open nature and ageing of the infrastructure a residual disinfectant is necessary to inactivate microbial pathogens that may enter the network through cross-connections, mains breaks, repairs and leaks. Also for smaller systems with limited resources, a disinfectant residual is a relatively simple and cheap solution to improve the microbial safety.
- The presence of a residual disinfectant controls the growth of microorganisms in the network. It limits non-compliance with microbial water quality standards such as total coliforms and heterotrophic plate counts. It is also argued that limiting the amount of nutrients in the water and materials is difficult and would require substantial investments in additional treatment (LeChevallier 1999) and focus on carbon-sources alone may not be enough (Haas 1999). Also, most networks now in use have been installed in past decades with the materials of that time and a control strategy should also consider this situation.
- The presence of a residual disinfectant may serve as sentinel for a breach of integrity of the system. When a distribution network is monitored with a consistent sampling strategy (or even online sensors) a reduced disinfectant residual concentration may signal that a contamination event has occurred.

The most important arguments against the use of a disinfectant residual are:

- Disinfectants react with organic and inorganic compounds in the water and this creates disinfection by-products (DBP) in small quantities. DBP formation depends on many factors, but DBP such as trihalomethanes (THM) are found in the vast majority of cases of chlorination. More than

600 DBP have been identified (Richardson *et al.* 2007). Several (groups of) by-products have been associated with illnesses in humans. Some of the DBP (bromate, N-nitrosodimethylamine (NDMA), benzaldehyde) show carcinogenic activity in long-term animal studies and several others are classified as possible carcinogens. A large body of epidemiological literature has accumulated over the years and meta-analyses have been conducted to assess the health risk of DBP. Lifetime exposure to chlorinated drinking water is associated with bladder cancer with a risk level of approximately 1 in 1000 (Hrudey and Charrois 2012). For colon and rectal cancer and also for reproductive health outcomes results are less univocal. A recent review (Nieuwenhuijsen 2009) indicated that there appears to be some evidence for an association between exposure to DBPs, specifically THMs, and congenital health, particularly in relation to the health effects for gestational age/intrauterine growth retardation and, to a lesser extent, pre-term delivery, but evidence for relationships with other outcomes such as low birth weight, stillbirth, congenital anomalies and semen quality is inconclusive and inconsistent.

- The reaction of disinfectants with organic compounds in the water may also yield compounds, such as halogenated phenols and anisoles that give rise to taste and odour complaints by consumers. Taste and odour complaints are the most frequent cause of consumer complaints and consumers have a negative opinion about chlorinous taste and odour, both in terms of aesthetics and safety (Crozes *et al.* 2007).
- The sensitivity of pathogens to the disinfectants used in the network differs. The disinfectants used are effective against bacterial pathogens, less so against viruses and even less so against parasitic protozoa (Payment 1999; Environmental Protection Agency [EPA] 2012). Chlorine and chloramine are not effective against *Cryptosporidium.* (LeChevallier and Au 2004).
- The use of a disinfectant residual may mask the failure of the integrity of the system and ingress of microbial contamination (Craun and Calderon 2001). Water quality testing for coliforms or *E. coli* that are very sensitive to chlorine may indicate that the water is not contaminated, while infectious pathogens that are more resistant may be present in the water. This is of particular importance in samples taken after repairs and maintenance.
- Disinfectant residuals are not very effective against microbes in biofilms on pipe walls or sediment. The disinfectants react with the biofilm matrix but do not reach the microbes (LeChevallier *et al.* 1988). Also microbes in biofilm particles that slough of the wall and enter the bulk water again are more difficult to reach and inactivate (Behnke *et al.* 2011).
- Disinfection is targeting the symptoms rather than the cause of the microbiological issues in the network. The cause of ingress is insufficient

hydraulic and structural integrity and hygiene. The cause of biofilm formation is the quality of the treated water and the materials used in the network. Targeting the cause is more effective and is also not sensitive to failures in the disinfection.

- In many settings, disinfectant residuals are not maintained throughout the entire network. That means that only a part of the network and consumers is protected by the presence of a residual and part is not (Payment 1999; Gauthier *et al.* 2001). In such settings, disinfectant residuals may even enhance regrowth as they react with the organic compounds in the water and produce compounds that are more readily biodegradable (Skadsen 1993; Zang and DiGiano 2002).
- The use of toxic chemicals such as chlorine and chlorine dioxide requires production and may require transportation of these chemicals with a risk of accidents and spills.

Looking at the pros and cons there is no simple answer to the question of whether a disinfectant residual in the distribution network is necessary/beneficial for health and a wholesome drinking-water quality. The answer certainly depends on the context, such as the quality of the water entering the network, hydraulic and structural integrity of the network and the ability to apply proper hygiene. In the context of drinking-water supply in the Netherlands, the answer is no. The next paragraphs describe the history and rationale of this answer and the consequences for safeguarding drinking-water safety, with the emphasis on protection of the distribution network against ingress of faecal contamination. Protection against regrowth is discussed in other chapters in this book.

5.1.2 The road to distribution without disinfectant residual in the Netherlands

In the 1970s, chlorination was applied in the Netherlands, particularly in surface water treatment. After the discovery of DBP of chlorination in the water of Rotterdam Water Works (Bellar *et al.* 1974; Rook 1974), a significant effort started to reduce the use of chlorine in water treatment. The reduction of chlorine use or replacement of chlorine was promoted: "the amount of chlorine to be added should, however, not be more than is absolutely necessary" (Kiwa 1978), but not without recognizing the significance of disinfection for public health "until … techniques have been found to be at least equivalent to chlorination, chlorine must from a public health viewpoint continue to be used as a disinfectant for potable water". In the late seventies, the amount of chlorine used in water treatment in the Netherlands was reduced significantly by limiting break-point chlorination and transport chlorination at low temperatures and using lower doses in summer (Kruithof 1986a, b). Also, post-treatment disinfection was re-examined and was shown to be an important contributor to the presence of

THMs in the network. Given the suggested relation between by-products of chlorination and several types of cancer and the observed mutagenicity (as seen with the Ames-test) of the water, it was recommended to reduce post-chlorination as much as possible. This meant in several water supplies that the residual was present only in the first part of the distribution network. The residual disappeared due to reaction of the disinfectant with reducing compounds in the water, biofilm, sediment and piping materials and no residual was present in the largest and more distant part of the network.

As an example, Amsterdam Water Supply started a full scale experiment in 1983 to reduce the chlorine dose stepwise and monitor what happened to the water quality in the distribution network (Schellart 1986). It turned out that when the chlorine was reduced to zero, the mutagenicity (Ames-test) of the water disappeared, total trihalomethanes were reduced from 12–22 μg/l to below the detection limit, no coliforms or enterococci were detected in the network (before and after stopping the chlorine dosing), the heterotrophic plate counts remained as low as they were (2–5 CFU/ml) and the assimilable organic carbon (AOC) content of the finished water was reduced by 40% (Schellart 1986). So the chlorine dosing was stopped permanently (it was kept on standby for emergency disinfection). Similar developments were seen at the other surface water utilities that used a chlorine residual. And even though the concentrations of chlorination by-products were below the 10^{-6} lifetime cancer risk level, the utilities abandoned the use of post-chlorination (van Genderen 1998).

In the 1990s, disinfection of surface water with chlorine was replaced in the Netherlands by ozone or membrane filtration as the primary barrier to pathogens, except for two water treatment plants. By 2006, these two chlorine-based disinfection plants changed to UV for primary disinfection, so chlorine is not used in water distribution or water treatment in the Netherlands any longer. Important arguments for abandoning chlorine also from water treatment were to: prevent DBP formation, improve taste and odour and hence consumer satisfaction, and apply sufficient disinfection, particularly of *Cryptosporidium*.

Today, there are two surface water utilities that use low doses of chlorine dioxide as post disinfection; not to create a residual in the network but to inactivate heterotrophic bacteria that may grow in activated carbon or sand filters in periods when the water temperature is high. In several other surface water, bank filtration and groundwater systems, UV is used for this purpose.

The Netherlands is not unique in the distribution of water without a disinfectant residual. Denmark and areas in other Nordic countries, areas of Germany (Hambsch 1998), Luxembourg and Switzerland (Klein and Foster 1998) distribute water without chlorination. These countries, and the European Union (EU), have no legal requirement for maintaining a disinfectant residual. Standards occur for microbial water quality and the owner of the network is obliged to ensure that the water meets these standards (with or without disinfectant residual).

5.1.3 Microbial safety in water legislation in the Netherlands

Water legislation in the Netherlands has adopted Quantitative Microbial Risk Assessment as the central approach to demonstrate the microbial safety of (surface) water supply systems (Anon. 2001, 2011). Water utilities have to monitor the presence of reference pathogens (*Campylobacter, Cryptosporidium, Giardia,* culturable Enteroviruses) in the source water and demonstrate the adequacy of the treatment processes to remove these pathogens to produce drinking water that has a quality that corresponds with a probability of infection of less than 1 in 10,000 persons per year. Although not identified specifically in the legislation, maintaining this high quality through the distribution network implies the same safety level applies to the water delivered to the consumer.

5.2 GOOD ENGINEERING PRACTICE

Good practices in construction, operation and maintenance of distribution networks are of paramount importance in protecting the water quality, both with and without a disinfectant residual in the network. When the network is in good condition and actively managed, this creates the proper conditions to consider distributing water without disinfectant residual.

The basic elements are:

- Physical integrity. The physical integrity prevents ingress in places or times where the hydraulic pressure is low or absent. Physical integrity is particularly important in places such as reservoirs that are not pressurized. The Netherlands has a very low leakage rate, generally <3% (Beuken *et al.* 2006). Compared to leakage rates in other European countries this is very low (Figure 5.1) and a comparative study identified several factors that contribute to this (United Kingdom Water Industry Research [UKWIR] 2006). The pressure in the network can be relatively low due to the flat terrain and high buildings are equipped with their own pump. The majority of the network is relatively young and produced of polyvinyl chloride (PVC) with relatively few joints and connections. Like many affluent nations, the Netherlands is facing ageing networks and need to actively assess and manage network condition and integrity to prevent leakage and ingress.
- Hydraulic integrity. Continuous maintenance of sufficiently high pressure in the network to prevent contaminants entering the network is required. Pressure fluctuations and surges are minimized by variable pumps, pressure dampening devices, valve closure procedures, automated distribution control to prevent large flow fluctuations and pressure zoning in (the few) hilly areas (Smeets *et al.* 2009).
- Protection against backflow. Use of break tanks before larger user-installations (industries, hospitals) and the use of backflow preventers in

the water meters of house connections, where >96% of the connections are metered.

- Strict hygiene during construction and maintenance. A national hygiene code was developed by the water utilities (Nobel 2001; van Lieverloo *et al.* 2002). Since then, this has become part of the quality assurance systems of the water utilities. It has recently been updated (Meerkerk and Kroesbergen 2010) and has become an integral part of the Dutch drinking-water legislation. The code lays down the principles of good hygiene during construction and maintenance and describes control measures for safe storage of materials, inspections, cleaning and disinfection of materials and pipes after construction or repair, personal hygiene and use of appliances such as fire hydrants. Also training, supervision and water quality monitoring to verify the efficacy of the hygiene code is described.
- Approval system for materials and valves and hydrants. Certification of materials and appendages that are to be used in distribution networks has been developed since the start of Kiwa in 1948. All materials have to be approved according to the national acceptance scheme.

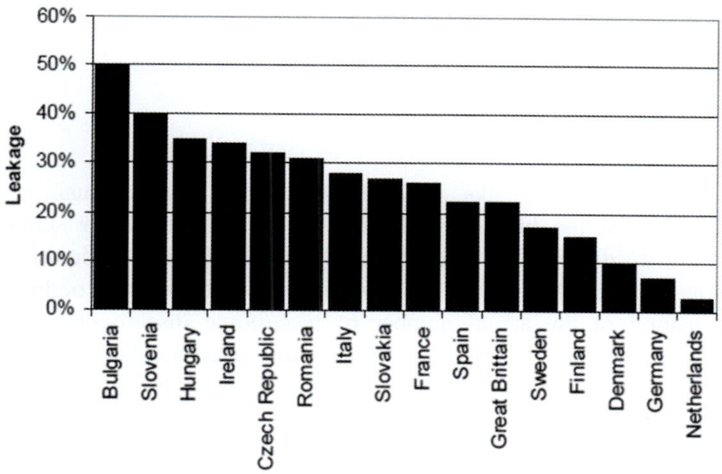

Figure 5.1 Leakage rates in European countries (Vereniging van waterbedrijven in Nederlands [VEWIN] 2009a; Deutcher Verein des Gas- und Wasserfaches [DVGW] 2008).

5.3 EVIDENCE OF SAFE DISTRIBUTION WITHOUT DISINFECTANT RESIDUAL

Noblesse oblige; with the absence of a disinfectant residual the water utilities have one barrier less in the distribution system and are obliged to ensure that the other barriers are in place and effective. Recognizing the difference in approach to safe distribution in the Netherlands and the United Kingdom, a collaborative research project was conducted around 2000 to evaluate the knowledge on distribution of drinking water with or without a disinfectant residual (van der Kooij *et al.* 2002). Both countries show a high compliance with microbial water quality standards and recognize that good engineering practice is the principal factor in protecting the distribution system against ingress. A chlorine residual appears to be a cheap way to reduce regrowth, but increases DBPs and taste and odour complaints. The UK-NL study indicated that a chlorine residual may offer at least some protection against faecal contamination of the network.

In the past decade, concerns were raised in North America and Europe over distribution system contaminations as a result of cross-connections, pressure transients and maintenance works and the association with illness (Payment *et al.* 1991, 1997; Craun and Calderon 2001; Karim *et al.* 2003; Hunter *et al.* 2005; Nygard *et al.* 2007; Besner *et al.* 2011; LeChevallier *et al.* 2011). In the Joint Research Programme (Bedrijfstakonderzoek [BTO]), of the water utilities in the Netherlands, research projects focused on the evaluation of the evidence that distribution without disinfectant residual offers sufficient protection against ingress of pathogens. The next paragraphs present the evidence coming from surveillance of waterborne outbreaks of disease, contamination events registered by the water utilities, water quality monitoring and operational monitoring.

5.3.1 Evidence from waterborne outbreaks

In affluent nations, outbreaks of waterborne illness are our main source of information about the burden of disease via community water supply, even though they are difficult to detect (Hunter *et al.* 2001). The difficulty to detect waterborne outbreaks means that the information we collect from outbreaks is an underestimation of the actual number of waterborne disease outbreaks (Ford 1999). Moreover, sporadic cases are thought to represent a larger proportion of waterborne disease than cases related to outbreaks (Nichols 2003) but go largely unnoticed. Such sporadic cases may be particularly the result of local contamination events in the distribution network. So, our picture is incomplete, but what is evident from outbreaks in public supplies is that harmful pathogens may spread to a large body of consumers resulting in substantial economic and health-related costs. This is illustrated by the April 1993 *Cryptosporidium* outbreak in Milwaukee (Mackenzie *et al.* 1994), where around 403,000 people suffered illness, 4,400 people were hospitalized and around 100 people died. Even though these figures have been disputed by others (Hunter and Syed 2001), it is clear from this and many other outbreaks that the health and economic consequences can be large.

Outbreaks do not only show that the impact can be substantial, but also inform the water and health community about the events, faults and circumstances, alone or in combination, that may lead to an outbreak. The next paragraph is a brief literature review about outbreaks that were caused by an event in the distribution network, to indicate the relative significance of the network and the causes of contamination events that have been implicated in outbreaks.

Hunter (1997) reported that 15 of the 57 outbreaks in public water supplies in the UK between 1911 and 1995 were associated with contamination within the distribution network. In the Nordic countries, 18–20% of the outbreaks through drinking-water between 1975 and 1991 were associated with cross-connections, both in community and private supplies (Stenström *et al.* 1994). In the European project MICRORISK, outbreaks through public water supplies in Europe between 1990–2004 were reviewed and subjected to a fault tree analysis (Risebro *et al.* 2007). Eighty-six outbreaks were reported, with a total of 72,546 cases, of which 341 were hospitalized and 1 died. In 33% of these outbreaks, contamination during distribution was the dominant cause of the outbreak. The fault tree analysis showed that distribution system events were usually solitary events, that is without a related event in source or treatment. Events that have contributed to outbreaks through contamination of distribution networks were:

- cross-connections/backflow
- construction or repair
- damaged/old mains
- low pressure
- cleaning of mains
- reservoir contamination

In the USA, 671 outbreaks reported in community water supplies from 1971 to 2002 were associated with distribution system deficiencies (Craun *et al.* 2006). Data on incidents collected by the US Environmental Protection Agency (USEPA) suggest that this is an underestimation of the actual number of identified outbreaks by a factor of 3–4 (Reynolds *et al.* 2008). Figure 5.2 illustrates the type of deficiencies that were identified as cause of the contamination.

The proportion of outbreaks associated with treatment deficiencies has declined over the years, and the deficiencies in the distribution network seem to be a proportionally increasing cause of waterborne outbreaks. From 1991–2000, 25% of the outbreaks were related to a contamination in the distribution system or in household plumbing. In the period 1920–1990, this was 11–18% (Craun and Calderon 2001). This increase of distribution deficiencies may be associated with ageing distribution networks. Why didn't the presence of a chlorine residual prevent these outbreaks? This is largely unknown. What is suggested (Craun *et al.* 2006) is that many pathogens associated with these outbreaks are not likely to be killed by the low level of disinfectant residual in the network. Another trend in the US that was associated with distribution-associated outbreaks was that they affect more people than other deficiencies.

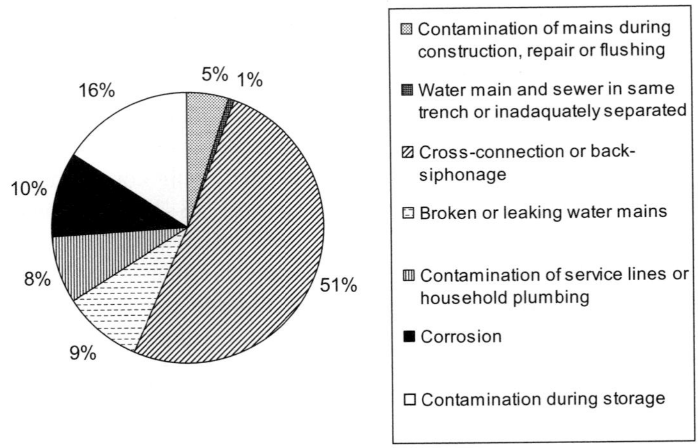

Figure 5.2 Distribution deficiencies associated with waterborne outbreaks in the United States, 1971–2000 ($n = 120$). (From Calderon 2004; Reynolds *et al.* 2008).

From the review of outbreaks through community drinking-water supplies (Hrudey and Hrudey 2004) it is clear that in many distribution-related outbreaks, lack of or non-compliance to adequate hygiene procedures to maintain the integrity of the network or to ensure safety during and after breaks and repairs has led to gross contamination of mains water which resulted in people falling ill and even to fatalities, such as in the Cabool and Gideon outbreaks in Missouri, USA. Hrudey and Hrudey (2004) state: "many of the most troubling cases have revealed no effort whatsoever at assuring distribution system integrity". In other cases, outbreaks resulted from cross-connections or open connections with contamination sources.

In the Netherlands, three outbreaks of waterborne diseases have been reported since 1945. All these three outbreaks were connected to contamination of drinking water in the distribution network. Six cases of typhoid in Amsterdam occurred in 1962, possibly due to sewage contamination of the drinking-water network. In 1981, 609 cases of enteric disease of multiple aetiology were reported because of wastewater inflow from a marine vessel by a cross-connection with the drinking-water network in Rotterdam (Huisman and Nobel 1981). At that time, water was still distributed with a (low) disinfectant residual. In 2001 an outbreak of gastroenteritis, with approx. 500 people experiencing symptoms, occurred due to a cross-connection in a dual reticulation system between a "household water" (partially treated river water) pipe and the drinking-water network (both without residual disinfectant) (Fernandes *et al.* 2007; Van Lieverloo *et al.* 2007b). Because of this outbreak, the use of dual reticulation systems was largely banned in the Netherlands.

Comparing the documented outbreaks in the Netherlands and the United States indicates that the number of outbreaks and disease cases in the Netherlands is low: in the period 1971–2002 there were 671 community water supply outbreaks in the US (roughly 22 outbreaks per year or 0.08 outbreaks per million consumers per year), compared to 2 (1 every 15 years or 0.004 outbreaks per million people per year) in the Netherlands. In Europe 86 outbreaks were recorded (roughly 6 per year or 0.01 outbreaks per million consumers per year) from 1990–2004. Though many differences in water supply and surveillance systems make a direct comparison difficult, the outbreak statistics certainly do not suggest that distribution of drinking water without a disinfectant residual increases the risk of a waterborne outbreak.

5.3.2 Evidence from contamination events

Outbreaks are the tip of the iceberg of contamination events (Craun *et al.* 2006) that may lead to illness among exposed consumers. Operational data from water utilities may provide a more extensive picture of contamination events that occur in the distribution net work. The absence of a registered outbreak of disease cases in these contamination events does not mean that no disease cases occurred. Health surveillance is not very sensitive in picking up waterborne outbreaks of intestinal illness.

Westrell *et al.* (2003) used operator logs of the Gothenburg water utility to estimate the frequency and magnitude of contamination events in the distribution network. Failures in the distribution system were derived from the incidence reports between 1980 and 2000 and personnel interviews. In a few events, cross-connections with pressurized sewage pipes were detected as the cause of the contamination, but for contamination events in the periphery of the distribution system or in reservoirs, no information on the contamination source was recorded. The events in the periphery and reservoirs were caused by leakage through cracks in concrete reservoir walls or through damage during maintenance of the network.

In the Netherlands, a large study was conducted to collect information about contamination events in distribution networks. Water utilities, representing the water supply to 11 million consumers, collated the information on events from the period 1994–2003 (Van Lieverloo *et al.* 2006, 2007b). Events were defined as repeated detection of faecal indicators. Fifty events were recorded or (roughly) 0.4 events per million people per year, 100-fold higher than the recorded outbreaks. One event was the recorded outbreak in 2001; in the other events no increased illness incidence was noted. The estimated affected population varied between 5 and 50,000 persons, with 9 events affecting over 1,000 consumers and a total affected population of 185,000 persons. The utilities emphasized that not all events were retrievable from the records, so this overview was an underestimation of the actual number of events. Data were compiled from the events on the duration (median 8 days, 95% 30d, Figure 5.3) and the level of contamination, described by the concentration of *E. coli*/thermotolerant coliforms detected (Figure 5.4).

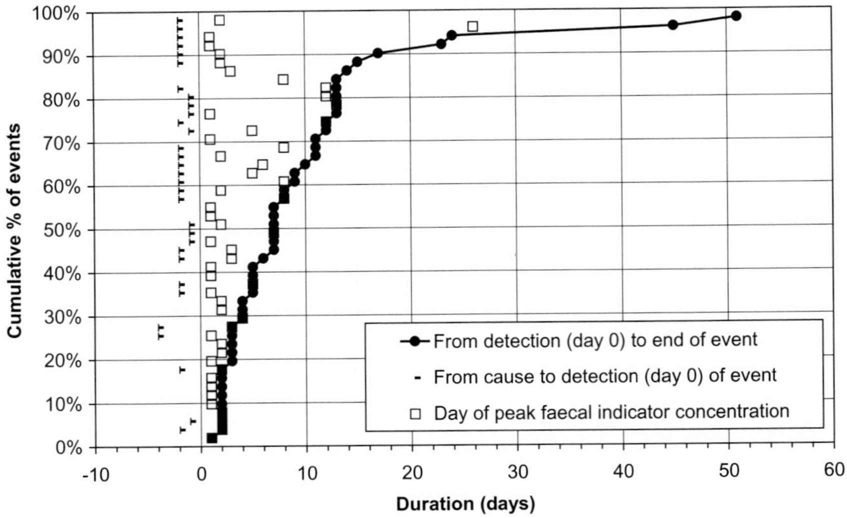

Figure 5.3 Duration of 50 faecal contamination events reported in the Netherlands from 1994 through 2003. The start was defined as the first day that faecal indicator bacteria were detected (from detection to end) or the day the suspected cause of the event occurred (from cause to detection). The end of the event was the second day when no indicator bacteria were detected anymore in 100 ml samples. (From Van Lieverloo *et al.* 2007b).

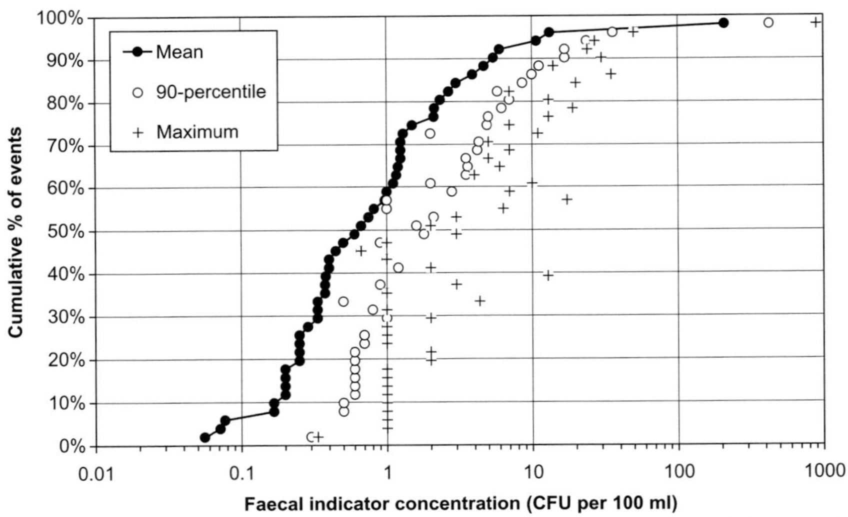

Figure 5.4 *E. coli* concentrations per 100 ml in water samples during 50 faecal contamination events. (From Van Lieverloo *et al.* 2007b).

The event data also provided information about the cause of the event, the suspected source of contamination and type of system failure that was involved, how the event was detected and what type of remedial actions were taken to restore safe water distribution. Table 5.1 provides a summary of the data.

Table 5.1 Characteristics of 50 faecal contamination events recorded from 1994–2003.

Phase of water supply system that was cause of the event

Distribution	Unknown	Source	Treatment	Before distribution
37	10	1	1	1

Site of water supply system that was cause of the event

Main	Unknown	Reservoir	Well	Treatment
34	13	2	1	0

Cause and type of failure

Unknown	Replacing mains/ hydrant	Repair mains	Damaged main	Leak in reservoir	Cross-connection	Flush-ing	Swabb-ing
19	18	8	1	1	1	1	1

Suspected source of contamination

Soil	Unknown	Sewage	Surface water	Roof material
26	19	3	2	0

Detection of faecal contamination event by

1st sample	2nd sample	3rd sample	4th sample	Complaints
25	7	1	1	3

Site of detection

Treatment plant	Distribution reservoir	Periodical tap water sample	After operations
2	12	7	29

Remedial measures taken

Flushing mains	Isolation of mains section	Unknown	Boiling advisory	Dosage of disinfectant
48	29	16	7	2

Only 3 of 50 events were reported to have occurred in wells or (groundwater) treatment plants, whereas 37 events occurred in distribution systems (Table 5.1). Over half of the reported events concerned contaminations that were detected after operations in mains (18 replacements, 8 repairs, 2 cleaning operations). In these cases, supply was commenced immediately after the operations, but not before the mains were flushed and in some cases disinfected. Standard procedure for operations is isolation of the distribution mains that were opened, until microbial safety has been verified by water quality testing. Most of the unknown causes were not well recorded as information was limited to sampling dates and concentrations of thermotolerant coliforms in laboratory databases. It is likely that the causes of these events in most cases also were also a result of mains operations. Isolation and flushing were the primary responses to prevent further spread of the contaminants and to remove the contaminants from the network. Evaluating the remedial actions of the utilities from a health perspective, more attention to issuing boiling water advisories (now issued in only 7 out of 50 events) and the use of remedial, targeted disinfection of the contaminated mains (used in only 2 out of 50 events) is advised.

What do these events mean for the consumer? The average annual probability of a consumer in the Netherlands to reside in an event-affected area is 1.7×10^{-3}. But what is the health risk of these events? Van Lieverloo *et al.* (2007b) used the event data (frequency, affected population, duration, *E. coli* concentration) in a Quantitative Microbial Risk Assessment (QMRA) to estimate the health risk associated with such events. This required a set of assumptions since the collected information did not disclose all the elements. One major assumption was the source of the contamination. To estimate the health risk, the information collected on duration and thermotolerant coliform concentrations had to be transformed into the exposure of consumers to faecal pathogens. To achieve such a transformation, pathogen-to-thermotolerant coliform ratios were derived from data on pathogen concentration (culturable thermotolerant *Campylobacter, Cryptosporidium* oocysts, *Giardia* cysts, culturable enteroviruses) and thermotolerant coliform concentration in sewage (Höller 1988; Medema *et al.* 2001). The calculated total risk of infection was dominated by the risk of *Campylobacter* because the *Campylobacter*-to-thermotolerant coliform ratios were the highest of the investigated faecal pathogens. Assuming surface water as the contamination source yielded a 95-percentile risk of infection per event of 0.16; when sewage was assumed as the source the 95-percentile risk of infection was 0.013; 12-fold lower, because the data showed that in sewage the average *Campylobacter*-to-thermotolerant coliform ratio was 12-fold lower than in the surface water. In any case, the best estimate of the risk of infection during events, even if low numbers of faecal indicator bacteria are detected, appears to be substantial. This is in line with recorded waterborne outbreaks where low numbers of *E. coli* were detected in the water (Hrudey and Hrudey 2004; Fernandes *et al.* 2006). Improvement of our ability to assess the health risk of such events can be achieved by collecting more data on the contamination level of the soil surrounding the water mains. The soil was identified as the main source of contamination in the events in the Netherlands and is the least well characterized in terms of microbial contamination (Van Lieverloo *et al.* 2007b).

Does the occurrence, frequency, magnitude and health risk of contamination events in the network imply that a disinfectant residual is necessary? Simulation studies indicate that chlorine (but not chloramine) disinfection is an effective barrier against ingress of microbial contaminants (Propato and Uber 2004; Teunis *et al.* 2010). Nevertheless, many outbreaks are reported via chlorinated systems and the epidemiological studies indicate that contamination events in chlorinated systems are associated with disease in the consumers (Payment *et al.* 1991, 1997; Hunter *et al.* 2005; Nygard *et al.* 2007). In the UK, the Drinking Water Inspectorate recorded between 14 and 47 microbiological contamination events per year in the period 1990–2005 (Gray 2008). They see an increasing number of events in the network of both microbiological contamination and discolouration and associate this with the increasing remedial activities of the ageing networks. These findings illustrate that a disinfectant residual in the network is certainly not an absolute protection. This was also concluded by Van der Kooij *et al.* (2002). Structural and hydraulic integrity and strict hygiene are preventative measures, while a disinfectant residual is a curative measure for failures. In the absence of a disinfectant residual, the water utility relies more heavily on the preventative measures, but also has a more sensitive monitoring system to detect faecal contamination. Several enteric pathogens, particularly viruses and protozoa are (much) more resistant to chlorine and related disinfectants than indicator bacteria such as *E. coli*. This means that infectious pathogens may still be present in contaminated water, where the indicator bacteria for faecal contamination have been inactivated. Several outbreaks of viral and protozoal illness have been reported from water in which no *E. coli* was detected (Anderson and Bohan 2001; Craun and Calderon 2001). Hence, considering the low frequency of events in the Netherlands, coupled with a rapid response to identify and repair these breaches, the limited protective value of a disinfectant residual is considered adequate protection of the safety of the consumer. In the context of the rapid response, improvements in consumer protection by proactive issuing of boil water advisories and in certain cases the use of targeted disinfection of the network to inactivate the contaminants is warranted.

5.3.3 Evidence from water quality monitoring

The statutory monitoring of faecal indicators by water utilities can also provide information about contamination events that occur in the distribution network. Bartram *et al.* 2002 accumulated the results of statutory drinking-water monitoring for faecal indicator bacteria in Europe. On average, 1–2% (range 0–12%) of the drinking-water samples show the presence of thermotolerant coliforms. This included countries in the (far) Eastern World Health Organization (WHO) Europe region and countries with a high proportion of small rural water supply systems. Studies from European countries with more centralized water supply systems indicate that only around 0.05–0.3% of the samples show the presence of *E. coli* (van der Kooij *et al.* 2002; Mendez *et al.* 2004; van Lieverloo *et al.* 2006, 2007b; Hambsch *et al.* 2007). In the Netherlands, the statutory monitoring for coliforms and *E. coli* is used by the water utilities in the water quality index of their benchmark. The statutory monitoring shows drinking water has a very high compliance with the drinking-water standards. On average there was 1 instance of incidental non-compliance with standards in 2008 for every 333 million m^3 supplied in the sector (VEWIN 2009b). This is a clear indication that water without a disinfectant residual can be delivered to the consumer without contamination by cross-connections or ingress. To study this in more detail, coliform and thermotolerant coliform monitoring data from 1996–1998 were collected from 8 Dutch water utilities. Table 5.2 shows the pooled data.

Table 5.2 Percentage of samples containing coliform and thermotolerant coliform; first samples and repeat samples taken the next day at the same or a nearby sampling site.

Location	n	Coliform positive	Repeated coliform positive	Thermotolerant coliform positive	Repeated thermotolerant coliform positive
Water leaving works	26656	0.70%	0.086%	0.10%	0.004%
Water in reservoirs	6682	2.02%	0.614%	0.31%	0.030%
Water in premises	54741	0.96%	0.037%	0.31%	0.0037%
Total	*88079*	*0.96%*	*0.095%*	*0.24%*	*0.0057%*

Both coliforms and thermotolerant coliforms were detected more often in premises and reservoirs than in water leaving the works, which was an indication that ingress of faecal contamination may occur at low levels during distribution. Repeated detection of (thermotolerant) coliforms was most frequently observed in the reservoirs. Given the residence time of water (and contaminants) in reservoirs, this is not surprising.

To try to understand what this low frequency and level of potential faecal contamination means in terms of safety, the sampling programme was assumed to be representative of the occurrence of faecal contamination in the network. The thermotolerant coliform drinking-water data were transformed to pathogen data, assuming sewage was the source of contamination and using concurrent data of pathogens and thermotolerant coliforms in sewage. The resulting pathogen data were used to calculate the probability of infection with enteric pathogens (van Lieverloo *et al.* 2003). The overall probability of infection (given the assumptions) was in the order of 10^{-6} per person per year, well below the 10^{-4} threshold. Incidents of repeated *E. coli* positive samples give a high probability of infection (see previous paragraph), but are very rarely seen. Given the transient nature of contamination events and the water flow in the network, the current practice of repeated sampling the following day at or around the same premises as the first positive sample was taken may not be very able to detect all incidents.

In a subsequent European study, 3 years of statutory monitoring data from the Netherlands, were compared to the data from France and Germany (Table 5.3, Hambsch *et al.* 2007). Overall, the percentage of samples containing *E. coli* was higher in France than in Germany or the Netherlands. The French data contained a high percentage of small rural water supplies, while the German and Dutch data represented larger systems. The larger German statutory data set of all systems does include small systems and yields higher percentages of *E. coli* positive samples (up to 0.23%). Also, the data from France indicated that no further deterioration of the water quality takes place in the network, while the German and Dutch data do show a higher percentage of positive samples in the distribution network compared to the water leaving the works. Comparing the data from the Netherlands from this period (2001–2003) to the previous suggests that the water quality in the network has improved (Table 5.3 vs Table 5.2).

Table 5.3 Incidence of *E. coli* in tap water.

Characteristic	Germany	France	The Netherlands
Represented # of consumers (million)	5.8	27.5	12.1
Water volume (million m^3)	400 (est)	2680	820
Water supply zones	13	1960	125
# of samples	42000	94000	149000
Chlorine residual	No	Yes	No
Leaving works with *E. coli*	0.005%	0.3%	0.04%
Distribution system with *E. coli*	0.1%	0.3%	0.09%

Source: Hambsch *et al.* (2007).

Although a direct comparison is hampered by the nature of the supply systems, the findings suggest that a high level of safety can be reached in the absence of a disinfectant residual. The data also indicate that water quality might be affected during distribution; compare the percentages of *E. coli* positive samples in water leaving the works and in the network in the data from Germany and the

Netherlands. In the same study, much larger volumes of distributed water (10–200 l) than the normal 100 ml were sampled and analysed to increase the sensitivity of detection. A total of 356 samples were collected in the United Kingdom, Germany and the Netherlands. None of these large volume samples showed the presence of *E. coli*. The study suggested that contaminations occur infrequently and during a short period.

But what is the level of protection that may be assigned to the statutory microbiological water quality monitoring? Not all water is tested all the time; the common approach is to take grab samples from predefined premises throughout the distribution network with a certain spatial and temporal distribution. The sample volume may vary, but the most common volume tested for *E. coli* is 100 ml. Contamination events due to cross-connections or ingress through leakage or orifices may occur throughout the system. Given the nature of the statutory monitoring, it will not be possible to detect all events that occur throughout the network and time. No information was available on the sensitivity of the statutory monitoring. We evaluated the sensitivity of a statutory monitoring system to detect an event of ingress of raw sewage into the distribution network in a simulation study (van Lieverloo *et al.* 2007a). The distribution network of a medium-sized city (72,000 inhabitants; 20,300 connections) was used. The network was fed with groundwater and did not contain a disinfectant residual. Contamination events were simulated, in which raw sewage entered the network with a rate of 1 l/h for a period of 16 hours. The *E. coli* concentration in raw sewage (1×10^8 cfu/l^{-1}) was the average of a sewage monitoring survey. Twelve different points of entry were simulated: at the treatment works, in a large transport main, in a reservoir and in nine (both central and peripheral) premises throughout the network.

A hydraulic model of the network, which was calibrated against flow and pressure measurements, was used to simulate the transport of *E. coli* in the network. Immediate and complete mixing of the *E. coli* from sewage into the mains and also in the reservoirs was assumed. *E. coli* was considered to behave as a conservative contaminant, be transported by convection and no inactivation or settling occurred. The hourly *E. coli* concentration at each sampling site was calculated for a period of 1,200 hours following the contamination event and the detection probability of the statutory monitoring programme was determined. The statutory monitoring programme was the actual programme that the water utility used in this city. The monitoring programme was designed in accordance with the EU Drinking Water Directive, but was more extensive than the minimum requirements of the Directive. It contained 47 sampling sites: one at the treatment works, two in reservoirs and 44 in premises throughout the network. The annual number of samples taken from the premises is four per site (176 in total, minimum requirement of Directive for this network: 39 check and 15 audit samples). The results of the simulations showed that the probability of detecting each of the nine contamination events throughout the network was on average 5.6% and varied from 0–13%. That means that on average 95% of such contamination events remain undetected by the statutory monitoring, even in the more expanded mode that was used here. This is in line with the reports of outbreaks of disease or contamination events that are reported by customer complaints rather than by water quality monitoring (Hrudey and Hrudey 2004; Fernandes *et al.* 2007). The low sensitivity is primarily determined by the transient nature of the contamination event and the infrequent water quality monitoring (over space and time). This appears to reflect the actual situation of contamination events (Van Lieverloo *et al.* 2007b; Besner *et al.* 2011; LeChevallier *et al.* 2011).

Several options to improve the detection probability were investigated (Blokker *et al.* 2009). Increasing the sample volume did not increase the detection probability. Increasing the sampling frequency at the sampling sites did increase the detection probability, a 3-fold increase yielded an increase of the detection probability to 10.9% and at weekly sampling (which is 13 times more frequent than the current programme) the average detection probability of all events was 31%, although the standard deviation was high. Selection of sampling sites based on the residence time of the water in the system or using the gate-keeper approach did not substantially improve the detection probability. We also investigated the (hypothetical) scenario of installing online *E. coli* sensors for continuous monitoring throughout the distribution network. Sensor locations were optimized with a genetic algorithm (Weickgenannt *et al.* 2010). With 5 sensors, 60% of all events were detected and with 25 sensors this increased to approx. 80% (Figure 5.5).

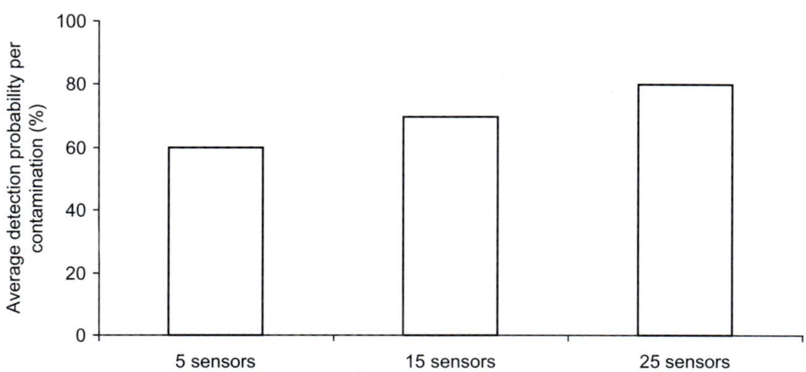

Figure 5.5 Probability of detection of a contamination of the distribution network with different (hypothetical) sensitive *E. coli* sensors.

5.3.4 Evidence from operational monitoring

The combination of physical and hydraulic integrity of the distribution network provides a double barrier against contamination of drinking water during distribution. For intrusion of contaminants to occur, both the physical and the hydraulic integrity of the network need to be lost at the same time and place, and contaminants should be present in the vicinity of the network. Loss of pressure is a breach in the protection that could result in backflow from cross-connections or intrusion of contaminants through leaks and orifices (LeChevallier *et al.* 2003, 2011; Besner *et al.* 2010). Causes of sustained pressure loss are mains breaks, repairs and power loss. Also short (seconds to minutes), transient events of low or negative pressures have been implicated as the cause of contamination of distributed drinking water. Such events can be caused by pump shutdowns but also by transmission main closures, use of hydrants, main repairs and pump power failures (Funk *et al.* 1992; Besner 2010b). The contaminants may enter the distribution network through (small) leaks in pipes, leaking couplings, cross-connections, joints or seals and submerged air valves. The health risk of contamination by low pressure events was suggested in two epidemiological studies in Europe. In England, a strong association between low water pressure, as noticed by the consumer, and self-reported diarrhoea was found (Hunter *et al.* 2005). The study was not designed to test this association (it was a case control study of sporadic cryptosporidiosis), but the authors hypothesized that 15% of intestinal illness in the community may be associated with low pressure events. In Norway, Nygard *et al.* (2007) studied 88 low pressure events in 7 distribution networks. The events were caused by main breaks and planned and unplanned repairs. Consumers in affected areas had a 1.6–2.0 times increased risk of intestinal illness. The authors hypothesized that if 20% of the consumers are exposed to a low pressure event every year, this would mean that these events caused 33,000 cases of intestinal illness annually in the Norwegian population of 4.5 million; if low pressure events are more frequent the number of cases will be higher. Flushing of the pipelines and use of chlorination after a repair episode yielded a reduction of the risk of illness. Guidelines in Norway recommend chlorination of distribution networks after maintenance, loss of water pressure or both and prior to repressurizing, to protect against contamination. However, these recommendations are specified to certain predefined risk incidents, where a contamination might have occurred and are often not followed. Only one of the seven participating water utilities chlorinated routinely in all episodes of work on the distribution network.

No outbreaks have been directly associated with short pressure transients. A number of outbreaks have been associated with low pressure events, leading to backsiphonage or ingress through cross-connections, including the 2001 outbreak in the Netherlands (Craun and Calderon 2001; Fernandes *et al.* 2007).

In the Netherlands, unplanned interruptions of supply are recorded, especially since the introduction of this parameter in the benchmark of the water utilities, to determine the level of service to the consumer. Not only the downtime but also the characteristics and causes of the unplanned supply interruptions are recorded and pooled by all water utilities. The primary objective of these records is to estimate the life expectancy of distribution network segments, as an aid to proactive evidence-based asset management (Vloerbergh *et al.* 2008). At the same time, they also provide information about the frequency of events that may be associated with ingress, such as low or no pressure. The interruptions are expressed as minutes of interrupted or inadequate supply. In 2009, consumers suffered an unplanned interruption of water supply of, on average, 7.5 min (range of average time between utilities was 2–13 min). This was an increase of 32% compared to 2006, probably due, at least in part, to improved registration.

The Water Services Regulation Authority (OFWAT) (2008) has collected data on leakage and supply interruptions from 2006–2007 from different countries. The data are not complete and not always directly comparable, but do reflect the good structural and hydraulic integrity of the distribution network in the Netherlands (Table 5.4).

Table 5.4 Operational data on distribution networks.

Country	Main bursts per 1000 km	Leakage $m^3/km/d$	Unplanned interruptions per 1000 properties	Properties at risk of low pressure
England/Wales	187	10.1	23	0.02%
Scotland	166	21.3	34	0.31%
Canada	66	11.9	21	No data
Australia	No data	4.4	50	No data
Portugal	67	7.0	0.4 (>12 hr)	No data
USA	629*	No data	3	No data
Netherlands	70**	1.6	***	No data

*LeChevallier *et al.* (2011) indicate an industry average of 23–27 breaks per 100 miles, or approx. 150 per 1000 km; **All failures, not only main bursts; ***Unplanned interruptions in the Netherlands are, on average, 7.5 min per property per year.
Source: Based on OFWAT (2008).

Researchers in Canada have monitored pressure in a distribution system of 1,600 km serving 380,000 consumers with an average water demand of 210,000 m^3/d (Besner *et al.* 2010). Pressure dropped to below 172 kPa at the treatment plant nine times in a period of 17 months and this resulted in four measured and five suspected negative pressure events throughout the distribution network. Another 4 negative pressure events were associated with the closure of a transmission main and 10 negative pressure events were recorded at only a single location. These lasted for minutes to hours and were associated with maintenance work in the same area of the network. The authors also confirmed the presence of faecal contamination in soil and water from pipe trenches, although both frequency and level of contamination was lower than previously reported in the USA (Karim *et al.* 2003).

Continuous pressure monitoring data from calibrated pressure monitors are limited in the Netherlands. In a dedicated study, the pressure was monitored for 50 days continuously at three locations along a limited part of a distribution network (boulevard of Zandvoort; Blokker *et al.* 2010). During these 50 days, the nominal pressure was 200–350 kPa and at two moments a pressure transient occurred in the network where the pressure dropped below 1 bar for several minutes. No negative pressures were observed; the minimum pressure was 0.2 bar. The movement of the low pressure transient through the network is shown in Figure 5.6. The duration of the low pressure event was short (Figure 5.7). The cause of the low pressure transients was not identified. This shows that pressure transients do occur in the network in the Netherlands. Because of the predominantly flat terrain, the network is probably less prone to negative pressures, but the evidence-base is limited.

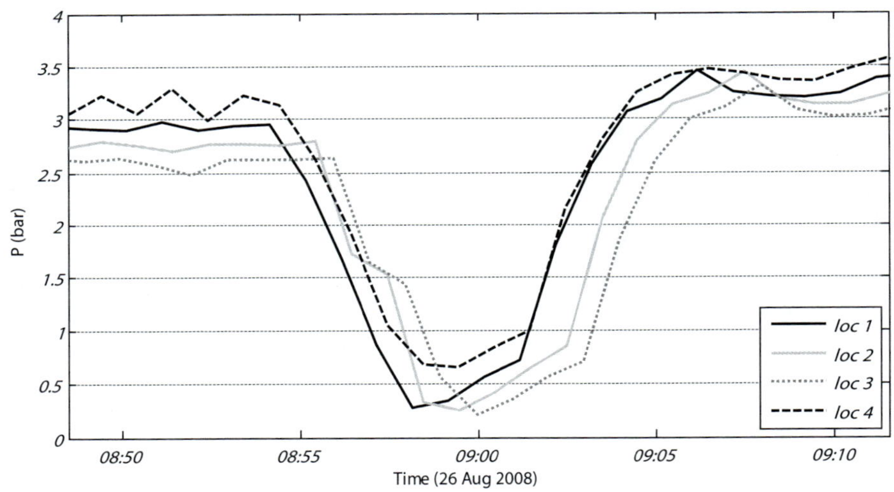

Figure 5.6 Low pressure transient in the distribution network of Zandvoort.

5.4 SYNOPSIS

The data collected in the Netherlands indicate that distribution networks with a high level of structural and hydraulic integrity and hygiene and without a disinfectant residual:

- have a low frequency of outbreaks;
- have a low frequency of contamination events;
- have a low frequency of detection of faecal indicator bacteria (while this system is more sensitive in non-disinfected waters);

- experience low but no negative pressure transients (based on limited data);
- have low leakage rates;
- have low rates/times of unplanned interruptions.

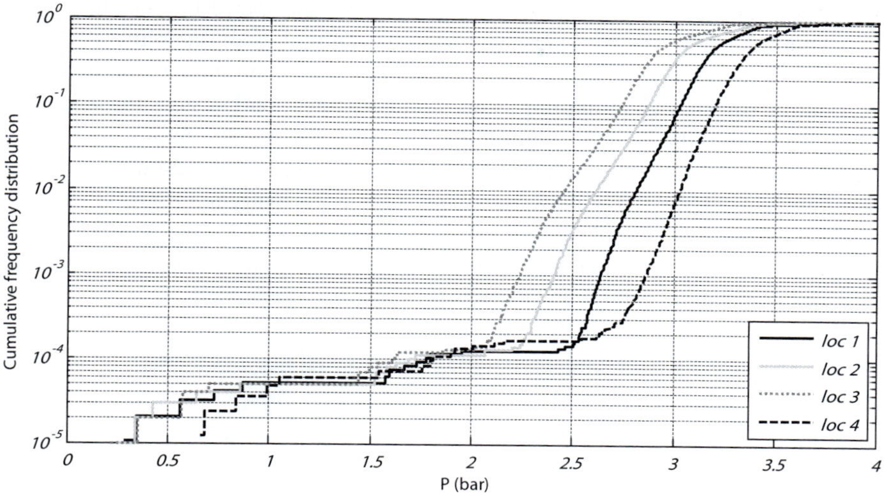

Figure 5.7 Frequency distribution of water pressure (x, in Bar) of the 50 day monitoring at three locations in the Zandvoort distribution network. A frequency of 10^{-3} in 50 day monitoring corresponds to 1.2 hours, 10^{-4} to 7.2 min.

This demonstrates the advantages of the focus on structural and hydraulic integrity and hygiene.

Is this sufficient proof of safety? The best available evidence is collected, representing the best state of the art, and the evidence clearly points to a high level of safety. Under these conditions disinfectant residuals, with DBP formation and taste and odour problems, are not needed to enhance the safety of the water network. It is clear that in drinking-water distribution networks, non-safety is more easily demonstrated than safety. The network is complex, open and underground and with our best efforts we currently monitor water quality, condition and integrity in a fraction of the network, both in space and in time. Are there ways to improve the assessment of the safety? New technologies and concepts emerge in the area of smart networks that may substantially improve our ability to assess the safety of the network.

5.5 OUTLOOK

5.5.1 Improved microbiological monitoring

The statutory monitoring of *E. coli* aims to verify the effectiveness of the measures to prevent contamination of the distribution network. Water utilities and health authorities use the level of compliance with the statutory monitoring of *E. coli* as an indication of the level of protection of the distribution network.

The level of protection that may be assigned to the statutory monitoring depends on the ability of the statutory monitoring to detect a faecal contamination event. Given the transient nature of contamination events and the results of our studies that showed a low recovery efficiency of the current grab sampling approach (see 5.3.3), we designed a continuous monitoring study to optimize the detection probability of events. Sampling sites were selected and tested for adequate placement with the genetic algorithm. The study uses continuous membrane filtration samplers that are installed on the tap and sample water 24 hours per day at a low flow to filter 10–100 l water. This study is currently in progress.

Rethinking of the sampling strategy upon detecting *E. coli* in the network, using the knowledge of the water (and contaminant) transport in the network between the first positive sample and the next samples, is also recommended.

5.5.2 Improved monitoring by combining sentinel sensors with rapid microbiological assays

The ability to monitor continuously for faecal contamination at dedicated sites in the network would mean a large improvement in the ability to demonstrate drinking-water safety and due diligence. There are currently developments in monitoring stations for *E. coli* (Brown *et al.* 2010; van de Vossenberg *et al.* 2010; Zibuschka *et al.* 2010) that indicate that this next generation of monitoring is becoming achievable, although true online monitoring of *E. coli* on many sites in the network is still “a bridge too far”. In the meantime, water utilities are exploring the application of non-microbiological sensor systems (turbidity, refractive light index and others) that are able to monitor online and are cheap and robust enough to install throughout the network (Storey *et al.* 2011; de Graaf *et al.* 2012). This is driven initially from the perspective of water security, but the developments are equally useful for water safety. The combination of sentinel sensors for detection of anomalies in the network and rapid, on site confirmation of a faecal contamination, is achievable given the current status of technology. Similarly, there are recent developments in the area of sensor placement strategies based on (improved) network modelling. This approach should also be applied to optimize the current (grab) sampling programmes.

5.5.3 Improved use of operational monitoring

The application of operational monitoring to aid the assessment of safety has been explored. Improved national registration of distribution system deficiencies (Vloerbergh *et al.* 2008) provides opportunities to upgrade the safety assessment even further. The first step is to extract information from this register (which is currently focused on the condition of network elements) that is indicative of contamination events.

In parallel, Geographical Information Systems are employed to evaluate the risk of main breaks for the surrounding area. This technology can also be used to identify spots in the network that are vulnerable to contamination, because of the presence of contamination sources, soil types, groundwater levels and pipe characteristics and condition etc.

5.5.4 Epidemiology to assess safety

Several epidemiological studies have been conducted in chlorinated systems in Canada (Payment *et al.* 1991, 1997), Australia (Hellard *et al.* 2001) and the USA (Colford *et al.* 2005) to determine the health burden associated with drinking water. The Canadian studies found a significant contribution of drinking water to intestinal illness in the population, and a link with events in the distribution network was suggested. No epidemiological study has been conducted in non-chlorinated systems to determine if there is a measurable contribution of drinking water to the intestinal illness in the community. Given the high background prevalence of intestinal illness, such a study requires a large population. An epidemiological study focused on the health effects of repairs and maintenance in the distribution network, in analogy with the Norwegian study (Nygard *et al.* 2007) is more appropriate.

5.5.5 QMRA to improve science-based management

The Netherlands has incorporated QMRA in the drinking-water legislation. Currently, QMRA is used to determine the safety of the surface water treatment systems. QMRA models have also been developed for intrusion of pathogens in the distribution network (McInnis 2004; van Lieverloo *et al.* 2007b; Teunis *et al.* 2010; Besner *et al.* 2011; LeChevallier *et al.* 2011). These QMRA models do suggest that contamination events do amount to a high probability of infection in the affected consumers. This, in combination with the detection probability of contamination events of the current statutory monitoring programmes for *E.coli,* is a strong incentive for QMRA-based safety management of the distribution network. The current state of knowledge implies that such a QMRA requires several assumptions. Key factors are the pathways of entry and the occurrence of pathogens in the material that enters the network via these pathways. The evidence from the Norwegian (Nygard *et al.* 2007) and the Canadian study (Besner *et al.* 2010) indicate that repair and maintenance work render the network vulnerable to ingress and may be associated with illness. Collection of information about the occurrence of pathogens in the trenches during repair and maintenance of the network, the amount of material and pathogens that can enter the network and the efficacy of the remedial actions (flushing, chlorination) in the Dutch setting is therefore recommended to evaluate the level of protection that is offered by current network operation practices.

Acknowledgements

This chapter has been prepared as part of the Joint Research Programme (BTO) of the water supply companies in the Netherlands and contains research conducted within the European FP6 project Microrisk.

5.6 REFERENCES

Anderson Y. and Bohan P. (2001). Disease surveillance and waterborne outbreaks. In: Water Quality: Guidelines, Standards and Health, L. Fewtrell and J. Bartram (eds), IWA Publishing, London, UK, pp. 115–133.

Anonymous (2001). Waterleidingbesluit. Staatsblad van het Koninkrijk der Nederlanden 2001, nr. 31.

Anonymous (2011). Drinkwaterbesluit. Staatsblad van het Koninkrijk der Nederlanden 2011, nr. 293.

Bartram J., Thyssen N. and Gowers A. (2002). Water and health in Europe. WHO Regional publications, European Series No. 93, Copenhagen, Denmark.

Behnke S., Parker A. E., Woodall D. and Camper A. K. (2011). Comparing the chlorine disinfection of detached biofilm clusters with those of sessile biofilms and planktonic cells in single- and dual-species cultures. *Appl. Environ. Microb.*, **77**(20), 7176–7184.

Bellar T. A., Lichtenberg J. J. and Kroner R. C. (1974). The occurrence of organohalides in chlorinated drinking water. *J. Am. Water Works Assoc.*, **66**, 703.

Besner M. C., Ebacher G., Jung B. S., Karney B., Lavoie J., Payment P. and Prévost M. (2010). Negative pressures in full-scale distribution system: field investigation, modelling, estimation of intrusion volumes and risk for public health. *Drinking Water Eng. Sci.*, **3**, 101–106.

Besner M. C., Prevost M. and Regli S. (2011). Assessing the public health risk of microbial intrusion events in distribution systems: conceptual model, available data and challenges. *Water Res.*, **45**, 961–979.

Beuken R., Lavooij K., Bosch A. and Schaap P. (2006). Low leakage in the Netherlands confirmed. Water Distribution System Analysis conference, Cincinnati, USA.

Blokker E. J. M., Vogelaar A. and Medema G. J. (2009). Optimalisatie meetprogramma Escherichia coli in distributienet. Report BTO 2009.008, KWR, Nieuwegein, Netherlands (in Dutch).

Blokker E. J. M., Vreeburg J. H. G., Beverloo H., Klein Arfman M. and van Dijk J. C. (2010). A bottom-up approach of stochastic demand allocation in water quality modelling. *Drinking Water Eng. Sci.*, **3**(1), 43–51.

Brown R. S., Marcotte E. J. P., Dunkinson C. E., Aston W. P., Gallant P. J. and Wilton D. (2010). An automated detection technology for on-site E. coli and coliform bacteria monitoring. Proceedings of the Water Environment Federation, WEFTEC 2010, pp. 7433–7442.

Calderon R. L. (2004). Measuring benefits of drinking water technology: ten years of drinking water epidemiology. NEWWA Water Quality Symposium, Boxborough, USA.

Colford J. M., Wade T. J., Sandhu S. K., Wright C. C., Lee S., Shaw S., Fox K., Burns S., Benker A., Brookhart M. A., van der Laan M. and Levy D. A. (2005). A randomized, controlled trial of in-home drinking water intervention to reduce gastrointestinal illness. *Am. J. Epidemiol.*, **161**(5), 472–482.

Craun G. F. and Calderon R. L. (2001). Waterborne disease outbreaks caused by distribution system deficiencies. *J. Am. Water Works Assoc.*, **93**(9), 64–75.

Craun M. F., Craun G. F., Calderon R. L. and Beach M. J. (2006). Waterborne outbreaks reported in the United States. *J. Water Health*, **4**(Suppl 2), 19–30.

Crozes G., Bruchet A. and Mackie E. (2007). Public perception of tap water chlorinous flavor. Water Research Foundation, Denver, USA.

de Graaf B. R., Williamson F., Koerkamp M. K., Verhoef J. W., Wuestman R., Bajema B., Trietsch E. and van Delft W. (2012). Implementation of an innovative sensor technology for effective online water quality monitoring in the distribution network. Proc. Singapore International Water week 2012, Singapore.

DVGW (2008). Branchenbild der deutschen Wasserwirtschaft. DVGW, Bonn, Germany.

EPA (2012). The effectiveness of disinfectant residuals in the distribution system. Total Coliform Rule issue paper. http://www.epa.gov/safewater/disinfection/tcr/regulation_revisions.html (accessed 27 November 2012).

Fernandes T. M., Schout C., De Roda Husman A. M., Eilander A., Vennema H. and van Duynhoven Y. T. (2007). Gastroenteritis associated with accidental contamination of drinking water with partially treated water. *Epidemiol Infect.*, **135**(5), 818–826.

Ford T. E. (1999). Microbiological safety of drinking water: United States and global perspectives. *Environ. Health Perspect.*, **107**(Suppl 1), 191–206.

Funk J. E., Wood D. J., Reddy L. S. and Denger D. C. (1992). Pressure surges due to rapid expulsion of air. In: Unsteady Flow and Fluid Transients, Bettess & Watts (eds), A. A. Balkema, Rotterdam, Holland.

Gauthier V., Besner M.-C., Trepanier M., Millette R. and Prevost M. (2001). Tracking the source for microbial contamination in distribution systems using an integrated approach. In: Proc. of the AWWA WQTC Conference, American Water Works Association, Denver, USA.

Gray J. (2008). Water contamination events in UK drinking-water supply systems. *J. Water Health*, **6**(Suppl 1), 21–26.

Haas C. N. (1999). Benefits of using a disinfectant residual. *J. Am. Water Works Assoc.*, **91**(1), 65–69.

Hambsch B. (1998). Change from chlorine residual distribution to no chlorine residual distribution in groundwater systems. Proc. Symposium Protecting water quality in the distribution system: what is the role of disinfection residuals. AWWA/IWA, April 26–28 1998, Philadelphia, USA.

Hambsch B., Bockle K. K. and van Lieverloo J. H. M. L. (2007). Incidence of faecal contaminations in chlorinated and non-chlorinated distribution systems of neighbouring European countries. *J. Water Health*, **5**(Suppl 1), 119–130.

Hellard M. E., Sinclair M. I., Forbes A. B. and Fairley C. K. (2001). A randomized, blinded, controlled trial investigating the gastrointestinal health effects of drinking water quality. *Environ. Health Perspect.*, **109**(8), 773–778.

Höller C. (1988). Quantitative und qualitative Untersuchungen an Campylobacter in der Kanalisation einer Grossstadt. *Zentralblatt für Bakteriologie, Mikrobiologie und Hygiene B*, **185**, 307–325.

Hrudey S. E. and Charrois J. W. A. (2012). Disinfection by-products and human health. IWA Publishing, London.

Hrudey S. E. and Hrudey E. J. (2004). Safe drinking water; lessons from recent outbreaks in affluent nations. IWA Publishing, London.

Huisman J. and Nobel P. J. (1981). *Enkele epidemiologische gegevens over de gevolgen van de faecale verontreiniging in het Scheepvaartkwartier te Rotterdam in maart 1981* [Some epidemiological data on the effects of the faecal contamination of drinking-water in the Scheepvaartkwartier residential area in Rotterdam in March 1981]. *H_2O*, **14**(26), 642–646 (in Dutch).

Hunter P. R. (1997). Water-borne Disease: Epidemiology and Ecology. John Wiley & Sons, Chichester, UK.

Hunter P. R. and Syed Q. (2001). Community surveys of self-reported diarrhoea can dramatically overestimate the size of outbreaks of waterborne cryptosporidiosis. *Water Sci. Technol.*, **43**, 27–30.

Hunter P. R., Syed Q. and Naumova E. N. (2001). Possible undetected outbreaks of cryptosporidiosis in areas of the North West of England supplied by an unfiltered surface water source. *Communicable Dis. Public Health*, **4**(2), 136–138.

Hunter P. R., Chalmers R. M., Hughes S. and Syed Q. (2005). Self-reported diarrhea in a control group: strong association with reporting of low-pressure events in tap water. *Clin. Infect. Dis.*, **40**(4), 32–34.

Karim M. R., Abbaszadegan M. and LeChevallier M. (2003). Potential for pathogen intrusion during pressure transients. *J. Am. Water Works Assoc.*, **95**(5), 134–146.

Kiwa (1978). Problematiek haloformen. Eindrapport van de ad hoc werkgroep Problematiek van de Haloformen. Kiwa mededeling nr. 57, Rijswijk, the Netherlands (in Dutch).

Klein H. P. and Forster R. (1998). Network operation without safety chlorination in Zurich. Proc. Symposium Protecting water quality in the distribution system: what is the role of disinfection residuals. AWWA/IWA, April 26–28 1998, Philadelphia, USA.

Kruithof J. C. (1986a). Side effects of chlorination. Kiwa report 74. Nieuwegein, the Netherlands.

Kruithof J. C. (1986b). Chlorination by-products: production and control. AWWAKiwa report, Nieuwegein, the Netherlands.

LeChevallier M. W. (1999). The case for maintaining a disinfectant residual. *J. Am. Water Works Assoc.*, **91**(1), 86–94.

LeChevallier M. W. and Au K. K. (2004). Water treatment and pathogen control. World Health Organization, Geneva, Switzerland.

LeChevallier M. W., Cawthon C. D. and Lee R. G. (1988). Inactivation of biofilm bacteria. *Appl. Environ. Microb.*, **54**(10), 2492–2499.

LeChevallier M. W., Gullick R. W., Karim M. R., Friedman M. and Funk J. E. (2003). The potential for health risks from intrusion of contaminants into the distribution system from pressure transients. *J. Water Health*, **1**(1), 3–14.

LeChevallier M. W., Xu M., Yang J., Tuenis P. and Fleming K. K. (2011). Managing distribution system low transient pressures for water quality. Water Research Foundation report, Denver, USA.

MacKenzie W. R., Hoxie N. J., Proctor M. E., Gradus M. S., Blair K. A., Peterson D. E., Kazmierczak J. J., Addiss D. G., Fox K. R., Rose J. B. and Davis J. P. (1994). A massive outbreak in Milwaukee of cryptosporidium infection transmitted through the public water supply. *New Engl. J. Medi.*, **331**(3), 161–167.

McInnis (2004). A relative-risk framework for evaluating transient pathogen intrusion in distribution systems. *Urban Water J.*, **1**(2), 113–127.

Medema G. J., Ketelaars H. A. M. and Hoogenboezem W. (2001). *Cryptosporidium* and *Giardia* in sewage, livestock manure and surface water. RIWA report 2000.035, RIWA, Amsterdam, the Netherlands.

Meerkerk M. A. and Kroesbergen J. (2010). Hygiënecode drinkwater; opslag, transport en distributie. BTO report 2001.175 second edition, KWR, the Netherlands (in Dutch).

Mendez J., Audicana A., Cancer M., Isern A., Llaneza J., Moreno B., Navarro M., Tarancón M. L., Valero F., Ribas F., Jofre J. and Lucena F. (2004). Assessment of drinking-water quality using indicator bacteria and bacteriophages. *J. Water Health*, **2**(3), 201–214.

Nichols G. (2003). Using existing surveillance-based data. In: Drinking Water and Infectious Disease: Establishing the Links, P. R. Hunter, M. Waite and E. Ronchi (eds), IWA Publishing, London, UK.

Nieuwenhuisen M. J., Grellier J., Smith R., Iszatt N., Bennett J., Best N. and Toledano M. (2009). The epidemiology and possible mechanisms of disinfection by-products in drinking water. *Philosophical transactions of the Royal Society A: Mathematical, Physical and Engineering Sciences*, **367**(1904), 4043–4076.

Nobel P. J. (2001). Bewaking van de microbiologische kwaliteit van het drinkwater in het distributienet na aanleg en ingrepen. BTO report 2001.113, Kiwa, the Netherlands (in Dutch).

Nygard K., Wahl E., Krogh T., Tveit O. A., Bohleng E., Tverdal A. and Aavitsland P. (2007). Breaks and maintenance work in the water distribution systems and gastrointestinal illness: a cohort study. *Int. J. Epidemiol*, **36**(4), 873–880.

OFWAT (2008). International comparison of water and sewerage service report. OFWAT, UK.

Payment P. (1999). Poor efficacy of residual chlorine disinfectant in drinking water to inactivate waterborne pathogens in distribution systems. *Can. J. Microbiol.*, **45**(8), 709–715.

Payment P., Richardson L., Siemiatycki J., Dewar R., Edwardes M. and Franco E. (1991). A randomized trial to evaluate the risk of gastrointestinal disease due to consumption of drinking water meeting current microbiological standards. *Am. J. Public Health*, **81**(6), 703–708.

Payment P., Siemiatycki J., Richardson L., Renaud G., Franco E. and Prevost M. (1997). A prospective epidemiological study of gastrointestinal health effects due to the consumption of drinking water. *Int. J. Environ. Heal. R.*, **7**(1), 5–31.

Propato M. and Uber J. G. (2004). Vulnerability of water distribution systems to pathogen intrusion: how effective is a disinfectant residual? *Environ Sci. Technol.*, **38**, 3713–3722.

Reynolds K. A., Mena K. D. and Gerba C. P. (2008). Risk of waterborne illness via drinking water in the United States. *Rev. Environ. Contaminant Toxicology*, **192**, 117–158.

Richardson S. D., Plewa M. J., Wagner E. D., Schoeny R. and Demarini D. M. (2007). Occurrence, genotoxicity, and carcinogenicity of regulated and emerging disinfection by-products in drinking water: a review and roadmap for research. *Mutat. Res.*, **636**(1–3), 178–242.

Risebro H. L., Doria M. F., Andersson Y., Medema G., Osborn K., Schlosser O. and Hunter P. R. (2007). Fault tree analysis of the causes of waterborne outbreaks. *J. Water Health*, **5**, 1–18.

Rook J. J. (1974). Formation of haloforms during chlorination of natural waters. *Water Treatment and Examination*, **23**, 234–243.

Schellart J. (1986). Disinfection and bacterial regrowth: some experiences of the Amsterdam water works before and after stopping the safety chlorination. *Water Supply*, **4**, 217–225.

Skadsen J. (1993). Nitrification in a distribution system. *J. Am. Water Works Assoc.*, **85**(7), 95–103.

Smeets P. W. M. H., Medema G. J. and van Dijk J. C. (2009). The Dutch secret: how to provide safe drinking water without chlorine in the Netherlands. *Drinking Water Eng. Sci.*, **2**, 1–14.

Stenström T. A., Boisen F., Georgson F., Lahti K., Lund V., Andersson Y. and Ormerod K. (1994). Waterborne infection in Nordic countries (Vattenburna infektioner I Norden). TemaNord 1994: 585 Nordiska Ministerrådet, (in Swedish).

Storey M., van der Gaag B. and Burns B. P. (2011). Advances in on-line drinking water quality monitoring and early warning systems. *Water Res.*, **45**(2), 741–747.

Teunis P. F. M., Xu M., Fleming K. K., Yang J., Moe C. L. and LeChevallier M. W. (2010). Enteric virus infection risk from intrusion of sewage into a drinking water distribution network. *Environ Sci. Technol.*, **44**, 8561–8566.

Trussel R. R. (1999). An overview of disinfectant residuals in drinking water distribution systems. *J. Water Supply Res. Technol. – Aqua*, **48**(1), 2–10.

UKWIR 2006. Comparison of Leakage Levels and Leakage Practice in the UK and the Netherlands. Report 06WM0834, UKWIR, United Kingdom.

Van de Vossenberg J. and Medema G. J. (2010). Rapid optical detection of Escherichia coli in drinking water. BTO report 2010.006, KWR, Nieuwegein, the Netherlands.

Van der Kooij D., van Lieverloo J. H. M. L., Schellart J. A. and Hiemstra P. (1999a). Distribution of drinking water without disinfectant: highest achievement or height of folly? *J. Water Supply Res. Technol. – Aqua*, **48**(1), 31–37.

Van der Kooij D., van Lieverloo J. H. M. L., Schellart J. A. and Hiemstra P. (1999b). Maintaining quality without a disinfectant residual. *J. Am. Water Works Assoc.*, **91**(1), 55–64.

Van der Kooij D., van Lieverloo J. H. M. L., Gale P. and Stanfield G. (2002). Distributing drinking water with a low or zero disinfectant residual. Operational and biological aspects. Kiwa report 127, Nieuwegein, the Netherlands.

Van Genderen J. (1998). Nevenproducten van chemische desinfectie in drinkwater: een toxicologische beoordeling op basis van literatuur- en praktijkgegevens. Kiwa report 96.220, Nieuwegein, the Netherlands (in Dutch).

Van Lieverloo J. H. M., Mesman G. A. M., Nobel P. J. and Kroesbergen J. (2002). Hygiënecode drinkwater; opslag, transport en distributie. BTO report 2001.175, Kiwa, Nieuwegein, the Netherlands (in Dutch).

Van Lieverloo J. H. M., Esveld-Amanatidou A., Hijnen W. A. M. and Groennou J. T. (2003). Microbiologische verontreiniging van drinkwater tijdens opslag, transport en distributie. Evaluatie en beheersing van risico's van eventuele afwijkingen en verontreinigingen. BTO report 2002.130, Kiwa Water Research, Nieuwegein, the Netherlands (in Dutch).

Van Lieverloo J. H. M., Medema G. and van der Kooij D. (2006). Risk assessment and risk management of faecal contamination in drinking water distributed without a disinfectant residual. *J. Water Supply Res. Technol. – Aqua*, **55**, 25–31.

Van Lieverloo J. H. M., Mesman G. A. M., Bakker G. L., Blokker E. J. M., Baggelaar P. K., Hamed A. and Medema G. J. (2007a). Probability of detecting and quantifying faecal contaminations of drinking water by periodically sampling for E. coli: A simulation model study. *Water Res.*, **41**, 4299–4308.

Van Lieverloo J. H. M., Blokker E. J. M. and Medema G. (2007b). Quantitative microbial risk assessment of distributed drinking water using faecal indicator incidence and concentrations. *J. Water Health*, **5**(Suppl 1), S131–S149.

VEWIN (2009a). Dutch drinking water statistics 2008. VEWIN, Rijswijk, the Netherlands.

VEWIN (2009b). Reflections on performance. VEWIN, Rijswijk, the Netherlands.

Vloerbergh I. N. (2008). USTORE: toelichting op en afsprake over uniforme storingsregistratie. BTO report 2008.057, KWR, Nieuwegein, the Netherlands (in Dutch).

Weickgenannt M., Kapelan Z., Blokker M. and Savic D. A. (2010). Risk-based sensor placement for contaminant detection in water distribution systems. *J. Water Res. Plan. Manage.*, **136**, 629–636.

Westrell T., Bergstedt O., Stenström T. A. and Ashbolt N. J. (2003). A theoretical approach to assess microbial risks due to failures in drinking water systems. *Int. J. Environ. Health Res.*, **13**(2), 181–197.

Zang W. and DiGiano F. A. (2002). Comparison of bacterial regrowth in distribution systems using free chlorine and chloramine: a statistical study of causative factors. *Water Res.*, **36**, 1469–1482.

Zibuschka F., Lendenfeld T. and Lindner G. (2010). Near real time monitoring of *E. coli* in water. *Österreichische Wasser- und Abfallwirtschaft*, **62**(11–12), 215–219.

Chapter 6

Legionella in drinking-water supplies

D. van der Kooij

6.1 INTRODUCTION

In the past decade, the number of reported cases of Legionnaires' disease (LD), a serious pneumonia, increased in Europe and the US and reached a total level of more than 10,000 cases in 2010 (European Centre for Disease Prevention and Control [ECDC] 2012; Centers for Disease Control and Prevention [CDC] 2012). This number is an underestimation for several reasons, for example, diagnosis is complicated and, although notification is mandatory in Europe, reporting to a national surveillance system is not always efficient. The disease is life-threatening, particularly in immunocompromised and elderly persons, with a current case-fatality rate in Europe of about 10% for hospitalized cases (ECDC 2012). Marston *et al.* (1997) estimated the annual number of LD cases in the US at 8000 to 18,000, based on a survey of persons with community-acquired pneumonia (CAP) requiring hospitalization. From a study in Germany, including both hospitalized and ambulatory CAP patients (von Baum *et al.* 2008), the total annual number of LD cases in Germany was estimated at 15,000 to 30,000, that is, more than the number reported for Europe and the US in 2010. LD has been observed world wide, which implies that the global number of cases is much higher than the total number reported in Europe and the US, but an estimate cannot be made due to a lack of data about incidences. The LD-causing organism, *Legionella pneumophila*, proliferates in freshwater environments at elevated temperatures and is transmitted by aerosols (Dondero *et al.* 1980; Baskerville *et al.* 1981). These environments, many of which are part of man-made water systems, obviously can pose a substantial public-health threat. After all the progress made in sanitary engineering and achieving safe drinking water, the public-health problem caused by *L. pneumophila* seems to be one of the major challenges in relation to using freshwater in man-made systems in our society.

The recognition of LD began with the large outbreak of pneumonia among members of the American Legion attending a conference in Philadelphia in July 1976, causing 182 cases of illness, 29 of which were fatal (Fraser *et al.* 1977). This outbreak led to the discovery of a previously unidentified pathogenic bacterium by the end of 1976. This organism was traced to a new bacterial family, the Legionellaceae, with one genus *Legionella* and one species *L. pneumophila* (McDade *et al.* 1977; Brenner *et al.* 1979). In the early 1980s it was found that *L. pneumophila* multiplied in potable-water systems and that this growth was associated with cases of LD (Tobin *et al.* 1980; Cordes *et al.* 1981). Also air-conditioning cooling towers and evaporative condensers appeared to be implicated in outbreaks of LD (Dondero *et al.* 1980; Cordes *et al.* 1980). The introduction of semi-selective culture methods facilitated the detection of *L. pneumophila* in aquatic environments (Edelstein *et al.* 1981; Wadowsky and Yee 1981).

Numerous surveys revealed that *L. pneumophila* is thriving in a diversity of man-made water systems, including: premise-plumbing systems, hot-spring spas, whirlpool spas, industrial cooling towers, rooftop-cooling towers, evaporative air-conditioning systems, medication nebulizers, eyewash stations, mist machines, fountains and waste-water treatment plants. At one or more locations, these systems have been implicated in cases of LD that were reported from all over the world. In 1999, the largest reported LD outbreak in terms of number of fatal cases (31) occurred in the Netherlands, where flower-show attendees were exposed to *L. pneumophila* from unchlorinated whirlpool spas on display (Den Boer *et al.* 2002; Nagelkerke *et al.* 2003). In the past decade, further outbreaks of the disease, which in most cases were associated with a cooling tower as the source of amplification and dissemination, have been reported. Examples of outbreaks in Europe are: in Murcia (Spain) in July 2001, caused by rooftop-cooling towers (Garcia-Fulgueiras *et al.* 2003); in Barrow-in-Furness (UK), July-August 2002 by a ventilation plant (Telford *et al.* 2006); at Pas-de Calais (France) in November 2003-January 2004 by industrial cooling towers (Nguyen *et al.* 2006); at Sarpsborg (Norway) in May 2005 by an industrial air scrubber (Nygard *et al.* 2008); in Amsterdam in July 2006 by a rooftop-cooling tower (van den Hoek *et al.* 2006); at Ulm in December 2009-January 2010, attributed to a wet cooling tower (von Baum *et al.* 2010); in Edinburgh (UK) in June 2012, attributed to cooling towers (McCormick *et al.* 2012). In Canada an outbreak occurred in Quebec in August 2012 that was attributed to a rooftop-cooling tower (CBC News 2012). Despite the large numbers of cases (more that 50 up to several hundreds) involved in these and other outbreaks, most LD cases are sporadic and community-acquired with an unidentified source.

Clearly, LD is a serious public-health problem that is not yet under control. The aim of this chapter is to describe the significance of community-water supplies and potable-water installations in buildings and homes as sources of LD. For this purpose, information is provided about the incidence of the disease in Europe and the US, the identity of the disease-causing organisms, their ecology and behaviour in natural and man-made water systems, and their presence in water supplies and potable-water installations. Measures for controlling the proliferation of *L. pneumophila* in these systems are briefly described. Finally, the role of drinking-water supply in LD is discussed.

6.2 INCIDENCE AND CAUSAL ORGANISMS

6.2.1 Legionnaires' disease in Europe and the US

The average rate of notified cases per million population in the reporting countries of the European Union (EU) increased from about 4 in 1994 to about 12 in 2008 (Figure 6.1). In 2009 and 2010, totals of 5518 and 6296 cases were reported by EU member states, giving an overall notification rate of 11.2 and 12.4 per million inhabitants, respectively (ECDC 2011, 2012). The crude notification rate showed large differences between countries in Europe and ranged from >20 per million in 2009 and/or 2010 in Slovenia, Spain, Denmark and the Netherlands to <1 in a number of countries in Eastern Europe. The crude notification rate in the Netherlands was about 3 per million in the period 1987–1998 and increased strongly after the outbreak in 1999, reaching 28 per million in 2010 (Bijkerk *et al.* 2011). The notified cases in the US were fairly stable at about 5 per million between 1991 and 2000 (Fields *et al.* 2002). Between 2000 and 2009, annually reported cases increased from 1110 in 2000 to 3522 in 2009, corresponding with a crude national notification rate of 3.9 per million inhabitants in 2000 and 11.5 in 2009 (Hicks *et al.* 2011). The highest numbers of cases were reported in the summer-early fall period, with the highest incidence in the Middle Atlantic region (New York, New Jersey and Pennsylvania).

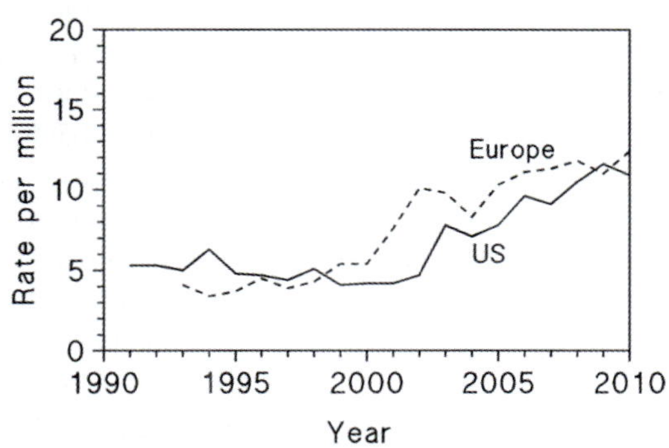

Figure 6.1 Overall notification rate of Legionnaires' disease in Europe and legionellosis (includes Pontiac fever) in the US in a period of about 20 years. Data from: Joseph and Ricketts (2010); ECDC (2012); CDC (2003), Hicks *et al.* (2011); CDC (2012). A few countries participating in the European Surveillance Scheme for Travel Associated Legionnaires' Disease (EWGLINET) (Joseph and Ricketts 2010) are not included in countries reporting to ECDC.

The observed increase in incidence in the US has been attributed to (i) improved diagnosis, (ii) expanding high risk populations (e.g. the elderly, transplant patients), and (iii) increased reporting (Editorial comment to Hicks *et al.* 2011). These factors probably also play a role in Europe (ECDC 2012). Travel-associated cases, including domestic travel, represented about 21% of the reported cases in the EU (Joseph and Ricketts 2010; ECDC 2011, 2012) and about 24% in the US in 2005–2009 (Hicks *et al.* 2011). Cases associated with healthcare facilities accounted for about 7.5% of the cases in Europe in 2009 and 2010 (ECDC 2012). The large majority of cases reported in Europe (62–71%) and the US (80%) have been classified as sporadic, community-acquired cases, that is, no source was identified.

The elevated number of cases in the warm season, both in the US and Europe, indicates that the environmental temperature has an impact on the LD incidence. The relatively high number of reported cases in the summer months in the metropolitan region of Philadelphia in 1995–2003 correlated with temperature, relative humidity and precipitation (Fisman *et al.* 2005). Elevated cases of LD in the Mid-Atlantic region of the US in 2003 correlated with temperature and rainfall as well (Hicks *et al.* 2007). In the first study, it was hypothesized that heavy rainfall had caused an increase of sediments in drinking-water supplies that subsequently promoted growth of *L. pneumophila*, leading to more cases of LD. An elevated rate in the UK in 2006 also correlated with warm and wet (increased humidity) weather during July to September (Ricketts *et al.* 2009). In the warm and wet summer of 2006 a significant increase in notified LD cases was observed in the Netherlands as well (Karagiannis *et al.* 2009). Statistical analysis of data from 2003–2006 revealed the impact on a weekly basis of temperature, relative humidity and precipitation intensity. Clustering for weather types showed that humid and wet conditions with little sunshine and a close to average temperature in summer corresponded with the highest incidence rate. An impact of rainfall on drinking water quality is unlikely in the Netherlands, because of the multiple barriers, including soil passage, in surface-water treatment and the use of groundwater in most supplies and no open surface reservoirs. A recent study in England and Wales showed that wet cooling systems may be an important source of sporadic community-acquired LD (Ricketts *et al.* 2012). Other potential sources of *Legionella* associated with rainfall include rainwater on roads (Sakamato *et al.* 2009) and windscreen-wiper fluid (Wallenstein *et al.* 2010).

6.2.2 Identity of clinical and environmental isolates

6.2.2.1 Clinical isolates

A total of 59 species and subspecies of the genus *Legionella* are recognized to date (Deutsche Sammlung von Mikroorganismen und Zellkulturen [DSMZ] 2012). Many of these species have been associated with disease (Fields *et al.* 2002). *L. pneumophila* represented over 95% of the isolates from culture-confirmed cases of LD in Europe in 2009–2010 (Table 6.1) with *L. pneumophila* serogroup 1 (SG1) in more than 83% of these cases. Other SGs of *L. pneumophila* together comprised less than 10% of the clinical isolates and *L.* non-*pneumophila* spp. played only a minor role with each <0.5–1%. The data reported by Yu *et al.* (2002) for community-acquired cases are based on a collection of isolates from the US and a few other countries. Also here, *L. pneumophila* was the predominating organism, with *L. pneumophila* SG1 as major representative. Most strains of *L. longbeachae* in this collection originated from Australia and New Zealand, where potting soil has been identified as a source of infection by this organism (Steele *et al.* 1990).

The introduction of the urine-antigen (UAg) test for LD diagnosis, which is currently the diagnostic tool for more that 80% of the cases in Europe (ECDC 2012), most likely is one reason for the increased notification rate in the EU and also the US (Editorial note to Hicks *et al.* 2011). This test is designed for the detection of serious cases of LD caused by *L. pneumophila* SG1 (Helbig *et al.* 2003). As a result, the presented distribution of species and serogroups in recent years is biased towards *L. pneumophila* SG1 and may not represent the actual distribution in all LD cases. A survey among CAP patients in Germany, using a UAg testing, culture and PCR, showed that 6.4% of the 94 LD cases were caused by *L.* non-*pneumophila* spp., including *L. anisa* (2.1%), *L. bozemanii* (2.1%), *L. longbeachae* (1%) and *L. erythra* (1%) (von Baum *et al.* 2008). A more recent study of 206 cases of community-acquired LD, also diagnosed with a variety of methods, gave the following distribution of *Legionella* species: *L. .pneumophila* (93.2%), *L. anisa* (1.5%), *L. bozemanii* (1.0%), *L. longbeachae* (1.0%), *L. maceachernii* (0.5), *L. erythra* (0.5%), *L. dumoffii* (0.5%) and four strains with an unclear sequence (von Baum and Lück 2011).

6.2.2.2 Environmental isolates

L. pneumophila SG2-16 and *L.* non-*pneumophila* species are more frequently isolated from environmental samples than from clinical samples (Table 6.1). However, the collections of environmental isolates may not represent the occurrence of these organisms in the aquatic environment. The main reasons include (i) *Legionella* spp. generally give a lower yield on Buffered Charcoal Yeast Extract (BCYE) medium with antibiotics than *L. pneumophila* (Lee *et al.* 1993), thereby reducing the relative proportion of these organisms in the collection of isolates, (ii) the collections of environmental isolates may not result from systematic surveys and (iii) molecular techniques show much larger diversities of *Legionella* spp. in the aquatic environment than culture methods (see Section 6.3.2.2). Still, certain proportional differences between environmental and clinical isolates reflect differences in the virulence of the various species and serogroups. For example, *L. anisa* is much more common among the environmental isolates than among clinical strains, and is much less infectious than *L. pneumophila* (Fields *et al.* 1990). This organism is frequently observed in

Table 6.1 Identity of clinical and environmental isolates of *L. pneumophila* and other *Legionella* species in different areas and time periods.

Identity	Clinical isolates (%)					Environmental isolates (%)		
	Canada 1978–1991 (Ref. 1)*	**Denmark 1994–2000 (Ref. 2)**	**US ** 2002 (Ref. 3)**	**the Netherlands 2003–2006 (Ref. 4)**	**Europe 2009–2010 (Ref. 5, 6)**	**France 2001–2002 (Ref. 7)**	**the Netherlands 2000–2006 (Ref. 4)**	**Korea 2008 (Ref. 8)**
L. pneumophila	71.1	97.5	91.5	99.8	97.5	75.5	74.8	85.5
L. pneumophila SG1	52.6	57	84.2	76.3	86.5	28.2	45.5	46.8
L. pneumophila SG 2-16	18.2	40.5	6.9	0.9	7.4	47.2	26.9	38.7
L. pneumophila SG 3	2.5	20	1.0	0.2	3.0	10.8	5.3	4.1
L. pneumophila SG 6	9.8	7.7	1.7	0.6	1.5	11.1	5.1	9.8
L. pneumophila SG ns#	0	2.6	0.6	20.2	3.6	18.7	1.6	0.3
Other *Legionella* species	28.8	2.5	8.5	0.2	1.2	24.5##	25.3	18.8$
L. anisa	0.2	0	0.2	nr#	nd#	13.8	nr	7.0
L. bozemanii	3.7	1	2.4	nr	0.4	nd	nr	3.0
L. dumoffii	2.5	0	0.6	nr	0.2	nd	nr	0.2
L. longbeachae	2.8	1	3.9	nr	0.5	nd	nr	nd
L. micdadei	5.1	0.5	0.6	nr	0.2	nd	nr	nd
Not specified/unknown	0	0	0.2	0	1.4	0	0	0
Total number of isolates	427	195	508	1091	1100	27447	742	560

*Ref.: 1, Tang and Krishnan (1993); 2, Uldum and Helbig (2002); 3, Yu *et al.* (2002); 4, Versteegh *et al.* (2007); 5,6, ECDC (2011, 2012); 7, Doleans *et al.* (2004); 8, Lee *et al.* (2010); **includes clinical strains from the US (72%), Italy, Switzerland, Australia and New Zealand; # ns, not specified; nr, not reported; nd, not detected; ## includes *L. tauriensis* (2.8%), *L. erythra* (1.5%), *L. rubrilucens* and L. *tucsonensis; $* includes also *L. rubrilucens* and *L. spiritensis.*

potable-water installations without causing disease (Stout *et al.* 2007), although an increased antibody titre for this organism has been reported for visitors of hot-spring spas (Bornstein *et al.* 1989). *L. anisa* was also implicated in outbreaks of Pontiac fever (Fenstersheib *et al.* 1990; Jones *et al.* 2003), a self-limited influenza-like illness.

6.2.2.3 Monoclonal antibody (MAb) types

MAb typing showed that *L. pneumophila* SG1 is highly heterogeneous and includes a large diversity of strains, differing in virulence (Joly *et al.* 1986). A pan-European study of 1335 unrelated *L. pneumophila* strains causing LD yielded 23 serogroups and monoclonal subgroups (Helbig *et al.* 2002). A total of 78.5% of these strains belonged to SG1, 66.8% were MAb3/1 positive and 11.7% of *L. pneumophila* SG1 strains were MAb3/1 negative. In strains originating from nosocomial cases, the percentages of non-serogroup 1 isolates (35.9%) and MAb3/1-negative isolates (53.5%) were significantly higher than these percentages for community-acquired cases, respectively 16.3% and 27.3%. The significant role of these organisms in healthcare-associated cases may be explained by the reduced resistance against infection of the exposed persons. In the US, 75% of 100 clinical isolates of *L. pneumophila* SG1 were positive for MAb2 (which reacts with the same *L. pneumophila* SG1 strains as MAb3/1) compared to 6% of 50 epidemiologically-unrelated environmental isolates (Kozak *et al.* 2009). In the UK, 91.6% of 167 clinical isolates were positive for MAb3/1, compared to 8.3% of 276 environmental isolates (Harrison *et al.* 2009).

6.2.2.4 Sequence-based types (STs)

The observations with MAbs were confirmed by molecular techniques of which multilocus sequence typing has become the tool for identification, enabling source tracing by determining the degree of genetic relatedness of clinical and environmental isolates of *L. pneumophila* (Gaia *et al.* 2005). To date, more than 1000 sequence types (STs) of *L. pneumophila* have been identified and MAb types include different STs. Certain STs are more common among clinical samples, whereas others are more common in samples from environments not related to disease (Borchard *et al.* 2008; Den Boer *et al.* 2008; Harrison *et al.* 2009; Kozak *et al.* 2009). Of a number of STs representing more than 60% of the clinical isolates in the UK, only *L. pneumophila* SG1 ST47, representing 25.7% of these isolates, was observed once in the environment (Harrison *et al.* 2009). *L. pneumophila* SG1 ST62, the causative organism of the outbreak at the flower show in 1999 in the Netherlands and the outbreak in Ulm in December 2009-January 2010 attributed to a cooling tower (von Baum *et al.* 2010), was isolated from the whirlpools at the flower show. This organism represented 9% of the clinical isolates in England and Wales (Harrison *et al.* 2009) and 8.2% in Ontario, Canada (Tijet *et al.* 2010), but was not represented in the environmental isolates in the UK (Harrison *et al.* 2009), the US (Kozak *et al.* 2009) and Korea (Lee *et al.* 2010).

Taken together, the studies in Germany (von Baum *et al.* 2008; von Baum and Lück 2011), confirm that more than 90% of the cases of LD are caused by *L. pneumophila*, despite the bias towards *L. pneumophila* SG1, caused by UAg testing. The relatively large proportion of other *Legionella* spp. among the clinical isolates collected in Canada (Table 6.1) might have been due to a large proportion of strains from nosocomial cases. Furthermore, the majority of clinical isolates of *L. pneumophila* in Europe and the US reacts with MAb3/1, which is equivalent to MAb2. These organisms represent only a small fraction of the environmental *L. pneumophila* strains. Also STs of *L. pneumophila* SG1 implicated in disease either represent only a small fraction of the environmental isolates or are not observed in the environment. The distributions of STs among the clinical isolates in Germany, UK, Portugal (Chasqueira *et al.* 2009), the US and Canada are different and include STs that are unique to the region or country. Geographical differences in the occurrence of STs are also observed for the environmental isolates. The diversity of environmental STs suggests that *L. pneumophila* proliferates in a large variety of microniches, with clinically-relevant STs multiplying under specific, not-yet clarified, environmental conditions.

6.2.3 Infectious dose

As for other pathogenic microorganisms, the infectious dose of Legionellae depends on the virulence of the involved organism, and the condition of the exposed person. Berendt *et al.* (1980) showed that aerosols containing *L. pneumophila* were infective for guinea pigs at inhaled dosages as low as 129 CFU. Based on the results of a number of exposure studies, Armstrong and Haas (2005), aiming at establishing the Quantitative Microbial Risk Assessment (QMRA) approach for exposure to *Legionella pneumophila*, derived dose-response relationships for the lethal dose (LD) and the infectious dose (ID) which gave an $LD_{50\%}$ of 6200 CFU ($LD_{1\%}$ 92 CFU) and $ID_{50\%}$ of 12 CFU ($ID_{1\%}$ 0.17 CFU), respectively. Intraperitoneal injection in guinea pigs, which requires higher dosages for infection and death than exposure via aerosols (Berendt *et al.* 1980), confirmed differences in pathogenicity between strains of *L. pneumophila* and also other species (Fields *et al.* 1986). Guinea pigs are considered suitable model organisms (Armstrong and Haas 2008), but LDs and IDs of specific *L. pneumophila* STs may be lower than those mentioned above, in particular for immunocompromised persons.

6.3 ECOLOGY OF *L. PNEUMOPHILA*

L. pneumophila is ubiquitous in freshwater environments at elevated temperatures, for example, in thermally-altered lakes in the US (Fliermans *et al.* 1981) and in tropical regions (Ortiz-Roque and Hazen 1987). A large diversity of other

Legionella spp. has been detected in all freshwater environments, including arctic regions (Brambilla *et al.* 2001; Carvalho *et al.* 2008). Legionellae generally represent only a small fraction (<1%) of the total number of bacteria. The ubiquitous presence of *Legionella* spp. in relatively low numbers in all types of freshwater environments indicates that these organisms proliferate in a specific, but common, niche.

6.3.1 Growth substrates and amoebae

Cultivation of *L. pneumophila* in defined media showed that certain amino acids, for example, cysteine, methionine, arginine, leucine and isoleucine, are essential for its growth and some, including alanine, aspartate, glutamate, serine and proline, can serve as a source of energy and carbon (Warren and Miller 1979; George *et al.* 1980; Franzus *et al.* 1984). Also proteolytic and phosphatase activities have been determined in the organism (Nolte *et al.* 1982). Furthermore, iron is needed for growth and (therefore) chelating agents are inhibitory. The use of the BIOLOG system, with a large diversity of substrates, showed that *L. pneumophila* can also utilize β-hydroxybutyrate, but no carbohydrates (Mauchline and Keevil 1991). Obviously, *L. pneumophila* is a fastidious microorganism that utilizes only a limited range of compounds as a source of energy. As a result, the organism cannot compete with oligotrophic aquatic bacteria that are able to utilize a range of biodegradable organic compounds at the μg/litre level. This conclusion is consistent with observations that pure cultures of *L. pneumophila* do not multiply in sterilized tap water (e.g. Wadowsky and Yee 1985). Colony counts of pure cultures added to sterilized water declined but the organisms remained detectable for a long time (Skaliy *et al.* 1979; Paszko-Kolva *et al.* 1992). The decline of the total cell number of cells of the pure cultures was much less than the decline of the colony count. Also the number of cells with intact membranes remained at the same level for several months (Ohno *et al.* 2003). The survival property of *L. pneumophila* has been attributed to its ability to produce poly-β-hydroxybutyrate during exponential growth as an intracellular energy source (James *et al.* 1999).

Experiments *in vitro* with pure cultures showed that a variety of amoebae, for example, *Acanthamoeba* spp., *Hartmannella* spp., *Naegleria* and *Valkampfia* can serve as a host for *L. pneumophila* (Rowbotham 1980, 1986; Tyndall and Domingue 1982; Holden *et al.* 1984). Also species of the ciliate *Tetrahymena* (Fields *et al.* 1984) and the mould *Dictylostelium discoideum* (Hägele *et al.* 2000) have been identified as host for *L. pneumophila.* Initially, it was generally accepted that *L. pneumophila* was a facultative intracellular parasite (Fields *et al.* 1986). However, Wadowsky *et al.* (1988) demonstrated that *L. pneumophila* was unable to multiply in sterilized tap water, with added associated microbiota, when the inoculum had been membrane filtered (1 μm). In the presence of the untreated inoculum, growth occurred and the amoebae *Hartmannella vermiformis* was identified as the host for *L. pneumophila*. Furthermore, *L. pneumophila* multiplied in a biofilm of three bacterial species only after the addition of *H. vermiformis*, although its survival was enhanced by the presence of these bacteria in the absence the amoeba (Murga *et al.* 2001). A biofilm of indigenous bacteria, grown at a high concentration on pieces of plasticized polyvinyl chloride (PVC) incubated in tap water, did not promote growth of *L. pneumophila* when the inoculum of *L. pneumophila* and associated microbiota had been membrane-filtered (3 µm) (Kuiper *et al.* 2004). However, *L. pneumophila* proliferated in the biofilm when the inoculum had not been filtered. Also in this study, *H. vermiformis* was identified as the host protozoan and *L. pneumophila* was observed in a large proportion of trophozoites. The high biofilm concentration on plasticized PVC corresponded with higher numbers of *H. vermiformis* and *L. pneumophila* than a biofilm on unplasticized PVC.

Incubation of water collected from man-made systems in a biofilm-batch test with pieces of polyethylene at 37°C resulted in the identification of a number of additional (candidate) hosts of *L. pneumophila* (Valster *et al.* 2010). This study also confirmed the prominent position of *H. vermiformis* as amoebal host in these environments, as had been reported previously by other investigators (Sanden *et al.* 1992; Rohr *et al.* 1998). Studies further showed that the protozoan hosts differ for different *Legionella* spp. and different *L. pneumophila* strains (e. g. Fields *et al.* 1986; Neumeister *et al.* 1997; Buse and Ashbolt 2011), demonstrating the complexity of the relationships between free-living protozoa and *Legionella* spp. in the environment.

A pure culture of *L. pneumophila*, present at 10^6 cells/ml in filter-sterilized tap water, multiplied after the addition of 10^8 to 10^9 cells/ml of a heat-killed *Pseudomonas putida* (Temmerman *et al.* 2006). The authors concluded that formation of dead microbial cells should be minimized to prevent the persistence of L. *pneumophila* in the environment. Piao *et al.* (2006) suggested that the ability of pure cultures of *L. pneumophila* to produce mycelial biofilms in 1:10 diluted buffered yeast extract broth is a strategy for its proliferation in the aquatic environment. However, both types of test environments, with a high concentration of dead cells or growth substrates, and absence of other (competitive) bacteria do not represent natural habitats and the significance of these observations for practice remains unclear.

6.3.2 Temperature effects

Information about the effects of temperature on *L. pneumophila* has been obtained in laboratory tests and by observations in natural freshwater environments and man-made water systems. Laboratory tests can be conducted with pure cultures and/or at strictly controlled temperatures. Observations in natural and man-made aquatic environments provide information about the combined impact of temperature and competition with other *Legionella* species on the presence and behaviour of *L. pneumophila*.

6.3.2.1 Laboratory systems

Naturally-occurring *L. pneumophila* multiplied in the presence of associated microbiota in tap water at temperatures ranging from 25 to 42°C, but not at 45°C (Yee and Wadowsky 1982; Wadowsky *et al.* 1985). Growth of *L. pneumophila* SG1 was observed in buffered charcoal yeast extract broth at temperatures ranging from 15°C to 40°C (States *et al.* 1993). However, in American Type Culture Collection medium 1034 (PYNFH) which supports axenic growth of *H. vermiformis*, *L. pneumophila* SG1 did not proliferate at 15°C and 20°C, grew well at temperatures up to 42°C, but not at a higher temperature (Figure 6.2a). The observation that *L. pneumophila* SG1 was digested *by Acanthamoeba castellani* at 15 and 20°C without infection of the amoeba, but proliferated by infection at 25°C (Ohno *et al.* 2008) is consistent with these findings. Digestion at 20°C was also observed in *Acanthamoeba palestinensis* (Anand *et al.* 1983).

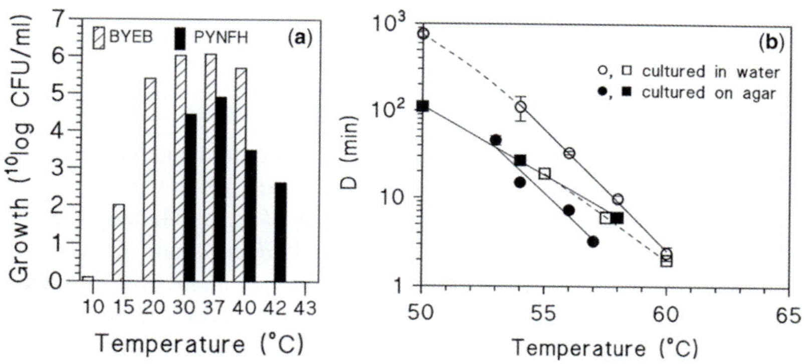

Figure 6.2 Effect of temperature on growth and inactivation of *L. pneumophila*. (a) Maximum colony count in Buffered Yeast Extract Broth (BYEB) and in PYNFH medium with *H. vermiformis* at temperatures ranging from 10 to 43°C (based on States *et al.* 1993); (b) Decimal reduction time (D) of *L. pneumophila* at different temperatures. *L. pneumophila* was: ● grown at 37°C on BCYE agar; ○ grown at 37°C in tap water in the presence of natural rubber (unpublished data); □ data from Schulze-Röbbecke *et al.* (1987); ■ data from Dennis *et al.* (1984).

Growth of *L. pneumophila*, measured as colony counts, in test systems with materials and a natural microbial community, was also limited at 20° (Rogers *et al.* 1994a), but relatively strong at 25°C (van der Kooij *et al.* 2002). A pure culture of *L. pneumophila* SG1 (Philadelphia) multiplied at 44°C in a chemostat (Berg *et al.* 1985). Also growth of *L. pneumophila* at 45°C, in association with cyanobacteria, has been reported (Tison *et al.* 1980). Tests with four pure cultures of *L. pneumophila* showed that growth at temperatures of 44 to 46°C was less than 10% of growth at 37°C (Kusnetsov *et al.* 1996). However, accelerated loss of cultivability was reported for a pure culture of *L. pneumophila* SG1 suspended in spring water at 42°C (Ohno *et al.* 2003). At temperatures above 50°C a more rapid decline of the number of cultivable *L. pneumophila* was observed, with a decimal reduction time (D) of a few minutes at 60°C (Figure 6.2B). Moreover, no thermotolerance was observed among strains of 10 *Legionella* species, including *L. pneumophila*, which all showed a D value of a few minutes at 60°C (Stout *et al.* 1986).

These observations demonstrate that representatives of *L. pneumophila* SG1 can multiply at relatively low temperatures as a pure culture in synthetic media and also at temperatures close to 45°C. Growth at a certain temperature (range) may differ between strains of *L. pneumophila* SG1 and between strains of different SGs and STs as well. Temperature also affects the communities of free-living protozoa, that is, the nature of the protozoan host, thereby also impacting the growth of *Legionella* spp. as well as SGs and STs of *L. pneumophila* (Valster 2011), because of specific bacterium-host relationships (Fields *et al.* 1986; Wadowsky *et al.* 1991; Buse and Ashbolt 2011).

6.3.2.2 Natural and man-made aquatic environments

In hydrothermal areas in Portugal, *L. pneumophila* predominated among isolates from water in hot springs of ≥45°C (Verissimo *et al.* 1991; Costa *et al.* 2005) and in biofilms at 35–42°C in hot springs in New Mexico (Marrao *et al.* 1993). *L.* non-*pneumophila* species, for example, *L. oakridgensis* and *L. sainthelensi*, predominated among the isolates from spring water at about 35°C. Polymerase Chain Reaction (PCR) amplification with *Legionella*-specific primers revealed that *L. sainthelensi* also was the predominating *Legionella* species in acidic water (pH 2.7) of 30°C from a hot spring in Yellowstone National Park (Sheehan *et al.* 2005). *L. micdadei* and *L. cherrii* were cultured from water of 38°C in which an unidentified *Legionella*-like amoebal pathogen (LLAP) predominated. In drinking water at about 30°C in the Caribbean, *L. pneumophila* comprised about 80% of the colony counts on BCYE agar (Valster *et al.* 2011). In an industrial cooling tower at a water temperature of 27°C, *Legionella* spp. including *L. falloni*, *L. lytica* and other LLAP-types were observed, but *L. pneumophila* was not detected by quantitative PCR (qPCR) (Wéry *et al.* 2008). Following a temperature increase to 29–30°C in July and August, *L. pneumophila* increased from 1% to 50% of the total *Legionella* population, along with a decrease in diversity of *Legionella* spp. *L. pneumophila* was not detected in water of 23–25°C abstracted from drinking water production wells, but a variety of other *Legionella* species, including *L. anisa*, *L. dumoffi*, *L. erythra* and *L. parisiensis*, were cultured (Pryor *et al.* 2004). Also, in biofilms and water (at 21–31°C) in the distribution system only *L.* non-*pneumophila* species were cultured. The predominating position of *L. anisa* in cold-water pipes of potable-water installations (Versteegh *et al.* 2007) is consistent with these observations, suggesting that this organism is adapted to growth at temperatures below the optimum for *L. pneumophila*.

By using molecular methods, *L. parisiensis*, *L. maceachernii*, *L. birminghamensis* and *L. erythra* were detected as the predominating *Legionella* spp. in slow sand filtrate at a temperature between 10 and 20°C, in which also *L. pneumophila* was observed (Calvo-Bado *et al.* 2003). In a pristine river, at a temperature ranging from 2 to 20°C *Legionella* spp. were observed at levels ranging from 7×10^3 to 9.4×10^5 genomic units (GU)/litre (Parthuisot *et al.* 2010). Most organisms were closely related to *Legionella* spp. including *Legionella*-like amoebal pathogens (LLAPs), but could not be identified further. *L. pneumophila* constituted 4.2% of the retrieved clones and was cultured in low numbers at two sites impacted by discharges from thermal baths and treated waste water. By using qPCR in a survey of the rivers Meuse and Rhine, brooks and lakes in the Netherlands, in spring and summer at temperatures ranging from 6 to 21°C, *Legionella* spp. were observed in all samples at levels of 10^5 to 10^7 GU/litre. *L. pneumophila* was only detected in a few of these samples (Wullings *et al.* 2009). Further testing showed that *L. pneumophila* originated from treated waste-water discharges. *L. pneumophila* constituted about 5% of the clones retrieved from samples of treated water and biofilms in drinking-water supplies in the Netherlands, at temperatures below 15°C, but all samples tested negative by culture (Wullings and van der Kooij 2006; Wullings *et al.* 2011). More than 25 different species, including LLAPs, were observed in these samples, together with more than 100 different operational taxonomic units (OTUs), which most likely represented not-yet described *Legionella* species. A number of clones retrieved from Arctic lakes at 0°C were closely related to some of these OTUs (Carvalho *et al.* 2008). In these lakes, located close to a scientific station, also *L. pneumophila* was detected with PCR and by culture.

The proliferation of a large variety of *Legionella* species in aquatic environments suggests that these organisms are all able to infect certain free-living protozoa. Such interactions have been observed for a number of *Legionella* species (Sanden *et al.* 1992) and for all LLAPs. *L. anisa* multiplied at temperatures of 30–35°C in *H. vermiformis* and *Acanthamoeba polyphaga*, but not at 37°C (Fields *et al.* 1990; Wadowsky *et al.* 1991). At 15°C, growth of unidentified *Legionella* spp. was observed in association with unidentified free-living protozoa (Valster *et al.* 2010).

The observations in natural and man-made freshwater environments demonstrate that (i) temperature affects the diversity and identity of *Legionella* spp. and (ii) the proportion of *L. pneumophila* in relation to other *Legionella* spp. decreases with decreasing temperature and *vice versa*. However, the information is rather fragmented and incomplete, because assessment of the effect of temperature on *L. pneumophila* and other *Legionella* spp. in the aquatic environments is complicated by a number of factors. These factors include: (i) culture methods do not give information about species that are viable but not cultivable (VBNC), (ii) PCR-based methods do not distinguish between living and dead cells and

(iii) measuring the true environmental temperature is hampered by fluctuations in time. For example, *L. pneumophila* cultured from water sampled at 60°C, most likely originates from a location where the water temperature was below the water temperature observed during sampling. Furthermore, temperature also affects the activity and diversity of free-living protozoa and (their interactions with) the bacterial community, for example, grazing activity.

Overall, the need of a protozoan host explains that the nutritionally fastidious bacterium *L. pneumophila*, as well as other *Legionella* spp., proliferate in oligotrophic aquatic natural and man-made environments, where bacteria in biofilms and sediments serve as food source for the protozoa. Multiplication of *L. pneumophila* depends on the accumulation of biofilms and/or sediments at a temperature range of 25 to 45°C. Temperatures below the optimum growth temperature may limit multiplication of *L. pneumophila* in the aquatic environment because of a decrease of its infectivity for host amoeba and competition with other *Legionella* species that are more adapted to growth at lower temperatures.

6.4 LEGIONELLAE IN DRINKING-WATER SUPPLIES

The presence and significance of *L. pneumophila* in relation to water supply was shown for the first time by the isolation of *L. pneumophila* from shower heads of a hot water installation in a hospital in the UK (Tobin *et al.* 1980). In this study, the concentrate of centrifuged water samples was injected intraperitoneally into guineapigs for the detection of *L. pneumophila*. Further investigations in plumbing systems initially were also conducted with guineapigs (Dennis *et al.* 1982), but the development of solid media for the detection of *L. pneumophila* (Edelstein 1981; Wadowsky and Yee 1981) facilitated numerous surveys in drinking-water distribution systems and in potable-water installations in hospitals, hotels and residential buildings. The introduction of molecular methods for detection and identification of *Legionella* was another major step forward because these methods enabled a more detailed assessment of the presence and identities of *Legionella* species, including culturable and VBNC bacteria as well as not-yet defined species, in freshwater environments.

Drinking-water supplies differ within and between countries by (i) the nature of the source water (groundwater, surface water), (ii) water treatment processes used for drinking water production, (iii) type of construction materials and (iv) the use and nature of a disinfectant during distribution. Furthermore, climate and season affect the temperature of water during treatment and distribution. As will be shown below, a number of these differences affect the presence and nature of Legionellae in these supplies. The observations of Legionellae in drinking-water supplies and potable-water installations are described in chronological order to show the developments in recognizing the significance of these organisms as waterborne pathogens.

6.4.1 Drinking water distribution systems

L. pneumophila was not or only sporadically cultured from water samples of chlorinated supplies in surveys conducted in the 1980s in the US. Tison and Seidler (1983) detected *Legionella* spp. by direct immunofluorescence in treated water of <20°C, but not by culture (n = 19). *L. pneumophila* and a few other species were isolated from 5 of 856 samples from chlorinated supplies (Hsu *et al.* 1984) and from 4 of 72 distribution sites (Voss *et al.* 1985), but not by States *et al.* (1987) (n = 50). Surveys in France (Desplaces *et al.* 1984), the UK (Colbourne and Trew 1985; Jones and Ashcroft 1988) and Canada (Tobin *et al.* 1986) also showed no or low percentages of *Legionella*-positive samples. In a survey in Germany, *L. pneumophila* was cultured from a few samples of treated groundwater and more frequently (7.5%) from water collected from cold-water taps in buildings (Seidel *et al.* 1986). In these studies, water temperature generally was not reported, but probably was <20°C in most cases. However, a temporary and/or periodically-elevated temperature may have promoted the growth of *L. pneumophila* at the sampling site. Drinking water collected from the bottom of finished-water reservoirs supported growth of *L. pneumophila*, when incubated at 35°C after pasteurisation and inoculation with water sampled from a hot-water tank containing *Legionella*. (States *et al.* 1987). This growth correlated with an elevated total organic carbon (TOC) concentration. Bezanson *et al.* (1992) reported that *L. pneumophila* strains isolated from the potable water of five different institutions in Halifax (Canada) served by one and the same reservoir of a surface water supply belonged to different monoclonal antibody subtypes. In contrast, a study conducted in Paris showed that a significant proportion of clinical isolates of nosocomial cases in a number of hospitals in Paris belonged to a specific genotype of *L. pneumophila* (Lawrence *et al.* 1999). This organism was also isolated from potable-water installations. Also Darelid *et al.* (2004) reported that a specific *L. pneumophila* genotype in a hospital water system was present in the municipal water supply.

In surveys in the Netherlands, using qPCR, *Legionella* spp. were observed in all drinking-water samples at temperatures <20°C (Wullings *et al.* 2006, 2011). *L. pneumophila* represented 2.5–6% of the retrieved clones, but this organism was not observed with the culture method. Diederen *et al.* (2007) sampled cold water taps of 250 public buildings in the Netherlands and, by using PCR analysis, detected *L. pneumophila* in 3.9% of these samples and *L.* non-*pneumophila* in 87.1%. Routine monitoring of drinking water in the Netherlands for *Legionella* by collecting samples from cold-water taps after heat sterilisation and flushing, showed approx. 5% *Legionella*-positive samples (Versteegh *et al.* 2011). The isolated strains were identified as *L. anisa*. In a survey of Caribbean water supplies, distributing drinking water of approx. 30°C without a disinfectant residual, *L. pneumophila* was cultivated in about 80% of the samples (Valster *et al.* 2011). In this study, biofilms and local accumulation of sediments were identified as growth-promoting conditions. In a recent study in the US, *L. pneumophila* was observed with qPCR in 4.4–5.6% of the samples from two supplies distributing drinking water at a temperature of about 20–27°C and a 2 ppm residual of monochloramine (Wang *et al.* 2012). Only one sample was culture-positive for *L. pneumophila*.

The sporadic detection of *L. pneumophila* in water supplies in temperate regions is in agreement with the observations showing that multiplication of this organism in the presence of protozoa is limited at temperatures below 25°C. Still, *L. pneumophila* may be present more frequently at levels below the detection limit of the testing methods and dissemination from the distribution system into potable-water systems has been reported. Furthermore, the ubiquitous presence of other *Legionella* species is consistent with the observations that drinking water has a potential for growth of *L. pneumophila* at elevated temperature.

6.4.2 Hospitals, hotels and homes

6.4.2.1 Hospitals and hotels

In a hospital with two to ten monthly cases of nosocomial LD in 1977–1980, *L. pneumophila* was detected (10^5-5×10^6 CFU/litre) in the potable water following a pressure shock (Shands *et al.* 1985). The decrease of cases after hyperchlorination was considered circumstantial evidence that the potable water system was implicated in the disease. Stout *et al.* (1982) reported that the water system of a hospital with nosocomial cases of LD was colonized by *L. pneumophila* and also *L. micdadei*, the Pittsburgh pneumonia agent, was isolated from a hospital shower (Brown et al. 1982; Stout *et al.* 1982). The organism was detected in water and sediments from hot-water tanks maintained at 30 to 54°C, but not at 71°C (Wadowsky *et al.* 1982). Also Fisher-Hoh *et al.* (1982) observed *L. pneumophila* in high numbers in the iron-containing deposits of hot-water cylinders in a hospital with LD cases. Plouffe *et al.* (1983) found colonization by *L. pneumophila* in four hospitals at a hot-water temperature of 43 to 45°C, but not in hospitals with hot-water storage reservoirs maintained at 58 to 60°C. Stagnation of water in hot-water tanks and obstruction in faucet aerators also promoted colonization (Ciesielsky *et al.* 1984). Growth of *L. pneumophila* in sediments from hot-water tanks correlated with elevated levels of iron and zinc in these sediments (States *et al.* 1985). Stout *et al.* (1985) observed that sediment from water systems promoted the growth of the indigenous microbiota, which in turn stimulated growth of *L. pneumophila*. A prospective study showed that 9 of 15 hospitals surveyed were *L. pneumophila* positive, with significant associations with elevated hot water temperature and concentrations of calcium and magnesium (Vickers *et al.* 1987).

In the Netherlands, *L. pneumophila*, later designated as SG10, was isolated from the warm water system in a hospital in association with nosocomial cases caused by the organism (Meenhorst *et al.* 1985). Guinea pigs exposed to aerosols of this water after concentration also developed pneumonia due to infection by the same

L. pneumophila type (Meenhorst *et al.* 1983). Maintaining a temperature of 60°C in the returning hot water eliminated *L. pneumophila* from a building water system, but this was not the case at a temperature of 54°C (Groothuis *et al.* 1985). A survey of 103 hospitals and 62 hotels in Germany yielded 70% *Legionella*-positive hospitals and 18% *Legionella*-positive hotels (Habicht and Müller 1988). Colonization was most frequently observed at water temperatures between 35 and 45°C and correlated with the concentration of iron and organic matter in drinking water. Another survey in Germany revealed that 61% of 75 buildings, including hospitals, were *Legionella* positive (Lück *et al.* 1993). Local heating, up to 38°C, of cold water was associated with two fatal cases of LD in a healthcare facility in the Netherlands (Hoebe *et al.* 1998).

More recently, it was reported that 75% of the hot-water systems of 40 hotels in Italy were colonized by *Legionella* spp. of which *L. pneumophila* was most frequently cultured (Borella *et al.* 2005). The presence of *Legionella* correlated with inorganic constituents (Zn, Ca, Mg, hardness) and organic content of the water and with temperature. A further study of hot-water systems of hotels and private homes in Italy yielded 47.5% *Legionella*-positive samples (up to 1.25×10^6 CFU/litre), mostly (92.8%) with *L. pneumophila* (Bargellini *et al.* 2011). The presence of *Legionella* correlated with concentrations of Mn, Zn and Fe, but not with the average water temperature (43–45°C) in these systems. In Greece, 20.8% of the water systems of 385 hotels were found colonized by *Legionella* spp. (Mouctouri *et al.* 2007). No *Legionella* were cultured from samples at <23.7°C or >60.3°C. Seasonal operation was identified as a risk factor but no LD cases were reported in association with the observation of *Legionella*. However, a recent study identified 10 different LD-associated accommodation sites across Corfu (Maini *et al.* 2012). In the Netherlands, in 2005–2006, the presence of *Legionella* (>100 CFU/litre) was reported in 39% of hospitals and other health-care facilities, 28% of swimming facilities, 23% of asylum-seekers' institutes and jails and 12% of hotels (Versteegh *et al.* 2007). Most (66–75%) samples with *Legionella* were from the cold-water pipes with *L. anisa* as the predominating organism. Also in Germany, cold-water pipes in healthcare facilities were found to be significantly more frequently colonized than the warm-water pipes (Arvand *et al.* 2011).

Overall, Legionellae were and still are quite common in hot-water installations in hospitals, hotels and other buildings, not only in warm but also in temperate climates. Water temperature is the most critical condition favouring growth of *L. pneumophila*, in combination with certain water quality aspects, which may be related to the formation of biofilms, sediments and scales, for example, organic matter, Fe, Ca and Mg.

6.4.2.2 Homes

The presence of *L. pneumophila* in a residential water system was reported for the first time in 1982 (Wadowsky *et al.* 1982). In 1987, LD was observed in two immunocompromised persons in association with the presence of *L. pneumophila* in the hot-water systems of two homes with a private well (Stout *et al.* 1987). In these systems, the hot-water temperature was 35–37°C at the bottom of the electric heater.

Table 6.2 shows the results of a number of surveys of water systems in residences, for example, apartments in buildings and family homes. Arnow *et al.* (1985) reported that systems with a maximum temperature <60°C were significantly more frequently colonized (42%) than systems with a maximum temperature >60°C (7%). *Legionella* was cultured from 1 of 93 samples from homes in different community water systems (Whitterel *et al.* 1988). Alari and Joly (1991) cultured *L. pneumophila*, mainly SG2 and 4, from samples of 39% of 178 houses with an electric heater, but none in 33 homes with an oil- or gas-heater. An extended survey in the Pittsburgh area showed 6% colonization with *L. pneumophila*, mostly SG1, without LD association nor elevated antibody titres to *Legionella* (Stout *et al.* 1992a). However, in another study the residential water system was identified as the source of *Legionella* in three of 20 LD cases, two of which lived in apartment buildings (Stout *et al.* 1992b). Surveys in Germany (Lück *et al.* 1993), Finland (Zacheus and Martikainen 1994) and Denmark (Pringler *et al.* 2002) revealed the presence of *L. pneumophila* in a substantial percentage of domestic water systems. The study in Denmark showed that counts were lower in systems with copper and steel pipes than in systems with iron pipes. The distribution of *L. pneumophila* serotypes in the systems differed from those in LD cases, but five of six serogroups implicated with LD in Denmark were observed in these systems.

In Germany no cases of LD were observed among persons (n = 53) in apartment buildings (house blocks) with *Legionella*-positive (40% > 1000 CFU/litre) systems and persons (n = 92) in family houses (3% > 1000 CFU/litre) (Heudorf *et al.* 2001). However, persons living in residences with an elevated number of *Legionella* in water had significantly higher antibody titres than those in residences with a low number of *Legionella*. A more recent study in Germany showed that hot water from instantaneous water heaters in single family residences did not contain *L. pneumophila*, but the organism was detected in 12% of the houses with a hot-water storage tank and water recirculation (Mathys *et al.* 2008). Systems with a water temperature below 46°C were most frequently colonized, but not newly constructed systems (<2 year). A survey in the Netherlands, including a total of 400 single-family homes in four different water supply areas, yielded 4% *Legionella*-positive systems (>100 CFU/litre). All isolates were identified as *L. anisa* (Oesterholt and van der Kooij 2003). Leoni *et al.* (2005) found 22% of systems in Italy *L. pneumophila* positive and 8.5% with *L.* non-*pneumophila*. Two LD cases were related to *L. pneumophila* in the potable-water system of two apartments in a newly built block of flats in Denmark (Krøjgaard *et al.* 2011). The hot water temperature was below 50°C and *Legionella* was cultured in high numbers (>10^5 CFU/litre) from the tap water. In the Netherlands, *L. pneumophila* was observed in the tap water of 7%

Table 6.2 *Legionellae* in domestic water systems.

Country/region	Period of investigation	Residence type	Number positive/tested	Range (CFU/litre)	LD	Environmental condition promoting occurrence	Ref. *
US	1981	ns**	1/5 (20%)	nr**	nr	ns	1
US/ Chicago	1985	apartment	30/95 (32%)	$1–10^4$	no	temperature <60°C	2
US/Pittsburgh	1988	ns	6/55 (11%)	nr	no	temperature	3
Canada	1991	ns	69/178 (37%)	ns	no	age of house; electric heater	4
Canada	1991	ns	0/33 (0%)	ns	no	gas heating	4
US/Pittsburgh	1992	apartment	14/218 (6%)	nr	no	temperature, Fe	5
Germany	1993	apartment	11/17 (65%)	ns	nr	nr	6
Finland	1994	apartment	20/67 (30%)	3×10^3– 3.5×10^5	no	organic carbon, Ca, microbial biomass, HPC##	7
Denmark	2002	apartment	41/49 (89%)	$>5 \times 10^4$	no	nr	8
the Netherlands	2003	family homes	16/400 (4%)	$<10^2–10^4$	no	variety of conditions	9
Italy	2005	apartments	18/52 (30.5%)	$25–9.8 \times 10^4$	no	centralized hot water system	10
Germany	2008	family homes	0/52 (0%)	–	no	nr	11
Germany	2008	family homes	12/400 (12%)	$390–10^5$	no	hot water circulation	11
the Netherlands	2002–2009	not specified	139/693 (20%)#	nr	yes#	ns	12
Denmark	2008–2009	apartment	2/2 (100%)	5.5×10^4– 6×10^5	yes#	Newly built installation	13

* References, 1, Wadowsky *et al.* (1982); 2, Arnow *et al.* (1985); 3, Lee *et al.* (1988); 4, Alary and Joly (1991); 5, Stout *et al.* (1992); 6, Lück *et al.* (1993); 7, Zacheus and Martikainen (1994); 8, Pringler *et al.* (2002); 9, Oesterholt and van der Kooij (2003); 10, Leoni *et al.* (2005); 11, Mathys *et al.* (2008); 12, Euser *et al.* (2012); 13, Krøjgaard *et al.* (2011); ** ns, not specified; nr, not reported: # see text; ## HPC, heterotrophic plate count.

of 693 homes of persons with LD in 2002 to 2010 and *L.* non-*pneumophila* in 13% (Euser *et al.* 2012). The residential water system was identified as the source of infection for a total of eight cases of LD (Euser *et al.* 2011).

The presented information shows that (i) colonization by *L. pneumophila* has been observed in substantial percentages of residential water systems in Europe and the US, (ii) a correlation with hot-water temperature was found in a number of studies, (iii) residential systems in apartment buildings seem more frequently colonized than family homes, (iv) no association between the presence of *L. pneumophila* and cases of LD was noticed in surveys, and (v) the water system in the home may serve as a source of infection for sporadic cases of community-acquired LD.

6.4.3 Critical concentration

Showers and faucets can generate *Legionella*-containing aerosols (e.g. Bollin *et al.* 1985). Schoen and Ashbolt (2011) developed a quantitative model for *Legionella* exposure during 15 minutes of showering and derived levels, necessary for infection, of 3.5×10^6–3.5×10^8 CFU/litre, with 3.5×10^7 CFU/litre as best estimate. These levels correspond with an infectious dose of 1–100 CFU, deposited in the alveolar region. Reported concentrations in water from potable-water systems generally are below 10^5 CFU/litre and less frequently between 10^5–5×10^5 CFU/litre (Table 6.2). Levels of 10^5–10^6 CFU/litre, associated with LD were observed in first flush samples of tap water in residential systems in Demark (Krøgaard *et al.* 2011). Levels $>10^6$ CFU/litre have been reported, for example 5×10^6 CFU/litre following a pressure shock in a potable-water system in a hospital with nosocomial LD in the US (Shands *et al.* 1985), and in one of 46 investigated hot water systems in Germany (Lück *et al.* 1993) and 10^5–10^6 CFU/litre in first flush samples of tap water in residential systems, associated with LD (Krøjgaard *et al.* 2011). Peak concentrations of 1.3×10^6 to 2.7×10^7 CFU/litre were observed in a model tap water system, after heat flushing followed by recirculation of the water (van der Kooij *et al.* 2005). In short, the *Legionella* concentrations in water, necessary for infection and derived by modelling, are much higher than the concentrations generally observed in hot water systems, but are close to values observed in exceptional situations. However, disease may require a higher dose than infection only. Hence, verification of the model is needed, for example, the conditions affecting the partitioning coefficient, and also the transmission of *Legionella* present in protozoa (Brieland *et al.* 1997) and vesicles, expelled by host protozoa (Berk *et al.* 1998), should be considered.

6.4.4 Disinfectant residual

Drinking water is distributed with a disinfectant residual in most countries, but this is not common practice in a few European countries (e.g. Germany and the

Netherlands). Many supplies in the US have converted from using chlorine to using monochloramine to comply with regulatory standards for disinfection by-products, including trihalomethanes (Kool *et al.* 1999b). Due to the lesser reactivity of monochloramine as compared to chlorine, this disinfectant is more stable in distribution systems and more effective in biofilms and sediments (LeChevallier *et al.* 1988). Quantification of the effect of chlorine and monochloramine on *L. pneumophila* suspended in tap water, showed a 99% reduction after 60 to 90 min at 0.25 mg of free residual Cl_2/litre (Kuchta *et al.* 1985) and after 15 min at 1 mg of monochloramine (Cunliffe 1990). From these observations a Ct_{99} of 15 to 20 mg.min/litre can be derived, that is, a residual of 0.1 mg/litre gives 99% inactivation after several hours.

In a study on nosocomial cases of LD, the presence of *L. pneumophila* in the hot-water system of 15 hospitals was assessed in relation to the type of disinfectant used by the water supply company (Kool *et al.* 1999a). *L. pneumophila* was not observed in four hospitals supplied with drinking water containing monochloramine, whereas the systems of 11 hospitals supplied with drinking water with a free chlorine residual all contained *L. pneumophila* and other *Legionella* spp. A case-control study was conducted across the US, including 32 hospitals with a history of reported nosocomial LD cases and 50 selected hospitals with no reported cases, to further investigate the effect of the residual disinfectant (Kool *et al.* 1999b). The disinfectant residual in the supplied drinking water ranged from 0 to 1.8 (median 0.55) mg $litre^{-1}$ free chlorine and 1.5–4.3 (median 2.7) mg/litre monochloramine. Statistical analysis of the observations showed that hospitals with LD cases were significantly more frequently supplied with drinking water containing free chlorine than monochloramine, although cases had been observed in one hospital supplied with water containing monochloramine. From these results it was concluded that the use of monochloramine instead of free chlorine would give a 90% reduction of nosocomial LD outbreaks (Kool *et al.* 1999b).

Conversion from free chlorine to monochloramine in a distribution system in Florida reduced the percentage of buildings colonized by *L. pneumophila* from 19.8% to 6.2% (Moore *et al.* 2006). Monochloramine was detected in 88% of samples from water heaters, whereas free chlorine was not detectable here or at the point of use. Similar observations were made in a study conducted in San Francisco, showing 60% colonized hot-water systems before and 4% colonization after conversion to monochloramine at a concentration of 0.15–3.4 (median 1.97) mg/litre (Flannery *et al.* 2006). These studies demonstrate that (i) a chlorine residual in drinking water is not effective against *Legionella* in hot-water systems and (ii) use of monochloramine in drinking-water distribution systems reduced colonization by *L. pneumophila* in nearly all potable-water systems.

The effect of free chlorine and monochloramine on *L. pneumophila* in a biofilm has been studied in a disk reactor, with surfaces of stainless steel colonized by a *Flavobacterium* sp, *Pseudomonas aeruginosa*, *Klebsiella pneumoniae*, *L. pneumophila* and *H. vermiformis* (Donlan *et al.* 2005). Exposure of the colonized disks to 0.5 mg/litre chlorine and 0.5 mg/litre monochloramine during 3 hrs had much less effect on biofilm-associated *L. pneumophila* in the presence of *H. vermiformis* than in the absence of this host (Figure 6.3). Obviously, the amoebae protected the organism against the biocidal properties of chlorine and monochloramine during the test period. However, the observations in distribution systems, at a higher concentration of monochloramine and a longer time of exposure, show that this disinfectant effectively limits multiplication of *L. pneumophila*.

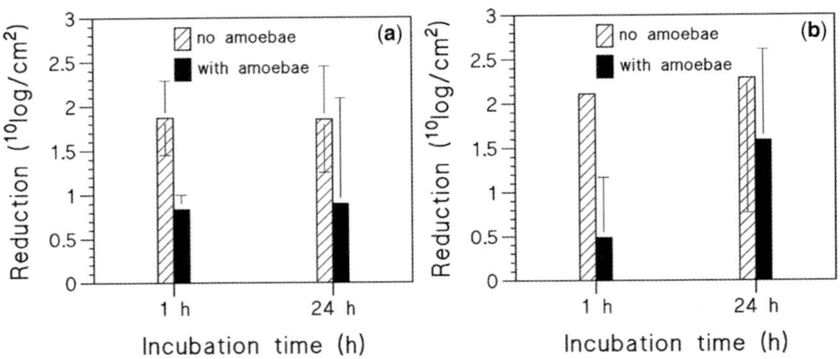

Figure 6.3 (a) Impact of free chlorine (0.5 mg/litre) and (b) monochloramine (0.5 mg/litre) on *L. pneumophila* in a biofilm on coupons of stainless steel in the presence and absence of *H. vermiformis* after 1 h and 24 h of exposure. (From Donlan *et al.* 2005, adapted.)

In another study, with a model system consisting of vertically-placed glass columns (i.d. 2.5 cm) containing segments of soft PVC tubing (Ø 1.6 cm, L 1.5–1.75 cm)), tap water (pH 7.9–8.1) of 37°C was circulated at 360 litre/h (0.2 m/s in empty column) with 15 litre/h replacement. In this system, a biofilm (c. 10.000 pg adenosine triphosphate [ATP]/cm^2) with *L. pneumophila* (c. 10^7 CFU/cm^2) developed within three weeks (Verhamme 2005). Subsequently, monochloramine, chlorine and chlorine dioxide were added to the water entering the system, in association with monitoring of the residual concentration in the water from the system. The disinfectants reduced the concentration of *L. pneumophila* in the water from c. 2×10^4 CFU/litre to below 1 CFU/litre within 1 h, but inactivation of *L. pneumophila* in the biofilm was less effective (Table 6.3). Monochloramine (0.35 mg/litre) and free chlorine (0.3 mg/litre) caused a larger reduction of *L. pneumophila* in the biofilm (with amoebae) on soft PVC than in the biofilm with amoebae on the steel disk (Figure 6.2). This difference may be due to the difference in flow rate, that is, 0.2 m/s vs. no flow with the disks. The effect of monochloramine at 1.8 mg/litre is consistent with the absence of *L. pneumophila* in potable-water systems after conversion from chlorine to monochloramine. However, the relatively high biofilm concentration, and its young age may have affected the level of reduction of *L. pneumophila* in the biofilms.

Table 6.3 Effect of monochloramine, free chlorine and chlorine dioxide in tap water at 37°C on the concentration of *L. pneumophila* in water and biofilms grown on soft PVC in a model system.

Type and concentration of disinfectant	Reduction (^{10}log *units*) of *L. pneumophila*		
	water	biofilm	
	1 h	4 h	24 h
Monochloramine			
0.35 mg/litre	>4	2	3
1.8 mg/litre	>4	>5	>5
Free chlorine			
0.3 mg/litre	>4	0.5	2
2 mg/litre	>4	2.5	4
Chlorine dioxide			
0.2 mg/litre	>4	<0.5	2
1.1 mg/litre	>4	2	>2

Source: Verhamme (2005).

From these experiments, the following estimations for Ct_{99} values were derived: monochloramine 70–90 mg.min/litre; Cl_2 150–250 mg.min/litre and ClO_2 300–500 mg · min/litre. These values are clearly higher than the Ct_{99} values for planktonic cells, confirming that biofilms protect against disinfectants. This protection was more pronounced for chlorine and chlorine dioxide than for monochloramine. Still, maintaining a disinfectant residual in the water ultimately will result in inactivation of the organism in the water, and, as the result of the long contact time, in the biofilm and sediments as well. However, the reactivity of free chlorine, chlorine dioxide, and to a lesser extent monochloramine, with natural organic matter (NOM), biofilm, and corrosion products in particular (LeChevallier *et al.* 1993), reduces the effect at a low residual. Another drawback is the regrowth-enhancing microbial decay of monochloramine at warm temperatures (Wilczak *et al.* 1996).

6.4.5 Plumbing materials

Enhancement of microbial growth by materials in contact with drinking water has been observed since the beginning of the 20th century (Schoenen and Schöler

1983). In 1984, Colbourne *et al.* (1984) reported that rubber materials in a plumbing system promoted growth of *L. pneumophila*. A study with a model hot-water system confirmed the stimulation of *L. pneumophila* colonisation by rubber components (Schofield *et al.* 1985). Natural isoprene elastomer and a range of synthetic rubber materials also promoted growth of *L. pneumophila* when incubated in a batch test at 37°C (Niedeveld *et al.* 1986). Schoenen *et al.* (1988) observed elevated concentrations of *L. pneumophila* (>10^3 CFU/ml) in water of 35–40°C flowing through small diameter tubes of plasticized PVC, polyethylene (PE), polytetrafluorethylene, natural rubber and silicone during a test period of 8 months. Under the same conditions, no growth of *L. pneumophila* was observed after contact with glass, stainless steel (SS), copper and polyamide. Enhanced growth, as compared to glass, was also observed on natural rubber (latex), ethylene-propylene, mild steel, chlorinated polyvinylchloride (PVC-C) and polypropylene (PP) in a two-stage chemostat at 30°C (Rogers *et al.* 1994b).

Assessment of the Legionella Growth Potential (LGP, CFU/cm^2) of a variety of materials, incubated in tap water at 25°C in the Biomass Production Potential (BPP) test, also demonstrated that most synthetic materials tested promoted growth of *L. pneumophila* (Figure 6.4). LGP correlated significantly with the Biomass Production Potential (BPP, pg ATP/cm^2) of the materials (van der Kooij *et al.* 2002).

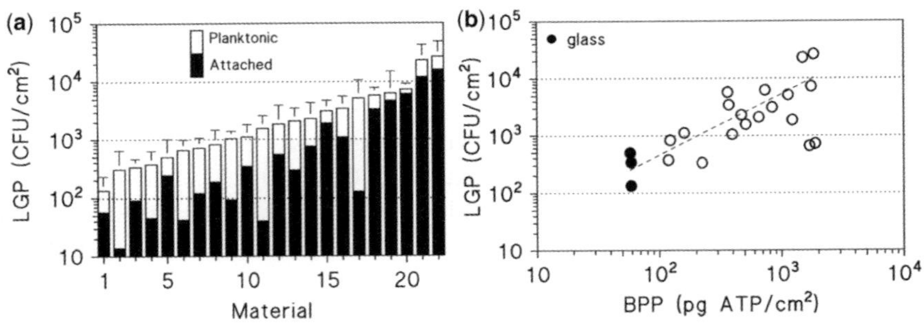

Figure 6.4 (a) Legionella Growth Potential (LGP, CFU/cm^2) of materials incubated in tap water at 25°C in a batch test at a surface to volume ration of 0.16 cm^2/cm^3. LGP is calculated from the *Legionella* concentrations on material pieces and in the water on days 56, 84 and 112, respectively. The upper limit of the standard deviation is indicated. Materials: 1, 2, glass; 3, PVC-C; 4, SS; 5, glass; 6, polyethylene cross-linked by the peroxide method (PE-Xa); 7, Silicone; 8, SS; 9, Copper; 10, SS; 11, polyethylene cross-linked by electron-beam processing (PE-Xc); 12, Silicone; 13, Copper; 14, PP; 15, polybutylene (PB); 16, PVC-C; 17, PE-Xc; 18, PB; 19, PP; 20, Silicone; 21, PE-Xa; 22, polyethylene cross-linked by silane (PE-Xb). (b) relationship between LGP and the Biomass Production Potential (BPP, pg ATP/cm^2) of the material. BPP is the average of the amount of active biomass (ATP) on the material and in the water at days 56, 84 and 112, respectively. *Source*: van der Kooij *et al.* 2002, with permission.

A study using a model hot-water system with pipes of copper, stainless steel (SS) and PE-Xc (polyethylene structurally cross-linked by electron-beam processing) with simulated domestic use of water at about 36°C, showed that *L. pneumophila* multiplied to a level (90 percentile) of 2.5×10^3 (copper); 8.2×10^3 (SS) and 1.5×10^4 (PE-Xc) CFU/cm^2, corresponding with 10^4-2.6×10^4 CFU/litre (van der Kooij *et al.* 2005). Under these conditions, the concentration of *L. pneumophila* in water from the PE-Xc pipes was not significantly higher than those from the pipes of stainless steel and copper. Hence, the microbial growth potential of the water used in this study was the main cause of the growth of *L. pneumophila.*

These observations demonstrate that a large variety of materials used in plumbing systems can enhance growth of *L. pneumophila*. This enhancement of growth most likely is associated with the release of biodegradable compounds, promoting biofilm formation and multiplication of protozoa that serve as hosts for *L. pneumophila*. Biofilm formation is also affected by the microbial growth potential of the water.

6.5 CONTROL MEASURES IN POTABLE-WATER SYSTEMS

6.5.1 Regulations, standards and guidelines

Water quality in potable-water installations must comply with the drinking-water quality standards defined in national drinking-water regulations. Compliance with these standards is the responsibility of the owner or contractor of the installation. In the EU water intended for human consumption 'shall be wholesome and clean if it is free from any micro-organisms and parasites and from any substance, which, in numbers or concentrations, constitute a potential danger to human health' (European Commission [EC] 1998). This principle is the basis of defining safe drinking water world-wide (World Health Organization [WHO] 2004a). Health-related drinking-water quality standards are defined for microorganisms of faecal origin (indicator bacteria) and organic and inorganic contaminants. A few countries in Europe have also defined standards for *Legionella* in potable-water installations, for example, <100 CFU/litre in the UK (Health and Safety Executive [HSE] 2000) and the Netherlands (Ministerie van Infrastructuur en Milieu [IM] 2011a, b) and <1000 CFU/litre in France (Ministère de la Santé et des Sports [MSS] 2010) and Germany (Bundesgesundheitsambt [BGA] 2011). In Flanders, attention for LD cases in healthcare-related facilities is prescribed at >1000 CFU/litre and intensified monitoring and control measures are required at >10,000 CFU/litre (Vlaamse Regering 2007). The standards for *Legionella* in potable water are not based on dose-response relationships (QMRA), but primarily aim at limiting the growth of and exposure to *Legionella.*

In the US a maximum contaminant level goal (MCGL) of 0/litre for *Legionella* has been regulated under the surface-water treatment rule, but neither a maximum tolerable concentration (MTC) nor regular monitoring are prescribed

(Environmental Protection Agency [EPA] 2012; CDC 2004). Routine monitoring of healthcare water supplies has been suggested (Yu *et al.* 1987; Lin *et al.* 1998) with 30% of *Legionella*-positive samples as criterion (Stout *et al.* 2007). Environmental investigations are recommended in healthcare facilities for transplant patients and health care facilities with continued nosocomial transmission (CDC 2004).

To facilitate compliance with drinking water quality standards in potable- water installations, guidance (Codes of Practice, or Building Codes) had been developed for the design, operation and maintenance of these systems. The observations of *Legionella* in potable-water systems in association with LD cases initiated adaptations of these guidance documents. As is shown in Sections 6.3 and 6.4, the proliferation of *L. pneumophila* depends on water temperature and the presence of biofilms and sediments. Hydraulic and operational conditions (e.g. retention time, periodic stagnation) and design (e.g. proximity of cold and warm water pipes, dead ends) affect water temperature, the formation of biofilms and sediments and the interactions between attached and planktonic cells. Therefore, a systematic analysis of the installation (risk assessment) is required to identify growth-promoting conditions and define control measures. Such an analysis generally includes a description of the installation, registration of the number of aerosol-generating devices, temperature of hot and cold water, dead ends, disinfectant residual, and so on. In the Netherlands potable-water installations in hospitals and other health care premises, hotels, bathing and swimming facilities, jails, asylum seekers' centres, etc., have been targeted for regulatory required risk assessment and monitoring (IM 2011a, b). Similar approaches are included in regulations in a number of other European countries, for example, Belgium (Vlaamse Regering 2007), France (MSS 2005), Germany (BGA 2011) and the UK (HSE 2000). In the US, also remedial action (superheating or hyperchlorination) is prescribed when cases of LD are diagnosed. (CDC 2004).

Preventive measures to control *Legionella* in potable-water installations include: (i) water temperature management, (ii) limitation of the formation of biofilms and sediments, (iii) hydraulic conditions to reduce stagnancy and obstruction and, (iv) techniques for elimination or inactivation of *L. pneumophila*. An overview of measures, aiming at preventing growth of and exposure to *Legionella* in potable-water systems, based on the Water Safety Plan approach, is given by the WHO (Bartram *et al.* 2007). Detailed guidance on risk assessment, control measures and monitoring is provided in regulations and standards in a number of European countries, for example, Germany (Deutscher Verein des Gas- und Wasserfaches [DVGW] 2004), the Netherlands (Instituut voor Studie en Stimulering van Onderzoek op het gebied van gebouwinstallaties [ISSO] 2005; IM 2011b) and the UK (HSE 2000). In the Netherlands, Codes of Practice have been developed as well for domestic installations and potable-water installations which are not included in regulations (ISSO 2003). Guidelines for control measures in potable-water installations in the US are described in American

Society of Heating, Refrigerating and Air-Conditioning Engineers (ASHRAE) Standards Guideline 12–2000 (ASHRAE 2000) and in the draft Standard 188P (ASHRAE 2011). Brief descriptions of the most commonly applied control measures are given below.

6.5.2 Water temperature management

L. pneumophila does not proliferate in aquatic habitats at a temperature ≤20°C and grows slowly at temperatures between 20 and 25°C (Section 6.3.2). Maintaining the cold-water temperature in potable-water installations at <25°C is only possible in temperate regions. *L. pneumophila* multiplies at temperatures up to 45°C but a clearly higher temperature is needed to achieve effective periodic thermal inactivation of the organism (Figure 6.2), for example, following a decrease of water temperature during stagnancy. Generally, a hot water temperature of ≥60°C is prescribed (ASHRAE 2000; DVGW 2004; IM 2012b) to prevent growth and ensure inactivation. According to ASHRAE (2011) the temperature should not be lower than 51°C at the coldest spot in the hot water system. Hence, maintaining a cold-water temperature at <25°C and a hot water temperature at >60°C is the basis of the *L. pneumophila* control policy in potable-water systems in buildings. This approach requires a series of measures to prevent cold water from increasing in temperature and to prevent hot water from decreasing in temperature, for example, during stagnation. Mixing water of <25°C with hot water (>50°C) at the point of use in a thermostatic device will prevent growth of *L. pneumophila*.

Hot-water flushing (>60°C) can be used as an emergency treatment. The effect is only temporary (Zacheus and Martikainen 1996) and periodic application is not attractive because it is laborious and requires safety measures to prevent scalding (Lin *et al.* 2011). Furthermore, this measure is not applicable in the cold water pipes of the premise-plumbing system.

6.5.3 Disinfection technologies

A number of disinfection techniques are available to control growth of *L. pneumophila* in hot-water installations. Some of these techniques may have a negative impact on water quality, for example, formation of disinfection by-products, and therefore, the application of such equipment requires approval. Disinfection-based methodologies to limit proliferation of *L. pneumophila* have been reviewed recently (Lin *et al.* 2011). Copper-silver ionization was found to be highly effective in most cases, provided that certain levels of copper (0.4 mg/litre) and silver (0.04 mg/litre) are maintained throughout the installation. These levels are below the concentrations of copper and silver included in drinking-water standards. An advantage of the method is the continuing inactivation by copper and silver at times when the installation is out of order and copper-silver ionization on a discontinuous basis has been suggested (Liu *et al.* 1998). The use of copper-silver ionisation is regulated in several European countries, for example, UK and the Netherlands. Application generally requires approval of the type of installation and a test, demonstrating the efficacy of the equipment and compliance with drinking-water standards. In the US no recommendations have been made regarding the structural use of copper-silver ionisation or another disinfection technique in potable water systems (CDC 2004) and application requires registration as a biocide for use in potable water (Lin *et al.* 2011). Recent cases of LD in a hospital with copper-silver ionisation revived the debate in the US about the best control method and monitoring for *Legionella* (Hamill 2012).

Anodic oxidation, with locally generated free chlorine from chloride in (or added to) the water, is allowed in the Netherlands, Belgium and the UK. Chlorine dioxide is another effective disinfectant, but a relatively long period may be needed after installation before the entire water system is free of *Legionella* (Srinivasan *et al.* 2003; Zhang *et al.* 2009; Lin *et al.* 2011). Rapid regrowth of *Legionella* when dosage is (temporarily) terminated may be a drawback of these techniques.

Physical disinfection technologies, apart from temperature, include UV disinfection and membrane filtration. UV disinfection effectively reduces the concentration of *Legionella* in water (Muraca *et al.* 1987) but has no effect on biofilms and sediments in the water system (Liu *et al.* 1995; Pozos *et al.* 2004), thus limiting its general application. Membranes (0.2 μm) at the point-of-use (e.g. near showerheads), particularly in hospitals (intensive-care units), are highly effective, not only to *L. pneumophila* but also to other opportunistic pathogens (Sheffer *et al.* 2005).

6.5.4 Limiting accumulation of biofilms and sediments

The accumulation of biofilms and sediments in potable-water systems depends on water composition, the type(s) of construction materials in contact with water and hydraulic conditions. Elevated concentrations of organic and certain inorganic compounds, including iron, calcium and magnesium, can promote growth of *L. pneumophila* in the installation (see Section 6.4). These compounds may enhance biofilm formation and/or sediment accumulation, but lack of quantitative data on these effects hampers the definition of guidelines for the involved parameters. Reducing the microbial growth potential of water may reduce the biofilm formation, but *L. pneumophila* can proliferate at very low biofilm concentrations (van der Kooij *et al.* 2005). Accumulation of sediments in the installation can be limited by proper hydraulic design and maintenance. EC regulations demand that construction materials in contact with drinking water should comply with specifications (EC 2002). Methods for testing the microbial growth enhancement of materials in contact with drinking water have been standardized in the UK, Germany and the Netherlands, but these methods differ from each other, as do pass-fail criteria (see Chapter 12).

6.6 DISCUSSION AND PERSPECTIVE

6.6.1 True incidence of LD

In the past decade the notification rate of LD in Europe and the US increased as did the knowledge about the properties of *L. pneumophila* and related species and the occurrence of these organisms in man-made water systems. This seems contradictory but the introduction of more effective diagnostic tools, for example, urine antigen (UAg) testing and PCR methods has facilitated diagnosis of LD. In the past 10 years, UAg has become the most frequently applied tool for diagnosis of LD in Europe and the US. This test effectively detects serious cases of LD caused by *L. pneumophila* SG1 (Helbig *et al.* 2003). Still, the detection of mild cases of LD and/or cases caused by other serogroups and other species is complicated, also because of the limitations of the culture methods. A more complete and rapid diagnosis can be achieved by the inclusion of PCR-based methods (Diederen *et al.* 2005; von Baum *et al.* 2008; Mentasi *et al.* 2012). Improvements in reporting and surveillance can contribute to a further increase of the notification rate as well.

Surveillance studies among hospitalized CAP patients yielded LD rates of 3% (n = 2776) in the US (Marston *et al.* 1997) and in the UK (n = 267) (Lim *et al.* 2001) and 3.8% (n = 1727) to 3.9% (n = 5238) in Germany (von Baum *et al.* 2008; von Baum and Lück 2011). By adjustment for age, sex and race, the total annual number of LP cases in the US was estimated at 8000 to 18,000 cases (Marston *et al.* 1997), that is, an overall incidence of 30 to 67 per million. These estimations are 2.6 to 5.8 times higher than the crude notification rate of LD in 2009, despite the observed increase in the past 10 years (Figure 6.1). The total annual number of hospitalised CAP cases has been estimated at about 16,000 (1/1000) in the Netherlands (Health Council 2003) and at 2/1000 in England (Trotter *et al.* 2008). A 4% proportion of LD in hospitalised CAP corresponds with overall LD incidences of 40/million in the Netherlands and 80/million in the UK, respectively. These estimates are 1.4× (the Netherlands) and 13× (UK) higher than the crude notification rates in these countries in 2010. The true incidence of LD differs from these estimations because the true LD percentage may be higher than those observed in surveys, for example, due to shortcomings in the diagnostic methods, demographic factors such as distribution of age and sex. Moreover, LD may occur in nonhospitalized CAP patients as well. In Germany, the percentage of LD diagnosed in nonhospitalized CAP patients (3.7%) was similar to the percentage in hospitalized CAP patients (von Baum *et al.* 2008). The number of nonhospitalized CAP cases is about 3 to 5 times higher than the number of hospitalized cases (Guest and Morris 1997; Marston *et al.* 1997; Brandsema *et al.* 2011). LD at a similar incidence in nonhospitalized CAP patients gives a 3 to 5 times higher incidence than those based on hospitalized CAP cases, although most of the additional cases do not require admission (von Baum *et al.* 2008). However, *Legionella* was not detected (by

PCR) in samples collected by general practitioners from the nose and throat of 230 persons with acute respiratory illness and 200 controls in the Netherlands (Diederen *et al.* 2005). Data about LD in nonhospitalized cases of CAP in other countries seem to be lacking.

The presented data demonstrate that the true LD incidence is unknown and most likely (much) higher than the notification rate. Assessment of the true LD incidence would require the use of optimized diagnostic methods in further surveys among CAP patients, both hospitalized and nonhospitalized. Finally, an increase of the true LD incidence over time cannot be excluded, because of: (i) the increasing proportions of elderly and immunocompromised persons in the populations in Europe and the US, (ii) an increase of travel to countries where control of *Legionella* is not yet implemented; (iii) an increase of the number of sources of infection due to intensified use of warm water, and (iv) an increase of the environmental temperature, due to climate change, leading to a more intensive use of cooling towers.

Overall, the notification rates and estimations of the incidence of LD demonstrate that the disease is not under control, despite the knowledge of the conditions favouring growth of *L. pneumophila*. The continuation of outbreaks, mainly caused by wet cooling towers, and clusters of travel-associated cases (ECDC 2012) show that effective implementation of control measures is complicated. Furthermore, most cases are sporadic and community-acquired without identified source of infection.

6.6.2 Water-supply associated LD

The number of *L. pneumophila* in drinking water in temperate climates is below the detection limit in the large majority of the samples (Section 6.4.1), but mains water has been identified as a source of dissemination of legionellae to potable-water systems. The role of drinking-water supply in LD therefore seems mainly related to the use of heated mains water in potable-water systems in hospitals and other healthcare premises, hotels, bathing and swimming facilities, residential water installations and other water systems with aerosol-generating devices. Travel-associated cases constitute a relatively large proportion of the reported LD cases in Europe and the US, and range from 0 to 100% in the European countries with 30% in Denmark and Germany, 31% in the Netherlands and 34% in the UK, respectively (ECDC 2012). In the afore-mentioned countries, the proportion of cases associated with travel abroad constitutes >80% of the travel-associated cases. In contrast, more than 85% of the travel-associated case in France, Italy and Spain are related to domestic travel. These data are consistent with results of surveys showing that the potable-water system of many hotels is colonized by *L. pneumophila* and other *Legionella* species (Section 6.4.2).

The proportion of nosocomial LD cases plus other healthcare-associated cases in Europe has decreased from about 30% to about 8%, although the number of notified nosocomial cases remained at the same level (Joseph *et al.* 2010; ECDC 2012). The notified cases related to domestic travel and healthcare premises account for 0.7 (UK) to 6.2 (France) cases/million (Table 6.4), but it is not clear to what extent these rates represent the notified LD cases related to water supply in these countries. Firstly, nosocomial cases may not all be potable-water related; cases associated with wet cooling towers near hospitals have been reported (O'Mahony *et al.* 1990). Secondly, travel-associated cases may not all be related to potable-water systems. However, an unknown proportion of sporadic community-associated cases may be related to potable-water usage. In the Netherlands, a total of 8 LD cases have been traced to the domestic system in a period of about 10 years (Euser *et al.* 2011), that is, an average yearly rate of about 0.05/million. In the US *Legionella* was responsible for 24 (28.6%) of the 84 drinking-water related outbreaks in 2001–2006, with 126 cases (12 deaths) of LD (Craun *et al.* 2010). This number corresponds with a yearly rate of approx. 0.07/million. These low rates may explain that LD cases were not observed in connection with surveys of residential systems, despite the detection of *L. pneumophila* (Table 6.2).

Table 6.4 Number and incidence (cases per million) of notified LD cases related with potable-water systems in a few countries in Europe in 2010 (data from ECDC 2012).

Country	Category				
	Nosocomial	Other healthcare	Domestic travel	Total domestic*	Travel abroad
Denmark	6 (1.1)	4 (0.7)	3 (0.5)	13 (2.4)	34 (6.2)
France	101 (1.56)	81 (1.25)	222 (3.43)	404 (6.2)	41 (0.7)
Italy	64 (1.06)	41 (0.68)	111 (1.84)	216 (3.6)	14 (0.2)
Spain	31 (0.67)	0 (0)	137 (2.98)	168 (3.7)	15 (0.3)
The Netherlands	2 (0.12)	7 (0.42)	26 (1.56)	35 (2.1)	144 (8.7)
UK	19 (0.31)	0(0)	26 (0.42)	45 (0.7)	129 (2.1)

*nosocomial plus other healthcare plus domestic travel.

In the Netherlands, an infection risk of $<10^{-4}$/year, associated with faecal pathogens, has been defined as a criterion for evaluating surface-water treatment (Ministerie van Volkshuisvesting, Ruimtelijke Ordening en Milieubeheer [VROM] 2001; IM 2011a). The proportion of exposed persons with an increased antibody titre in the 'flower-show' outbreak of LD in the Netherlands was estimated at 20 to 60× the LD rate (Nagelkerke *et al.* 2003). From this ratio, it can be derived that an annual infection risk of $<10^{-4}$ corresponds with an annual LD rate of <1.7/million to <5/million. In several countries, these levels are exceeded by the notification rate related to domestic travel or travel abroad (Table 6.4). The (assumed) potable-water related notification rate of LD in 2010 (ECDC 2012), due to exposure within the country, is above the level of 1.6/million in five of the countries listed, and above 5/million in France. The notification rate related to residential water systems in private homes and apartments is estimated at 0.05 LD/million in the Netherlands and 0.07 LD/million in the US. Another reference value for evaluating the LD rate is the 10^{-5} excess death risk over a life span (70 years). At a case fatality ratio for LD of 10%, the potable-water associated mortality rates derived from the notified number of cases are: 4.3×10^{-5} (France); 2.6×10^{-5} (Spain), 2.5×10^{-5} (Italy), 1.7×10^{-5} (Denmark), 1.5×10^{-5} (the Netherlands) and 4.9×10^{-6} (UK), that is, mostly above the mentioned reference level of 10^{-5}. In a few countries, this level was also exceeded in association with travel abroad. Assessment of the impact of LD on the basis of disability-adjusted life years (DALY) approach would also enable an evaluation in comparison with the burden of disease (WHO 2004a). A DALY value has not (yet) been defined for LD, however, it would be relatively high as the result of the severity of the disease and the high mortality risk.

The estimations described above indicate that the health impact of potable-water-associated cases of LD is relatively high in comparison with reference risk values. A true incidence rate higher than the notified rate would give higher risk levels. Prevention of the colonisation of potable-water systems by *L. pneumophila* therefore requires adequate control measures and continuous vigilance.

6.6.3 Drinking-water quality

Studies in the US show that monochloramine is more effective than free chlorine in limiting colonization of potable water systems by *L. pneumophila* and reducing the number of LD cases in healthcare premises (Section 6.4.4). The policy of the water supplies in the Netherlands is to prevent microbial growth in the distribution system by distributing biologically stable drinking water (van der Kooij *et al.* 1999). The issue of applying post disinfection emerged in the Netherlands after the 'flower-show' outbreak of LD in 1999 and the observations of *L. pneumophila* in potable-water systems thereafter. However, the Health Council of the Netherlands concluded that post disinfection is not the appropriate measure to control growth of *L. pneumophila* in potable-water systems in the Netherlands (Health Council 2003). The main arguments were (i) prevention of multiplication of *L. pneumophila* in these systems is the responsibility of the owner of the system and (ii) adding a chemical disinfectant to finished water would impair drinking water quality. The drawbacks of using monochloramine include (i) the (potential for) formation of undesirable by-products, (ii) decay resulting in the growth of nitrifying bacteria and (iii) toxic properties of monochloramine itself (WHO 2004b). Also *Mycobacterium* spp. have been observed more frequently (42.2%

vs. 19.1%) in the water systems after the introduction of monochloramine (Pryor *et al.* 2004). These disadvantages may be less relevant when the disinfectant is needed to prevent faecal contamination and regrowth (e.g. coliforms) in distributed water (LeChevallier 1999).

Distribution of biologically stable water effectively prevents high HPCs and growth of coliforms, but noncultivable, mostly yet undefined, *Legionella* spp. have been observed in all water supplies by using PCR-based methods (Wullings *et al.* 2006). Hence, even drinking water with a high degree of biostability has a potential for growth of *Legionella* and a water temperature of <25°C is a precondition to prevent growth of *L. pneumophila* in drinking water distributed without disinfectant residual.

6.6.4 Temperature and virulence

LD is most frequently caused by *L. pneumophila* but large differences in virulence have been demonstrated between *L. pneumophila* strains, that is, serogroups and sequence types (Section 6.2.2). Generally, *L. pneumophila* is not cultured from water at <25°C and *L.* non-*pneumophila* spp. seem to predominate at temperatures up to 30°C (Section 6.3.2). It is tempting to hypothesize that the proportion of virulent genotypes of *L. pneumophila* is highest at a water temperature ≥37°C. The consequence of this hypothesis is that exposure to aerosols of water from hot water systems in temperate climates (and also cooling towers) poses a higher health risk than exposure to water at 30°C in the tropics. The lack of information about the occurrence of the most virulent genotypes of *L. pneumophila* in the environment may be related to rapid growth at a local, and/or temporarily increased, temperature of 37–42°C, the optimal temperature for growth of *L. pneumophila*, followed by washout after temperature decline. Further research should therefore focus on the effect of temperature and other environmental conditions, for example, concentration and composition of biofilms and sediments, including the presence of free-living protozoa, on the proliferation of virulent *L. pneumophila* SG1, MAb3/1 +ve strains, for example, STs 23, 47, 62. The development of PCR-based methods to detect these organisms may be helpful in identifying the heath risks associated with growth of *L. pneumophila* in water-supply related systems and other man-made water installations.

6.6.5 Perspective

The presented data show that LD is a serious public health threat related to aerosol-generating man-made water systems. The true incidence of LD is unknown and is (much) higher than the notification rate. Still, implementation of measures to limit growth of *L. pneumophila* in potable-water systems, and in cooling towers as well, is a slow process. Improved surveillance may lead to more pressure on the implementation of control measures, but most likely outbreaks and clusters of cases will continue to play a major role in such implementation. The use of PCR-based methods for LD diagnosis, the detection of virulent *L. pneumophila* and the elucidation of the environmental niches of these organisms, may enhance the development and implementation of effective control methods.

Acknowledgements

The author is much indebted to Carol A. Joseph and Ans F. M. Versteegh for valuable comments made on the draft of this chapter.

6.7 REFERENCES

Alary M. and Joly J. R. (1991). Risk factors for contamination of domestic hot water systems by Legionellae. *Appl. Environ. Microbiol.*, **57**(8), 2360–2367.

Anand C. M., Skinner A. R., Malic A. and Kurtz J. B. (1983). Interaction of *L. pneumophila* and a free-living amoeba (*Acanthamoeba palestinensis*). *J. Hyg. Camb.*, **91**, 167–178.

Armstrong T. W. and Haas C. N. (2005). Quantitative microbial risk assessment model for Legionella: summary of methods and results. In: Legionella, state of the art 30 years after its recognition, N. P. Cianciotto et al. (eds), ASM Press, Washington DC, US, pp. 486–488.

Armstrong T. W. and Haas C. N. (2008). Legionnaires' disease: evaluation of a quantitative microbial risk assessment model. *J. Wat. Health*, **6**(2), 149–165.

Arnow P. M., Weil D. and Para M. F. (1985). Prevalence and significance of *Legionella pneumophila* contamination of residential hot-tap water systems. *J. Infect. Dis.*, **152**(1), 145–151.

Arvand M., Jungkind K. and Hack A. (2011). Contamination of the cold water distribution system of health care facilities by *Legionella pneumophila*: do we know the true dimension? *Euro Surveill.*, **16**(16), 19844.

ASHRAE (2000). Minimizing the risk of legionellosis associated building water systems. ASHRAE Guideline 12-2000, American Society of Heating, Refrigerating and Air–Conditioning Engineers, Inc. Atlanta GA.

ASHRAE (2011). Proposed new standard 188, prevention of legionellosis associated with building water systems. BSR/ASHRAE Standard 188P, American Society of Heating, Refrigerating and Air–Conditioning Engineers, Inc. Atlanta GA.

Bargellini A., Marchesi I., Right E., Ferrari A., Cencetti S., Borella P. and Rovesti S. (2011). Parameters predictive of *Legionella* contamination in hot water systems: association with trace elements and heterotrophic plate counts. *Water Res.*, **45**, 2315–2321.

Bartram J., Chartier Y., Lee J. V., Pond K. and Surman-Lee S. (2007). Legionella and the prevention of legionellosis. World Health Organization, Geneva, Switzerland.

Baskerville A., Broster M., Fitzgeorge R. B., Hambleton P. and Dennis P. J. (1981). Experimental transmission of Legionnaires' disease by exposure to aerosols of *Legionella pneumophila*. *Lancet*, December 19/26, 1389–1390.

Berendt R. F., Young H. W., Allen R. G. and Knutsen G. L. (1980). Dose-response of guinea pigs experimentally infected with aerosols of *Legionella pneumophila*. *J. Infect. Dis.*, **141**(2), 186–192.

Berg J. D., Hoff J. C., Roberts P. V. and Matin A. (1985). Growth of *Legionella pneumophila* in continuous culture. *Appl. Environ. Microbiol.*, **49**(6), 1534–1537.

Berk S. G., Ting R. S., Turner G. W. and Ashburn R. J. (1998). Production of respirable vesicles containing live *Legionella pneumophila* cells by two *Acanthamoeba* spp. *Appl. Environ. Microbiol.*, **64**(1), 279–286.

Bezanson G., Burbridge S., Haldane D., Yoell C. and Marrie T. (1992). Diverse populations of *Legionella pneumophila* present in the water of geographically clustered institutions served by the same water reservoir. *J. Clin. Microbiol.*, **30**(3), 570–576.

BGA (2011). Erste Verordnung zur Änderung der Trinkwasserverordnung vom 3. Mai 2011. *Bundesgesundheitsblatt 2011*, **1**(21), Bonn.

Bijkerk P., van Lier E. A., Swaan C. M. and Kretzschmar M. E. E. (2011). Staat van infectieziekten in Nederland 2010. Rapport 21022110072011 Rijksinstituut voor Volksgezondheid en Milieu Bilthoven, the Netherlands.

Bollin G. E., Plouffe J. E., Para M. F. and Hackman B. (1985). Aerosols containing *Legionella pneumophila* generated by shower heads and hot-water faucets. *Appl. Environ. Microbiol.*, **50**(5), 1128–1132.

Borchard J., Helbig J. H. and Luck P. C. (2008). Occurrence and distribution of sequence types among *Legionella pneumophila* strains isolated from patients in Germany: common features and differences to other regions in the world. *Eur. J. Clin. Infect. Dis.*, **27**, 29–36.

Borella P., Montagna M. T., Stampi S., Stancanelli G., Romao-Spica V., Triassi M., Marchesi I., Bargellini A., Tatò D., Napoli C., Zanetti F., Leoni E., Moro M., Scaltriti S., D'Alcala G. R., Santarpia R. and Boccia S. (2005). *Legionella* contamination in hot water of Italian hotels. *Appl. Environ. Microbiol.*, **71**(10), 5805–5813.

Bornstein N., Marmet D., Surgot M., Nowicki M., Arslan A., Esteve J. and Fleurette J. (1989). Exposure to Legionellaceae at a hot spring spa: a prospective clinical and serological study. *Epidemiol. Inf.*, **102**, 31–36.

Brambilla E., Hippe H., Hagelstein A., Tindall B. J. and Stackebrandt E. (2001). 16SrDNA diversity of cultured and uncultured prokaryotes of a mat sample from Lake Fryxell, McMurdo Dry Valleys, Antarctica. *Extremophiles*, 23–33.

Brandsema P. S., Dijkstra F., Euser S. M., van Gageldonk-Lafeber A. B., de Lange M. M. A., Meijer A., Slump E., Snijders B. E. P. and van der Hoek W. (2011). Jaarrapportage surveillance respiratoire infectieziekten 2010. Projectgroep respiratoire infecties RIVM rapport 210231008201 1. Bilthoven, the Netherlands.

Brieland J. K., Fantone J. C., Remick D. G., LeGendre M., McClain M. and Engleburg N. C. (1997). The role of *Legionella pneumophila*-infected Hartmannella vermiformis as an infectious particle in a murine model. *Infect. Immun.*, **65**(12), 5330–5333.

Brenner D. J., Steigerwald A. G. and McDade J. E. . (1979). Classification of the Legionnaires' disease bacterium: *Legionella pneumophila*, genus novum, species nova, of the family *Legionellaceae*, family nova. *Ann. Internal Med.*, **90**, 656–658.

Brown A., Yu V. L., Magnussen M. H., Vickers R. M., Garrity G. M. and Elder E. M. (1982). Isolation of Pittsburgh pneumonia agent from a hospital shower. *Appl. Environ. Microbiol.*, **43**(3), 725–726.

Buse H. Y. and Ashbolt N. J. (2011). Differential growth of *Legionella pneumophila* strains within a range of amoebae at various temperatures associated with in-premise plumbing. *Let. Appl. Microbiol.*, **53**, 217–224.

Calvo-Bado L. A., Morgan J. A. W., Sergeant M., Pettitt T. R. and Whipps J. M. (2003). Molecular characterization of *Legionella* populations present within slow sand filters used for fungal plant pathogen suppression in horticulture crops. *Appl. Environ. Microbiol.*, **69**(1), 533–541.

Carvalho F. R. S., Nastasi F. R., Gamba R. C., Foronda A. S. and Pellizari V. H. (2008). Occurrence and diversity of Legionellaceae in polar lakes of the Antarctic peninsula. *Curr. Microbiol.*, **57**, 294–300.

CBC News (2012). Quebec legionnaires' outbreak found. CBC News Canada, 19 September 2012.

CDC (2003). Summary of notifiable diseases—United States, 2001. MMWR *Morb. Mortal. Weekly Report.*, **50**(53), 90.

CDC (2004). Guidelines for preventing health-care-associated pneumonia, 2003. Recommendations of CDC and the Healthcare Infection Control Practices Advisory Committee. *MMWR Morb. Mortal. Weekly Report*, **53**(RR-3), 1–36.

CDC (2012). Summary of notifiable diseases-United States, 2010. *MMWR Morb. Mortal. Weekly Report*, **59**(53), June 1, 2012.

Chasqueira M. J., Rodrigues L., Nascimento M. and Marques T. (2009). Sequence-based and monoclonal antibody typing of *Legionella pneumophila* isolated from patients during 1987–2008. *Euro Surveill.*, **14**(28), 1–4.

Ciesielski C., Blaser M. J. and Wang W. L. L. (1984). Role of stagnation and obstruction of water flow in isolation of *Legionella pneumophila* from hospital plumbing. *Appl. Environ. Microbiol.*, **48**(5), 984–987.

Colbourne J. S. and Trew R. M. (1985). Presence of *Legionella* in London's water supplies. *Isr. J. Med. Sci.*, **22**, 633–639.

Colbourne J. S., Pratt D. J., Smith M. G., Fisher-Hoch S. P. and Harper D. (1984). Water fittings as sources of *Legionnella pneumophila* in a hospital plumbing system. *Lancet*, January i, 210–213.

Cordes L. G., Fraser D. W., Skaliy P., Perlindo C. A., Elsea W. R., Mallison G. F. and Hayes P. S. (1980). Legionnaires' disease outbreak at an Atlanta, Georgia, country club: evidence for spread from an evaporative condenser. *Am. J. Epidemiol.*, **111**(4), 425–431.

Cordes L. G., Wiesenthal A. M., Gorman G. W., Phair J. P., Sommers H. M., Brown A., Yu V. L., Magnussen M. H., Meyer R. D., Wolf J. S., Shands K. N. and Fraser D. W. (1981). Isolation of *Legionella pneumophila* from hospital shower heads. *Ann. Internal Med.*, **94**, 195–197.

Costa J., Tiago I., da Costa M. S. and Verissimo A. (2005). Presence and persistence of *Legionella* spp. in groundwater. *Appl. Environ. Microbiol.*, **71**(2), 663–671.

Craun G. F., Brunkard J. M., Yoder J. S., Roberts V. A., Carpenter J., Wade T., Calderon R. L., Roberts J. M., Beach M. J. and Roy S. L. (2010). Causes of outbreaks associated with drinking water in the United States from 1971 to 2006. *Clin. Microbiol. Rev.*, **23**(3), 507–528.

Cunliffe D. A. (1990). Inactivation of *Legionella pneumophila* by monochloramine. *J. Appl. Bacteriol.*, **68**, 453–459.

Darelid J., Bernander S. S. and Löfgren S. S. (2004). The presence of a specific genotype of *Legionella pneumophila* serogroup 1 in a hospital and municipal water distribution system over a 12-year period. *Scand. J. Infect. Dis.*, **36**(6–7), 417–423.

Den Boer J. W., Yzerman E. P. F., Schellekens J., Lettinga K. D., Boshuizen H. C., van Steembergen J. E., Bosman A., van den Hof S., van Vliet H. A., Peeters M. C., van Ketel R. J., Speelman P., Kool J. L. and Conyn-van Spaendonck M. A. E. (2002). A large outbreak of Legionnaires' disease at a flower show, The Netherlands, 1999. *Emerg. Infect. Dis.*, **8**(1), 37–43.

Den Boer J. W., Bruin J. P., Verhoef L. P. B., Van der Zwaluw K., Jansen R. and Yzerman E. P. F. (2008). Genotype comparison of clinical *Legionella* isolates and patient-related environmental isolates in The Netherlands. *Clin. Microbiol. Infect.*, **14**, 459–466.

Dennis P. J., Fitzgeorge R. B., Taylor J. A., Bartlett C. L. R. and Barrow G. I. (1982). *Legionella pneumophila* in water plumbing systems. *Lancet*, April 24, 949–951.

Dennis P. J., Green D. and Jones B. P. C. (1984). A note on the temperature tolerance of *Legionella*. *J. Appl. Bacteriol.*, **56**, 349–350.

Desplaces N., Nahapetian K. and Dournon E. (1984). Inventaire des Legionella dans l'environnement; implications practiques. *La Presse Medicale*, **13**(3), 1875–1879.

Diederen B. M. W., de Jong C. M. A., Aarts I., Peeters M. F., van Gageldonk,-Lafeber A. B., Wilbrink B. and van der Zee A. (2005). No evidence of *Legionella* species infection in general practice patients presented with acute respiratory infections in the Netherlands. *Clin. Microbiol. Infect.*, **11**, 401–412.

Diederen B. M. W., de Jong C. M. A., Aarts I., Peeters M. F. and van der Zee A. (2007). Molecular evidence for the ubiquitous presence of *Legionella* species in Dutch tap water installations. *J. Wat. Health*, **5**(3), 375–383.

Doleans A., Aurell H., Reyrolle M., Lina G., Freney J., Vandenesch F., Etienne J. and Jarraud S. (2004). Clinical and environmental distributions of *Legionella* strains in France are different. *J. Clin. Microbiol.*, **42**(1), 458–460.

Dondero T. J., Rendtorff R. C., Mallison G. F., Weeks R. M., Levy J. S., Wong E. W. and Schaffner W. (1980). An outbreak of Legionnaires' disease associated with a contaminated air-conditioning cooling tower. *N. Eng. J. Med.*, **302**(7), 365–370.

Donlan R. M., Forster T., Murga R., Brown E., Lucas C., Carpenter J. and Fields B. (2005). *Legionella pneumophila* associated with the protozoan *Hartmannella vermiformis* in a model multi-species biofilm has reduced susceptibility to disinfectants. *Biofouling*, **21**(1), 1–7.

DSMZ (2012). Leibnitz Institut DSMZ-Deutsche Sammlung von Mikroorganismen und ZellkulturenGmbH. http://old.dsmz.de/microorganisms/bacterial_nomenclature_info.php?genus=Legionella.

DVGW (2004) Arbeitsblatt W 551: Trinkwasserverwarmings- und Trinkwasserleitungs-anlagen; Technische Massnamen zur Verminderung des Legionellenwachstums; Planung, Errichtung, Betreib und Sanierung von Trinkwasser-Installations. Regelwerk des Deutscher Verein des Gas-und Wasserfachs e.V., http://www.dvgw.de/.

EC (1998). Council Directive 98/83/EC of November 1998 on the quality of water intended for human consumption. *Off. J. Eur. Comm.*, **L330**, 32–54.

EC (2002). Commission decision of 13 May 2002 on the procedure for attesting the conformity of construction products in contact with water intended for human consumption, pursuant to Article 20(2) of Council Directive 89/106/EEC. *Off. J. Eur. Comm.*, **L127**, 16–18.

ECDC (2011) *Legionnaires' disease in Europe, 2009; Surveillance report*, European Centre for Disease Prevention and Control, Stockholm.

ECDC (2012) *Legionnaires' disease in Europe, 2010; Surveillance report*, European Centre for Disease Prevention and Control, Stockholm.

Edelstein P. H. (1981). Improved semiselective medium for isolation of *Legionella pneumophila* from contaminated clinical and environmental specimens. *J. Clin. Microbiol.*, **14**(3), 298–303.

EPA (2012) 2012 Edition of the drinking water standards and health advisories. Report EPA 822-S-12-001, Office of Water, U.S. Environmental protection Agency, Washington DC, US.

Euser S. M., Brandsema P., Ruijs W. M. L. and den Boer J. W. (2011). *Legionella* bronopsporing in Nederland 2009–2010. *Infectieziekten Bulletin*, **22**(5), 160–163.

Euser S. M., Bruin J. P., van der Hoek W., Schop W. A. and Boer J. W. (2012). Wellness centres: an important but overlooked source of Legionnaires' disease. Eight years of source investigation in the Netherlands. 1 August 2002-1 August 2010. *Euro Surveill.*, **17**(8), 20097. Available online: http:/www.eurosurveillance.org/View Article.aspx? Articleid=20097

Fenstersheib M. D., Miller M., Diggings C., Liska S., Detwiler L., Werner S. B., Lindquist D., Thacker W. L. and Benson R. F. (1990). Outbreak of Pontiac fever due to *Legionella anisa*. *Lancet*, **330**, 35–37.

Fields B. S., Shotts E. B., Feeley J. C., Gorman G. W. and Martin W. T. (1984). Proliferation of *Legionella pneumophila* as an intracellular parasite of the ciliated protozoan *Tetrahymena pyriformis*. *Appl. Environ. Microbiol.*, **47**(3), 467–471.

Fields B., Barbaree J., Shotts E., Feeley J., Morilll W., Sanden G. and Dykstra M. (1986). Comparison of guinea pig and protozoan models for determining virulence of *Legionella* species. *Infect. Immun.*, **53**(3), 553–559.

Fields B., Barbaree J., Sanden G. and Morrill W. (1990). Virulence of a *Legionella anisa* strain associated with Pontiac fever: an evaluation using protozoan, cell culture and guinea pig models. *Infect. Immun.*, **58**(9), 3139–3142.

Fields B. S., Benson R. F. and Besser R. (2002). Legionella and Legionnaires' disease: 25 years of investigation. *Clin. Microbiol. Rev.*, **15**(3), 506–526.

Fisher-Hoch S. P., Smith M. G. and Colbourne J. S. (1982). *Legionella pneumophila* in hospital hot water cylinders. *Lancet*, May, 1073.

Fisman D. N., Lim S., Wellenius G. A., Johnson C., Britz P., Gaskins M., Maher J., Mittleman A., Spain C. V., Haas C. N. and Newbern C. (2005). It's not the heat, it's the humidity: wet weather increases legionellosis risk in the greater Philadelphia metropolitan area. *JID*, **192**, 2066–2072.

Fliermans C. B., Cherry W. B., Orrison L. H., Smith S. J., Tison D. L. and Pope D. H. (1981). Ecological distribution of *Legionella pneumophila*. *Appl. Environ. Microbiol.*, **41**(1), 9–16.

Flannery B., Gelling L. B., Vugia D. J., Weintraub J. M., Salerno J. J., Conroy M. J., Stevens V. A., Rose C. E., Moore M. R., Fields B. S. and Besser R. R. (2006). Reducing *Legionella* colonization of water systems with monochloramine. *Emerg. Infect. Dis.*, **12**(4), 588–596.

Franzus M., Malcolm G. B. and Pine L. (1984). Taxonomic evaluation of amino acid metabolism in *Legionella*. *Curr. Microbiol.*, **11**, 73–80.

Fraser D. W., Tsai T. R., Orenstein W., Parkin W. E., Beecham H. J., Sharrar R. G., Harris J., Mallison G. F., Martin S. M., McDade J. E., Shepard C. C., Brachman P. S. the field investigation team (1977). Legionnaires' disease; description of an epidemic of pneumonia. *N. Eng. J. Med.*, **297**(22), 1189–1197.

Gaia V., Fry N. K., Afshar B., Luck P. C., Meugnier H., Etienne J., Peduzzi R. and Harrison T. G. (2005). Consensus sequence-based scheme for epidemiological typing of clinical and environmental isolates of *Legionella pneumophila*. *J. Clin. Microbiol.*, **43**(5), 2047–2052.

Garcia-Fulgeras A., Navarro C., Fenoll D., Garcia J., Gonzalez-Diego P., Pacheco T., Ruiz J., Segovia M., Baladron B. and Pelaz C. (2003). Legionnaires' disease outbreak in Murcia, Spain. *Emerg. Infect. Dis.*, **9**(8), 915–921.

George J. R., Pine L., Reeves M. W. and Harrell W. K. (1980). Amino acid requirements of *Legionella pneumophila*. *J. Clin. Microbiol.*, **11**(3), 286–291.

Groothuis D. G., Veenendaal H. R. and Dijkstra H. L. (1985). Influence of temperature on the number of *Legionella pneumophila* in hot water systems. *J. Appl. Bacteriol.*, **59**, 529–536.

Guest J. F. and Morris A. (1997). Community-acquired pneumonia: the annual cost to the National Health Service in the UK. *Eur. Respir. J.*, **10**, 1530–1534.

Habicht W. and Muller H. E. (1988). Occurrence and parameters of frequency of *Legionella* in warm water systems of hospitals and hotels in Lower Saxony. *Zbl. Backt. Hyg.*, **B 186**, 79–88.

Hagele S., Kohler R., Merkert H., Sleicher M., Hacker J. and Steinert M. (2000). *Dictyostelium discoidum*: a new host model system for intracellular pathogens of the genus *Legionella*. *Cell Microbiol.*, **2**, 165–171.

Hamill S. D. (2012). Legionnaires' prevention under dispute. Pittsburgh Post Gazette. December 24, 2012. Pittsburgh, US.

Harrison T. G., Asfar B., Doshi N. and Fry N. K. (2009). Distribution of *Legionella pneumophila* serogroups, monoclonal antibody subgroups and DNA sequence types in recent clinical and environmental isolates from England and Wales (2000–2008). *Eur. J. Clin. Microbiol. Infect. Dis.*, **28**, 781–791.

Health Council (2003) *Controlling legionnaire's disease* (in Dutch), Publication no. 2003/12. Health Council of the Netherlands, The Hague.

Helbig J. H., Bernander S., Castellani Pastoris M., Etienne J., Gaia V., Lauwers S., Lindsay D., Lück P. C., Marques T., Mentula S., Peeters M. F., Pelaz C., Struelens M., Uldum S. A., Wewalka G. and Harrison T. G. (2002). Pan-European study on culture-proven Legionnaires' disease: distribution of Legionella pneumophila serogroups and monoclonal subgroups. *Eur. J. Clin. Microbiol. Infect. Dis.*, **21**, 710–716.

Helbig J. H., Uldum S. A., Bernander S., Luck P. C., Wewalka G., Abraham B., Gaia V. and Harrison T. G. (2003). Clinical utility of urinary antigen detection for diagnosis of community-acquired, travel-associated and nosocomial Legionnaires' disease. *J. Clin. Microbiol.*, **41**(2), 838–840.

Heudorf U., Hentschel W., Hoffmann M., Luck C. and Schubert R. (2001). Legionella im hauseigenen Warmwasser – Auswirkungen auf die Gesundheit der Bewohner. *Gesundheitswezen*, **63**, 326–334.

Hicks L. A., Rose C. E., Fields B. S., Drees M. L., Engel J. P., Jenkins P. R., Rouse B. S., Blythe D., Khalifah A. P., Feikin D. R. and White C. C. (2007). Increased rainfall is associated with increased risk for legionellosis. *Epidemiol. Infect.*, **135**, 811–817.

Hicks L. A., Garrison L. E., Nelson G. E. and Hampton L. M. (2011). Legionellosis – United States, 2000–2009. *Morbid. Mortal. Weekly Report*, **60**, 1083–1086.

Hoebe C. J. P., Cluitmans J. J. M. and Wagenvoort J. H. T. (1998). Two fatal cases of nosocomial *Legionella pneumophila* pneumonia associated with a contaminated cold water supply. Brief report. *Eur. J. Clin. Microbiol. Infect. Dis.*, **17**, 740–749.

Holden E. P., Winkler H. H., Wood D. O. and Leinbach E. D. (1984). Intracellular growth of *Legionella pneumophila* within *Acanthamoeba castellanii* Neff. *Infect. Immun.*, **45**(1), 18–24.

HSE (2000). The control of legionella bacteria in water systems. Health and Safety Commission L8 – Legionnaire's disease. Approved Code of Practice and Guidance. Health and Safety Executive, London, UK.

Hsu S. C., Martin R. and Wentworth B. B. (1984). Isolation of *Legionella* species from drinking water. *Appl. Environ. Microbiol.*, **48**(4), 830–832.

IM (2011a). Besluit van 23 mei 2011, houdende bepaling inzake de productie en distributie van drinkwater en de organisatie van de openbare drinkwatervoorziening (drinkwaterbesluit). *Staatsblad*, **293**, 1–90. Ministerie van Infrastructuur en Milieu, Den Haag, Nederland.

IM (2011b). Regeling van de Staatssecretaris van Infrastructuur en Milieu van 27 juni 2011, nr. BJZ2011046957, houdende nadere regels met betrekking tot enige onderwerpen inzake legionellapreventie in drinkwater en warm tapwater (Regeling legionellapreventie in drinkwater en warm tapwater). *Staatscourant* 10828.

ISSO (2003). Legionella Code for Domestic Installations (in Dutch). Report ISSO/ Uneto-VNI richtlijn 30.5. ISSO, Rotterdam.

ISSO (2005). Manual Legionella Prevention in Potable Water (in Dutch). Report ISSO 55.1, ISSO, Rotterdam.

James B. W., Mauchline W. S., Dennis P. J., Keevil C. W. and Wait R. (1999). Poly-3-hydroxybutyrate in *Legionella pneumophila*, an energy source for survival in low-nutrient environments. *Appl. Environ. Microbiol.*, **65**(2), 822–827.

Joly J. R., McKinney R. M., Tobin J. O., Bibb W. F., Watkins I. D. and Ramsay D. (1986). Development of a standardized subgrouping scheme for *Legionella pneumophila* serogroup 1 using monoclonal antibodies. *J. Clin. Microbiol.*, **23**, 768–771.

Jones F. and Ashcroft C. (1988). Survey to detect *Legionella pneumophila* in potable waters in North West England. *J. IWEM*, **2**, 460–464.

Jones T. F., Benson R. F., Brown E. W., Rowland J. R., Crosier S. C. and Schaffner W. (2003). Epidemiologic investigation of a restaurant-associated outbreak of Pontiac fever. *Clin. Infect. Dis.*, **37**, 1291–1297.

Joseph C. A. and Ricketts D. (2010). Legionnaires' disease in Europe 2007–2008. Surveillance and outbreak reports. *Euro Surveill.*, **15**(8), 1–8.

Karagiannis I., Brandsema P. and van der Sande M. (2009). Warm wet weather associated with increased Legionnaires' disease incidence in The Netherlands. *Epidemiol. Infect.*, **137**(2), 181–187.

Kool J. L., Bergmire-Sweet D., Butler J. C., Brown E. W., Peabody D. J., Massi D. S., Carpenter J. C., Pruckler J. M., Benson R. F. and Fields B. S. (1999a). Hospital characteristics associated with colonization of water systems by *Legionella* and risk of nosocomial Legionnaires' disease: a cohort study of 15 hospitals. *Infect. Control Hosp. Epidemiol.*, **20**, 798–805.

Kool J. L., Carpenter J. C. and Fields B. S. (1999b). Effect of monochloramine disinfection of municipal drinking water on the risk of nosocomial Legionnaires' disease. *Lancet*, **353**, 272–277.

Kozak N. A., Benson R. F., Brown E., Alexander N. T., Taylor T. H., Shelton B. G. and Fields B. S. (2009). Distribution of lag-1 alleles and sequence-based types among *Legionella pneumophila* serogroup 1 clinical and environmental isolates in the United States. *J. Clin. Microbiol.*, **47**(8), 2525–2535.

Krøjgaard L. H., Krogfelt K. A., Albrechtsen H. J. and Uldum S. A. (2011). Cluster of Legionnaires' disease in a newly built block of flats, Denmark, December 2008 – January 2009. *Euro Surveill.*, **16**(1), 11–17.

Kuchta J. M., States S. J., McGlaughlin J. E., Overmeyer J. H., Wadowsky R. M., McNamara A. A., Wolford R. S. and Yee R. B. (1985). Enhanced chlorine resistance of tap water adapted *Legionella pneumophila* as compared with agar medium-passaged strains. *Appl. Environ. Microbiol.*, **50**(1), 21–26.

Kuiper M. W., Wullings B. A., Akkermans A. D. L., Beumer R. and van der Kooij D. (2004). Intracellular proliferation of *Legionella pneumophila* in *Hartmannella vermiformis* in aquatic biofilms grown on plasticized polyvinyl chloride. *Appl. Environ. Microbiol.*, **70**(11), 6826–6833.

Kusnetsov J. M., Ottoila E. and Martikainen P. J. (1996). Growth, respiration and survival of *Legionella pneumophila* at high temperatures. *J. Appl. Bacteriol.*, **81**, 341–347.

Lawrence C., Reyrolle M., Dubrou S., Forey F., Decludt B., Goulvestre C., Matsiota-Bernards P., Etienne J. and Nauciel C. (1999). Single clonal origin of a high proportion of Legionella pneumophila serogroup 1 isolates from patients and the environment in the area of Paris, France, over a 10-year period. *J. Clin. Microbiol.*, **37**(8), 2652–2655.

LeChevallier M. W. (1999). The case for maintaining a disinfectant residual. *J. Am. Wat. Works Assoc.*, **91**, 86–94.

LeChevallier M. W., Cawthon C. D. and Lee R. G. (1988). Inactivation of biofilm bacteria. *Appl. Environ. Microbiol.*, **54**(10), 2492–2499.

LeChevallier M. W., Lowry C. D., Lee R. G. and Gibbon D. L. (1993). Examining the relationship between iron corrosion and the disinfection of biofilm bacteria. *J. Am. Wat. Works Assoc.*, **85**, 111–123.

Lee T. C., Vickers R. M., Yu V. L. and Wagener M. M. (1993). Growth of 28 Legionella species on selective culture media: a comparative study. *J. Clin. Microbiol.*, **31**(10), 2764–2768.

Lee H. K., Shim J. I., Kim H. E., Yu J. Y. and Kang Y. H. (2010). Distribution of *Legionella* species from environmental water sources of public facilities and genetic diversity of *L. pneumophila* serogroup 1 in South Korea. *Appl. Environ. Microbiol.*, **76**(19), 6547–6554.

Leoni E., DeLuca G., Legnant P. P., Sacchetti R., Stampi S. and Zanetti F. (2005). *Legionella* waterline colonization: detection of *Legionella* species in domestic, hotel and hospital hot water systems. *J. Appl. Microbiol.*, **98**, 373–379.

Lim W. S., MacFarlane J. T., Boswell T. C. J., Harrison T. G., Rose D., Leinonen M. and Saikku P. (2001). Study of community acquired pneumonia aetiology in adults admitted to hospital: implications for management guidelines. *Thorax*, **56**, 296–301.

Lin Y. E., Vidic R. D., Stout J. E. and Yu V. L. (1998). Legionella in water distribution systems. *J. Am. Water Works Assoc.*, **90**(9), 112–121.

Lin Y. E., Stout J. E. and Yu VF.L. (2011). Controlling *Legionella* in hospital drinking water supplies: an evidence based review of disinfection methods. *Infect. Control Hosp. Epidemiol.*, **32**(2), 166–173.

Liu Z., Stout J. E., Tedesco L., Boldin M., Hwang C. and Yu V. L. (1995). Efficacy of ultraviolet light in preventing *Legionella* colonization of a hospital water distribution system. *Water Res.*, **29**(10), 2275–2280.

Liu Z., Stout J. E., Boldin M., Rugh J., Diven W. E. and Yu V. L. (1998). Intermittent use of copper-silver ionisation for *Legionella* control in water distribution systems: a potential option in buildings housing individuals at low risk of infection. *Clin. Infect. Dis.*, **26**, 138–140.

Lück P. C., Leupold I., Hlawitschka M., Helbig J. H., Carmienke I., Jatzwauk L. and Guderitz T. (1993). Prevalence of Legionella species, serogroups and monoclonal antibody subgroups in hot water systems in south-eastern Germany. *Zbl. Hyg.*, **193**, 450–460.

Maini R., Naik F., Harrisson T. G., Mentasi M., Spala G., Velonakis E., Hadjichristodoulou C., de Jong B., Vatapoulos A. and Phin N. (2012). Travel-associated Legionnaires' disease in residents from England and Wales travelling to Corfu, Greece, August to October 2011. *Euro Surveill.*, **17**(32), 20240. Available online: http://www.eurosurveillance.org/ViewArticle.aspx?Articled=20240.

Marrao G., Verissimo A., Bowker R. G. and da Costa M. S. (1993). Biofilms as major sources of *Legionella* spp. in hydrothermal areas and their dispersion into stream water. *FEMS Microbiol. Ecol.*, **12**, 25–33.

Marston B., Plouffe J. E., File T. M., Hackman B. A., Salstrom S. J., Lipman H. B., Kolczak M. S., Breiman R. F. for the Community based pneumonia-based incidence group (1997). Incidence of community-acquired pneumonia requiring hospitalization. *Arch. Intern. Med.*, **157**, 1799–1718.

Mathys W., Stanke J., Harmuth M. and Junge-Mathys E. (2008). Occurrence of *Legionnella* in hot water systems of single-family residences in suburbs of two German cities with special reference to solar and district heating. *Int. J. Hyg. Environ. Health*, **211**, 179–185.

Mauchline W. S. and Keevil C. W. (1991). Development of the BIOLOG substrate utilization system for identification of *Legionella* spp. *Appl. Environ. Microbiol.*, **57**(11), 3345–3349.

McCormick D., Thorn S., Milne D., Evans C., Stevenson J., Llano M., Donaghy M. on behalf of the Incident Management Team (2012). Public health response to an outbreak of Legionnaires' disease in Edinburgh, United Kingdom, June 2012. *Euro Surveill.*, **17** (28), 1–4.

McDade J. E., Shepard C. C., Fraser D. W., Tsai T. R., Redus M. A., Dowdle W. R. the Laboratory Investigation team (1977). Isolation of a bacterium and demonstration of its role in upper respiratory disease. *N. Eng. J. Med.*, **297**(22), 1197–1203.

Meenhorst P. L., Reingold A. L., Gorman G. W., Feeley J. C., van Cronenburg B. J., Meyer C. L. M. and van Furth R. (1983). *Legionella pneumophila* in guinea pigs exposed to aerosols of concentrated potable water from a hospital with nosocomial Legionnaires' disease. *J. Inf. Diseases*, **147**, 129–132.

Meenhorst P. L., Reingold A. L., Groothuis D. G., Gorman G. W., Wilkinson H. W., McKinney R. M., Feeley J. C., Brenner D. J. and van Furth R. (1985). Water-related nosocomial pneumonia caused by *Legionella pneumophila* serogroups 1 and 10. *J. Infect. Dis.*, **132**(2), 356–364.

Mentasi M., Fry N. K., Afshar B., Palepou-Foxley C., Naik T. G. and Harrison N. C. (2012). Application of *Legionella pneumophila*-specific quantitative real-time PCR combined with direct amplification and sequence-based typing in the diagnosis and epidemiological investigation of Legionnaires' disease. *Eur. J. Clin. Microbiol. Infect. Dis.*, **31**, 2017–2028.

Moore M. R., Pryor M., Fields B., Lucas C., Phelan M. and Besser R. (2006). Introduction of monochloramine into a municipal water system: impact on colonization of buildings by *Legionella* spp. *Appl. Environ. Microbiol.*, **72**(1), 378–383.

Mouchtouri V., Velonakis E., Tsakalov A., Kapoula C., Goutziana G., Vatopoulos A., Kremastinou J. and Hadjichristodoulou C. (2007). Risk factors for contamination of hotel water distribution systems by *Legionella* species. *Appl. Environ. Microbiol.*, **73**(5), 1489–1492.

MSS (2005) *L' eau dans les Établissements de Santé; guide technique*. Ministère de la Santé et des Solidarité, France. (http:/www.sante.gouv.fr/htm/dossiers/eau_etabs/ accueil.hm).

MSS (2010) *Arrêté du 1er février 2010 relatif à la surveillance des légionelles dans les installations de production, de stockage et de distribution d'eau chaude sanitaire*. Ministere de la Santé et des Sports. J. Off. de la Republique Française. Texte 17 sur 122.

Murga R., Forster T. S., Brown E., Prickler J. M., Fields B. S. and Donlan R. M. . (2001). Role of biofilms in the survival of *Legionella pneumophila* in a model potable- water system. *Microbiology*, **147**, 3121–3126.

Nagelkerke N. J. D., Boshuizen H. C., de Melder H. E., Schellekens J. F. P., Peters M. F. and Conyn-van Spaendonk M. (2003). Estimating the incidence of subclinical infections with *Legionella pneumophila* using data augmentation: analysis of an outbreak in The Netherlands. *Statist. Med.*, **22**, 3713–3724.

Neumeister B., Schöniger S., Faigle M., Eichner M. and Dietz K. (1997). Multiplication of different *Legionella* species in Mono Mac 6 cells and in *Acanthamoeba castellani*. *Appl. Environ. Microbiol.*, **63**(4), 1219–1224.

Niedeveld C. J., Pet F. M. and Meenhorst P. L. (1986). Effects of rubbers and their constituents on proliferation of *Legionella pneumophila* in naturally contaminated hot water. *Lancet*, July, 180–184.

Nguyen T. M., Jarraud S., Rouil L., Campese C., Che D., Haeghebaert S., Ganiayre E., Marcel F., Etienne J. and Desenclos J. C. (2006). A community-wide outbreak of legionnaires' disease linked to industrial cooling towers- how far can contaminated aerosols spread? *J. Infect. Dis.*, **193**(1), 102–111.

Nolte F. S., Hollick G. E. and Robertson R. G. (1982). Enzymatic activities of *Legionella pneumophila* and *Legionella*-like organisms. *J. Clin. Microbiol.*, **15**(1), 175–177.

Nygard K., Werner-Johansen Ø., Rønsen S., Caugant D. A., Simonsen Ø., Kanestrom A., Ask E., Ringstad J., Ødegard R., Jensen T., Krogh T., Høiby E. A., Ragnhilstveit E., Aaberge I. S. and Aavitsland P. (2008). An outbreak of Legionnaires' disease caused by long distance spread from an industrial air scrubber in Sarpsborg, Norway. *Clin. Infect. Dis.*, **46**, 61–69.

Oesterholt F. and van der Kooij D. (2003). *Legionella* in Household plumbing systems (in Dutch). *H_2O*, **38**(4), 19–20.

Ohno A., Kato N., Yamada K. and Yamaguchi K. (2003). Factors influencing survival of *Legionella pneumophila* serotype 1 in hot spring water and tap water. *Appl. Environ. Microbiol.*, **69**, 2540–2547.

Ohno A., Kato N., Sakamoto R., Kimura S. and Yamaguchi K. (2008). Temperature dependent parasitic relationship between *Legionella pneumophila* and a free-living amoeba (*Acantanmoeba castellani*). *Appl. Environ. Microbiol.*, **74**(14), 4585–4588.

O'Mahoney M. C., Stanwell-Smith R. E., Tillett H. E., Harper D., Hutchinson J. G. P., Farrell I. D., Hutchinson D. N., Lee J. V., Dennis P. J., Duggal H. V., Scully J. A. and Denne C. (1990). The Stafford outbreak of Legionnaires' disease. *Epidemiol. Infect.*, **104**, 361–380.

Ortiz-Roque C. M. and Hazen T. C. (1987). Abundance and distribution of Legionellaceae in Puerto Rican waters. *Appl. Environ. Microbiol.*, **53**(9), 2231–2236.

Parthuisot N., West N. J., Lebaron P. and Baudari J. (2010). High diversity and abundance of *Legionella* spp. in a pristine river and impact of seasonal and anthropogenic effects. *Appl. Environ. Microbiol.*, **76**(24), 8201–8210.

Paszko-Kolva C., Shahamat M. and Colwell R. R. (1992). Long term survival of *Legionella pneumophila* serogroup 1 under low-nutrient conditions and associated morphological changes. *Microbiol. Ecol.*, **102**, 45–55.

Piao Z., Sze C. C., Barysheva O., Iida K. and Yoshida S. (2006). Temperature-regulated formation of mycelial mat-like biofilms by *Legionella pneumophila*. *Appl. Environ. Microbiol.*, **72**(2), 1613–1622.

Plouffe J. F., Webster L. R. and Hackman B. (1983). Relationship between colonization of hospital buildings with *Legionella pneumophila* and hot water temperatures. *Appl. Environ. Microbiol.*, **46**(1), 769–770.

Pozos N., Scow K., Wuertz S. and Darby J. (2004). UV disinfection in a model distribution system: biofilm growth and microbial community. *Water Res.*, **38**, 3083–3091.

Pringler N., Brydov P. and Uldum S. A. (2002). Occurrence of Legionella in Danish hot water systems. In: Legionella, Marre et al. (ed.), ASM Press, Washington DC, pp. 298–301.

Pryor M., Springthorpe S., Riffard S. and Broks T. (2004). Investigations of opportunistic pathogens in municipal drinking water under different supply and treatment regimes. *Wat. Sci. Technol.*, **50**, 83–90.

Ricketts K. D., Charlett A., Geld D., Lane C., Lee J. V. and Joseph C. A. (2009). Wheater patterns and Legionnaires' disease: a meteorological study. *Epidemiol. Infect.*, **137**, 1003–1012.

Ricketts K. D., Joseph C. A., Lee J. V. and Wilkinson P. (2012). Wet cooling systems as a source of sporadic Legionnaires' disease: a geographical analysis of data for England and Wales, 1996–2006. *J. Epidemiol. Community Health*, **66**, 616–623.

Rogers J., Dowsett A. B., Dennis P. J., Lee J. V. and Keevil C. W. (1994a). Influence of temperature and plumbing material selection on biofilm formation and growth of *Legionella pneumophila* in a model potable water system containing complex microbial flora. *Appl. Environ. Microbiol.*, **60**(5), 1585–1592.

Rogers J., Dowsett A. B., Dennis P. J., Lee J. V. and Keevil C. W. (1994b). Influence of plumbing materials on biofilm formation and growth of *Legionella pneumophila* in potable water systems. *Appl. Environ. Microbiol.*, **60**(6), 1842–1851.

Rohr U., Weber S., Michel R., Selenka F. and Wilhelm M. (1998). Comparison of free-living amoebae in hot water systems of hospitals with isolates from moist sanitary areas by identifying genera and determining temperature tolerance. *Appl. Environ. Microbiol.*, **64**(5), 1822–1824.

Rowbotham T. J. (1980). Preliminary report on the pathogenicity of *Legionella pneumophila* for freshwater and soil amoebae. *J. Clin. Pathol.*, **33**, 1179–1183.

Rowbotham T. J. (1986). Current views on the relationships between amoebae, Legionellae and man. *Isr. J. Med. Sci.*, **22**, 678–689.

Sakamato R., Ohno A., Nakahara T., Satomura K., Iwanaga S., Kouyama Y., Kura F., Kato N., Matsubayashi K., Okumiya K. and Yamaguchi K. (2009). *Legionella penumophila* in rainwater on roads. *Emerg. Infect. Dis.*, **15**(8), 1295–1297.

Sanden G. N., Morrill W. E., Fields B. S., Breiman R. F. and Barbaree J. M. (1992). Incubation of water samples containing amoebae improves detection of Legionellae by the culture method. *Appl. Environ. Microbiol.*, **58**(6), 2001–2004.

Schoen M. E. and Ashbolt N. J. (2011). An in-premise model for Legionella exposure during showering events. *Water Res.*, **45**, 5826–5836.

Schoenen D. and Schöler H. F. (1983). Trinkwasser und Werkstoffe: Praxisbeobachtungen und Untersuchungsverfahren; Praxisbeobachtungen und Untersuchungsverfahren. DVGW-Schriftenreihe, Wasser Nr. 37. Gustav Fischer Verlag, Stuttgart.

Schoenen D., Schulze-Röbbecke R. and Schirdewahn N. (1988). Microbial contamination of water by materials of pipes and hoses. 2nd Communication: growth of *Legionella pneumophila. Zbl. Bakt Hyg.*, **B 186**, 326–332.

Schofield G. M. and Locci R. (1985). Colonization of components of a model hot water system by *Legionella pneumophila. J. Appl. Bacteriol.*, **58**, 151–162.

Schulze-Röbbecke R., Rödder M. and Exner M. (1987). Multiplication and inactivation of naturally occurring Legionellae (in German). *Zbl. Bakt.*, **154**, 495–500.

Seidel K., Bornert W., Baz G., Blankenburg A. and Alexander I. (1986). Presence of *Legionella pneumophila* in ground water, cold and warm drinking waters. *Vom Wasser*, **69**, 39–48.

Shands K., Ho J. L., Meyer R. D., Gorman G. W., Edelstein P. H., Mallison G. F., Finegold S. M. and Fraser D. W. (1985). Potable water as a source of Legionnaires' disease. *JAM*, **253**(10), 1412–1416.

Sheehan K. B., Henson J. M. and Ferris M. J. (2005). Legionella species diversity in an acidic biofilm community in Yellowstone National Park. *Appl. Environ. Microbiol.*, **71**(1), 507–511.

Sheffer P. J., Stout J. E., Wagener M. M. and Muder R. R. (2005). Efficacy of new point-of-use water filter for preventing exposure to *Legionella* and waterborne bacteria. *Am. J. Infect. Control*, **33**, 20–25.

Skaliy P. and McEachern H. V. (1979). Survival of the Legionnaires' disease bacterium in water. *Ann. Internal Med.*, **90**, 662–663.

Srinivasan A., Bova G., Ross T., Mackie K., Paquette N., Merz W. and Perl T. M. (2003). A 17-month evaluation of a chlorine dioxide water treatment system to control *Legionella pneumophila* in a hospital water system. *Infect. Control Hosp. Epidemiol.*, **24**, 575–579.

States S. J., Conley L. F., Ceraso M., Stephenson T., Wolford R. S., Wadowsky R. M., McNamara A. M. and Yee R. B. (1985). Effects of metals on *Legionella pneumophila* growth in drinking water plumbing systems. *Appl. Environ. Microbiol.*, **50** (5), 1149–1154.

States S. J., Conley L. F., Kuchta J. M., Oleck B. M., Lipovich M. J., Wolford R. S., Wadowsky R. M., McNamara A. M., Sykora J. L., Keleti G. and Yee R. B. (1987). Survival and multiplication of *Legionella pneumophila* in municipal drinking water systems. *Appl. Environ. Microbiol.*, **53**(5), 979–986.

States S. J., Podorski J. A., Conley L. F., Young W. D., Wadowsky R. M., Dowling J. N., Kuchta J. M., Navratil J. S. and Yee R. B. (1993). Temperature and the survival and multiplication of *Legionella pneumophila* associated with Hartmannella vermiformis. In: Legionella, current status and emerging perspectives, J. M. Barbaree, R. F. Breiman and A. P. Dufour (eds), ASM, Washington DC, US, pp. 147–149.

Steele T. W., Moore C. V. and Sangster N. (1990). Distribution of *Legionella longbeachae* serogroup 1 and other legionellae in potting soils in Australia. *Appl. Environ. Microbiol.*, **56**, 2984–2988.

Stout J. E., Yu V. L., Vickers R. M., Zuravleff J., Best M., Brown A., Yee R. B. and Wadowsky R. (1982). Ubiquitousness of *Legionella pneumophila* in the water supply of a hospital with endemic legionnaires' disease. *N. Eng. J. Med.*, **306**, 466–468.

Stout J. E., Yu V. L. and Best M. G. (1985). Ecology of *Legionella pneumophila* within water distribution systems. *Appl. Environ. Microbiol.*, **49**(1), 221–228.

Stout J. E., Best M. G. and Yu V. L. (1986). Susceptibility of members of the family Legionellaceae to thermal stress; implications for heat eradication methods in water distribution systems. *Appl. Environ. Microbiol.*, **52**(2), 396–399.

Stout J. E., Yu V. L. and Muraca P. (1987). Legionnaires' disease acquired within the home of two patients. Link to the home water supply. *J. Am. Med. Assoc.*, **257**, 1215–1217.

Stout J. E., Yu V. L., Lee Y. C., Vaccarello S., Diven W. and Lee T. C. (1992a). *Legionella pneumophila* in residential water supplies: environmental surveillance with clinical assessment for Legionnaires' disease. *Epidemiol. Infect.*, **109**, 49–57.

Stout J. E., Yu V. L., Muraca P., Joly J., Troup N. and Tompkins L. S. (1992b). Potable water as a cause of sporadic cases of community-acquired Legionnaires' disease. *N. Eng. J. Med.*, **326**(3), 151–155.

Stout J. E., Muder R. R., Mietzner S., Wagener M. M., Perri M. B., DeRoos K., Goodrich D., Arnold W., Williamson T., Ruark O., Treadway C., Eckstein E. C., Marshall D., Rafferty M. E., Sarro K., Page J., Jenkins R., Oda G., Shimoda K. J., Servos M. J., Bittner M., Camhi S. L., Panwalker A. P., Donskey C. J., Nguyen M. H., Holodniy M., Yu V. L. the Legionella Study Group (2007). Role of environmental surveillance in determining the risk of hospital acquired legionellosis: a national surveillance study with clinical correlations. *Infect. Control Hosp. Epidemiol.*, **28**(7), 818–824.

Tang P. and Krishnan C. (1993). Legionellosis in Ontario, Canada: laboratory aspects, In: Legionella, current status and emerging perspectives, J. M. Barbaree, R. F. Breiman and A. P. Dufour (eds), American Society for Microbiology, Washington DC, pp. 6–17.

Teford D., Partidge S., Cumming I., Smith A. and Calvert N. (2006). The legionnaires' outbreak in Barrow-in Furness, summer 2002. *J. Epidemiol. Community Health*, **60**, 464–466.

Temmerman R., Vervaeren H., Noseda B., Boon N. and Verstraete W. (2006). Necrotrophic growth of *Legionella pneumophila*. *Appl. Environ. Microbiol.*, **72**(6), 4323–4328.

Tijet N., Tang P., Romilowych M., Duncan C., Ng V., Fisman D. N., Jamieson F., Low D. E. and Guyard C. (2010). New endemic *Legionella pneumophila* serogroup 1 clones, Ontario, Canada. *Emerg. Infect. Dis.*, **16**(3), 447–454.

Tison D. L. and Seidler R. (1983). *Legionella* incidence and density in potable drinking water supplies. *Appl. Environ. Microbiol.*, **45**(1), 337–339.

Tison D. L., Pope H., Cherry W. B. and Fliermans C. B. (1980). Growth of *Legionella pneumophila* in association with blue-green algae (cyanobacteria). *Appl. Environ. Microbiol.*, **39**, 456–459.

Tobin J. O'. H., Beare J. and Dunell M. S. (1980). Legionnaires' disease in a transplant unit: isolation of the causative organism from shower baths. *Lancet*, **ii**, 118–121.

Tobin R. S., Ewan P., Walsh K. and Dutka B. (1986). A survey of *Legionella pneumophila* in water in 12 Canadian cities. *Water Res.*, **4**, 495–501.

Trotter C. L., Stuart J. M., George R. and Miller E. (2008). Increasing hospital admissions for pneumonia, England. *Emerg. Infect. Dis.*, **14**(5), 727–733.

Tyndall R. L. and Domingue E. L. (1982). Cocultivation of *Legionella pneumophila* and free-living amoebae. *Appl. Environ. Microbiol.*, **55**(4), 954–959.

Uldum S. and Helbig J. H. (2002). Legionella serogroup and subgroup distribution among patients with Legionnaires' disease in Denmark. In: Legionella R. Marre et al. (eds), ASM Press, Washington DC, pp. 200–203.

Valster R. M. (2011). Free-living Protozoa in Drinking-water Supplies. Community Composition and Their Role as Hosts for Legionella Pneumophila. Thesis, Wageningen University, Wageningen, the Netherlands.

Valster R. M., Wullings B. A. and van der Kooij D. (2010). Detection of protozoan hosts for *Legionella pneumophila* in engineered water systems by using a biofilm batch test. *Appl. Environ. Microbiol.*, **76**(21), 7144–7153.

Valster R. M., Wullings B. A., van den Berg R. and van der Kooij D. (2011). Free-living protozoa and *Legionella* in three Caribbean water supplies. *Appl. Environ. Microbiol.*, **77**(20), 7321–7328.

van den Hoek J. A. R., Yzerman E. P. F. and Coutinho R. A. (2006). Legionella outbreak in Amsterdam: cooling tower as source (in Dutch). *Ned. Tijdschr. Geneeskd*, **150**(33), 1808–1811.

van der Kooij D., van Lieverloo J. H. M., Schellart J. and Hiemstra P. (1999). Maintaining quality without a disinfectant residual. *J. Am. Water Works Assoc.*, **91**(1), 55–64.

van der Kooij D., Veenendaal H. R., Slaats N. P. G. and Vonk D. (2002). Biofilm formation and multiplication of *Legionella* on synthetic pipe materials in contact with treated water under static and dynamic conditions, In: Legionella, R. Marre et al. (eds), ASM Press, Washington DC, pp. 176–180.

van der Kooij D., Veenendaal H. R. and Scheffer W. J. H. (2005). Biofilm formation and multiplication of *Legionella* in a model warm water system with pipes of copper, stainless steel and cross-linked polyethylene. *Water Res.*, **39**, 2789–2798.

Verhamme D. (2005) Effect of Chlorine, Chlorine Dioxide and Monochloramine on Legionella in Biofilms at Continuous Flow *(in Dutch)*. Report BTO 2005.029. Kiwa Water Research, Nieuwegein, the Netherlands.

Verissimo A., Marrao G., Gomes da Silva F. and da Costa M. S. (1991). Distribution of *Legionella* spp. in hydrothermal areas in continental Portugal and the island of Sao Miguel, Azores. *Appl. Environ. Microbiol.*, **57**(10), 2921–2927.

Versteegh J. F. M., Brandsema P. S., van der Aa N. G. F. M., Dik H. H. J. and de Groot G. M. (2007) Evaluation of the Dutch Water Supply Act: Legionella prevention *(in Dutch)*. Report 7037190202007 RIVM, Bilthoven, the Netherlands.

Versteegh J. F. M. and Dik H. H. J. (2011) The quality of drinking water in the Netherlands in 2010, RIVM report nr. 7037190812011(in Dutch). Ministerie van Infrastructuur en Milieu, The Hague, the Netherlands.

Vickers R. M., Yu V. L., Hanna S. S., Muraca P., Diven W., Carmen N. and Taylor F. B. (1987). Determinants of *Legionella pneumophila* contamination of distribution systems: 15-hospital prospective study. *Infect. Control*, **8**(9), 357–363.

Vlaamse Regering (2007) Besluit van de Vlaamse regering betreffende de preventie van de veteranenziekte op publiek toegankelijke plaatsen. Belgisch Staatsblad 4. V 2007.

Von Baum H. and Lück C. (2011). Ambulant erworbene Legionellenpneumonie. *Bundesgesundheitsbl.*, **54**, 688–692.

von Baum H., Ewig S., Marre R., Suttopr N., Gonschior S., Welte T., Lück C. for the Competence Network for Community Acquired Pneumonia study group (2008). Community-acquired Legionella pneumonia: new insights from the German competence network for community acquired pneumonia. *Clin. Inf. Dis.*, **46**, 1356–1364.

Von Baum H., Harter G., Essig A., Lück C., Gonser T., Embacher A. and Brockmann S. (2010). Preliminary report: outbreak of Legionnaires' disease in the cities of Ulm and Neu-Ulm in Germany, December 2009-January 2010. *Euro Surveill.*, **15**(4), 1–2.

Voss L., Button K. S. and Tuovinen O. H. (1985) *Legionella pneumophila* in a metropolitan water distribution system. *Environmental Technology Letters*, **6**, 429–438.

VROM (2001). Besluit van 9 januari 2001 tot wijziging van het Waterleidingbesluit in verband met de richtlijn betreffende de kwaliteit van voor menselijke consumptie bestemd water (Ministry of Housing, Spatial Planning and Environment). *Staatsblad*, **31**, 1–26.

Wadowsky R. M. and Yee R. B. (1981). A glycine-containing selective medium for isolation of Legionellaceae. *J. Clin. Microbiol.*, **13**, 380–382.

Wadowsky R. M. and Yee R. B. (1985). Effect of non-Legionellaceae bacteria on the multiplication of *Legionella pneumophila* in potable water. *Appl. Environ. Microbiol.*, **49**(5), 1206–1210.

Wadowsky R. M., Yee R. B., Mezmar L., Wing E. J. and Dwoling J. N. (1982). Hot water systems as sources of *Legionella pneumophila* in hospital and nonhospital plumbing fixtures. *Appl. Environ. Microbiol.*, **43**, 1104–1110.

Wadowsky R. M., Wolford R., McNamara A. M. and Yee R. (1985). Effect of temperature, pH and oxygen level on the multiplication of naturally occurring *Legionella pneumophila* in potable water. *Appl. Environ. Microbiol.*, **49**(5), 1197–1205.

Wadowsky R. M., Butler L. J., Cook M. K., Verma S. M., Paul M. A., Fields B. S., Keleth G., Sukora J. L. and Yee R. B. (1988). Growth-supporting activity for *Legionella pneumophila* in tap water cultures and implication of hartmannellid amoeba as growth factors. *Appl. Environ. Microbiol.*, **54**(11), 2677–2682.

Wadowsky R. M., Wilson T. M., Kapp N. J., West A. J., Kuchta J. M., States S. J., Dowling J. N. and Yee R. B. (1991). Multiplication of *Legionella* spp. in tap water containing *Hartmannella vermiformis*. *Appl. Environ. Microbiol.*, **57**(7), 1950–1955.

Wallenstein A., Oliver I., Ricketts K., Kafatos G., Stuart J. M. and Joseph J. A. (2010). Windscreen wiper fluid without added screen wash in motor vehicles: a newly identified risk factor for Legionnaires' disease. *Eur. J. Epidemiol.*, **25**(9), 661–665.

Wang H., Edwards M., Falkinham J. O. and Pruden A. (2012). Molecular survey of the occurrence of *Legionella* spp., *Mycobacterium* spp., *Pseudomonas aeruginosa*, and amoeba hosts in two chloraminated drinking water distribution systems. *Appl. Environ. Microbiol.*, **78**(17), 6285–6294.

Warren W. J. and Miller R. D. (1979). Growth of Legionnaires' disease bacterium (*Legionella pneumophila*) in chemically defined medium. *Appl. Environ. Microbiol.*, **10**(1), 50–55.

Wéry N., Bru-Adan V., Minervini C., Delgénes J. P., Garrelly L. and Godon J. J. (2008). Dynamics of *Legionella* spp. and bacterial populations during the proliferation of *L. pneumophila* in a cooling tower facility. *Appl. Environ. Microbiol.*, **74**(10), 3030–3037.

WHO (2004a) Guidelines for drinking-water quality. Volume 1, Recommendations. 3rd Edition, World Health Organization, Geneva.

WHO (2004b) Monochloramine in drinking-water. Background document for development of WHO Guidelines for Drinking water quality. WHOSDEWSH03.0483.

Wilczak A., Jackangelo J. G., Marcinko J. P., Odell L. H., Kimeyer G. J. and Wolfe R. L. (1996). Occurrence of nitrification in chloraminated distribution systems. *J. Am. Wat. Works Assoc.*, **88**, 74–85.

Witherell L. E., Duncan R. W., Stoen K. M., Statton L. J., Orciari L., Kappel S. and Jillson D. A. (1988). Investigation of *Legionella pneumophila* in drinking water. *J. Am. Wat. Works Assoc.*, **80**, 87–93.

Wullings B. A. and van der Kooij D. (2006). Occurrence and genetic diversity of uncultured *Legionella* spp. in drinking water treated at temperatures below 15°C. *Appl. Environ. Microbiol.*, **72**(1), 157–166.

Wullings B., Veenendaal H. R. and van der Kooij D. (2009). *Legionella pneumophila* komt sporadisch voor in Nederlands oppervlaktewater. *H_2O*, **7**, 44–46.

Wullings B. A., Bakker G. and van der Kooij D. (2011). Concentration and diversity of uncultured *Legionella* spp. in two unchlorinated drinking water supplies with different concentrations of natural organic matter. *Appl. Environ. Microbiol.*, **77**(3), 634–641.

Yee R. and Wadowsky R. M. (1982). Multiplication of *Legionella pneumophila* in unsterilized tap water. *Appl. Environ. Microbiol.*, **43**(6), 1330–1334.

Yu V. L., Beam T. R., Lumish R. M., Vickers R. M., Fleming J., McDermott C. and Romano J. (1987). Routine culturing for *Legionella* in the hospital environment may be a good idea: a three-hospital prospective study. *Am. J. Med. Sci.*, **294**(2), 97–99.

Yu V. L., Plouffe J. E., Pastoris M. C., Stout J. E., Schousboe M., Widmer A., Summersgill J., File T., Heath C. M., Patterson D. L. and Chereshsky A. (2002). Distribution of *Legionella* species and serogroups isolated by culture in patients with sporadic community-acquired legionellosis: an international collaborative study. *J. Inf. Dis.*, **186**, 127–128.

Zacheus O. M. and Martikainen P. J. (1994). Occurrence of legionellae in hot water systems of Finnish apartment buildings. *Can. J. Microbiol.*, **40**, 993–999.

Zacheus O. M. and Martikainen P. J. (1996). Effect of heat flushing on the concentrations of Legionella pneumophila and other heterotrophic microbes in hot water systems of apartment buildings. *Can. J. Microbiol.*, **42**, 811–818.

Zhang Z., McCann C., Hanrahan J., Jencson A., Joyce D., Fyffe S., Piesczynski S., Hawks R., Stout J., Yu V. L. and Vidic R. D. (2009). *Legionella* control by chlorine dioxide in hospital water systems. *J. AWWA*, **101**(5), 117–127.

Chapter 7

Opportunistic pathogens in drinking water in the Netherlands

P. W. J. J. van der Wielen, R. Italiaander, B. A. Wullings, L. Heijnen and D. van der Kooij

7.1 INTRODUCTION

Most of the microorganisms on earth do not cause disease in humans. In contrast to these non-pathogenic species are virulent strains of certain pathogenic microbial species that can cause disease in humans. Some of these pathogenic species have highly virulent strains (e.g. *Vibrio cholerae*, *Salmonella typhi* and *Yersinia pestis*), which can cause disease in both healthy humans and humans with a less active or inactive immune system (i.e. immunocompromised individuals). Other pathogenic organisms have more mildly virulent strains, which mainly causes disease in immunocompromised humans. Microorganisms that can cause disease in immunocompromised individuals are named opportunistic pathogens. It is known that several of these opportunistic pathogens are capable of growing in drinking water or drinking-water-related biofilms (Bartram *et al.* 2003).

7.1.1 Climate change

The climate on earth has remained relatively stable over the last millennia and has a strong temperate central tendency (Solomon *et al.* 2007). However, numerous studies have indicated that the climate has changed over the last decades (reviewed in Solomon *et al.* 2007). In 2007, the Intergovernmental Panel on Climate Change (IPCC) reported that the average temperature on earth has increased by 0.6 to 0.9°C over the last 100 years. Moreover, climate models predict that by 2100 the average temperature on earth will have increased by 1.1 to 6.4°C (Solomon *et al.* 2007).

It has been reported that climate change will have a negative impact on our society, which is caused by several different mechanisms (reviewed in Haines *et al.* 2000; Frumkin *et al.* 2008). An aspect of climate change that has not received much attention in the literature is enhanced growth of opportunistic pathogens in drinking water when temperatures rise (Levin *et al.* 2002). Increasing temperatures, due to climate change, will result in an increase in surface water and drinking water temperatures. There are several clues that these higher (drinking) water temperatures will increase the growth of pathogenic microorganisms in water. First, the number of *Legionella pneumophila* cases is higher in summer (Fisman *et al.* 2005). Second, an increased risk of severe melioidosis, an infectious disease caused by *Burkholderia pseudomallei*, is observed when the number of cyclones increases (Currie and Jacups 2003). Third, the optimum temperature for growth of many pathogenic microorganisms is between 30 and 40°C. Consequently, it is expected that increased temperatures in drinking water distribution systems might result in increased numbers of opportunistic pathogens in drinking water and drinking-water-related biofilms.

7.1.2 Demographic change

During the last decades, the number of elderly in our society has increased and will continue to increase till approximately 2040. When people age, their immune system will be less active, which makes them more vulnerable to infection with opportunistic pathogens (Morris and Potter 1997). Besides the trend of increasing age, the medical care has improved as well in the last decades (Morris and Potter 1997). Improved health care has resulted in longer life expectation for people with serious underlying diseases like cancer, HIV/AIDS or rheumatic fever. However, patients suffering from these diseases have a less active or inactive immune system, because of their disease or the medication they use. These trends have resulted in an increase of immunocompromised individuals in our western society and it is expected that these numbers will continue to rise in the next few decades (Glasmacher *et al.* 2003). Consequently, the number of people that are vulnerable to opportunistic pathogens, which can be present in drinking water, will increase as well.

7.2 LITERATURE SURVEY

The expected increase of opportunistic pathogens in drinking water and the number of immunocompromised persons in our western society press the need to study the occurrence of opportunistic pathogens in drinking water. Therefore, it is essential to know which opportunistic pathogenic organisms might be important in relation to drinking-water (biofilms). Such information was obtained by performing a literature survey in relation to opportunistic microorganisms that can occur and grow in drinking water and might be of significance in the Netherlands.

In a first step, literature was screened to determine opportunistic pathogens that can grow in drinking water and this resulted in a list of organisms shown in

Table 7.1. Subsequently, four different aspects about each of the opportunistic pathogens listed in Table 7.1 were obtained from the scientific literature:

(1) What is the number of (reported) cases in the Netherlands?
(2) Is there an epidemiological link between patients and drinking water?
(3) Is the organism observed in drinking water in the Netherlands?
(4) What is the effect of temperature on growth of the organism?

Table 7.1 Opportunistic pathogens capable of growing in drinking water (in alphabetical order).

Acinetobacter baumannii complex	*Legionella (pneumophila)*
Aeromonas (hydrophila)	Methylobacteria
Afipia spp.	Nontuberculous mycobacteria
Bosea spp.	*Pseudomonas aeruginosa*
Burkholderia cepacia complex	*Stenotrophomonas maltophilia*
Burkholderia pseudomallei	*Yersinia enterocolitica*
Chlamydia-like bacteria	Pathogenic fungi
Elizabethkingia meningoseptica	Pathogenic protozoa

Based on the obtained information, the need for further research in the Netherlands for each of the opportunistic pathogen listed in Table 7.1 was prioritized.

7.2.1 (Reported) cases in the Netherlands

In 2010, 476 cases of legionellosis were reported to the Dutch health authorities (Bijkerk *et al.* 2011; Table 7.2). In 334 cases patients obtained the *Legionella* infection in the Netherlands, whereas in the other cases patients obtained the infection in a foreign country (Bijkerk *et al.* 2011).

For some of the other opportunistic pathogenic organisms estimates have been made for the number of disease-related cases in the Netherlands. Estimates by van Ingen (van Ingen 2009) indicated that approximately 270 cases of nontuberculous *Mycobacterium* (NTM)-related disease occur in the Netherlands every year (Table 7.2). In addition, it was observed that the number of NTM isolates from clinical material increased during the last decade (van Ingen 2009). Around 150 to 200 cases of gastrointestinal disease caused by *Yersinia enterocolitica* were reported in the Netherlands annually in the period 1990–1999 (Goverde 1999) (Table 7.2). *Acanthamoeba* can cause granulomatous amoebic encephalitis (GAE) and keratitis (Ma *et al.* 1990). Cases of GAE caused by *Acanthamoeba* have not been reported in the Netherlands. However, estimates have shown that the incidence of keratitis caused by *Acanthamoeba* is around 3 out of 1 million contact lens wearers per year in the Netherlands (Cheng *et al.* 1999), which makes the estimation for keratitis cases caused by *Acanthamoeba* under contact lens wearer around 6 per year in the Netherlands (Table 7.2).

Table 7.2 Reported cases and hospital-acquired infection cases in the Netherlands caused by the different opportunistic pathogens that are capable of growing in drinking water.

	Reported cases[a]	**Reported cases of hospital-acquired infection**[b]
Legionella (pneumophila)	476	0
Nontuberculous mycobacteria	±270	0
Yersinia enterocolitica	±250	0
Acanthamoeba spp.	±6	0
Pseudomonas aeruginosa	Unknown	7.2%
Stenotrophomonas maltophilia	Unknown	1.2%
Acinetobacter baumannii complex	Unknown	0.6%
Aeromonas (hydrophila)	Unknown	0.1%
Burkholderia cepacia complex	Unknown	0
Aspergillus fumigatus	Unknown	Unknown
Fusarium oxysporum	Unknown	Unknown
Afipia spp.	0	0
Bosea spp.	0	0
Burkholderia pseudomallei	0	0
Simkania negevensis	0	0
Elizabethkingia meningoseptica	0	0
Methylobacteria	0	0
Neigleria fowleri	0	0

[a]Data from Cheng *et al.* (1999); Goverde (1999); van Ingen (2009); Bijkerk *et al.* (2011).
[b]Based on data from 83 Dutch hospitals in the period 1998–2007 (PREZIES 2008a).

It is known that disease caused by several other organisms (*Pseudomonas aeruginosa*, *Stenotrophomonas maltophilia*, *Acinetobacter baumannii* complex, *Aeromonas*, *Burkholderia cepacia* complex, *Aspergillus fumigatus* and *Fusarium oxysporum*) occurs in the Netherlands, but the number of cases is unknown. Most of these organisms can cause hospital-acquired infections. A survey in the period 1997 till 2007 among 83 different hospitals in the Netherlands showed that *P. aeruginosa* and *S. maltophilia* were responsible for a reasonable number of hospital-acquired infections (Table 7.2) (Preventie van ziekenhuisinfecties door surveillance [PREZIES] 2008a). In 2007, *P. aeruginosa* was even the most isolated organism in hospital acquired pneumonia (PREZIES 2008b). Disease caused by *B. cepacia* complex occurs mainly in cystic fibrosis (CF) patients. Until 1995 around 3% of the CF patients in the Netherlands were infected by *B. cepacia* complex (Horrevorts *et al.* 1995), but the incidence has most likely increased during the last decade as has been observed in other countries (Mahenthiralingam *et al.* 2005). Pathogenic filamentous fungi like *A. fumigatus* and *F. oxysporum* also cause disease in the Netherlands (ter Maaten *et al.* 1995; Verduyn Lunel *et al.* 2006), but information about the incidence of these organisms in disease in the Netherlands could not be found in literature.

No reported cases or hospital-acquired infection cases caused by *Afipia* spp., *Bosea* spp., *Burkholderia pseudomallei*, *Simkania negevensis*, *Elizabethkingia meningoseptica*, methylobacteria or *Neigleria fowleri* in the Netherlands have been found in the literature (Table 7.2). However, this does not necessarily mean that these organisms do not cause disease in the Netherlands, but their incidence is probably very low.

Overall, the number of patients with legionellosis in the Netherlands is probably higher than the number of patients that obtained disease caused by the other opportunistic pathogens that are listed in Table 7.1. However, it is difficult to make this comparison, because diseases caused by the other opportunistic pathogens do not have to be reported to the health authorities.

7.2.2 Epidemiological link between organisms from drinking water and patients

The opportunistic pathogenic organisms shown in Table 7.1 are capable of growing in drinking water or drinking-water-related biofilms, but that does not necessarily mean that strains observed in drinking water are identical to strains isolated from patients. For instance *Aeromonas* strains isolated from patients that suffer from diarrhoea are genotypically and phenotypically different from strains isolated from drinking water (Havelaar *et al.* 1992; Moyer *et al.* 1992; Hanninen and Siitonen 1995; Borchardt *et al.* 2003). This demonstrates that, although there is still debate whether certain *Aeromonas* strains are the causative agents for gastrointestinal disease, there is no proof that *Aeromonas* strains in drinking water are involved in diarrhoeal disease. In addition to *Aeromonas*, studies have shown that *Y. enterocolitica* strains from water environments are genotypically different from strains that cause disease in patients (Aleksic and Bockemuhl 1988; Sachdeva and Virdi 2004). Furthermore, no studies were found in the literature that have demonstrated that drinking water plays a role in the epidemiology of *A. baumannii* complex, *Afipia* spp., *Bosea* spp. or methylobacteria.

Other studies have demonstrated that drinking-water strains of *L. pneumophila*, certain nontuberculous mycobacterial species, *Acanthamoeba* that cause keratitis and *B. pseudomallei* can be genotypically identical to strains isolated from patients (Engel *et al.* 1980; von Reyn *et al.* 1994; Chadha *et al.* 1998; Aronson *et al.* 1999; Currie *et al.* 2001; Kilvington *et al.* 2004; Garcia-Nunez *et al.* 2008; Luck *et al.* 2008). In addition, *F. oxysporum* and *A. fumigatus* strains isolated from drinking water in a hospital in the Netherlands were identical to strains isolated from patients (pers. comm. Prof. Dr. P. Verweij, Radboud Hospital Nijmegen). Consequently, it can be assumed that drinking water plays a role in the epidemiology of these bacterial, protozoan and fungal species.

Strains from *B. cepacia* complex, *P. aeruginosa*, *S. maltophilia* and *E. meningoseptica* have been isolated from sinks, taps, shower heads, bathtubs and bathing toys. These strains can be genotypically identical to strains isolated from patients (Nelson *et al.* 1991; Fisher *et al.* 1993; Doring *et al.* 1996; Buttery *et al.* 1998; Denton *et al.* 1998; Weber *et al.* 1999; Hoque *et al.* 2001; Vochem *et al.* 2001). Although these results suggest that drinking water may play a role in the epidemiology of these organisms, definite evidence that these strains came from distributed drinking water instead of the surrounding environment is lacking. Moreover, results from studies where *P. aeruginosa* strains isolated from drinking water were compared to strains isolated from patients, demonstrated that drinking-water strains were genotypically different from patient strains (Rudnick *et al.* 1996; Reuter *et al.* 2002; Trautmann *et al.* 2005). *S. negevensis* and *N. fowleri* have been isolated from drinking water that was used by patients who were infected with these organisms (Marciano-Cabral *et al.* 2003; Kahane *et al.* 2007). However, it was not investigated in these studies whether the strains isolated from drinking water were phenotypically or genotypically identical to the strains isolated from patients.

Overall, it is concluded that the epidemiological role of drinking water is proven for some opportunistic pathogens that can grow in drinking-water distribution and/or plumbing systems. For others, definite proof is lacking that drinking water is a source for strains that can infect patients, and additional epidemiological studies are required to investigate the role of drinking water in the epidemiology of these opportunistic pathogens.

7.2.3 Occurrence of opportunistic pathogens in drinking water in the Netherlands

Water supply companies in the Netherlands have to monitor the presence of *Aeromonas* spp. and *Legionella* spp. in drinking water, because both parameters are included in the Dutch drinking-water decree (Anonymous 2001). *Aeromonas* spp. is often observed in distributed drinking water in the Netherlands, especially in the summer when numbers range from 0 to 1000 cfu 100 ml^{-1}. *Legionella* spp. is detected sporadically in distributed drinking water in the Netherlands with numbers normally ranging between 0 and 1000 cfu l^{-1}. In almost all cases the observed *Legionella* spp. belongs to non-pathogenic species, like *Legionella anisa*. *L. pneumophila* is normally not observed in the drinking-water distribution system, but the organism is sometimes cultivated from plumbing systems of large buildings (e.g. hotels, hospitals etc).

In 1979 research was conducted to investigate the occurrence of *Mycobacterium kansasii* in drinking water from the Rotterdam area (Engel *et al.* 1980). The results demonstrated that *M. kansasii* was observed frequently in drinking water from household installations (38 out of 78 samples), but the organism was only detected in 1 out of 112 drinking-water samples from the distribution system (Engel *et al.* 1980). These results demonstrate that growth of *M. kansasii* occurred mainly in the household installation. *P. aeruginosa*, *F. oxysporum* and *Aspergillus versicolor* have been observed in Dutch drinking water (van der Kooij *et al.* 1982; Valster *et al.* 2009), but the occurrence of these organisms in drinking water was studied sporadically and locally. As a result, it remains unknown how often these organisms occur in distributed drinking water in the Netherlands. *P. aeruginosa*, *F. oxysporum* and *A. fumigatus* are observed in drinking water in other countries as well (Papapetropoulou *et al.* 1994; Gottlich *et al.* 2002; Emtiazi *et al.* 2004; Hageskal *et al.* 2006; Lee *et al.* 2006), which shows that these organisms can be a part of the microbial community in drinking water. The occurrence of *S. maltophilia* in drinking-water distribution systems in the Netherlands has not been studied. However, *S. maltophilia* was observed in drinking water sampled from a tap in a hospital in the eastern part of the Netherlands (Verweij *et al.* 1998). The presence of *Acanthamoeba* spp. in distributed drinking water in the Netherlands has been studied with quantitative polymerase chain reaction (qPCR), but the organism was not detected in drinking water (Valster 2011).

The occurrence of *B. cepacia* complex, *B. pseudomallei*, *N. fowleri*, *Chlamydia*-like bacteria (including *S. negevensis*), *A. baumannii* complex, *Afipia* spp., *Bosea* spp., *E. meningoseptica* and methylobacteria in drinking water in the Netherlands has not been studied. However, studies in Italy and the US have shown that *B. cepacia* complex was present in 3.5 to 5% of the drinking water samples in numbers ranging from 1 to 40 cfu ml^{-1} (Edberg *et al.* 1996; Zanetti *et al.* 2000). *B. pseudomallei*, *N. fowleri* and *S. negevensis* have also been observed in drinking water samples from other countries; mainly countries with a warm climate (e.g. South East Asia, Australia, the southern states of the US and Israel) (Anderson and Jamieson 1972; Zanetti *et al.* 2000; Currie *et al.* 2001; Kahane *et al.* 2004).

7.2.4 Effect of temperature on growth of opportunistic pathogens

The opportunistic pathogens discussed in this paper have a growth optimum at 25 to 40°C (Garrity 2001–2011). In addition, most of these organisms are unable to grow at temperatures lower than 15°C. *B. pseudomallei*, *P. aeruginosa* and *N. fowleri* occur more often in drinking water in regions with a warm/tropical climate than in regions with a more temperate climate (Reitler and Seligmann 1957; Anderson and Jamieson 1972; Schubert and Scheiber 1975; Zanetti *et al.* 2000; Currie *et al.* 2001). Consequently, it is expected that when climate change results in elevated drinking water temperatures in the distribution system in the Netherlands, the occurrence of these opportunistic pathogens in Dutch drinking water will increase. An exception is *Y. enterocolitica*, which is capable of growing at low temperatures (1 to 10°C) as well (Goverde 1999).

7.2.5 Priority for further research

The literature survey demonstrated that drinking water temperatures of 25°C or higher are advantageous for all opportunistic pathogens discussed and based on that criterion it cannot be concluded which organism has a higher priority for additional research in the Netherlands. In contrast, the number of cases in the Netherlands, the epidemiological link between patients and drinking water strains and the occurrence of the organisms in Dutch drinking water differ between the opportunistic pathogens. Thus, based on those criteria the opportunistic pathogens can be prioritized for further research (Table 7.3).

Table 7.3 Priority for further research on different opportunistic pathogens that can grow in drinking water. Priority is based on the number of cases in the Netherlands, the epidemiological link between patient and drinking-water strains and the occurrence in drinking water in the Netherlands.

Organism	Cases	Epidemiology	Occurrence	Priority
L. pneumophila	+++	++	+	Very high
NTM	++	++	+	High
P. aeruginosa	+	+	+	High
Pathogenic fungi	+	++	+	High
S. maltophilia	+/−	+	+	Moderate
Acanthamoeba spp.	+/−	++	−	Moderate
B. cepacia complex	+/−	+	Unknown	Low
A. baumannii complex	+/−	−	Unknown	Low
Aeromonas spp.	+/−	−	+	Low
Y. enterocolitica	++	−	Unknown	Low
Afipia spp.	−	−	Unknown	Low
Bosea spp.	−	−	Unknown	Low
B. pseudomallei	−	++	Unknown	Low
S. negevensis	−	+/−	Unknown	Low
E. meningoseptica	−	+/−	Unknown	Low
Methylobacteria	−	−	Unknown	Low
N. fowleri	−	+/−	Unknown	Low

L. pneumophila has the highest priority for further research, because the number of legionellosis cases is relatively high in the Netherlands, an epidemiological link with drinking water has been proven and the organism has been detected in Dutch drinking water (Table 7.3). Because this high priority emerged after the large outbreak of legionellosis in 1999 in the Netherlands (Den Boer *et al.* 2002), a number of studies involving *Legionella* and *L. pneumophila* have already been conducted in the Netherlands (Kuiper *et al.* 2004; van der Kooij *et al.* 2005; Wullings and van der Kooij 2006; Wullings *et al.* 2011). Nontuberculous mycobacteria, *P. aeruginosa* and pathogenic fungi have also a high priority for additional research, because there are reported cases of patients that are infected with these organisms in the Netherlands, the epidemiological link with drinking water has been demonstrated and the organisms have been observed in Dutch drinking water (Table 7.3).

S. maltophilia, *Acanthamoeba* spp, and *B. cepacia* complex have a moderate priority for further research. The lower priority compared to the organisms mentioned above is caused by the low number of reported cases in the Netherlands, although the epidemiological link with drinking water has been established (Table 7.3). Currently, the other organisms have a low priority for additional research. This low priority was given because no reported cases exist in the Netherlands for certain organisms (*Afipia* spp., *Bosea* spp., *B. pseudomallei*, *S. negevensis*, *E. meningoseptica*, methylobacteria and *N. fowleri*) or an epidemiological link with drinking water has not been demonstrated (*A. baumannii* complex, *Aeromonas* spp., *Y. enterocolitica*, *Afipia* spp., *Bosea* spp. and methylobacteria) (Table 7.3).

7.3 DETECTION OF OPPORTUNISTIC PATHOGENS IN DRINKING WATER

Based on the priority for further research, it was decided to use quantitative methods for the specific detection of *L. pneumophila*, nontuberculous mycobacteria, *P. aeruginosa*, fungi, *S. maltophilia* and *Acanthamoeba* spp. Fast, specific, reliable and quantitative methods are necessary to be able to determine the number of these microorganisms in drinking water. However, specific methods are not available for the detection of some of these organisms in drinking water, whereas for others selective culture media are available (Neumann *et al.* 1997; Hageskal 2007; International organisation for standardisation [ISO] 16266 2006; Nederlands normalisatie instituut [NEN] 6265 2007). Some of these agar-based methods are time consuming, because the incubation time is long (1 week to 3 months) (Neumann *et al.* 1997; NEN 6265 2007). In addition, agar media will not detect organisms that cannot be cultivated. Therefore, quantitative PCR (qPCR) methods for the specific detection of these opportunistic pathogens in drinking water were developed, since qPCR methods can be fast, specific, reliable and quantitative.

In the past, qPCR methods for the specific detection of *L. pneumophila* and *Acanthamoeba* spp. have been developed and can be used to detect these organisms in drinking water (Wullings *et al.* 2011; Qvarnstrom *et al.* 2006; Valster *et al.* 2011). For the other organisms of interest, PCR methods have been described (Telenti *et al.* 1993; Smit *et al.* 1999; Challier *et al.* 2004; da Silva Filho *et al.* 2004; Shannon *et al.* 2007; Feazel *et al.* 2009), but these methods were not used in water-related environments. Consequently, we investigated whether these PCR methods can be used in a fast, reliable and quantitative manner to specifically detect the different opportunistic pathogens in drinking water.

7.3.1 Fungi and *Aspergillus fumigatus*

Smit *et al.* (1999) described primer pair EF4 and Fung5 for the specific detection of 18S rRNA genes of fungi in the wheat rhizosphere using PCR. We tested this primer pair on DNA isolated from drinking water sampled from four different distribution systems in the Netherlands and the formation of one PCR product of the expected length was observed. Subsequently, PCR products were cloned in *E. coli* and the DNA sequence of 48 clones per sample were determined to verify the specificity of the PCR primers for the 18S rRNA gene of fungi.

The closest relative from the Genbank database for all obtained 18S rRNA gene sequences was a 18S rRNA gene sequence from a fungal species (Table 7.4), which demonstrates that the PCR primers EF4 and Fung5 can be used for the specific detection of fungi in drinking water. Some 18S rRNA gene sequences from drinking water had high similarities to 18S rRNA gene sequences of cultivated fungal species (98–100%), whereas other 18S rRNA gene sequences from drinking water had low similarities (84–91%) (Table 7.4). In addition, it was observed that some relatively dominant 18S rRNA gene sequences in drinking water were closely related to fungal species that have been reported to be involved in human disease (e.g. *Aspergillus restrictus*, *Cladosporium cladosporioides*) (Pitt 1994; Stevens *et al.* 2000; Vieira *et al.* 2001) or fish disease (*Exophiala salmonis*) (Richards *et al.* 1978). These findings indicate the need for further characterization of the fungal diversity in drinking-water samples.

The PCR method was further optimized for quantitative detection (qPCR assay). SYBRgreen was included in the PCR reaction mixture and the formation of a PCR product was detected using a CFX96 multicolour real-time PCR detection system (Biorad). Quantification was based on comparison of the threshold cycle (C_T) of the sample with C_T values of a calibration curve, which was based on tenfold dilution of known copy numbers of the 18S rRNA gene of *Penicillium brevicompactum.* A log-linear relationship between gene copies and C_T values was observed ($R^2 = 0.986$) when gene copies were between 26 and 2.6×10^5. In addition, a relatively high PCR efficiency (92.0%) was detected. It was concluded that this qPCR assay can be used to quantify 18S rRNA gene copy numbers of fungi in drinking water.

Table 7.4 Nearest 18S rRNA gene sequence of cultivated fungal species in the GenBank database.

Distribution system	OTU	# seqs	Nearest relative Genbank	Similarity (%)
A[1]	1	8	*Exophiala salmonis* (EF413608)	100
	2	17	*Ramichloridium anceps* (NG_013193)	99
	3	15	*Cladosporium cladosporioides* (GQ458030)	100
	4	3	*Dothiora cannabinae* (DQ479933)	100
	5	1	*Coniosporium apollinis* (GU250919)	98
	6	1	*Rhexocercosporidium carotae* (AF487897)	100
	7	1	*Cosmospora coccinea* (AY489702)	99
B[1]	1	17	*Phlebia radiata* (EF491866)	99
	2	1	*Phlebia radiata* (EF491866)	98
	3	18	*Aspergillus restrictus* (EU723495)	100
	4	8	*Spizellomyces pseudodichotomus* (GQ499384)	87
	5	1	*Nowakowskiella* sp. (EU828479)	84
C[2]	1	6	*Allochytridium luteum* (EU828473)	87
	2	22	*Triparticalcar arcticum* (DQ536480)	90
	3	11	*Allochytridium luteum* (EU828473)	87
	4	1	*Tolypocladium parasiticum* (EF468993)	99
	5	1	*Rhizophlyctis rosea* (GQ160456)	91
	6	1	*Phlyctochytrium africanum* (FJ827649)	91
	7	4	*Rhizophlyctis rosea* (AF164251)	95
D[2]	1	1	*Megacollybia platphylla* (AY786053)	88
	2	4	*Elaphocordyceps inegoensis* (AB027322)	96
	3	7	*Rhizophlyctis rosea* (GQ160456)	92
	4	3	*Megacollybia platphylla* (AY786053)	88
	5	2	*Nowakowskiella sp.* (EU828479)	87

[1]Drinking water in distribution system A and B was produced from groundwater.
[2]Drinking water in distribution system C and D was produced from surface water.

A qPCR assay for the specific detection of the 28S rRNA gene of *A. fumigatus* in blood serum has been published by Challier *et al.* (2004) and is based on the use of a specific primer pair and a specific Taqman probe. *In silico* analysis of the primer and probe sequences with sequences from the GenBank database indicated that the 28S rRNA gene of *A. fumigates* and *Aspergillus phialisepticus* was detected with the primers and probe. However, *A. phialisepticus* is an *Aspergillus* species that belongs to the *A. fumigatus* lineage (Peterson 2008), has been involved in human disease and is equally pathogenic to *A. fumigatus* in mice (Kwon-Chung 1975). Thus, the qPCR specifically detects opportunistic fungi that belong to the *A. fumigatus* lineage.

DNA from drinking water sampled at different locations in a distribution system and surface water were tested with the qPCR assay. PCR resulted in the formation of a PCR product of the expected length. Subsequently, PCR products were cloned in *E. coli* and the 28S rRNA gene sequences of all analysed clones (n = 54) were 100% identical to 28S rRNA gene sequences from *A. fumigatus* strains that have been deposited in the GenBank database. A calibration curve with known 28S rRNA gene copies of *A. fumigatus* was developed to be able to quantify 28S rRNA gene copies of *A. fumigatus* in drinking water. Gene copies between 25 and 2.5×10^5 showed a log-linear relationship with the C_T value ($R^2 = 0.996$), whereas the efficiency of the qPCR was 98.0%. Consequently, this qPCR assay can be used to quantify 28S rRNA gene copies of *A. fumigatus* in drinking water.

7.3.2 Mycobacteria and *Mycobacterium avium* complex

A new qPCR method to quantify the 16S rRNA genes of mycobacteria was developed. The primers and probe specific for the 16S rRNA gene of bacteria belonging to the genus *Mycobacterium* were constructed using the ARB program and database. The specific forward primer Myco637f (5′GGGCGATACGG GCAGAC3′) and specific reversed primer Myco841r (5′GAAACCGACAC CTAGTACC3′) resulted in a PCR product of 208 bp. The qPCR assay was further complemented with the specific probe Myco662p (5′CAGGGGAGA CTGGAATTCCTGGTGTA3′). The qPCR program for this assay was as follows: 5 min 95°C; 43 cycles of 20 sec at 95°C and 60 sec at 60°C.

The specificity of the primers and probe were tested on 15 different species of the genus *Mycobacterium* and *L. pneumophila*. DNA from all mycobacterial species resulted in the formation of a PCR product of the expected length, whereas DNA from *L. pneumophila* did not result in the formation of a PCR product. DNA was isolated from drinking-water samples that were obtained from eight different distribution systems in the Netherlands. The 16S rRNA gene of mycobacteria was amplified using primers Myco635f and Myco841r and the PCR-product was cloned in *E. coli*. Thereafter, the 16S rRNA gene of 9 to 48 clones per drinking water sample was sequenced (resulting in a total of 258 16S rRNA gene sequences). Subsequently, the 16S rRNA gene sequences were compared with sequences deposited in the GenBank database. The results from this comparison revealed that all 258 16S rRNA gene sequences had a sequence from a cultivated *Mycobacterium* species as nearest relative, with an identity of 99 to 100%. The probe sequence Myco662p was present in 257 of the 258 analyzed 16S rRNA gene sequences. It is concluded that this qPCR is selective for the detection of *Mycobacterium* spp. in unchlorinated drinking water.

A calibration curve with the 16S rRNA gene of *M. avium* subsp. *avium* was constructed and a log-linear relationship between the 16S rRNA gene copies and C_T values was observed between 2.5 and 2.5×10^4 gene copies ($R^2 = 0.992$). In addition, the PCR efficiency was 93.2%, making this qPCR assay useful for the quantification of mycobacterial 16S rRNA genes in drinking water.

Feazel *et al.* (2009) recently published a qPCR assay for the quantitative detection of the 16S rRNA gene of bacteria belonging to the *M. avium* complex. The primers used in their assay are primer MAV199R, a reverse primer that is specific for the 16S rRNA gene of *M. avium* complex, and a forward primer that is non-specific and anneals to the 16S rRNA gene of all bacterial species. We developed a more specific forward primer (MAV77F, 5′GCCTCTTCGGAG GTACTCG3′) and a specific Taqman probe (MavPrC, 5′CAATCTGCCCTGCA CTTCGGGATAAG3′) to increase the specificity of the PCR assay. The new primers and probe were tested on DNA of four different *M. avium* complex strains, four other mycobacterial species, *E. coli* and *L. pneumophila*. Only DNA from the four *M. avium* complex strains resulted in the formation of a PCR product, which demonstrates the specificity of the primers for the detection of bacteria belonging to the *M. avium* complex.

A calibration curve with the 16S rRNA gene of *M. avium* subsp. *avium* was constructed for the quantification of 16S rRNA genes of *M. avium* complex bacteria in drinking water. The calibration curve showed a log-linear correlation between gene copies of the 16S rRNA gene of *M. avium* subsp. *avium* (range: 25 to 2.5×10^5 gene copies) and C_T values ($R^2 = 0.998$). Furthermore, the PCR specificity was observed to be 92.7%. These values indicate that this qPCR assay can be used to quantify 16S rRNA gene copies of *M. avium* complex bacteria in drinking water.

7.3.3 *Pseudomonas aeruginosa*

A qPCR assay based on the amplification and detection of the *regA* gene of *P. aeruginosa* using specific PCR primers Paer-F and Paer-R and a Taqman probe (Paer-Pr) has been described in the literature (Shannon *et al.* 2007). However, primer dimers were formed when these primers were tested on DNA from drinking water. To avoid the formation of primer dimers, both primers and probe sequences were slightly altered, resulting in the primers Paer-Fb (5′ATCGA GTACCTGAACCGGC3′) and Paer-Rb (5′TGGTGCAGTTCCTCATTGTC3′) and the probe Paer-PrA (5′CCAGATGCTTTGCCTCAAC3′). The specificity of the primers was investigated with DNA from drinking water sampled from a distribution network in the Netherlands. The PCR product obtained was of the expected size and the PCR product was cloned in *E. coli* and subsequently the *regA* genes were sequenced. The results demonstrated that all *regA* gene sequences showed 100% similarity to the *regA* gene of *P. aeruginosa* deposited in the Genbank database. In addition the obtained *regA* gene sequences contained the probe sequence as well.

Quantification of the *regA* gene of *P. aeruginosa* was achieved by the development of a calibration curve using the *regA* gene of a *P. aeruginosa* strain. A strong log-linear relationship was observed between *regA* gene copies and C_T values when gene copies ranged from 25 to 2.5×10^5 ($R^2 = 0.997$). The specificity of the PCR was observed to be 99.9%. It was concluded that this qPCR assay is suitable for the quantification of *regA* gene copies from *P. aeruginosa* in drinking water.

7.3.4 *Stenotrophomonas maltophilia*

Da Silva Filho *et al.* (2004) published a PCR method for the specific detection of the *chitA* gene of *S. maltophilia* in sputa material from patients. We tested this PCR method on several strains of *S. maltophilia* and related bacterial species. A PCR product was only detected with DNA from *S. maltophilia*, but not with DNA from related bacterial species. Subsequently, the PCR primers were also tested on DNA from drinking water sampled from two different distribution systems in the Netherlands and resulted in the formation of a PCR product of the expected size. PCR products were further characterized by cloning the PCR product in *E. coli* and sequencing the *chitA* gene of 30 to 50 clones for each drinking water sample. The obtained *chitA* gene sequences were compared with sequences deposited in the Genbank database. In all cases the nearest relative was a *chitA* gene from a *S. maltophilia* strain. The different *chitA* gene sequences obtained from drinking water showed 89 to 95% similarity to *chitA* gene sequences of *S. maltophilia* deposited in the Genbank database. This similarity of 89% was within the *S. maltophilia* species level, since the sequence similarity between the *chitA* gene of *S. maltophilia* strain 34S1 (Genbank number: AF014950) and the *chitA* gene of *S. maltophilia* strain R551-3 (Genbank number: CP001111) was 89% as well. A phylogenetic tree, based on the neighbour-joining method, was constructed (Figure 7.1) and demonstrated that most *chitA* gene sequences from drinking water formed a separate cluster.

The method was further optimized to a qPCR assay. SYBRgreen was included in the PCR reaction mixture and the formation of a PCR product was detected using a CFX96 real-time PCR detection system (Biorad). Quantification was based on comparison of the C_T of the sample with C_T values of a calibration curve. The calibration curve was based on tenfold dilution of known gene copies of the *chitA* gene from *S. maltophilia*. A log-linear relationship between gene copies, in the range of 21 to 2.1 × 10^5, and C_T values was observed ($R^2 = 0.992$) as well as a relatively high PCR efficiency (96.8%). It was concluded that this qPCR assay can be used to specifically quantify *chitA* gene copy numbers of *S. maltophilia* in drinking water.

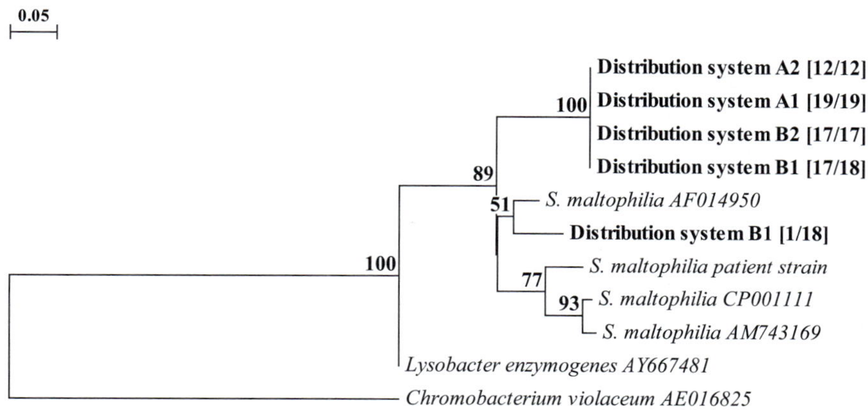

Figure 7.1 Phylogenetic tree of *chitA* gene sequences of *S. maltophilia*, obtained from DNA isolated from distributed drinking water samples of two different treatment plants. Between brackets is the number of sequences belonging to that operational taxonomic unit and the total number of analysed sequences.

7.4 OPPORTUNISTIC PATHOGENS IN UNCHLORINATED DRINKING WATER

The presence of opportunistic pathogens in drinking water was investigated by sampling drinking water from the distribution systems of two production plants in the Netherlands. These production plants produce drinking water from surface water, which is treated with multiple treatment processes and distributed without a residual disinfectant. In September 2009 and February 2010, drinking water samples from the distribution systems were taken from the taps of 13 (distribution system A) or 11 (distribution system B) houses. Before samples were taken, the tap was flushed until the water temperature remained stable for 30 sec.

In total 500 ml (September 2009) or 3000 ml (February 2010) drinking water was sampled from the taps. The temperature of the drinking water samples was determined immediately after the sample was taken from the tap, but the other parameters were measured after transportation at 4°C to the laboratory. Adenosine triphosphate (ATP) was determined as described by van der Wielen and van der Kooij (2010), cell numbers were determined by flow cytometry as previously described (van der Wielen and van der Kooij 2013). For DNA isolation, 500 ml (September 2009) or 2000 ml (February 2010) water was filtered over 25 mm polycarbonate filter (0.22 μm pore size). Subsequently, DNA was isolated from the filters as described by van der Wielen *et al.* (2009). The copies of the *mip* gene of *L. pneumophila* and the 18S rRNA gene of *Acanthamoeba* spp. were determined by previously described qPCR assays (Qvarnstrom *et al.* 2006; Valster *et al.* 2011; Wullings *et al.* 2011). The copy numbers of the 28S rRNA gene of fungi, the 18S rRNA gene of *A. fumigatus*, the 16S rRNA gene of mycobacteria, the 16S rRNA gene of *M. avium* complex, the *regA* gene of *P. aeruginosa* and the *chitA* gene of *S. maltophilia* were quantified with the qPCR assays described in Section 7.3.

7.4.1 Temperature, ATP and cell numbers

The average temperatures measured in drinking water sampled from distribution system A and B in September 2009 and February 2010 are visualized in Figure 7.2. The average drinking-water temperatures measured in the summer were significantly higher than in the winter ($p < 0.01$; t-test). ATP levels ranged from 1.7 to 15.2 ng l^{-1} in drinking water from distribution system A, with average concentrations of 6.3 ± 3.4 and 3.6 ± 2.0 ng l^{-1} in September 2009 and February 2010, respectively (Figure 7.3). In distribution system B, ATP levels ranged from <1 to 13.6 ng l^{-1} with average values of 7.4 ± 2.6 ng l^{-1} in September 2009 and <1 ng l^{-1} in February 2010 (Figure 7.3). The differences in ATP levels between September and February were statistically significant ($p < 0.01$; t-test). The colder temperature in the winter month could be responsible for the lower active biomass (ATP) in drinking water sampled in February 2010. The average cell numbers in drinking water from distribution system B were also significantly ($p < 0.01$; t-test) lower in February 2010 compared to September 2009 (Figure 7.3) and correspond to the lower ATP concentrations. In contrast, the average cell numbers in drinking water from distribution system A were significantly higher ($p < 0.01$; t-test) in February 2010 compared to September 2009, which is opposite to the ATP concentrations and unexpected at the lower water temperatures in February 2010. However, it is known that cell numbers are an inappropriate parameter for biomass, because cells can be active/non-active and can differ in size (Schlegel 1993). Consequently, the higher cell numbers in distribution system A in February 2010 might be caused by non-active or smaller cells in the winter than in the summer.

7.4.2 Opportunistic pathogens

The volume of drinking water sampled and filtered in September 2009 for DNA analyses was 500 ml, which was not enough for all qPCR assays. Therefore, the number of 16S rRNA gene copies of mycobacteria and the 28S rRNA gene copies of *A. fumigatus* were not determined in drinking water sampled in September 2009. All qPCR assays could be performed on DNA isolated from drinking waters sampled in February 2010, because the drinking water volume was increased from 500 to 2000 ml.

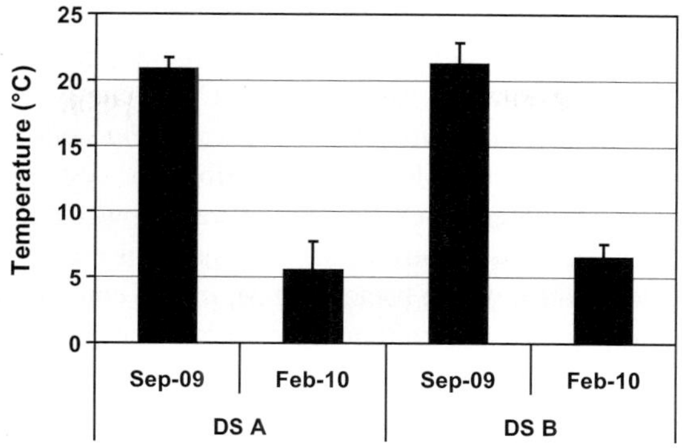

Figure 7.2 The average temperature (±standard deviation) measured in the drinking-water samples from distribution system A (DS A) and distribution system B (DS B) in September 2009 and February 2010.

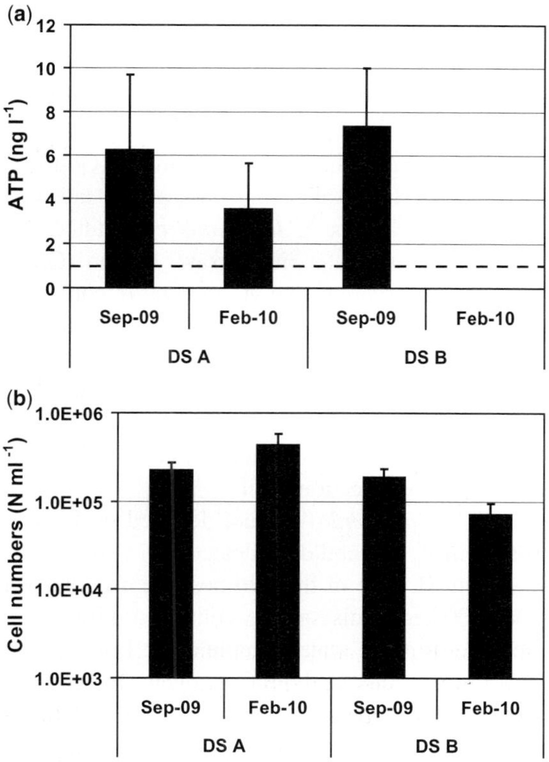

Figure 7.3 Average ATP levels and cell numbers (±standard deviation) in drinking water from distribution system A (DS A) and B (DS B) sampled in September 2009 and February 2010. Dashed line is detection limit ATP method (1 ng l^{-1} ATP).

M. avium complex, *Acanthamoeba* spp. and *A. fumigatus* were not detected in any of the drinking water samples analysed, meaning that copies of the 16S rRNA gene of *M. avium* complex and the 28S rRNA gene of *A. fumigatus* were below 200 copies l^{-1} and the number of *Acanthamoeba* cells were lower than 1 $N l^{-1}$ in drinking water sampled from distribution system A and B. The observation that *M. avium* complex bacteria were below the detection limit in drinking water from these two distribution systems is in contrast to results from studies in the United States, where bacteria of *M. avium* complex have often been detected in drinking water using cultivation or molecular methods (Carson *et al.* 1988; von Reyn *et al.* 1993; Aronson *et al.* 1999; Falkinham *et al.* 2001; Pryor *et al.* 2004; Feazel *et al.* 2009). In the US, drinking water is distributed with a disinfectant residual like chlorine or monochloramine, whereas drinking water in the Netherlands is distributed without such a disinfectant residual. Studies have shown that *M. avium* complex bacteria can be resistant to chlorine or chloramine (Taylor *et al.* 2000; Berry *et al.* 2010). The use of these disinfectants in drinking water might, therefore, result in a competitive advantage for *M. avium* complex bacteria over other bacterial species that are vulnerable to these disinfectant residuals in drinking water, although other factors (e.g. temperature, easily assimilable organic carbon) play a role as well (Falkinham *et al.* 2001).

The occurrence of *A. fumigatus* in drinking water has also been studied in other countries. *A. fumigatus* was not or only rarely encountered in drinking water samples from 29 different water localities in Germany (Gottlich *et al.* 2002), 14 water supplies in Norway (Hageskal *et al.* 2006) and at taps of 60 homes in the US (Vesper *et al.* 2007), whereas *A. fumigatus* constituted 50% of the mould species isolated from drinking water in Warsaw, Poland (Grabinska-Loniewska *et al.* 2007). Thus, the observation that *A. fumigatus* could not be detected in drinking water from distribution system A and B is consistent with results on *A. fumigatus* in drinking water from Germany, Norway and the US. However, distributed drinking water from other production plants in the Netherlands have to be analysed for *A. fumigatus*, *M. avium* complex and *Acanthamoeba*, before it can be concluded that the number of these opportunistic pathogens in Dutch drinking water are below the detection limit.

The *mip* gene of *L. pneumophila* was not detected in drinking water sampled from distribution system A, but could be detected in two drinking water samples from distribution system B. One of the two positive drinking water samples was taken in September 2009 and this sample contained 1100 *mip* gene copies per litre. The other positive drinking water sample came from another tap location in February 2010; this sample contained 3300 *mip* gene copies per litre. Despite the positive PCR results, *L. pneumophila* could not be cultivated from the same water sample on a selective agar medium (buffered charcoal yeast extract agar) for *Legionella*. The discrepancy between the qPCR and cultivation results can be caused by several factors: (i) the observed *mip* gene of *L. pneumophila* came from dead *L. pneumophila* cells, (ii) not all *L. pneumophila* strains in drinking water can be cultivated on the selective agar medium or (iii) the agar plates were overgrown with other *Legionella* species, preventing the detection of *L. pneumophila*. The consequence of these findings for public health remains uncertain, also because not all *L. pneumophila* strains are pathogenic (Yzerman *et al.* 2010).

The *regA* gene of *P. aeruginosa* was not detected in drinking water from distribution system A or B sampled in September 2009. In February 2010 three of thirteen (23.1%) drinking water samples from distribution A and one of eleven (9.1%) drinking water samples from distribution B were positive for the *regA* gene of *P. aeruginosa*. Copy numbers of the *regA* gene ranged from 440 to 4400 $N\ l^{-1}$ in the four *P. aeruginosa* positive samples. Only a low number of studies have reported the numbers of *P. aeruginosa* in distributed drinking water, which makes it difficult to compare our data with data from other countries. However, the number of positive drinking-water samples observed in our study was consistent with the number of *P. aeruginosa* positive drinking-water samples found with cultivation methods in Germany (7.6%) and Greece (9%) (Schubert and Blum 1974; Papapetropoulou *et al.* 1994).

The difference between the occurrence of *P. aeruginosa* in distributed drinking water from distribution system A and B and the higher number of *P. aeruginosa* in the winter might indicate that certain factors (e.g. temperature, water composition, treatment processes used for the production of drinking water, pipe materials) might influence the growth of *P. aeruginosa* in drinking water, but additional research is required to determine the influence of these factors on the occurrence of *P. aeruginosa*.

The *chitA* gene of *S. maltophilia* was detected in only one distributed drinking-water sample. This drinking-water sample was taken in September 2009 from distribution system A and a relatively high number of *chitA* gene copies ($3.0 \times 10^4\ N\,l^{-1}$) was observed in this sample. In all other drinking-water samples, *chitA* gene copies were below the detection limit of 412 gene copies per litre. Thus, *S. maltophilia* seems to be only sporadically present in the two sampled drinking-water distribution systems and the occurrence of the organism is too low to conclude whether temperature or other factors influence the presence of *S. maltophilia* in drinking water. Others have also reported the occurrence of *S. maltophilia* in drinking water (reviewed in (Rusin *et al.* 1997)), but none of those studies have specifically searched for *S. maltophilia* in drinking water and *S. maltophilia* was obtained on general agar media for heterotrophic bacteria.

The 18S rRNA gene of fungi was detected in all drinking-water samples, indicating that fungi are a normal part of the drinking-water microflora in the two sampled distribution systems. The average fungal 18S rRNA gene copies ranged from 1.0×10^5 to $1.3 \times 10^7\ N\,l^{-1}$ (Figure 7.4). A significant difference in gene copies was observed between drinking water sampled in September 2009 and drinking water sampled in February 2010 ($p < 0.01$; t-test). This indicates that fungal numbers are higher in summer than in winter, which is probably caused by the higher drinking water temperatures in summer (Figure 7.2) and corresponds to the higher level of active microbial biomass (ATP) in September 2009 (Figure 7.3). No statistical significant difference was observed between the fungal 18S rRNA gene copies in drinking water from distribution system A and B in September 2009 ($p > 0.01$; t-test), but in February 2010 fungal 18S rRNA gene copy numbers were higher in drinking water sampled from distribution system A than from distribution system B (Figure 7.4; $p < 0.01$; t-test). It cannot be deduced from this small study which factor(s) might be responsible for this statistical difference.

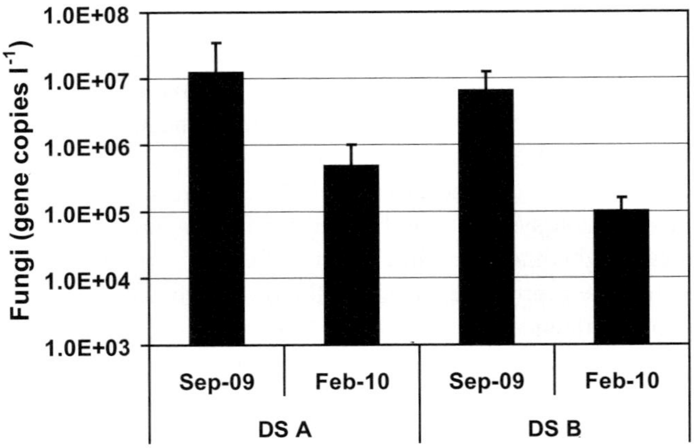

Figure 7.4 Average 18S rRNA gene copy numbers ($\pm$standard deviation) of fungi in drinking water from distribution system A (DS A) and B (DS B) sampled in September 2009 and February 2010.

The occurrence of fungi in drinking water has also been studied in other countries, but these studies have been based on fungi that can be cultivated on agar media. The results from these studies showed in general that fungi could be detected in 43 to 70% of the drinking-water samples (Hinzelin and Block 1985; Frankova and Horecka 1995; Anaissie *et al.* 2003; Hageskal *et al.* 2006; Pedro-Botet *et al.* 2007). However, Gottlich *et al.* (2002) detected fungi in only 8% of drinking-water samples from 29 different water localities in Germany, whereas 86 to 95% of drinking water sampled in Finland and Greece contained cultivable fungi (Niemi *et al.* 1982; Arvanitidou *et al.* 1999). The number of fungi observed in these studies was between 1 and 500 cfu l^{-1}, which is considerably lower than the number of 18S rRNA gene copies of fungi observed in our study (Figure 7.4). There are a number of possible explanations for this apparent discrepancy: (i) not all fungal species can be cultivated, (ii) the qPCR assay also detects dead fungal cells, (iii) fungi are often not present as separate cells and, therefore, one colony might be formed by more than one fungal cell and (iv) the number of 18S rRNA gene copies per fungal cell can be up to 100.

The importance of the observation that fungal species are present in drinking water for public health remains unclear, because most fungal species are not pathogenic to humans. However, as shown in paragraph 7.3.1 some fungal species detected in drinking water in the Netherlands are related to fungal species involved in human, animal or plant disease. Additional research on the identification of fungal species and strains from drinking water is necessary to determine the role fungal species from drinking water may play in causing disease in (immunocompromised) humans (Anaissie *et al.* 2003).

The 16S rRNA gene of bacteria belonging to the genus *Mycobacterium* was only determined on DNA from drinking-water samples taken in February 2010. The 16S rRNA gene was detected in all drinking water samples from distribution system A and B. The average 16S rRNA gene copies in drinking water from distribution system A was 1.3 ($\pm$1.3) $\times$ 10^5 N l^{-1}. In drinking water from distribution system B gene copies were 2.6 ($\pm$5.3) $\times$ 10^5 N l^{-1}. These differences between distribution system A and B were not statistically significant ($p > 0.01$; t-test on log-transformed data).

Previous studies in the Netherlands have shown that between 78 and 100% of drinking-water samples taken directly from the tap contained cultivated mycobacteria (Engel *et al.* 1980; van Ingen *et al.* 2010), which is similar to the number of positive drinking-water samples observed in our study. Mycobacteria were also detected in drinking-water samples in other Western European countries and the US, but the percentage of positive samples varied considerably between countries (15 to 83%) and within a country (US; 15 to 83%) (Carson *et al.* 1988; Fischeder *et al.* 1991; Covert *et al.* 1999; Tsintzou *et al.* 2000; Falkinham *et al.* 2001; Le Dantec *et al.* 2002; Pryor *et al.* 2004; Torvinen *et al.* 2004; Hussein *et al.* 2009). However, the number of mycobacteria in drinking water was determined by using selective agar media in all these studies. Hussein *et al.* (2009) used a different qPCR assay for the specific detection of the 16S rRNA gene of mycobacteria in drinking water from taps in four different hospitals in Germany. They also detected mycobacteria in all drinking-water samples with the qPCR assay, but could only cultivate mycobacteria from 62% of the samples. This discrepancy between cultivation and qPCR studies is most likely caused by mycobacterial species that cannot be cultivated or the detection of dead mycobacteria with the qPCR assay.

Overall, the results from our studies show (i) that certain opportunistic pathogens seem to be able to multiply in drinking-water distribution systems and (ii) that some opportunistic pathogens were not detected in distributed drinking water from two treatment plants, other pathogens were sporadically detected and some broader groups of microorganisms that harbour opportunistic pathogenic species (i.e. fungi and mycobacteria) were detected in all drinking-water samples. It is difficult to draw general conclusions from this preliminary study, since only drinking water from two distribution systems was investigated. Consequently, it remains to be determined whether these opportunistic pathogens are capable of growing in drinking-water distribution and plumbing systems and whether water source, treatment processes, water composition, temperature and distribution materials affect the occurrence and growth of opportunistic pathogens. Still, the observation that several opportunistic pathogens can sporadically be detected in drinking water, has resulted in a more comprehensive study on the occurrence of opportunistic pathogens in drinking-water distribution and plumbing systems in the Netherlands (van der Wielen and van der Kooij 2013).

Acknowledgements

This study was financed by the water supply companies in The Netherlands as part of the Joint Research Programme (Bedrijfstakonderzoek [BTO]).

7.5 REFERENCES

Aleksic S. and Bockemuhl J. (1988). Serological and biochemical characteristics of 416 *Yersinia* strains from well water and drinking water plants in the Federal Republic of Germany: lack of evidence that these strains are of public health importance. *Zentralblatt für Bakteriologie, Mikrobiologie und Hygiene [B]*, **185**, 527–533.

Anaissie E. J., Stratton S. L., Dignani M. C., Lee C. K., Summerbell R. C., Rex J. H., Monson T. P. and Walsh T. J. (2003). Pathogenic molds (including *Aspergillus* species) in hospital water distribution systems: a 3-year prospective study and clinical implications for patients with hematologic malignancies. *Blood*, **101**, 2542–2546.

Anderson K. and Jamieson A. (1972). Primary amoebic meningoencephalitis. *Lancet*, **299**, 902–903.

Anonymous (2001). Dutch Drinking Water Decree. Staatsblad van het Koninkrijk der Nederlanden 31 (in Dutch).

Aronson T., Holtzman A., Glover N., Boian M., Froman S., Berlin O. G., Hill H. and Stelma G. (1999). Comparison of large restriction fragments of *Mycobacterium avium* isolates recovered from AIDS and non-AIDS patients with those of isolates from potable water. *J. Clin. Microbiol.*, **37**, 1008–1012.

Arvanitidou M., Kanellou K., Constantinides T. C. and Katsouyannopoulos V. (1999). The occurrence of fungi in hospital and community potable waters. *Lett. App. Microbiol.*, **29**, 81–84.

Bartram J., Cotruvo C., Exner M., Fricker C. and Glasmacher A. (2003). Heterotrophic Plate Counts and Drinking-water Safety. IWA Publishing on behalf of the WHO, London, UK.

Berry D., Horn M., Xi C. and Raskin L. (2010). *Mycobacterium avium* infections of *Acanthamoeba* strains: host strain variability, grazing-acquired infections, and altered dynamics of inactivation with monochloramine. *Appl. Environ. Microbiol.*, **76**, 6685–6688.

Bijkerk P., van Lier E. A., Swaan C. M. and Kretzschmar M. E. E. (2011). State of infection diseases in the Netherlands 2010. RIVM report 210211007/2011, Bilthoven, the Netherlands (in Dutch).

Borchardt M. A., Stemper M. E. and Standridge J. H. (2003). *Aeromonas* isolates from human diarrheic stool and groundwater compared by pulsed-field gel electrophoresis. *Emerg. Infect. Dis.*, **9**, 224–228.

Buttery J. P., Alabaster S. J., Heine R. G., Scott S. M., Crutchfield R. A., Bigham A., Tabrizi S. N. and Garland S. M. (1998). Multiresistant *Pseudomonas aeruginosa* outbreak in a pediatric oncology ward related to bath toys. *Pediatr. Infect. Dis. J.*, **17**, 509–513.

Carson L. A., Bland L. A., Cusick L. B., Favero M. S., Bolan G. A., Reingold A. L. and Good R. C. (1988). Prevalence of nontuberculous mycobacteria in water supplies of hemodialysis centers. *Appl. Environ. Microbiol.*, **54**, 3122–3125.

Chadha R., Grover M., Sharma A., Lakshmy A., Deb M., Kumar A. and Mehta G. (1998). An outbreak of post-surgical wound infections due to *Mycobacterium abscessus*. *Pediatr. Surg. Int.*, **13**, 406–410.

Challier S., Boyer S., Abachin E. and Berche P. (2004). Development of a serum-based Taqman real-time PCR assay for diagnosis of invasive aspergillosis. *J. Clin. Microbiol.*, **42**, 844–846.

Cheng K. H., Leung S. L., Hoekman H. W., Beekhuis W. H., Mulder P. G., Geerards A. J. and Kijlstra A. (1999). Incidence of contact-lens-associated microbial keratitis and its related morbidity. *Lancet*, **354**, 181–185.

Covert T. C., Rodgers M. R., Reyes A. L. and Stelma G. N. (1999). Occurrence of nontuberculous mycobacteria in environmental samples. *Appl. Environ. Microbiol.*, **65**, 2492–2496.

Currie B. J. and Jacups S. P. (2003). Intensity of rainfall and severity of melioidosis, Australia. *Emerg. Infect. Dis.*, **9**, 1538–1542.

Currie B. J., Mayo M., Anstey N. M., Donohoe P., Haase A. and Kemp D. J. (2001). A cluster of melioidosis cases from an endemic region is clonal and is linked to the water supply using molecular typing of *Burkholderia pseudomallei* isolates. *Am. J. Trop. Med. Hyg.*, **65**, 177–179.

da Silva Filho L. V., Tateno A. F., de Velloso L. F., Levi J. E., Fernandes S., Bento C. N., Rodrigues J. C. and Ramos S. R. (2004). Identification of *Pseudomonas aeruginosa*, *Burkholderia cepacia* complex, and *Stenotrophomonas maltophilia* in respiratory samples from cystic fibrosis patients using multiplex PCR. *Pediatr. Pulmonol.*, **37**, 537–547.

Den Boer J. W., Yzerman E. P., Schellekens J., Lettinga K. D., Boshuizen H. C., van Steenbergen J. E., Bosman A., van den Hof S., van Vliet H. A., Peeters M. F., van Ketel R. J., Speelman P., Kool J. L. and Conyn van Spaendonck M. A. E. (2002). A large outbreak of Legionnaires' disease at a flower show, the Netherlands, 1999. *Emerg. Infect. Dis.*, **8**, 37–43.

Denton M., Todd N. J., Kerr K. G., Hawkey P. M. and Littlewood J. M. (1998). Molecular epidemiology of *Stenotrophomonas maltophilia* isolated from clinical specimens from patients with cystic fibrosis and associated environmental samples. *J. Clin. Microbiol.*, **36**, 1953–1958.

Doring G., Jansen S., Noll H., Grupp H., Frank F., Botzenhart K., Magdorf K. and Wahn U. (1996). Distribution and transmission of *Pseudomonas aeruginosa* and *Burkholderia cepacia* in a hospital ward. *Pediatr. Pulmonol.*, **21**, 90–100.

Edberg S. C., Gallo P. and Kontnick C. (1996). Analysis of the virulence characteristics of bacteria isolated from bottled, water cooler, and tap water. *Microb. Ecol. Health Dis.*, **9**, 67–77.

Emtiazi F., Schwartz T., Marten S. M., Krolla-Sidenstein P. and Obst U. (2004). Investigation of natural biofilms formed during the production of drinking water from surface water embankment filtration. *Wat. Res.*, **38**, 1197–1206.

Engel H. W., Berwald L. G. and Havelaar A. H. (1980). The occurrence of *Mycobacterium kansasii* in tapwater. *Tubercle*, **61**, 21–26.

Falkinham J. O. III, Norton C. D. and LeChevallier M. W. (2001). Factors influencing numbers of *Mycobacterium avium*, *Mycobacterium intracellulare*, and other mycobacteria in drinking water distribution systems. *Appl. Environ. Microbiol.*, **67**, 1225–1231.

Feazel L. M., Baumgartner L. K., Peterson K. L., Frank D. N., Harris J. K. and Pace N. R. (2009). Opportunistic pathogens enriched in showerhead biofilms. *PNASUSA*, **106**, 16393–16399.

Fischeder R., Schulze-Robbecke R. and Weber A. (1991). Occurrence of mycobacteria in drinking water samples. *Zentralbl. Hyg. Umweltmed.*, **192**, 154–158.

Fisher M. C., LiPuma J. J., Dasen S. E., Caputo G. C., Mortensen J. E., McGowan K. L. and Stull T. L. (1993). Source of *Pseudomonas cepacia*: ribotyping of isolates from patients and from the environment. *J. Pediatr.*, **123**, 745–747.

Fisman D. N., Lim S., Wellenius G. A., Johnson C., Britz P., Gaskins M., Maher J., Mittleman M. A., Spain C. V., Haas C. N. and Newbern C. (2005). It's not the heat, it's the humidity: wet weather increases legionellosis risk in the greater Philadelphia metropolitan area. *J. Infect. Dis.*, **192**, 2066–2073.

Frankova E. and Horecka M. (1995). Filamentous soil fungi and unidentified bacteria in drinking water from wells and water mains near Bratislava. *Microbiol. Res.*, **150**, 311–313.

Frumkin H., Hess J., Luber G., Malilay J. and McGeehin M. (2008). Climate change: the public health response. *Am. J. Public Health*, **98**, 435–445.

Garcia-Nunez M., Sopena N., Ragull S., Pedro-Botet M. L., Morera J. and Sabria M. (2008). Persistence of *Legionella* in hospital water supplies and nosocomial Legionnaires' disease. *FEMS Immunol. Med. Microbiol.*, **52**, 202–206.

Garrity G. M. (2001–2011). Bergey's Manual of Systematic Bacteriology. Springer, New York.

Glasmacher A., Engelhart S. and Exner M. (2003). Infections from HPC organisms in drinking-water amongst the immunocompromised. In: Heterotrophic Plate Counts and Drinking-Water Safety, J. Bartram, J. Cotruvo, M. Exner, C. Fricker and A. Glasmacher (eds), WHO, Cornwall, UK, pp. 137–145.

Gottlich E., van der Lubbe W., Lange B., Fiedler S., Melchert I., Reifenrath M., Flemming H. C. and de Hoog S. (2002). Fungal flora in groundwater-derived public drinking water. *Int. J. Hyg. Environ. Health*, **205**, 269–279.

Goverde R. (1999). *Yersinia enterocolitica*. Genes Involved in Cold-Adaptation. PhD thesis, Utrecht University, Utrecht.

Grabinska-Loniewska A., Konillowicz-Kowalska T., Wardzynska G. and Boryn K. (2007). Occurrence of fungi in water distribution system. *Polish J. Environ. Stud.*, **16**, 539–547.

Hageskal G. (2007). Moulds in Norwegian Drinking Water. Occurrence and Distribution of Mould Species in Drinking Water Distribution Systems. PhD thesis, University of Oslo, Oslo, Norway.

Hageskal G., Knutsen A. K., Gaustad P., de Hoog G. S. and Skaar I. (2006). Diversity and significance of mold species in Norwegian drinking water. *Appl. Environ. Microbiol.*, **72**, 7586–7593.

Haines A., McMichael A. J. and Epstein P. R. (2000). Environment and health: 2. Global climate change and health. *CMAJ*, **163**, 729–734.

Hanninen M. L. and Siitonen A. (1995). Distribution of *Aeromonas* phenospecies and genospecies among strains isolated from water, foods or from human clinical samples. *Epidemiol. Infect.*, **115**, 39–50.

Havelaar A. H., Schets F. M., van Silfhout A., Jansen W. H., Wieten G. and van der Kooij D. (1992). Typing of *Aeromonas* strains from patients with diarrhoea and from drinking water. *J. Appl. Bacteriol.*, **72**, 435–444.

Hinzelin F. and Block J. C. (1985). Yeasts and filamentous fungi in drinking water. *Environ. Technol. Lett.*, **6**, 101–106.

Hoque S. N., Graham J., Kaufmann M. E. and Tabaqchali S. (2001). *Chryseobacterium (Flavobacterium) meningosepticum* outbreak associated with colonization of water taps in a neonatal intensive care unit. *J. Hosp. Infect.*, **47**, 188–192.

Horrevorts A. M., Heijerman H. G., Moller A. V., Dankert-Roelse J. E., Mouton J. W., van der Laag J. and Dankert J. (1995). Cystic fibrosis and *Burkholderia* (formerly *Pseudomonas*) *cepacia*: an increasing clinical and psychosocial problem. *Ned. Tijdschr. Geneeskd.*, **139**, 2343–2346.

Hussein Z., Landt O., Wirths B. and Wellinghausen N. (2009). Detection of non-tuberculous mycobacteria in hospital water by culture and molecular methods. *Int. J. Med. Microbiol.*, **299**, 281–290.

Kahane S., Platzner N., Dvoskin B., Itzhaki A. and Friedman M. G. (2004). Evidence for the presence of *Simkania negevensis* in drinking water and in reclaimed wastewater in Israel. *Appl. Environ. Microbiol.*, **70**, 3346–3351.

Kahane S., Greenberg D., Newman N., Dvoskin B. and Friedman M. G. (2007). Domestic water supplies as a possible source of infection with *Simkania*. *J. Infect.*, **54**, 75–81.

Kilvington S., Gray T., Dart J., Morlet N., Beeching J. R., Frazer D. G. and Matheson M. (2004). *Acanthamoeba* keratitis: the role of domestic tap water contamination in the United Kingdom. *Invest. Ophthalmol. Vis. Sci.*, **45**, 165–169.

Kuiper M. W., Wullings B. A., Akkermans A. D. L., Beumer R. R. and van der Kooij D. (2004). Intracellular proliferation of *Legionella pneumophila* in *Hartmannella vermiformis* in aquatic biofilms grown on plasticized polyvinyl chloride. *Appl. Environ. Microbiol.*, **70**, 6826–6833.

Kwon-Chung K. J. (1975). A new pathogenic species of *Aspergillus* in the *Aspergillus fumigatus* series. *Mycologia*, **67**, 770–779.

Le Dantec C., Duguet J. P., Montiel A., Dumoutier N., Dubrou S. and Vincent V. (2002). Occurrence of mycobacteria in water treatment lines and in water distribution systems. *Appl. Environ. Microbiol.*, **68**, 5318–5325.

Lee D. G., Kim S. J. and Park S. J. (2006). Effect of reservoirs on microbiological water quality in a drinking water distribution system. *J. Microbiol. Biotechnol.*, **16**, 1060–1067.

Levin R. B., Epstein P. R., Ford T. E., Harrington W., Olson E. and Reichard E. G. (2002). US drinking water challenges in the twenty-first century. *Environ. Health Persp.*, **110** (Suppl 1), 43–52.

Luck P. C., Schneider T., Wagner J., Walther I., Reif U., Weber S. and Weist K. (2008). Community-acquired Legionnaires' disease caused by *Legionella pneumophila* serogroup 10 linked to the private home. *J. Med. Microbiol.*, **57**, 240–243.

Ma P., Visvesvara G. S., Martinez A. J., Theodore F. H., Dagget P. M. and Sawyer T. K. (1990). *Naegleria* and *Acanthamoeba* infections: review. *Rev. Infect. Dis.*, **12**, 490–513.

Mahenthiralingam E., Urban T. A. and Goldberg J. B. (2005). The multifarious, multireplicon *Burkholderia cepacia* complex. *Nature Rev. Microbiol.*, **3**, 144–156.

Marciano-Cabral F., MacLean R., Mensah A. and LaPat-Polasko L. (2003). Identification of *Naegleria fowleri* in domestic water sources by nested PCR. *Appl. Environ. Microbiol.*, **69**, 5864–5869.

Morris J. G. Jr. and Potter M. (1997). Emergence of new pathogens as a function of changes in host susceptibility. *Emerg. Infect. Dis.* **3**, 433–441.

Moyer N. P., Luccini G. M., Holcomb L. A., Hall N. H. and Altwegg M. (1992). Application of ribotyping for differentiating aeromonads isolated from clinical and environmental sources. *Appl. Environ. Microbiol.*, **58**, 1940–1944.

Nelson J. W., Doherty C. J., Brown P. H., Greening A. P., Kaufmann M. E. and Govan J. R. (1991). *Pseudomonas cepacia* in inpatients with cystic fibrosis. *Lancet*, **338**, 1525.

Neumann M., Schulze-Robbecke R., Hagenau C. and Behringer K. (1997). Comparison of methods for isolation of mycobacteria from water. *Appl. Environ. Microbiol.*, **63**, 547–552.

Niemi R. M., Knuth S. and Lundstrom K. (1982). Actinomycetes and fungi in surface waters and in potable water. *Appl. Environ. Microbiol.*, **43**, 378–388.

Papapetropoulou M., Iliopoulou J., Rodopoulou G., Detorakis J. and Paniara O. (1994). Occurrence and antibiotic-resistance of *Pseudomonas* species isolated from drinking water in southern Greece. *J. Chemother.*, **6**, 111–116.

Pedro-Botet M. L., Sanchez I., Sabria M., Sopena N., Mateu L., Garcia-Nunez M. and Rey-Joly C. (2007). Impact of copper and silver ionization on fungal colonization of the water supply in health care centers: implications for immunocompromised patients. *Clin. Infect. Dis.*, **45**, 84–86.

Peterson S. W. (2008). Phylogenetic analysis of *Aspergillus* species using DNA sequences from four loci. *Mycologia*, **100**, 205–226.

Pitt J. I. (1994). The current role of *Aspergillus* and *Penicillium* in human and animal health. *J. Med. Vet. Mycol.*, **32**(Suppl 1), 17–32.

PREZIES (2008a). Referentiecijfers 1998–2007. Landelijke surveillance netwerk ziekenhuisinfecties. http://www.prezies.nl/ (accessed 29 September 2011)

PREZIES (2008b). Resultaten prevalentiestudie maart 2007. Landelijke surveillance netwerk ziekenhuisinfecties. http://www.prezies.nl/ (accessed 29 September 2011)

Pryor M., Springthorpe S., Riffard S., Brooks T., Huo Y., Davis G. and Sattar S. A. (2004). Investigation of opportunistic pathogens in municipal drinking water under different supply and treatment regimes. *Wat. Sci. Technol.*, **50**, 83–90.

Qvarnstrom Y., Visvesvara G. S., Sriram R. and Da Silva A. J. (2006). Multiplex real-time PCR assay for simultaneous detection of *Acanthamoeba* spp., *Balamuthia mandrillaris*, and *Naegleria fowleri*. *J. Clin. Microbiol.*, **44**, 3589–3595.

Reitler R. and Seligmann R. (1957). *Pseudomonas aeruginosa* in drinking water. *J. Appl. Bacteriol.*, **20**, 145–150.

Reuter S., Sigge A., Wiedeck H. and Trautmann M. (2002). Analysis of transmission pathways of *Pseudomonas aeruginosa* between patients and tap water outlets. *Crit. Care Med.*, **30**, 2222–2228.

Richards R. H., Holliman A. and Helgason S. (1978). *Exophiala salmonis* infection in Atlantic salmon *Salmo salar L. J. Fish Dis.*, **1**, 357–368.

Rudnick J. R., Beck-Sague C. M., Anderson R. L., Schable B., Miller J. M. and Jarvis W. R. (1996). Gram-negative bacteremia in open-heart-surgery patients traced to probable tap-water contamination of pressure-monitoring equipment. *Infect. Control Hosp. Epidemiol.*, **17**, 281–285.

Rusin P. A., Rose J. B., Haas C. N. and Gerba C. P. (1997). Risk assessment of opportunistic bacterial pathogens in drinking water. *Rev. Environ. Contam. Toxicol.*, **152**, 57–83.

Sachdeva P. and Virdi J. S. (2004). Repetitive elements sequence (REP/ERIC)-PCR based genotyping of clinical and environmental strains of *Yersinia enterocolitica* biotype 1A reveal existence of limited number of clonal groups. *FEMS Microbiol. Lett.*, **240**, 193–201.

Schlegel H. G. (1993). General Microbiology. Cambridge University Press, Cambridge, UK.

Schubert R. and Blum U. (1974). Zur Frage der Erweiterung der hygienischen Wasserkontrolle auf den Nachweis von *Pseudomonas aeruginosa*. *Gas- und Wasserfach*, **115**, 224–226 (in German).

Schubert R. and Scheiber P. (1975). Das Vorkommen von *Pseudomonas aeruginosa* im Grundwasser, Oberflächenwasser und in Wasserversorgungsleitungen in tropischen Gebieten. *Gas- und Wasserfach*, **116**, 413–415 (in German).

Shannon K. E., Lee D. Y., Trevors J. T. and Beaudette L. A. (2007). Application of real-time quantitative PCR for the detection of selected bacterial pathogens during municipal wastewater treatment. *Sci. Total Environ.*, **382**, 121–129.

Smit E., Leeflang P., Glandorf B., van Elsas J. D. and Wernars K. (1999). Analysis of fungal diversity in the wheat rhizosphere by sequencing of cloned PCR-amplified genes encoding 18S rRNA and temperature gradient gel electrophoresis. *Appl. Environ. Microbiol.*, **65**, 2614–2621.

Solomon S., Qin D., Manning M., Marquis M., Averyt K., Tignor M. M. B. and Miller H. L. (2007). Climate Change 2007. The Physical Science Basis. Cambridge University Press, Cambridge, England.

Stevens D. A., Kan V. L., Judson M. A., Morrison V. A., Dummer S., Denning D. W., Bennet J. E., Walsh T. J., Patterson T. F. and Pankey G. A. (2000). Practice guidelines for diseases caused by *Aspergillus*. *Clin. Infect. Dis.*, **30**, 696–709.

Taylor R. H., Falkinham J. O. III, Norton C. D. and LeChevallier M. W. (2000). Chlorine, chloramine, chlorine dioxide, and ozone susceptibility of *Mycobacterium avium*. *Appl. Environ. Microbiol.*, **66**, 1702–1705.

Telenti A., Marchesi F., Balz M., Bally F., Bottger E. C. and Bodmer T. (1993). Rapid identification of mycobacteria to the species level by polymerase chain reaction and restriction enzyme analysis. *J. Clin. Microbiol.*, **31**, 175–178.

ter Maaten J. C., Golding R. P., Strack van Schijndel R. J. and Thijs L. G. (1995). Disseminated aspergillosis after near-drowning. *Neth. J. Med.*, **47**, 21–24.

Torvinen E., Suomalainen S., Lehtola M. J., Miettinen I. T., Zacheus O., Paulin L., Katila M. L. and Martikainen P. J. (2004). Mycobacteria in water and loose deposits of drinking water distribution systems in Finland. *Appl. Environ. Microbiol.*, **70**, 1973–1981.

Trautmann M., Lepper P. M. and Haller M. (2005). Ecology of *Pseudomonas aeruginosa* in the intensive care unit and the evolving role of water outlets as a reservoir of the organism. *Am. J. Infect. Control*, **33**, S41–49.

Tsintzou A., Vantarakis A., Pagonopoulou O., Athanassiadou A. and Papapetropoulou M. (2000). Environmental mycobacteria in drinking water before and after replacement of the water distribution network. *Wat. Air Soil Poll.*, **120**, 273–282.

Valster R. M., Wullings B. A., Bakker G., Smidt H. and van der Kooij D. (2009). Free-living protozoa in two unchlorinated drinking water supplies, identified by phylogenic analysis of 18S rRNA gene sequences. *Appl. Environ. Microbiol.*, **75**, 4736–4746.

Valster R. M., Wullings B. A., van den Berg R. and van der Kooij D. (2011). Relationships between free-living protozoa, cultivable *Legionella* spp., and water quality characteristics in three drinking water supplies in the Caribbean. *Appl. Environ. Microbiol.*, **77**, 7321–7328.

van der Kooij D., Oranje J. P. and Hijnen W. A. (1982). Growth of *Pseudomonas aeruginosa* in tap water in relation to utilization of substrates at concentrations of a few micrograms per liter. *Appl. Environ. Microbiol.*, **44**, 1086–1095.

van der Kooij D., Veenendaal H. R. and Scheffer W. J. (2005). Biofilm formation and multiplication of *Legionella* in a model warm water system with pipes of copper, stainless steel and cross-linked polyethylene. *Wat. Res.*, **39**, 2789–2798.

van der Wielen P. W. J. J. and van der Kooij D. (2010). Effect of water composition, distance and season on the adenosine triphosphate concentration in unchlorinated drinking water in the Netherlands. *Wat. Res.*, **44**, 4860–4867.

Van der Wielen P. W. J. J. and van der Kooij D. (2013). Nontuberculous mycobacteria, fungi and opportunistic pathogens in unchlorinated drinking water in the Netherlands. *Appl. Environ. Microbiol.*, **79**, 825–834.

van der Wielen P. W. J. J., Voost S. and van der Kooij D. (2009). Ammonia-oxidizing bacteria and archaea in groundwater treatment and drinking water distribution systems. *Appl. Environ. Microbiol.*, **75**, 4687–4695.

van Ingen J. (2009). Nontuberculous Mycobacteria. From Gene Sequences to Clinical Relevance. PhD thesis, Radboud University Nijmegen, Nijmegen, The Netherlands.

van Ingen J., Blaak H., de Beer J., de Roda Husman A. M. and van Soolingen D. (2010). Rapidly growing nontuberculous mycobacteria cultured from home tap and shower water. *Appl. Environ. Microbiol.*, **76**, 6017–6019.

Verduyn Lunel F., Koeleman J. G., Spanjaard L., VandenBroucke-Grauls C., Schultz C., Verbrugh H. A., Vos G., Troelstra A., Mascini E., Verweij P. E. and Voss A. (2006). Trends in fungaemia and antifungal susceptibility in the Netherlands. *Neth. J. Med.*, **64**, 236–242.

Verweij P. E., Meis J. F., Christmann V., van der Bor M., Melchers W. J., Hilderink B. G. and Voss A. (1998). Nosocomial outbreak of colonization and infection with *Stenotrophomonas maltophilia* in preterm infants associated with contaminated tap water. *Epidemiol. Infect.*, **120**, 251–256.

Vesper S. J., Haugland R. A., Rogers M. E. and Neely A. N. (2007). Opportunistic *Aspergillus* pathogens measured in home and hospital tap water by quantitative PCR (QPCR). *J. Wat. Health*, **5**, 427–431.

Vieira M. R., Milheiro A. and Pacheco F. A. (2001). Phaeohyphomycosis due to *Cladosporium cladosporioides*. *Med. Mycol.*, **39**, 135–137.

Vochem M., Vogt M. and Doring G. (2001). Sepsis in a newborn due to *Pseudomonas aeruginosa* from a contaminated tub bath. *N. Engl. J. Med.*, **345**, 378–379.

von Reyn C. F., Waddell R. D., Eaton T., Arbeit R. D., Maslow J. N., Barber T. W., Brindle R. J., Gilks C. F., Lumio J., Lahdevirta J., Ranki A., Dawson D. and Falkinham J. O. III. (1993). Isolation of *Mycobacterium avium* complex from water in the United States, Finland, Zaire, and Kenya. *J. Clin. Microbiol.*, **31**, 3227–3230.

von Reyn C. F., Maslow J. N., Barber T. W., Falkinham J. O. III and Arbeit R. D. (1994). Persistent colonisation of potable water as a source of *Mycobacterium avium* infection in AIDS. *Lancet*, **343**, 1137–1141.

Weber D. J., Rutala W. A., Blanchet C. N., Jordan M. and Gergen M. F. (1999). Faucet aerators: A source of patient colonization with *Stenotrophomonas maltophilia*. *Am. J. Infect. Control*, **27**, 59–63.

Wullings B. A. and van der Kooij D. (2006). Occurrence and genetic diversity of uncultured *Legionella* spp. in drinking water treated at temperatures below 15°C. *Appl. Environ. Microbiol.*, **72**, 157–166.

Wullings B. A., Bakker G. and van der Kooij D. (2011). Concentration and diversity of uncultured *Legionella* spp. in two unchlorinated drinking water supplies with different concentrations of natural organic matter. *Appl. Environ. Microbiol.*, **77**, 634–641.

Yzerman E., den Boer J., Caspers M., Almal A., Worzel B., van der Meer W., Montijn R. and Schuren F. (2010). Comparative genome analysis of a large Dutch Legionella pneumophila strain collection identifies five markers highly correlated with clinical strains. *BMC Genomics*, **11**, 433.

Zanetti F., De Luca G. and Stampi S. (2000). Recovery of *Burkholderia pseudomallei* and *B. cepacia* from drinking water. *Int. J. Food Microbiol.*, **59**, 67–72.

Chapter 8

The last meters before the tap: where drinking water quality is at risk

H.-C. Flemming, B. Bendinger, M. Exner, J. Gebel, T. Kistemann, G. Schaule, U. Szewzyk and J. Wingender

8.1 INTRODUCTION

Drinking water is most rigorously and systematically surveyed throughout the entire supply chain from raw water abstraction through treatment and distribution to the water meter. In most parts of Europe, the quality of drinking water usually is unobjectionable – but only up to the water meter. Behind, a twilight area begins: the household drinking water installation. Plumbing systems in buildings can be a significant source of contamination, and poor management of these systems has contributed to sporadic illness or even outbreaks of diseases, as was already reported decades ago (Wadowski *et al.* 1982). This situation was the focus of a collaborative research project in Germany that was carried out by five research institutions and supported by 17 industry partners between 2006 and 2010. A key issue of this research project was to pay more attention to the household installation, because here, the best drinking water can lose its quality. Useful suggestions to detect, sanitize and prevent such problems were generated. On the other hand, the project also revealed further research and regulation demand – not only concerning the materials but also the examination methods. The last meters to the tap are critical.

In this chapter, the most relevant results are presented in an overview, based on the final project report (Flemming 2010) and completed by most recent findings.

8.1.1 Biofilms in drinking water systems

It is long known that in drinking water distribution systems, all surfaces in contact with water can be colonized by microorganisms (Servais *et al.* 1995; Kilb *et al.*

2003; Wingender and Flemming 2004, 2011; Bressler *et al.* 2009; Wingender 2011). Certain types of plastic and elastomeric materials can promote biofilm formation due to the release of biodegradable compounds such as monomers or additives like plasticizers, anti-oxidants, providing favorable nutrient conditions for microorganisms (Schoenen and Schoeler 1983; van der Wende *et al.* 1989; Keevil 2003; Kilb *et al.* 2003; Rogers *et al.* 1994b; Anonymous 2007). It has been estimated that about 95% of all microbial cells present in drinking water distribution systems exist as biofilms on pipe surfaces, only 5% occur in the water phase (Flemming *et al.* 2002). Similarly, in a domestic hot water system, most of the culturable bacteria (72%) were found to be surface-associated (Bagh *et al.* 2004). Drinking water biofilms are formed predominantly by microorganisms of the autochthonous aquatic microflora without any relevance to human health. However, biofilms have the potential to harbor microbial pathogens which can harm human health, especially in immunocompromised persons (Keevil 2003). Once integrated in a drinking water biofilm, these organisms can tolerate external stresses such as the action of disinfectants much better and can thus persist and possibly multiply inside the biofilm. Contamination of drinking water occurs when hygienically relevant pathogens are released from a biofilm as a consequence of physical disturbance or active detachment of infectious cells, which then pose a potential threat to human health (Szewzyk *et al.* 2000; Flemming *et al.* 2002).

8.1.2 The problem of pathogen detection

Traditionally, pathogenic bacteria are detected and quantified by culture methods. However, some of these organisms may enter a viable but non-culturable (VBNC) state (Oliver 2005). Bacteria in the VBNC state do not grow on conventional microbiological media on which they would normally develop into colonies, but are still alive and are characterized by low levels of metabolic activity (Bosian and Borneuf 2001; Oliver 2010). Their viability and infectivity is subject of current debate (Nyström *et al.* 2001; Harvey 2011; Rochelle *et al.* 2011).

The conversion to the VBNC state is supposed to be a response to adverse environmental conditions (Oliver 2005) such as lack of nutrients, unfavorable water temperature, the presence of disinfectants or metal ions at toxic concentrations such as copper (Dwidjosiswojo *et al.* 2011). VBNC bacteria can become culturable again upon resuscitation under more favorable conditions. Oliver (2010) provided a list of pathogens known to enter the VBNC state, in which all relevant water-associated bacterial pathogens are present. Investigations of the VBNC state are usually performed with planktonic cells. For biofilms, there is still considerable research demand.

To circumvent the shortcomings of non-culturability of biofilm organisms, culture-independent methods are increasingly used in order to characterize the composition and diversity of microbial biofilm communities, and to detect VBNC cells in biofilms of drinking water systems (e.g. Gião *et al.* 2009a, b).

Some hygienically relevant bacterial species have been detected in real or experimental drinking water systems, using culture-independent methods, alone or in combination with conventional culture methods. Significantly higher cell numbers were frequently detected by culture-independent methods compared to colony counts determined on culture media, suggesting that viable cells are present and may persist in a VBNC state in drinking water biofilms. However, the biofilm-associated non-culturable state and the potential of resuscitation of these bacteria have not yet been elucidated. From a health perspective, the relevance of pathogens in the VBNC state may be underestimated, because it has to be taken into account that they can regain their virulence and are able to initiate infection when they revert to the culturable state (Oliver 2010; Dwidjosiswojo *et al.* 2011). Thus, pathogens in the VBNC state might represent an infectious potential when present in biofilms of drinking water systems (Wingender and Flemming 2011). Therefore, particular attention was put on that aspect in this study.

8.2 DRINKING WATER INSTALLATIONS IN BUILDINGS VS. PUBLIC NETWORKS

There are significant and important differences between drinking water installations as opposed to the public distribution network. Some characteristics of drinking water installations are:

- The range of water temperatures is much higher than in the distribution network. Drinking water systems in buildings usually comprise hot and cold water. It is not uncommon that even the cold water temperature exceeds 20 °C and more, particularly if the drinking water pipe is insufficiently insulated, located close to that of warm water, or if pipes run through heated rooms.
- The retention time of the water can be quite long (days to weeks) with stagnation periods, in particular during holidays or other periods of low consumption.
- Drinking water installations tend to include dead ends and areas with particularly low flow through or irregular flow regimes. Water outlets such as taps can be contaminated and act as sources for retrograde contamination.
- The range of materials employed in drinking water installations is partially different from that of drinking water distribution networks (e.g. synthetic polymers, copper, stainless steel, elastomers); possibly, they may be unapproved for use in drinking water. Occasionally, materials are employed which support microbial growth and, thus, biofilm formation.
- The surface-volume ratio is much higher, providing a large surface area ready to be colonized. Thus, biofilms play a more important role in drinking water installations.

- Auxiliary equipment can be part of the installation such as softeners, filters, phosphate dosage devices or others. It has been shown that such equipment raises the risk of elevated numbers of colony forming units in the water (Völker *et al.* 2010).
- The range of hygienically relevant organisms is different: in temperate climates, *Legionella pneumophila*, *Pseudomonas aeruginosa* and non-tuberculous mycobacteria are organisms that play a minor role in public water distribution systems compared to drinking water installations in buildings.
- Surveillance is more difficult and less frequent than in public networks. Experience shows that in particular, private apartment blocks are very rarely surveyed.
- In hospitals and other large buildings, frequently an unclear situation of piping system is encountered due to continuous reconstruction measures.
- The design of plumbing systems in buildings is highly variable and influenced by the diversity of water uses and building types (e.g. schools, hotels, residential buildings, sport facilities, office blocks as well as hospitals, nursing homes and other health-care facilities).
- Drinking water systems in buildings may be connected to water-based devices (e.g. cooling towers, boilers, swimming pools) or point-of-use equipment (e.g. washing machines, water dispensers) (World Health Organization [WHO] 2011).

These factors may enhance biofilm formation and elevate the risk of water contamination in drinking water installations. No epidemics related to these points are reported so far, except in hospitals. Here, water systems have been identified as reservoirs and sources of contamination by *L. pneumophila* and *P. aeruginosa* (e.g. Anaissie *et al.* 2002, 2003).

8.2.1 Surveillance – the twilight zone

A nationwide survey in Germany showed that in 2007 only 13% of the public health authorities had completed the monitoring of drinking water installations in public buildings in compliance with legal obligations (Völker *et al.* 2009). This was mainly due to over-commitment and understaffing of the health authorities which have to carry out the surveillance. The methods applied for the investigations differ widely and do not always lead to comparable results. But even if this is taken into account, the data allow for some conclusions (Kistemann *et al.* 2010).

For the survey, data from a total of 4600 public buildings that have been inspected by German health authorities with 30,000 water samples during 7 years (2003–2009) were available. Furthermore, 16,000 data points in over 1000 buildings concerning the specific technical drinking water installation construction could be collected. A statistical evaluation of the data showed an exceedance of the technical action value for *Legionella* spp. (100 cfu in 100 mL) in 12.8% of the samples and an exceedance of the guideline value for *P. aeruginosa* (0 cfu in 100 mL, as recommended by the Federal Environmental Agency) in 2.9% of the samples. Although *Legionella* could be found mainly in the warm water systems, which was not very surprising (American Society of Heating, Refrigerating and Air-Conditioning Engineers [ASHRAE] Standard 2000; Arnow *et al.* 1985; Codony *et al.* 2002; Darelid *et al.* 2002; Mathys *et al.* 2008), the organism was also found in 5.4% of the cold water samples, exceeding the technical action value. This demonstrates that *Legionella* spp. can find suitable survival and persistence conditions in cold water systems. Recently, high *Legionella* spp. numbers ($\geq$1000 cfu 100 mL^{-1}) were reported in cold water samples at temperatures as low as 11 °C in health-care facilities (Arvand *et al.* 2011). This underlines the importance of assessing cold water plumbing systems as potential contamination sources of *Legionella.* Pseudomonads (not further differentiated) could be found in cold and warm water systems to approximately equal percentage (Völker *et al.* 2010). The most relevant results on exceedance of limit and guidelines values respectively are summarized in Table 8.1.

Table 8.1 Exceedance of limit/guideline values of microbiological parameters of water in drinking water installations.

Microbiological parameters	Limit/Guideline values	Samples	Exceedance	
			Absolute	%
Legionella spp.	100 cfu 100 mL^{-1}	22,786	2908	12.8
Colony count 36 °C	100 cfu mL^{-1}	10,928	380	3.5
Colony count 22 °C	100 cfu mL^{-1}	10,869	129	1.8
Coliform bacteria	100 cfu 100 mL^{-1}	8652	152	1.8
Escherichia coli	100 cfu 100 mL^{-1}	8330	25	0.3
P. aeruginosa	0 cfu 100 mL^{-1}	3468	102	2.9

[1]Control and action levels that were proven of practical value for *Legionella* spp. and *P. aeruginosa*, respectively.
Source: Völker *et al.* (2010).

Frequently, the generally recognized standards of good practice were not met, such as regular maintenance of the drinking water installation (35%, n = 498), dismantled dead legs (65%, n = 327), or flushing after extended interruption of operation, for example, after summer holidays in schools (56%, n = 290). The statistical evaluation revealed that a daily temperature rise of the warm water reservoirs to at least 60 °C reduced the relative risk for positive reports of *Legionella* by approximately 30% (Kistemann *et al.* 2010). Any additional device for water treatment – be it softeners, filters, phosphate dosage or others – will at least double the risk (Kistemann *et al.* 2010), mainly because of lacking maintenance or other non-compliance to generally recognized standards of good practice. The survey showed clearly that the following points should be harmonized:

- Prioritization of institutions to be surveyed;
- Extent and intensity of inspections;
- Harmonization of methods;
- Documentation and archiving of data;
- Expertise of personnel.

Furthermore, the data showed without any doubt that determination of heterotrophic plate counts (HPC) according to the analytical methods defined by the Council Directive 98/83/EC (1998) and the German Drinking Water Ordinance (2001) only is not fully sufficient to assess the hygienic status of drinking water: no correlation could be found between HPC numbers and occurrence of *Legionella* after evaluation of more than 2200 data points. The proportion of *Legionella* and pseudomonads in objectionable samples was much higher than that of *E. coli*. In addition, no significant correlation could be found between the presence of *E. coli* and coliform bacteria or *Legionella* or pseudomonads (Ahmed *et al.* 2010; Kistemann *et al.* 2010).

Especially in cases of doubt with recurring contaminations, it is strongly recommended to refer to advanced methods for determination and verification of *Legionella*, such as fluorescence in situ hybridization (FISH) (Buchbinder *et al.* 2002; Wullings *et al.* 2002; Emtiazi *et al.* 2004; Whiley *et al.* 2011) or polymerase-chain-reaction (PCR)-based methods (Lye *et al.* 1997; Shen *et al.* 2006). Thus, hidden contamination sources with cells that survive disinfection treatment within the system can be detected. For differentiation between local and systemic contamination sources, it is recommended to employ pulsed-field gel electrophoresis (genetic fingerprinting) of bacterial isolates (Bender *et al.* 1990).

8.3 THE ROLE OF MATERIALS

Drinking water is not sterile and does not have to be so. The very successful strategy of water works for production of safe drinking water is based on nutrient limitation and, thus, prevents microbial growth and results in "biologically stable drinking water" (van der Kooij 1992). This is a way to avoid or minimize chlorination. The effect of plumbing materials on biofilm formation as well as on the persistence of pathogens in biofilms has mainly been studied on materials which make up drinking water distribution systems, while domestic plumbing materials usually differ from those and have been considered less frequently (Rogers *et al.* 1994a, b; Eboigbodin *et al.* 2008). Common plumbing materials are copper and plastics, but other material types such as elastomers can also be found as components of plumbing systems (WHO 2011).

If starved drinking water bacteria meet materials which release nutrients, they settle, form biofilms and grow (Rogers *et al.* 1994b; Keevil 2003; van der Kooij

2005). Usual suspect materials in such cases are synthetic polymers which are not approved for use in drinking water. For example, shower tubes or even small rubber fittings may become hospitable for bacteria settling on those surfaces. They will proliferate on the expense of small molecular weight plasticizers, anti-oxidants, residual monomers and other low-molecular weight components of polymers which are bioavailable directly at the surface where biofilms form. Here, facultative pathogens can settle and even proliferate if the conditions are favorable (Kilb *et al.* 2003). This was investigated and verified with seeding experiments with facultative pathogens in testing systems run under conditions close to real drinking water installations. Furthermore, it was shown that countermeasures such as application of disinfectants could improve the situation temporarily but did not lead to essential sanitation (Benölken *et al.* 2010; Schaule *et al.* 2010). Therefore, the quality of materials in drinking water installations is crucial. Materials that comply with the requirements of the German technical guidelines and promote the development of only small amounts of biofilms cannot prevent naturally occurring potentially pathogenic bacteria in water (e.g. *L. pneumophila* and *P. aeruginosa*) from settling in the biofilms. However, in this case the materials with low amounts of biofilm provide the best preconditions for a successful cleaning and disinfection of a drinking water installation (Benölken *et al.* 2010). Available materials – in particular those at the lower end of the market – might lead straight to enhanced biofilm formation and, thus, increase the risk for pathogen multiplication. A problem is that in drinking water installations, the use of approved materials is indeed mandatory, but the recognized standards of good practice are not always followed and difficult to monitor.

The choice of materials in the project was made concerning the most commonly used polymers in drinking water installations and, copper as a frequently employed material in drinking water installations in Germany. The materials employed in the project are listed in Table 8.2.

It is known that ethylene-propylene-diene-monomer (EPDM) without certificate for use in drinking water supports microbial growth and has led to microbial contamination problems in practice, although it is still frequently in use. Therefore, this was chosen as a “worst case material”. EPDM with certificate can also cause problems with microbial contamination. Therefore, this was also included in the study. Under conditions close to practice and after exposure to drinking water for 1–2 weeks, a biofilm forms on all materials. After 6–10 weeks, a quasi-stationary state is reached. The colonization density depends upon the supporting material. The highest densities were observed on EPDM which is not approved for use in drinking water but has been employed in older drinking water installations frequently. The average bacterial density for three different water qualities and at 12 °C to 15 °C and 37 °C ranged between $10^6 - 10^8$ cells cm^{-2} for EPDM materials, $10^5 - 10^7$ cells cm^{-2} for PE-Xb and $10^4 - 10^6$ cells cm^{-2} for copper and PE-Xc (Benölken *et al* 2010).

Table 8.2 Materials employed in the study.

Material	Usage	Application	Compliance to DVGW W 270[5]	KTW category[7]
Copper	Metal pipe	Drinking water installation	–[6]	–
HD-PE-Xb[1]	Composite pipe	Drinking water installation	Yes	A[8]
HD-PE-Xc[2]	Polymer pipe	Drinking water installation	Yes	A
EPDM with certificate[3] (=w cert.)	Reinforced hoses	Connection to faucets (before 2007) and devices for example, (dish) washing machines	Yes	C[9]
EPDM without certificate[4] (=w/o cert.)	Reinforced hoses	Fittings and connections in drinking water installations (before 1995)	No	C

[1]HD-PE-Xb: high density silane-cross-linked polyethylene.
[2]HD-PE-Xc: high density electron-ray cross-linked polyethylene.
[3]Ethylene-propylene-diene-monomer which has passed the admission by DVGW W 270.
[4]Ethylene-propylene-diene-monomer which has not passed the admission by DVGW W 270.
[5]DVGW worksheet W 270: German standard procedure for admission of polymeric materials in drinking water systems (Anonymous 2007).
[6]Not applicable.
[7]German commission for drinking water. In order to assess the health risk of organic materials in contact with drinking water different categories were defined (KTW-Empfehlung 1977).
[8]Category A applies for pipes with a surface to volume ratio of 1:1 $cm^2\ mL^{-1}$. After a triple migration test for 72 h each, among other parameters to be tested, the total organic carbon (TOC) content of test water should not exceed 2.5 $mg\ m^{-2}\ day^{-1}$.
[9]Category C applies for pieces of equipment and rigid jointing sealants with a surface to volume ratio of 1:6 $cm^2\ mL^{-1}$. After a triple migration test for 72 h each, among other parameters to be tested, the TOC content of test water should not exceed 15.0 $mg\ m^{-2}\ day^{-1}$.

Of course, the drinking water quality also plays a role in biofilm formation and pathogen multiplication. In close-to-practice test rigs quasi-stationary biofilms on different materials were inoculated with a suspension of starved *L. pneumophila*. With increasing amount of biofilm the number of *L. pneumophila* in the biofilm increased, which again resulted in an increasing dissemination of *L. pneumophila* into stagnating water in warm water systems (37 °C) (Figures 8.1 and 8.2). In most cases the technical action value (100 cfu mL^{-1}) was exceeded. A combination of low material quality, which provides high amounts of biodegradable organic carbon (EPDM without certificate), with unfavorable water quality (12 mg L^{-1} nitrate and 1 mg L^{-1} phosphate) lead to the highest observed biofilm formation and *Legionella* concentration in biofilm and stagnating water (Benölken *et al.* 2010).

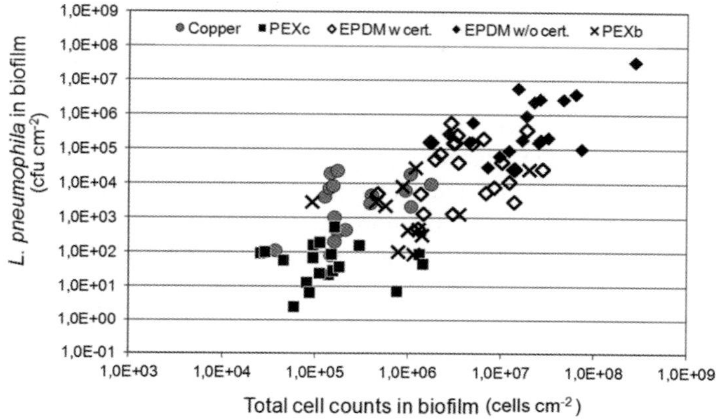

Figure 8.1 Colony counts of *L. pneumophila* in biofilms vs. total cell counts (DAPI) in biofilms on five different materials. Data are from three different water qualities at 37 °C in the close-to-practice test rigs 8 to 28 weeks after inoculation with *L. pneumophila*. *L. pneumophila* were quantified according to ISO 11731 on buffered charcoal yeast extract with α-ketoglutarate (BCYEα) agar without heat or acid pre-treatment at 36 °C.

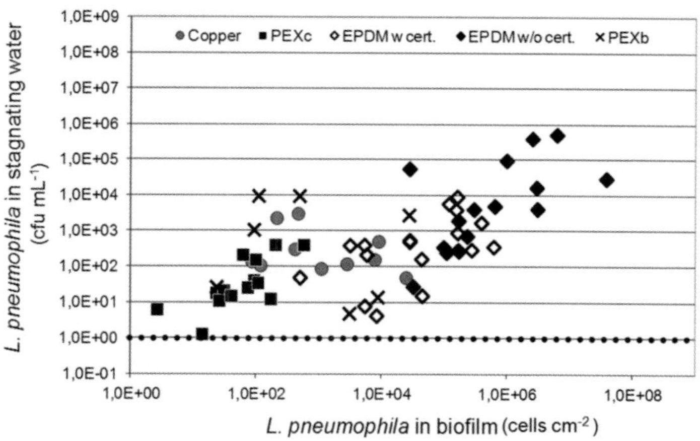

Figure 8.2 Colony counts of *L. pneumophila* in 8 h stagnating drinking water vs. colony counts of *L. pneumophila* in biofilms on five different materials. Data are from three different water qualities at 37 °C in the close-to-practice test rigs 8 to 28 weeks after inoculation with *L. pneumophila*. *L. pneumophila* were quantified according to ISO 11731 on BCYEα agar without heat or acid pretreatment at 36 °C. The threshold value for action in Germany is 1000 cfu L^{-1}.

8.3.1 Biofilm formation of new and aged materials

An important aspect concerning the role of materials is that they can change their properties during use, in particular in response to oxidative stress as applied by oxidizing disinfectants and by elevated temperatures ("ageing"). Therefore, it was interesting to compare the biofilm formation of new to used and, thus, aged materials. Changes in hydrophobicity, surface structure, roughness and even the formation of cracks and crevices can be the consequence of ageing (Schaule *et al.* 2010).

Artificial ageing was performed according to the standard test methods for evaluating the oxidative resistance of polyethylene (PE) pipe to chlorinated water (American Society for Testing and Materials [ASTM] 2008). The hypothesis was that ageing processes based on oxidation processes (disinfection) in combination with elevated temperature may increase the release of bioavailable substances from PE-Xc and EPDM materials. It is known that the oxidation of organic substances in water may enhance regrowth and biofilm formation because of new reaction products which are biodegradable. It was shown earlier that oxidizing agents such as chlorine may increase the assimilable organic carbon (AOC) and enhance regrowth potential. Artificial ageing was performed, for example, by exposure of the materials to a pressure of 4 bar at 40 °C in presence of 2.5 mg L^{-1} sodium hypochlorite or 4 mg L^{-1} chlorine dioxide respectively for a period of 4 weeks. Chlorine dioxide showed the highest effect on the materials under comparable conditions. It was obvious that ageing procedures may lead to the formation of cracks and crevices in all synthetic materials but it could not be shown that these cracks are preferred settling areas for bacteria (Schaule *et al.* 2010).

Biofilm formation was assessed in a multifactorial experimental approach in which the most relevant parameters influencing biofilm growth were selected and their effect was assessed. The effect was expressed as bacteria cell numbers per square centimeter. Different drinking water compositions were considered such as temperature, AOC, hardness, humic substances. This approach revealed that aged EPDM and aged PE-Xc, which was depleted of low molecular weight components, showed a tendency for supporting less biofilm development. The reason is therefore probably exhaustive leaching of biologically degradable substances at the interface and no further growth support. In practice, this indicates a lower biofilm formation potential for aged polymers with lower disinfectant demand.

The experiments were carried out with drinking water from the IWW Water Centre laboratory drinking water installation which contained no disinfectants and which was supplemented to simulate different types of drinking water, for example, oligotrophic (AOC equivalent to 5 µg L^{-1} acetate) and copiotrophic (AOC equivalent to 150 µg L^{-1} acetate). The results of the multi-parameter experiments showed that all chosen factors influence biofilm formation, but they do so to a different extent:

- The amount of organic nutrients represented the highest single influencing factor;

- If materials leached biodegradable substances, the dissolved organic carbon (DOC) or AOC respectively of water was of minor relevance;
- In situations where materials leach out low amounts of nutrients, the nutrient content of the water played a more important role;
- Temperature played a stronger role under copiotrophic conditions, but to a lesser extent under oligotrophic conditions;
- Material age also significantly influenced biofilm formation with aged materials having lower biofilm formation than new materials;
- Water hardness showed a low influence on biofilm formation potential.

8.4 THE RESULT OF QUANTIFICATION OF BACTERIA IN DRINKING WATER AND BIOFILMS DEPENDS UPON THE ANALYTICAL METHOD

For surveillance of the general microbiological drinking water quality, heterotrophic plate count (HPC) methods are usually employed. The routine application of these culture-dependent methods successfully contributes to ensure the provision of safe drinking water as a process parameter. As mentioned earlier, it is well known that only a minor fraction of the actual number of bacteria in drinking water is detected by these methods, a phenomenon which has been termed as the "Great Plate Count Anomaly" (Staley and Konopka 1985). For the purpose of quantifying the total number of microorganisms in the biofilms, culturing methods are unsuitable, because these methods usually underestimate the actual number of cells. Widely used methods for the determination of total cells counts in drinking water environments are based on the staining of bacteria with DNA-binding fluorescent dyes such as 4′,6-diamidino-2-phenylindole (DAPI) or SYBR Green I and the subsequent enumeration of cells by epifluorescence microscopy (Kempner and Pratt 1976) or flow cytometry (Berney *et al.* 2007; Hammes and Egli 2010).

In drinking water environments, the fraction of culturable bacteria frequently varies between approximately 0.01% and a few percent of the total cell counts (Kalmbach *et al.* 1997; Wingender and Flemming 2004). Low culturability is a typical property of bacteria in oligotrophic water environments, and seems to be characteristic of bacteria in months to years old drinking water biofilms. However, observations from field studies suggest that in drinking water biofilms an increase in the proportion of (culturable) HPC bacteria in relation to total cell counts can occur under favorable nutrient conditions on the surface of materials which release biodegradable substances (Kilb *et al.* 2003; Wingender and Flemming 2004). A high proportion of culturable organisms indicates better nutrition conditions. The ratio between colony counts of HPC bacteria and total cells counts can be used as a sign for elevated nutrient levels and, thus, for possible nutrient sources in the system. Therefore, in cases of contamination problems, it is recommended to include the determination of both total cell counts and HPC in the analysis of drinking water biofilms.

In the current project, the gap between total cell counts and HPC was also observed on materials relevant for plumbing systems in Germany (Moritz *et al.* 2010). Drinking water biofilms were grown in flow-through reactors on coupons of the materials as listed in Table 8.2. In the beginning, differences in colonization density were observed among the materials, but after 43 days, only a minor difference in the total cell counts between all materials was observed (Figure 8.3).

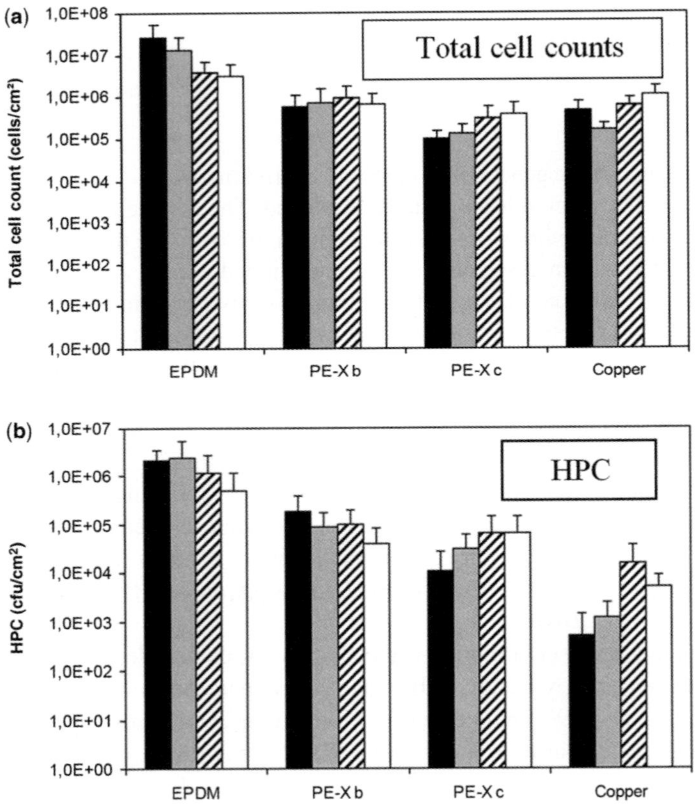

Figure 8.3 Total cell counts (a) and HPC (b) of ■ 14, ▫ 16, ▪ 29 and □ 43 day old drinking water biofilms grown on EPDM with certificate, PE-Xb, PE-Xc and copper under constant flow conditions at a mean temperature of approximately 19 °C. Total cell counts after 14 days were determined in four independent experiments and total cell counts after 16, 29 and 43 days were determined in three independent experiments. Total cell counts of DAPI-stained bacteria were determined by epifluorescence microscopy, and HPC was determined on R2A medium (20 °C, 7 d). (Moritz *et al.* 2010, with permission).

The concentration of HPC bacteria (Figure 8.3b) was always about 1 order of magnitude lower than the total cell counts in biofilms on EPDM with certificate, PE-Xb and PE-Xc. Biofilms on copper showed the lowest concentrations of culturable bacteria compared to the biofilms on the other materials; the HPC values were 2 to 3 log units lower than the total cell counts. In the period between 14 and 43 days the variation in total cell counts and HPC was never more than one log unit on the single materials, indicating that after 14 days of drinking water flow-through the biofilms had already reached relatively high and stable cell densities.

An exception were biofilms on copper, which showed an increase in HPC of more than 1 log unit between 14 and 29 days, indicating a delayed growth of biofilm bacteria on copper material that could be cultivated on agar media. In other studies, biofilm formation on copper analyzed by colony counts or adenosine triphosphate (ATP) content has also been observed to be limited or delayed compared to alternative materials (Lehtola *et al.* 2004; van der Kooij *et al.* 2005; Tsvetanova and Hoekstra 2009; Yu *et al.* 2010).

8.4.1 *L. pneumophila* and *P. aeruginosa* incorporated into drinking water biofilms as a potential water contamination source

All surfaces in drinking water distribution networks and domestic plumbing systems are colonized by microorganisms (Wingender and Flemming 2004; Eboigbodin *et al.* 2008). Such biofilms can serve as a habitat for pathogenic organisms which can be released from the biofilms, resulting in the contamination of drinking water systems (for review, see Wingender 2011; Wingender and Flemming 2011).

In the current project, seeding experiments in flow-through laboratory reactors were performed to investigate the fate of the facultative pathogens *L. pneumophila* (serogroup 1) and *P. aeruginosa* incorporated into drinking water biofilms on materials relevant to plumbing systems (Moritz *et al.* 2010). For this purpose, 14 day-old biofilms grown on EPDM with certificate, PE-Xb, PE-Xc and copper coupons (Table 8.2) were spiked with a mixture of *L. pneumophila* and *P. aeruginosa* cells that had been starved in drinking water for 24 h. After inoculation, the presence of the bacteria was monitored with standard cultural methods as well as with fluorescence in situ hybridization (FISH) over a period of 4 weeks under continuous flow-through with cold drinking water (mean water temperature: 18.8 °C). Prior to inoculation, *P. aeruginosa* and *L. pneumophila* serogroup 1 could not be cultivated from 14 day-old drinking water biofilms and 100-mL volumes of influent drinking water.

L. pneumophila and *P. aeruginosa* were detected in biofilms on EPDM with certificate, PE-Xb and PE-Xc materials and persisted there for up to 28 days (end of experimental period) as shown for representative experiments in Figures 8.4 and 8.5. *L. pneumophila* was also detected in biofilms on copper by FISH; in contrast, *P. aeruginosa* was not detectable either by culture or by FISH on copper (Figure 8.4), although the biofilms contained significant numbers of culturable HPC bacteria, indicating that there was no complete inhibition of heterotrophic biofilm organisms on the copper surface.

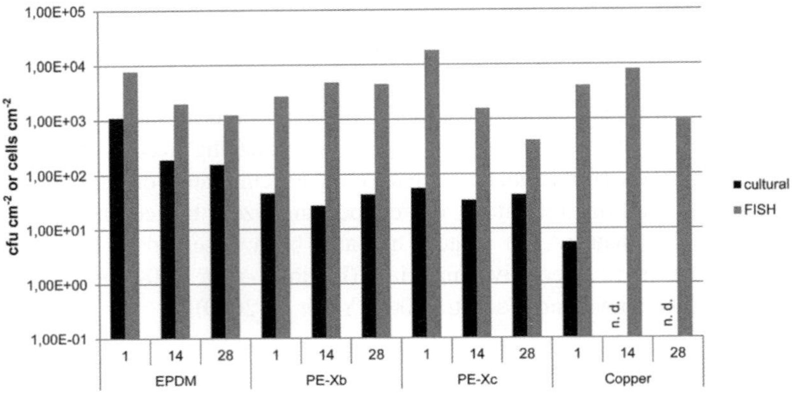

Figure 8.4 Persistence of *L. pneumophila* in drinking water biofilms on EPDM with certificate, PE-Xb, PE-Xc and copper under constant flow conditions. At days 1, 14, and 28 after inoculation, *L. pneumophila* was quantified by cultivation on glycin-vancomycin-polymyxin-B-cycloheximid (GVPC) agar (36 °C, up to 10 d) and FISH using Cy3-labeled probe LEGPNE1 (Moritz *et al.* 2010). *L. pneumophila* was not detected by cultivation methods at days 14 and 28 on copper material. n.d. = not detected.

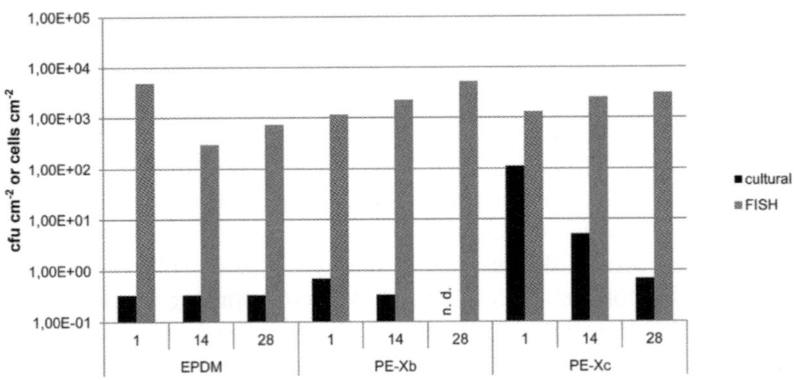

Figure 8.5 Persistence of *P. aeruginosa* in drinking water biofilms on EPDM with certificate, PE-Xb and PE-Xc under constant flow conditions. At days 1, 14 and 28 after inoculation, *P. aeruginosa* was quantified by cultivation on cetrimid-nalidixinic acid (CN) agar (36 °C, 2 d) and FISH using Cy3-labeled probe Psae16S-182 (Moritz *et al.* 2010). n.d. = not detected.

The concentration of *P. aeruginosa* and *L. pneumophila* determined by FISH was frequently several orders of magnitude higher than those determined by culturing methods. On some occasions, the bacteria could not be detected using the culture method but only by FISH. It is not known if all bacteria observed by FISH using oligonucleotide probes which are targeted at intact 16S rRNA are still viable. But based on the assumption that detection of rRNA with fluorescent oligonucleotide probes indicates viability, the low ratio of culturable cells to FISH-positive cells suggests that at least a fraction of *L. pneumophila* and *P. aeruginosa* cells may occur here in a viable state without being culturable. In 100 mL-volumes of drinking water sampled at the outlet of the biofilm reactors, *L. pneumophila* and *P. aeruginosa* were detected at least as long as the organisms were found culturally or by FISH in the corresponding biofilms, indicating that these organisms were released from the biofilms and contaminated the water phase as a result.

A number of other studies have considered the fate of *L. pneumophila* either as a component of naturally occurring mixed-population biofilms or following artificial spiking of drinking water biofilms, however, under experimental conditions that differed from those of our study. The inclusion and persistence of naturally occurring *L. pneumophila* at cold-water temperatures and multiplication of these bacteria at warm-water temperatures in biofilms of autochthonous water bacteria have been observed on various plumbing materials including ethylene-propylene, PE-X, non-plasticized PVC and copper (Keevil 2003; Rogers *et al.* 1994a, b; van der Kooij *et al.* 2002, 2005; Gião *et al.* 2009a, b). Seeding experiments indicated that *L. pneumophila* could become integrated in pre-existing biofilms and survive or even multiply in these biofilms, depending on the environmental conditions (Armon *et al.* 1997; Långmark *et al.* 2005; Lehtola *et al.* 2007; Benölken *et al.* 2010). In contrast to *L. pneumophila*, relatively little information exists regarding the occurrence of *P. aeruginosa* in biofilms on materials which are relevant to plumbing systems. In real drinking water systems, culturable *P. aeruginosa* was only sporadically found in biofilms on galvanized iron surfaces (Lee and Kim 2003), on steel coupons (Emtiazi *et al.* 2004), and on EPDM-coated valves (Kilb *et al.* 2003). In mixed-population drinking water biofilms cultivated in the laboratory on polybutylene, polyvinyl chloride (PVC), PE and steel (Rogers *et al.* 1994b) and on EPDM (Bressler *et al.* 2009), *P. aeruginosa* was found at relatively low concentrations using culturing methods. Thus, it appears that *P. aeruginosa* may occur in a culturable state in drinking water biofilms, mostly at relatively low concentrations compared to the general biofilm microflora, on materials which are employed in plumbing systems. As shown in this study, quantification of *P. aeruginosa* using FISH revealed orders of magnitude higher concentrations of the bacteria on EPDM with certificate, PE-Xb and PE-Xc, suggesting that culture-based methods may significantly underestimate the actual occurrence of *P. aeruginosa* in drinking water biofilms. However, the FISH method alone based on 16S rRNA detection does not provide definitive proof of cell viability (Cenciarini *et al.* 2008), so supplementary viability assays have to be applied to confirm a possible VBNC state of *P. aeruginosa* and *L. pneumophila* in drinking water biofilms.

Taken together, the experiments demonstrate that *L. pneumophila* and *P. aeruginosa* are capable of attaching to and becoming integrated into mixed-population drinking water biofilms of high cell density, where these can persist without significant growth for several weeks in a culturable state, indicating successful competition with the resident biofilm microflora. In addition, in laboratory and pilot scale experiments a single contamination event with *L. pneumophila* and *P. aeruginosa* was sufficient for their establishment in existing biofilms and succeeding long-term dissemination of the pathogens into the water which demonstrates the vulnerability of the system by external contamination. The observations indicate that drinking water biofilms grown under cold water conditions on domestic plumbing materials, especially EPDM and PE-X in the present study, can be a reservoir for *L. pneumophila* and *P. aeruginosa* that seem to persist in these habitats mostly in a non-culturable state (Moritz *et al.* 2010). Comparable results were obtained in parallel experiments performed under cold and warm water conditions in close-to-practice test rigs simulating a drinking water installation with a representative water consumption profile (Benölken *et al.* 2010). Thus, the results might be extrapolated to actual drinking water systems. From a health perspective, established drinking water biofilms can represent a reservoir for *L. pneumophila* and *P. aeruginosa*, and thus might act as a source of contamination in plumbing systems of buildings.

8.4.2 The influence of copper ions

Sources of copper ions in drinking water can be copper materials which are widely used for pipes or fixtures of plumbing systems (WHO 2011). During our study, it was observed that *P. aeruginosa* lost culturability when exposed to copper-containing drinking water ($0.021-0.066\ mg\ L^{-1}$ Cu) from different building plumbing systems (Dwidjosiswojo *et al.* 2011). For *P. aeruginosa* inactivation of planktonic bacteria as well as of biofilms by copper ions had been shown in a number of studies (Hoekstra *et al.* 1984; Huang *et al.* 2008; Teitzel and Parsek 2003; Teitzel *et al.* 2006) based exclusively on cultural methods, but it was unknown if copper ions actually killed the bacteria or possibly induced a VBNC state.

For the current study, the LIVE/DEAD *Bac*Light Bacterial Viability Kit was employed to differentiate between membrane-intact and membrane-disrupted cells. Using a mixture of SYTO 9 and propidium iodide, bacteria with intact membranes fluoresce bright green, whereas in bacteria with damaged membranes, propidium iodide abolishes the green fluorescence of SYTO 9 by displacing this dye from complexes with DNA and causes these cells to fluoresce red (Haugland 2005). The ratio of SYTO 9 and propidium iodide was adapted to the target organisms. Thus, using a combination of SYTO 9 and propidium iodide allows distinguishing between membrane-intact and membrane-damaged cells. Exposure of *P. aeruginosa* to deionized water with or without copper sulfate supplement revealed an almost equal level of green fluorescent cells to the total cell counts during incubation at 20 °C for 24 h, whereas culturability decreased by several orders of magnitude in the presence of copper ions (Figure 8.6). Using a *P. aeruginosa*-specific oligonucleotide probe Psae16S-182 targeted at 16S rRNA, no significant decrease of the concentration of FISH-positive cells occurred upon exposure to 10 µM copper for 24 h (Dwidjosiswojo *et al.* 2011).

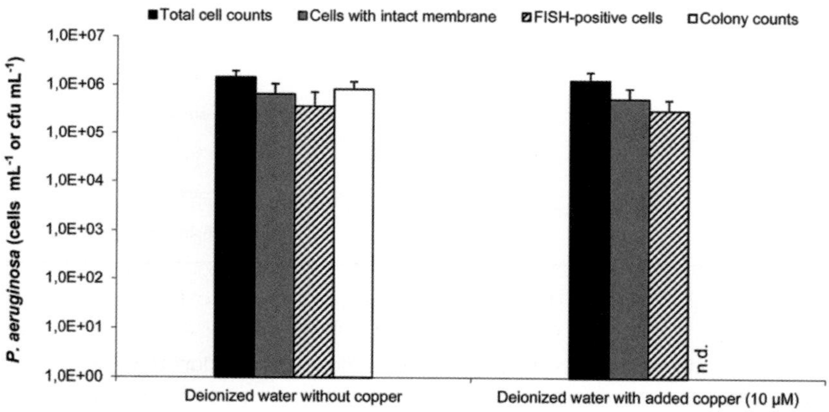

Figure 8.6 Viability of *P. aeruginosa* AdS after 24 h exposure to deionized water with or without 10 µM copper sulfate ($CuSO_4 \times 5\ H_2O$). Epifluorescence microscopy was used to determine total cell counts of DAPI-stained bacteria, the concentration of cells with intact membranes (LIVE/DEAD® *Bac*Light™ Bacterial Viability Kit) and the number of FISH-positive cells. Colony counts were determined on nutrient agar after incubation at 36 °C for 2 d. Error bars indicate standard deviation (n = 3). n. d., not detected (Dwidjosiswojo *et al.* 2011, with permission).

To investigate the potential restoration of culturability by abolishing copper stress, suspensions of *P. aeruginosa* which had been exposed to 10 µM copper for 24 h were incubated at 20 °C for 14 days in the presence or absence of the copper chelator diethyldithiocarbamate trihydrate (DDTC), and total cell counts and colony counts were monitored over 14 days. In control suspensions without added copper, the total cells counts and colony counts did not change significantly over the whole experimental period, indicating that *P. aeruginosa* did not grow in the deionized water used as the test medium (Figure 8.7). In suspensions of copper-treated cells, bacteria were not detected culturally, while the total cell counts of approximately 10^6 cells mL^{-1} remained constant. When the chelator DDTC had been added to suspensions of copper-stressed cells, a slow increase in colony counts was observed and after 14 days the concentrations of culturable cells of approximately 10^6 cfu mL^{-1} corresponded to those in bacterial suspensions without added copper (Figure 8.7, Dwidjosiswojo *et al.* 2011). The increase in culturability did not seem to be due to multiplication, since the total cells counts (approximately 10^6 cells mL^{-1}) remained unchanged over the whole incubation period.

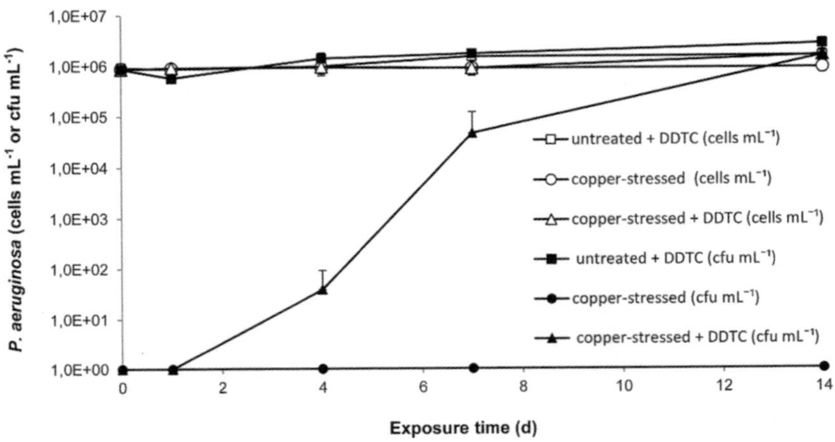

Figure 8.7 Fate of copper-stressed *P. aeruginosa* AdS induced by the chelator DDTC. After 24-h pre-incubation of the bacteria in deionized water with copper sulfate (10 µM), DDTC (100 µM) was added at time zero to abolish copper stress. Controls were copper-untreated bacteria with added DDTC and copper-stressed bacteria in the absence of DDTC. During subsequent incubation of the bacterial suspensions at 20 °C over a period of 14 d, total cell counts of DAPI-stained bacteria (open symbols) were determined by epifluorescence microscopy and colony counts (closed symbols) were determined on nutrient agar after incubation at 36 °C for 2 d. Error bars indicate standard deviation ($n = 3$). n. d., not detected (after Dwidjosiswojo *et al.* 2011).

Another interesting question is whether the organisms are infectious after relief of the copper stress. The infectious potential of *P. aeruginosa* cells in the culturable state, in the copper-induced non-culturable condition and in the culturable state after abolishing copper stress was investigated by the determination of their cytotoxicity towards Chinese hamster ovary (CHO-9) cells. Untreated culturable bacterial suspensions caused a cytotoxic effect with a complete loss of viability of CHO-9 cells within 24 h (of incubation). Copper-stressed bacteria did not affect viability of CHO-9 cells, while copper-stressed cells that had been incubated with DDTC treatment were cytotoxic again (Figure 8.8, Dwidjosiswojo *et al.* 2011). The chelator DDTC alone had no cytotoxic effect on CHO-9 cells (data not shown).

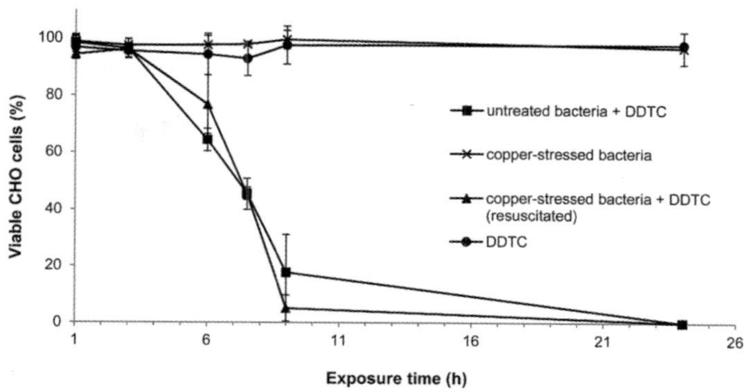

Figure 8.8 Cytotoxicity of *P. aeruginosa* AdS towards Chinese hamster ovary (CHO-9) cells. The bacteria had been incubated for 14 d in deionized water in the absence or presence of copper sulfate (10 μM), or were incubated in the presence of the chelator DDTC (100 μM) for 14 days (see Figure 8.7), before they were added to CHO-9 cells. The CHO-9 cell cultures were incubated at 36 °C for 24 h, before the viability of CHO-9 cells was determined by trypan blue staining. Viability of bacteria-exposed CHO-9 cells was calculated as the percentage of intact cells in relation to control cell cultures that were incubated in the absence of bacteria (100% viable CHO cells). Error bars indicate standard deviation (n = 3) (after Dwidjosiswojo *et al.* 2011).

Although many studies have demonstrated the inhibitory effect of copper ions on *P. aeruginosa* by cultural methods monitoring the decrease of colony counts on agar media, the possible copper-induced non-culturable state has not been considered in these investigations. However, copper ions have been reported to induce the VBNC state in a number of other gram-negative bacteria, including soil and phytopathogenic species such as *Agrobacterium tumefaciens* and *Rhizobium leguminosarum* (Alexander *et al.* 1999), *Xanthomonas campestris* pv. *campestris* (Ghezzi and Steck 1999), *Ralstonia solanacearum* (Grey and Steck 2001a), *Erwinia amylovora* (Ordax *et al.* 2006) and *Xanthomonas axonopodis* pv. *citri* (Del Campo *et al.* 2009), and the fecal bacterium *Escherichia coli* (Grey and Steck 2001b; Aurass *et al.* 2011).

The results indicate that *P. aeruginosa* might enter the VBNC state in the presence of copper ions at concentrations relevant to drinking water environments of plumbing systems. The hygienic relevance of copper ions at concentrations relevant to drinking water environments in plumbing systems on the culturability and viability of *P. aeruginosa* in water and biofilms has still to be established. The current study gives preliminary indications that *P. aeruginosa* may possibly occur in a VBNC state in copper plumbing systems and then are not detectable by culture-based methods routinely employed for monitoring drinking water quality (Dwidjosiswojo *et al.* 2011).

8.5 DISINFECTION

The efficacy of current disinfection procedures against biofilms has been investigated. The model system is based on the use of silicon tubes and allowed for defined conditions, representing a "worst case" situation as the silicon material supports biofilm growth (Exner *et al.* 2010). Again, it was shown that the bacterial numbers as obtained by culture methods only reflect a minor fraction of the total cell numbers as obtained by fluorescence microscopy. Both methods were used to determine the efficacy of a range of currently employed disinfection procedures which shed a differentiated light on their efficacy. These included the use of chlorine as generated by the anolyte procedure (Kraft *et al.* 1999), chlorine dioxide, ozone, and hydrogen peroxide with fruity acid or with silver (Exner *et al.* 2010), at various dosing regimens. The summary of the results is shown in Table 8.3.

Table 8.3 Efficacy of disinfection vs. cleaning.

Disinfectant	Dosing regime	Conc. [mg L^{-1}]	Duration of Treatment	Reduction (cfu cm^{-2})	Reduction TCC cm^{-1}
Cl_2 (ECA)[2]	Continuous	0.3	70 days	6 log	2 log
Cl_2 (ECA)	Shock	25	6 hours	6 log	6 log
ClO_2	Continuous	0.2	70 days	6 log	3 log
ClO_2	Shock	10	6 hours	6 log	1 log
H_2O_2 + fruity acid	Shock	1000	4 hours	6 log	1 log
$H_2O_2 + Ag^+$	Shock	10,000	4 hours	6 log	1 log
O_3	Discontinuous	0.1	14 days	6 log	6 log

[1]TCC = total cell count by fluorescence microscopy.
[2]ECA = electrolytically generated chlorine.
Source: (Exner *et al.* 2010).

It is evident that all procedures led to a sufficient reduction of the culturable bacterial numbers as determined by plate counts (colony forming units, cfu). However, none of them with the only exception being high shock dosage of chlorine or ozone did lead to biomass removal shown as reduction in total cell counts. Therefore regrowth of the microorganisms 14 days after the disinfection measurement can be explained. The remaining biofilms offered a nutrient source for surviving biofilm organisms and those which enter the system after the disinfection procedure. It may sound trivial, but it may have quite some practical implication: disinfection does not necessarily mean cleaning. Particularly in industrial biofouling, these two are frequently considered equivalent (Flemming 2002, 2011) – a misapprehension which leads to rapid regrowth and recurrent biofouling problems. From this point of view, the rule to apply cleaning methods prior to disinfection is confirmed. Air-water flushing can support cleaning efficacy, although the biofilm may not be removed completely (Lehtola *et al.* 2004).

A mobile biofilm device was designed, termed as a "Hygiene Monitor" which basically consists of an apparatus with silicon tubing (or other tubing materials) which can be implemented in water systems in order to assess the biofilm formation potential of waters. It represents the "worst case" situation and cleaning/disinfection success in this device is considered to be sufficient for extrapolation to a drinking water installation. This has been shown to be quite useful, for example, if there are problems with recurring growth events, for example, in hospitals (Brensing *et al.* 2009).

Furthermore, it is obvious that the effect of a time-limited disinfection measurement does not last long (less than two weeks) if the conditions remain the same under which the contamination occurred, particularly in the case of materials which support microbial growth. This may also sound trivial, but it has serious practical consequences, for example, the replacement of such materials by others which do not support microbial growth.

8.6 POPULATION ANALYSIS

The population composition that developed in biofilms on the materials used in the project was investigated with 16S rDNA clone libraries, denaturing gradient gel electrophoresis (DGGE), FISH and quantitative PCR (qPCR) in order to find any regular patterns.

Biofilm communities grown on materials which do not release considerable amounts of easily degradable organic compounds such as glass and PE consisted of some predominating groups and a background population of higher diversity, as determined by DGGE and 16S rDNA clone libraries (Szewzyk and Röder 2010). Both, the highest cell density and biodiversity were found on EPDM. Interestingly, in these biofilms, higher occurrence of *Sphingomonas* and flavobacteria was detected, both organisms known for the wide spectrum of substrates they can use (Leys *et al.* 2004; Bissett *et al.* 2008). This is attributed to the nutrients provided by the material, supporting biofilm growth. Possibly, the occurrence of these organisms can be interpreted as an indicator for growth-promoting plumbing materials. Biofilms generated on the test materials using the local drinking water of three different locations showed no regular pattern of bands.

Figure 8.9 shows the diversity of biofilm populations and their similarities on different materials exposed both in cold and warm water systems as determined by DGGE. The biofilm samples were taken from (i) the surface of the materials exposed for 6 weeks (6w) and (ii) the surface of the materials 24 h after contamination of the water system with *P. aeruginosa*, *L. pneumophila*, *Enterobacter amnigenus* and *Citrobacter freundii* in week 8 (8w).

In warm water, a higher biodiversity was observed on PE-Xb, PE-Xc and copper than in cold water. After contamination of 8-week-old biofilms, the diversity on copper and PE-Xb and PE-Xc biofilms increased strongly. In cold water biofilms, diversity remained similar before and after the contamination. In biofilms on

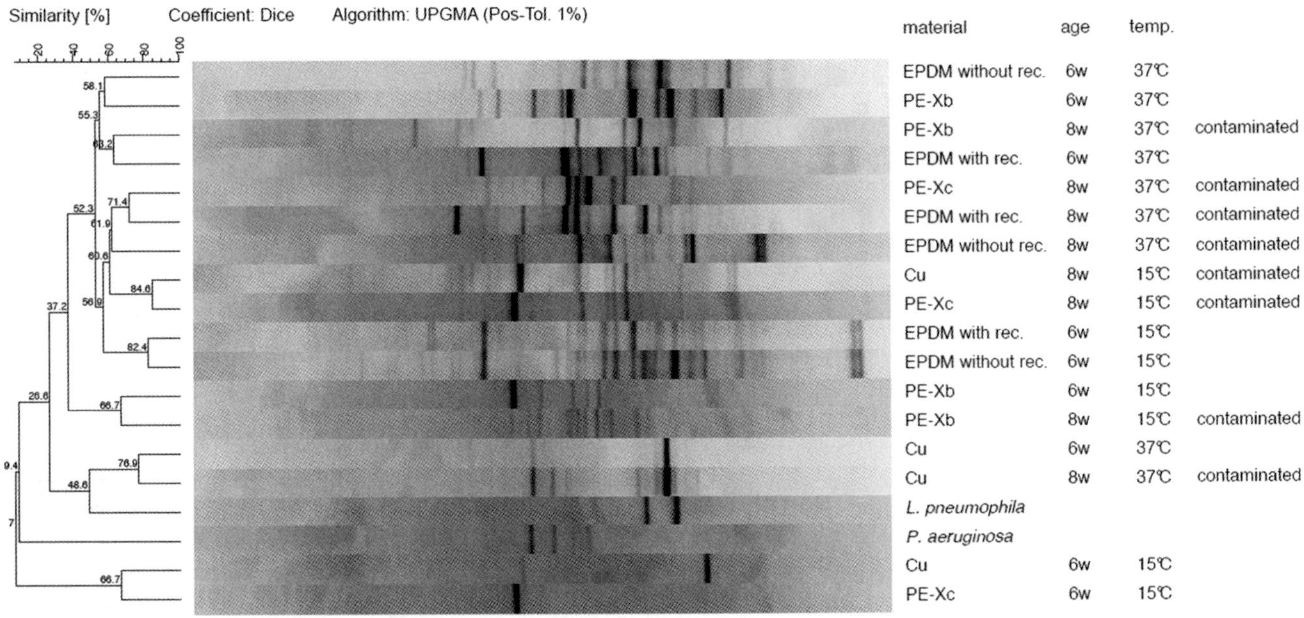

Figure 8.9 Similarity of biofilms from location H2 before (6w) and after (8w) contamination with *L. pneumophila*, *P. aeruginosa*, *E. amnigenus* and *C. freundii* as determined by DGGE (Röder *et al.* 2010a; Szewzyk and Röder 2010).

copper at 37 °C, bands occurred at the same positions as *L. pneumophila* and *P. aeruginosa* respectively in control experiments, indicating that *these organisms* had settled. The populations on various materials were very diverse.

As a general result it could be observed, that after 6 weeks the biofilms on the same type of material in warm and cold water showed a higher overall similarity than biofilm populations from different materials in either warm or cold water. Exceptions were the biofilms on the two different EPDM materials (one with and one without certificate) which showed a high similarity. The biofilms on copper and PE-Xc showed less diversity than biofilms on the other materials and therefore the similarity calculation was less conclusive. In total the diversity of the biofilm populations decreased in dependence of the respective support material in warm and cold water in the following order: EPDM > copper > PE-Xc. Diversity of the warm water biofilms from copper and PE-Xc was, however, slightly higher than from their cold water counterparts.

After contamination of the water systems with different pathogens in week 8, the diversity of the biofilms on copper and PE-Xc in cold water strongly increased. Similarity of the biofilms on these two materials was high at around 80%. Obviously, the influence of the contamination on the diversity and similarity was predominant over the existing biofilm population before contamination which had shown a low diversity on both materials. This may not only be the result of the added bacteria but also a population shift due to unintentionally added nutrients. Since the biofilm samples were taken only 24 h after contamination this may reflect an adaptation phase of the biofilms to the new conditions. In contrast to the cold water biofilms, diversity of biofilms exposed to warm water remained more or less constant after contamination. For the copper biofilm, bands were observed in the DGGE which indicated the presence of *P. aeruginosa* and *L. pneumophila* in the biofilm. Bands at the same positions were obtained from pure culture experiments and bands from population analysis were cut out and analyzed by 16s rDNA clone library. No indication was found for the presence of *E. amnigenus* and *C. freundii.*

It can be concluded that disinfection in the long term does not necessarily reduce the cell density of a biofilm but causes changes in population composition. Part of the biofilm population will be killed by the disinfectant and this results in the availability of easily degradable nutrients and can favor fast growing for opportunistic pathogenic microorganisms (Röder *et al.* 2010b).

8.7 CONCLUSIONS

Drinking water installations represent a potential contamination risk due to the range of materials, the flow pattern and the complexity of the system. Biofilms occur on all surfaces in contact with water, particularly, if the materials support microbial growth by leaching of biodegradable substances. Opportunistic pathogens can be incorporated into biofilms, persist and proliferate at elevated nutrient levels. As a consequence, material quality is crucial in limiting regrowth of microorganisms in the system. Compliance with recognized standards of good practice is effective in prevention of microbial contamination. However, the situation is similar to traffic: if everybody would follow the rules, there would be no accidents. Enhanced surveillance, harmonization of the methods and training of personnel are good measures for improvement of safety (van der Kooij *et al.* 2003).

Pathogen detection by cultivation-dependent methods can be impaired by a transition into the VBNC state. This is considered a stress response. Then, potentially harmful remain undetected. However, they can return to culturability and can even become infective again if the stress is suspended. An example is the response to copper stress and the resuscitation after stress abolishment. It is not recommended to abandon culture-depending methods, as they are a key element for providing drinking water safety. However, in critical cases, for example, with recurring and recalcitrant contaminations, they should be completed by advanced molecular biological methods. This should definitely be considered in risk assessments.

Disinfection remains a standard tool in drinking water safety but it has to be kept in mind that it does not necessarily lead to removal of biomass, that is, cleaning. Furthermore, a population shift towards fast growing species is observed. Particularly in situations with strong biofilm growth, as encountered upon employment of unsuitable construction materials, rapid regrowth must be expected after disinfection. Therefore, careful choice of materials used in drinking water installations is a key factor for drinking water safety.

8.8 SUMMARY

The best drinking water quality can be compromised if plumbing systems of the drinking water installation in buildings support biofilm growth. Unfortunately, this is the least monitored part in the supply chain. A nationwide survey in Germany showed that in 2007, more than five years since inspection had become mandatory by the new national drinking water regulation, only 13% of the public health authorities had completed the monitoring of drinking water installations in public buildings in compliance with legal obligations. This was not due to inactivity of the authorities but rather to overburden and understaffing. Furthermore, sampling and analytical methods applied were not homogeneous and did not always yield comparable results. But even under these conditions, statistical evaluation of more than 20,000 measurements showed that 12% of the water samples contained *Legionella* spp. and 3% of about 3500 samples contained *Pseudomonas* spp. Implementation of devices in the installation system such as water filters, softeners or dosage of phosphate significantly increased the incidence of positive results. It remains beyond doubt that under compliance with the generally recognized standards of good practice, hygienic risks associated with drinking water will be minimized. One of the results of the inspections was that infrastructure and operation conditions as well as maintenance frequently did not comply with these rules.

Drinking water is not sterile and does not have to be so. A very successful strategy of water works is based on nutrient limitation and, thus, prevents microbial growth and produces "biologically stable drinking water" as a way to avoid or minimize chlorination. However, if drinking water bacteria meet materials which release nutrients, they settle, form biofilms and grow. Usual suspect materials in such cases are synthetic polymers which are not approved for use in drinking water. In these biofilms, hygienically relevant organisms can settle and even proliferate if the conditions for their growth are favorable. They can be partially released into the water and then represent a hygienic risk. This was investigated and verified in testing systems run under conditions close to real drinking water installation systems. Furthermore, it was shown that countermeasures such as application of disinfectants could improve the situation but did not lead to essential sanitation: disinfection is not necessarily equivalent to cleaning. It was observed that disinfection led to population shifts in a direction towards fast growing organisms. Clearly, the quality of materials in drinking water installations is crucial. Many available materials – in particular those on the lower end of the market – are leading straight to enhanced biofilm formation.

L. pneumophila and *P. aeruginosa* can become integrated into drinking water biofilms, and may proliferate and subsequently contaminate the water phase. Common detection methods are based on cultivation. However, hygienically relevant microorganisms can enter a "viable-but-non-culturable state" (VBNC), for example, if they are stressed by disinfectants, toxic metal ions, temperature shocks or as a general result of adaptation to limiting conditions. An example is the observation that copper ions at concentrations common in copper-material-based drinking water installations possibly turned *P. aeruginosa* into such a VBNC state. However, they may remain viable and can be resuscitated by application of a copper chelator (sodium diethyldithiocarbamate, DDTC) to full culturability. Such processes may be one of the reasons for cases with recurrent contamination in water systems as occasionally found in public buildings such as hospitals, schools, military barracks, hotels and so on. If the contaminating organisms enter and leave the VBNC state, erratic patterns will be seen with cultivation methods. Organisms in this state can only be found by cultivation-independent methods such as fluorescence-in-situ-hybridization (FISH), PCR-based techniques or others.

Investigation on cytotoxicity revealed that the putative VBNC form of *P. aeruginosa* was not infectious for Chinese hamster ovary (CHO-9) cells while the DDTC-resuscitated organisms were as infectious as the original strain. This indicates that a transition into the VBNC state and back to culturability is not associated with a loss of infectivity.

In conclusion, it is strongly recommended (i) to ensure that household installation plumbing materials with low biofilm formation potential are employed and, (ii) not to rely on cultivation methods only, when contamination sources have to be identified. In particular, in problematic cases, cultivation-independent methods are suggested; otherwise, only the "tip of the iceberg" is detected and the occasional false-negative findings lead to the observation of "recurring contaminations".

Acknowledgements

Authors' main contribution to sections: Surveillance: Thomas Kistemann; Materials: Bernd Bendinger; Materials ageing: Gabriela Schaule Disinfection: Jürgen Gebel and Martin Exner; Population analysis: Ulrich Szewzyk; All others: Jost Wingender and Hans-Curt Flemming.

The valuable contributions of the Ph. D. students and postdocs in this project are gratefully acknowledged: Dr. Jens Benölken, Prof. Dr. Elke Dopp, Zenyta Dwidjosiswojo, Dr. Susanne Grobe, Dr. Johannes Lenz, Stefan Linke, Dr. Miriam Moritz, Dr. Jessica Richard, Dr. Rosemarie Röder, Dr. Christiane Schreiber, Dr. Simone Schulte, Sebastian Völker.

The project was supported by our industrial partners, that is, ActiDes, Deutsche Messe AG, European Copper Institute, Franke Aquarotter AG, Freudenberg GmbH & Co. KG, Geberit International AG, Grünbeck Wasseraufbereitung, Grundfos GmbH, Hammann GmbH, Jati Products, Luftfahrtgeräte Gauting GmbH, Pall GmbH, ProMinent ProMaqua GmbH, Sanosil Service GmbH, Scheideler Verfahrenstechnik, Stadtwerke Duisburg, Viega GmbH & Co. KG.

The main funding was supplied by the German Federal Ministry of Education and Research, grant No. 02WT0832–02WT0837. The kind and patient support of the Project Management Agency Karlsruhe by Vera Horak and Sylvia Gräber is very gratefully appreciated.

8.9 REFERENCES

Ahmed W., Goonetilleke A. and Gardner T. (2010). Implications of faecal indicator bacteria for the microbiological assessment of roof-harvested rainwater quality in southeast Queensland, Australia. *Can. J. Microbiol.*, **56**, 471–479.

Alexander E., Pham D. and Steck T. R. (1999). The viable-but-nonculturable condition is induced by copper on *Agrobacterium tumefaciens* and *Rhizobium leguminosarum*. *Appl. Environ. Microbiol.*, **65**, 3754–3756.

Anaissie E. J., Penzak S. R. and Dignani M. C. (2002). The hospital water supply as a source of nosocomial infections. *Arch. Internal Med.*, **162**, 1483–1492.

Anaissie E. J., Stratton S. L., Dignani M. C., Lee C., Summerbell R. C., Rex J. H., Monson T. P. and Walsh T. J. (2003). Pathogenic molds (including *Aspergillus* species) in hospital water distribution systems: a 3-year prospective study and clinical implications for patients with hematologic malignancies. *Blood*, **101**, 2542–2546.

Anonymous (2007). Technische Regel Arbeitsblatt W 270. Vermehrung von Mikroorganismen auf Werkstoffen für den Trinkwasserbereich – Prüfung und Bewertung. DVGW Deutscher Verein des Gas und Wasserfaches, Josef Wirmer Str. 1–3, D-53123 Bonn, Germany.

Armon R., Starosvetzky J., Arbel T. and Green M. (1997). Survival of *Legionella pneumophila* and *Salmonella typhimurium* in biofilm systems. *Wat. Sci. Technol.*, **35** (11–12), 293–300.

Arnow P. M., Weil D. and Para M. F. (1985). Prevalence and significance of *Legionella pneumophila* contamination of residential hot tap water systems. *J. Infect. Dis.*, **153**, 145–151.

Arvand M., Jungkind K. and Hack A. (2011). Contamination of the cold water distribution system of health care facilities by *Legionella pneumophila*: do we know the true dimension? *Euro Surveill.*, **16**, 9–14.

ASHRAE Standard (2000). Minimizing the Risk of Legionellosis Associated with Building Water Systems. ASHRAE Guideline 12–2000. American Society of Heating, Refrigeration and Air-Conditioning Engineers, Inc., Atlanta.

ASTM F2263 03 (2008). Standard Test Method for Evaluating the Oxidative Resistance of Polyethylene (PE) Pipe to Chlorinated Water, ASTM International, West Conshohocken, PA.

Aurass P., Prager R. and Flieger A. (2011). EHEC/EAEC O104:H4 strain linked with the 2011 German outbreak of haemolytic uremic syndrome enters into the viable but non-culturable state in response to various stresses and resuscitates upon stress relief. *Environ. Microbiol.*, **13**, 3139–3148.

Bagh L. K., Albrechtsen H.-J., Arvin E. and Ovesen K. (2004). Distribution of bacteria in a domestic hot water system in a Danish apartment building. *Wat. Res.*, **38**, 225–235.

Bender L., Ott M., Marre R. and Hacker J. (1990). Genome analysis of *Legionella* -spp. by orthogonal field alternation gel electrophoresis (OFAGE). *FEMS Microbiol. Lett.*, **72**, 253–257.

Benölken J. K., Dorsch T., Wichmann K. and Bendinger B. (2010). Praxisnahe Untersuchungen zur Kontamination von Trinkwasser in halbtechnischen Trinkwasser-Installationen. In: Vermeidung und Sanierung von Trinkwasserkontaminationen durch hygienisch relevante Mikroorganismen aus Biofilmen der Hausinstallation, H.-C. Flemming (ed.), IWW Schriftenreihe 54, Mülheim, pp. 101–180.

Berney M., Hammes F., Bosshard F., Weilenmann H.-U. and Egli T. (2007). Assessment and interpretation of bacterial viability using the LIVE/DEAD BacLight kit in combination with flow cytometry. *Appl. Environ. Microbiol.*, **73**, 3283–3290.

Bissett A., Bowman J. and Burke C. (2008). Flavobacterial response to organic pollution. *Aquat. Microb. Ecol.*, **51**, 31–43.

Bogosian G. and Bourneuf E. (2001). A matter of bacterial life and death. *EMBO Rep.*, **21**, 770–774.

Brensing K. A., Neugebauer E., Heidkamp P., Raab P., Pöge U., Gerhardt T., Dedenbach M., Kluth R., Speuser W., Kistemann T., Linke S., Lenz J., Exner M. and Gebel J. (2009). Intermittierende Ozonierung des Wasserringsystems in einem ambulanten Hämodialyse-Zentrum. *Hyg. Med.*, **34**, 176–182.

Bressler D., Balzer M., Dannehl A., Flemming H.-C. and Wingender J. (2009). Persistence of *Pseudomonas aeruginosa* in drinking-water biofilms on elastomeric material. *Wat. Sci. Technol., Wat. Supply*, **9**, 81–87.

Buchbinder S., Trebesius K. and Heesemann J. (2002). Evaluation of detection of *Legionella* spp. in water samples by fluorescence in situ hybridization, PCR amplification and bacterial culture. *Int. J. Med. Microbiol.*, **292**, 241–245.

Cenciarini C., Courtois S., Raoult D. and La Scola B. (2008). Influence of longtime storage in mineral water on RNA stability of *Pseudomonas aeruginosa* and *Escherichia coli* after heat inactivation. *PLoS One*, **3**, e3443.

Codony F., Alvarez J., Oliva J. M., Ciurana B., Company M., Camps N., Torres J., Minguell S., Jové N., Cirera E., Admetlla T., Abós R., Escofet A., Pedrol A., Grau R., Badosa I. and Vila G. (2002). Factors promoting colonization by Legionellae in residential water distribution systems: an environmental case–control survey. *Eur. J. Clin. Microbiol. Infect. Dis.*, **21**, 717–721.

Council Directive 98/83/EC of 3 November 1998 on the quality of water intended for human consumption. *Official Journal L 330*, 5.12.1998, 32–54.

Darelid J., Löfgren S. and Malmvall B.-E. (2002). Control of nosocomial Legionnaires' disease by keeping the circulating hot water temperature above 55 °C: experience from a 10-year surveillance programme in a district general hospital. *J. Hosp. Infect.*, **50**, 213–219.

Del Campo R., Russi P., Mara P., Mara H., Peyrou M., Ponce de León I. and Gaggero C. (2009). *Xanthomonas axonopodis* pv. *citri* enters the VBNC state after copper treatment and retains its virulence. *FEMS Microbiol. Lett.*, **298**, 143–148.

Dwidjosiswojo Z., Richards J., Moritz M. M., Dopp E., Flemming H.-C. and Wingender J. (2011). Influence of copper ions on the viability and cytotoxicity of *Pseudomonas aeruginosa* under conditions relevant to drinking water. *Int. J. Hyg. Environ. Health*, **214**, 485–492.

Eboigbodin K. E., Seth A. and Biggs C. A. (2008). A review of biofilms in domestic plumbing. *J. Am. Wat. Works Assoc.*, **100**, 131–138.

Emtiazi F., Schwartz T., Marten S. M., Krolla-Sidenstein P. and Obst U. (2004). Investigation of natural biofilms formed during the production of drinking water from surface water embankment filtration. *Wat. Res.*, **38**, 1197–1206.

Exner M., Gebel J. and Lenz J. (2010). Überprüfung der Wirksamkeit von Desinfektionsmaßnahmen und Erarbeitung von Handlungsanweisungen in Schadensfällen. In: Vermeidung und Sanierung von Trinkwasserkontaminationen durch hygienisch relevante Mikroorganismen aus Biofilmen der Hausinstallation, H.-C. Flemming (ed.), IWW Schriftenreihe 54, Mülheim 2010, pp. 181–230.

Flemming H.-C. (2002). Biofouling in water systems – cases, causes, countermeasures. *Appl. Environ. Biotechnol.*, **59**, 629–640.

Flemming H.-C. (ed.) (2010). Vermeidung und Sanierung von Trinkwasser-Kontaminationen durch hygienisch relevante Mikroorganismen aus Biofilmen der Hausinstallation. IWW-Schriftenreihe Band 54, Mülheim 2010, IWW Water Centre Mülheim, Moritzstr. 26, D-45476 Mülheim.

Flemming H.-C. (2011). Microbial biofouling – unsolved problems, insufficient approaches and possible solutions. In: Biofilm Highlights, H.-C. Flemming, J. Wingender and U. Szewzyk (eds), Springer Int. Heidelberg, New York, pp. 81–109.

Flemming H.-C., Percival S. L. and Walker J. T. (2002). Contamination potential of biofilms in water distribution systems. *Wat. Sci. Technol., Wat. Supply*, **2**, 271–280.

German Drinking Water Ordinance (2001). Verordnung über die Qualität von Wasser für den menschlichen Gebrauch. Trinkwasserverordnung vom 21. Mai 2001 (BGBl, I, 959), die durch Artikel 363 der Verordnung vom 31. Oktober 2006 (BGBl I, 2047) geändert worden ist (in German).

Ghezzi J. I. and Steck T. R. (1999). Induction of the viable but non-culturable condition in *Xanthomonas campestris* pv. *campestris* in liquid microcosms and sterile soil. *FEMS Microbiol. Ecol.*, **30**, 203–208.

Gião M. S., Wilks S. A., Azevedo N. F., Vieira M. J. and Keevil C. W. (2009a). Comparison between standard cultures and peptide nucleic acid 16S rRNA hybridization quantification to study the influence of physico-chemical parameters on *Legionella pneumophila* survival in drinking water biofilms. *Biofouling*, **25**, 335–343.

Gião M. S., Wilks S. A., Azevedo N. F., Vieira M. J. and Keevil C. W. (2009b). Comparison between standard culture and peptide nucleic acid 16S rRNA hybridization quantification to study the influence of physico-chemical parameters on *Legionella pneumophila* survival in drinking water biofilms. *Biofouling*, **25**, 343–351.

Grey B. and Steck T. R. (2001a). The viable but nonculturable state of *Ralstonia solanacearum* may be involved in long-term survival and plant infection. *Appl. Environ. Microbiol.*, **67**, 3866–3872.

Grey B. and Steck T. R. (2001b). Concentrations of copper though to be toxic to *Escherichia coli* can induce the viable but nonculturable condition. *Appl. Environ. Microbiol.*, **67**, 5325–5327.

Hammes F. and Egli T. (2010). Cytometric methods for measuring bacteria in water: advantages, pitfalls and applications. *Anal. Bioanal. Chem.*, **397**, 1083–1109.

Harvey H. (2011). Life, death, and in-between: meanings and methods in microbiology. *Appl. Environ. Microbiol.*, **78**, 5571–5576.

Haugland R. P. (2005). A Guide to Fluorescent Probes and Labeling Technologies, 10th edn. Invitrogen, Carlsbad.

Hoekstra A. C., van der Kooij D., Visser A. and Hijnen W. A. M. (1984). Bacteriological, chemical and physical characteristics of samples from two hot water systems containing *Legionella pneumophila*, compared with drinking water from municipal waterworks. In: Legionella, Proceedings of the 2nd International Symposium, C. Thornsberry, A. Balows, J. C. Feeley and W. Jakubovsky (eds), Am. Soc. Microbiol., Washington, DC, pp. 343–346.

Huang H.-I., Shih H.-Y., Lee C.-M., Yang T. C., Lay J.-J. and Lin Y. E. (2008). In vitro efficacy of copper and silver ions in eradicating *Pseudomonas aeruginosa*, *Stenotrophomonas maltophilia* and *Acinetobacter baumannii*: implications for on-site disinfection for hospital infection control. *Wat. Res.*, **42**, 73–80.

Kalmbach S., Manz W. and Szewzyk U. (1997). Dynamics of biofilm formation in drinking water: phylogenetic affiliation and metabolic potential of single cells assessed by formazan reduction and in situ hybridization. *FEMS Microbiol. Ecol.*, **22**, 265–279.

Keevil C. W. (2003). Pathogens in environmental biofilms. In: Encyclopedia of environmental microbiology, volume 4, G. Bitton (ed.), J. Wiley & Sons, New York, pp. 2339–2356.

Kempner R. L. and Pratt J. R. (1976). Use of fluorochromes for direct enumeration of total bacteria in environmental samples. *Microbiol. Rev.*, **58**, 603–615.

Kilb B., Lange B., Schaule G., Flemming H.-C. and Wingender J. (2003). Contamination of drinking water by coliforms from biofilms grown on rubber-coated valves. *Int. J. Hyg. Environ. Health*, **206**, 563–573.

Kistemann T., Schreiber C. and Völker S. (2010). Vorkommen mikrobieller Trinkwasserkontaminationen in Hausinstallationen. In: Vermeidung und Sanierung von Trinkwasserkontaminationen durch hygienisch relevante Mikroorganismen aus Biofilmen der Hausinstallation, H.-C. Flemming (ed.), IWW Schriftenreihe 54, Mülheim 2010, pp. 31–100.

Kraft A., Stadelmann M., Blaschke M., Kreysig D., Sandt B., Schröder F. and Rennau J. (1999). Electrochemical water disinfection Part I: Hypochlorite production from very dilute chloride solutions. *J. Appl. Electrochem.*, **29**, 859–866.

Långmark J., Storey M. V., Ashbolt N. J. and Stenström T. A. (2005). Accumulation and fate of microorganisms and microspheres in biofilms formed in a pilot-scale water distribution system. *Appl. Environ. Microbiol.*, **71**, 706–712.

Lee D.-G. and Kim S.-J. (2003). Bacterial species in biofilm cultivated from the end of the Seoul water distribution system. *J. Appl. Microbiol.*, **95**, 317–324.

Lehtola M. J., Torvinen E., Kusnetzov J., Pitkanen T., Maunula L., von Bonsdorff C. H., Martikainen P. J., Wilks S. A., Keevil C. W. and Meittinen I. T. (2007). Survival of *Mycobacterium avium*, *Legionella pneumophila*, *Escherichia coli* and caliciviruses in drinking water-associated biofilms grown under high-shear turbulent flow. *Appl. Environ. Microbiol.*, **73**, 2854–2859.

Leys N. M. E. J., Ryngaert A., Bastiaens L., Verstraete W., Top E. M. and Springael D. (2004). Occurrence and phylogenetic diversity of *Sphingomonas* strains in soils contaminated with polycyclic aromatic hydrocarbons. *Appl. Environ. Microbiol.*, **70**, 1944–1955.

Lye D., Fout G. S., Crout S. R., Danielson R., Thio C. L. and Paszko-Kolva C. M. (1997). Survey of ground, surface, and potable waters for the presence of *Legionella* species by EnviroAmp^R PCR *Legionella* Kit, culture, and immunofluorescent staining. *Wat. Res.*, **31**, 287–293.

Mathys W., Stanke J., Harmuth M. and Junge-Mathys E. (2008). Occurrence of *Legionella* in hot water systems of single-family residences in suburbs of two German cities with special reference to solar and district heating. *Int. J. Hyg. Environ. Health*, **211**, 179–185.

Moritz M., Flemming H.-C. and Wingender J. (2010). Integration of *Pseudomonas aeruginosa* and *Legionella pneumophila* in drinking water biofilms grown on domestic plumbing materials. *Int. J. Hyg. Environ. Health*, **213**, 190–197.

Nyström T. (2001). Not quite dead enough: On bacterial life, culturability, senescence and death. *Arch. Microbiol.*, **176**, 159–164.

Oliver J. D. (2005). The viable but nonculturable state in bacteria. *J. Microbiol.*, **43**, 93–100.

Oliver J. D. (2010). Recent findings on the viable but nonculturable state in pathogenic bacteria. *FEMS Microbiol. Rev.*, **34**, 415–425.

Ordax M., Marco-Noales E., López M. M. and Biosca E. G. (2006). Survival strategy of *Erwinia amylovora* against copper: induction of the viable-but-nonculturable state. *Appl. Environ. Microbiol.*, **72**, 3482–3488.

Rochelle P. A., Camper A. K., Nocker A. and Burr M. (2011). Are they alive? Detection of viable organisms and functional gene expression using molecular techniques. In: Environmental Microbiology, K. Sen and N. Ashbolt (eds), Caister Acad. Press, Norfolk, UK, pp. 179–202.

Röder R. S., Heeg K., Tarne P., Benölken J. K., Schaule G., Bendinger B., Flemming H.-C. and Szewzyk U. (2010a). Influence of materials, water qualities and disinfection methods on the drinking water biofilm community. *Wat. Pract. Technol.*, **5**, 1–12.

Röder R. S., Lenz J., Tarne P., Gebel J., Exner M. and Szewzyk U. (2010b). Long-term effects of disinfectants on the community composition of drinking water biofilms. *Int. J. Hyg. Environ. Health*, **213**, 183–189.

Rogers J., Dowsett A. B., Dennis P. J., Lee J. V. and Keevil C. W. (1994a). Influence of temperature and plumbing material selection on biofilm formation and growth of *Legionella pneumophila* in a model potable water system containing complex microbial flora. *Appl. Environ. Microbiol.*, **60**, 1585–1592.

Rogers J., Dowsett A. B., Dennis P. J., Lee J. V. and Keevil C. W. (1994b). Influence of plumbing materials on biofilm formation and growth of *Legionella pneumophila* in potable water systems. *Appl. Environ. Microbiol.*, **60**, 1842–1851.

Schaule G., Schulte S., Grobe S. and Flemming H.-C. (2010). Untersuchung der Langzeitwirkung von Reinigungs- und Desinfektionsmaßnahmen auf Werkstoffe hinsichtlich erneuter Biofilmbildung. In: Vermeidung und Sanierung von Trinkwasserkontaminationen durch hygienisch relevante Mikroorganismen aus Biofilmen der Hausinstallation, H.-C. Flemming (ed.), IWW Schriftenreihe 54, Mülheim 2010, pp. 231–264.

Schoenen D. and Schöler H. F. (1983). Trinkwasser und Werkstoffe: Praxisbeobachtungen und Untersuchungsverfahren DVGW-Schriftenreihe Wasser nr. 37.

Servais P., Laurent P. and Randon G. (1995). Comparison of the bacterial dynamics in various French distribution systems. *J. Wat. Supply Res. Technol. AQUA*, **44**, 10–17.

Shen H., Rogelj S. and Kieft T. L. (2006). Sensitive, real-time PCR detects low-levels of contamination by Legionella pneumophila in commercial reagents. *Mol. Cell. Probes*, **20**, 147–153.

Staley J. T. and Konopka A. (1985). Measurement of in situ activities of nonphotosynthetic microorganisms in aquatic and terrestrial habitats. *Ann. Rev. Microbiol.*, **39**, 321–346.

Szewzyk U. and Röder R. (2010). Untersuchung der Auswirkung von Desinfektionsmaßnahmen auf die Biofilm-Population unter besonderer Berücksichtigung der Ansiedlung und Vermehrung hygienisch relevanter Mikroorganismen. In: Vermeidung und Sanierung von Trinkwasserkontaminationen durch hygienisch relevante Mikroorganismen aus Biofilmen der Hausinstallation, H.-C. Flemming (ed.), IWW Schriftenreihe 54, Mülheim 2010, pp. 329–388.

Szewzyk U., Szewzyk R., Manz W. and Schleifer K.-H. (2000). Microbiological safety of drinking water. *Annu. Rev. Microbiol.*, **54**, 81–127.

Teitzel G. M. and Parsek M. R. (2003). Heavy metal resistance of biofilm and planktonic *Pseudomonas aeruginosa*. *Appl. Environ. Microbiol.*, **69**, 2313–2320.

Teitzel G. M., Geddie A., De Long S. K., Kirisits M. J., Whiteley M. and Parsek M. R. (2006). Survival and growth in the presence of elevated copper: transcriptional profiling of copper-stressed *Pseudomonas aeruginosa*. *J. Bacteriol.*, **188**, 7242–7256.

Tsvetanova Z. G. and Hoekstra E. J. (2009). A study on assessment of biomass production potential of pipe materials in contact with drinking water. *Wat. Sci. Technol., Wat. Supply*, **9**, 423–429.

Van der Kooij D. (1992). Assimilable organic carbon as an indicator of bacterial regrowth. *J. Am. Wat. Works Assoc.*, **84**, 57–66.

Van der Kooij D. (2003). Managing regrowth. In: Heterotrophic plate counts and drinking water safety, J. Bartram, J. Cotruvo, M. Exner, C. Fricker and A. Glasmacher (eds), IWA Publishing, London, UK, pp. 199–232. ISBN: 1 84339 025 6.

Van der Kooij D., Veenendaal H. R., Slaats N. P. G. and Vonk D. (2002). Biofilm formation and multiplication of *Legionella* on synthetic pipe materials in contact with treated water under static and dynamic conditions. In: R. Marre et al. (eds), Legionella. ASM Press, Washington DC, pp. 176–180.

Van der Kooij D., Veenendaal H. R. and Scheffer W. J. H. (2005). Biofilm formation and multiplication of Legionella in a model warm water system with pipes of copper, stainless steel and cross-linked polyethylene. *Wat. Res.*, **39**, 2789–2798.

Van der Wende E., Characklis W. G. and Smith D. B. (1989). Biofilms and bacterial drinking water quality. Wat. Res., 23, 1313–1322.

Völker S., Schreiber C. and Kistemann T. (2009). Umsetzung des § 18 TrinkwV 2001 in der behördlichen Praxis. *Umweltmed. Forsch. Prax.*, **14**, 265–266.

Völker S., Schreiber C. and Kistemann T. (2010). Drinking water quality in household supply infrastructure – a survey of the current situation in Germany. *Int. J. Hyg. Environ. Health*, **213**, 204–209.

Wadowski R. M., Yee R. B., Mezmar R., Wing E. J. and Dowling J. N. (1982). Hot water systems as sources of *Legionella pneumophila* in hospital and nonhospital plumbing fixtures. *Appl. Environ. Microbiol.*, **43**, 1104–1110.

Whiley H., Taylor M. and Bentham R. (2011). Detection of *Legionella* species in potting mixes using fluorescent *in situ* hybridisation (FISH). *J. Microb. Meth.*, **86**, 304–309.

WHO, World Health Organization (2011). Guidelines for Drinking-Water Quality. 4th edn. Geneva, World Health Organization.

Wingender J. (2011). Hygienically relevant microorganisms in man-made water systems. In: Biofilm Highlights, H.-C. Flemming, J. Wingender and U. Szewzyk (eds), Springer, Heidelberg, New York, pp. 189–238.

Wingender J. and Flemming H.-C. (2004). Contamination potential of drinking water distribution network biofilms. *Wat. Sci. Technol.*, **49**, 277–286.

Wingender J. and Flemming H.-C. (2011). Biofilms in drinking water and their role as reservoir for pathogens. *Int. J. Hyg. Environ. Health*, **214**, 417–423.

Wullings B., Remoc Voogt H., Veenendaal R. V. H. and Kooij D. (2002). The fluorescent in situ hybridization test in comparison with culture for detection of *Legionella pneumophila* in water samples. In: R. Marre et al. (eds), Legionella. ASM Press, Washington DC, 263–266.

Yu J., Kim D. and Lee T. (2010). Microbial diversity in biofilms on water distribution pipes of different materials. *Wat. Sci. Technol.*, **61**, pp. 163–171.

Chapter 9

Invertebrates in drinking water distribution systems

J. H. M. van Lieverloo, W. Hoogenboezem, G. Veenendaal and D. van der Kooij

9.1 INTRODUCTION

9.1.1 Invertebrates in drinking water

Few people realise that their drinking water is delivered by distribution systems predominantly resembling clean, pristine mountain brooks: carrying water that usually is not just clear and fresh, but that is also flowing over surfaces supporting communities of invertebrate life. Invertebrates such as worms and crustaceans are likely to be found in drinking water distribution systems (Anonymous 1996; van Lieverloo *et al.* 2002; Evins 2004). Representatives of the taxa most commonly reported in literature are presented in Figure 9.1. These organisms crawl, slide and swim over biofilms and sediments in the mains, along with many free-living protozoans and bacteria, presumably serving as their prey. The organisms visible to the naked eye are too large to be easily swept from the mains into the service lines and even if they actively invade these lines, most water meters are equipped with a filter that stops most of them. The small invertebrates that do pass the water meter filter and the aerator mounted on modern taps are not likely to be observed.

Consequently, no one ever notices one of these invertebrates in a glass of water or a bathtub. People who do usually are shocked and disgusted, falsely assuming that these organisms are indicators of low hygiene, as invertebrates are in many other domestic situations.

9.1.2 Significance

Bichai *et al.* (2008, 2009) suggested, as have others before them (summarised by van Lieverloo *et al.* 2002), that Nematoda and other invertebrates may protect pathogenic microorganisms from disinfection in surface water treatment. No study, however, has ever been able to show in a quantitative microbial risk assessment that these processes significantly contribute to a diminished microbial safety. In the Netherlands, with multiple barrier surface water treatment (van der Kooij *et al.* 1999), it is very unlikely that these processes have a notable influence.

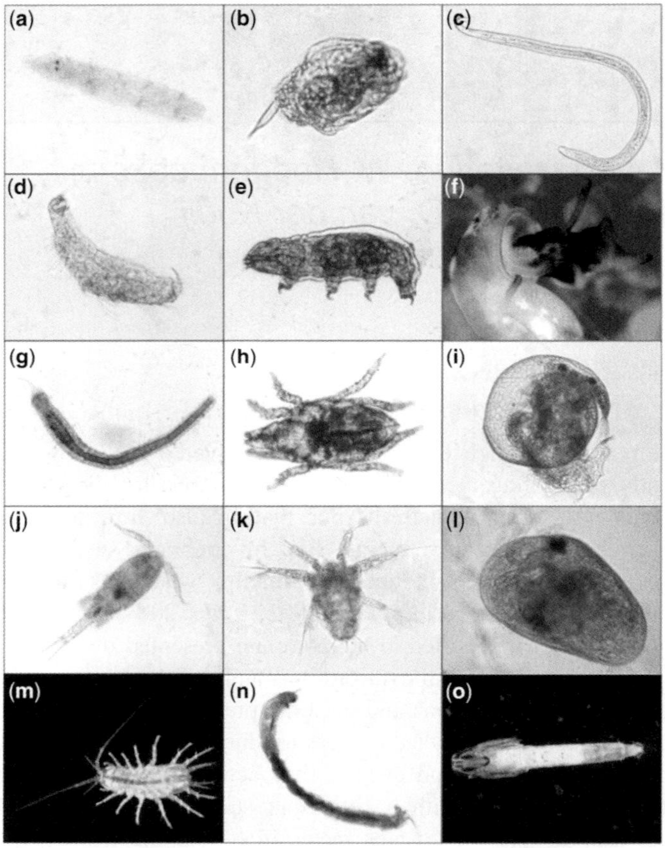

Figure 9.1 Examples of invertebrate taxa commonly found in drinking-water distribution systems. (a) Turbellaria (flatworms, c. 0.5–20 mm); (b) Rotifera (0.1–1 mm); (c) Nematoda (roundworms, 0.5–2 mm); (d) Gastrotricha (0.1–1 mm); (e) Tardigrada (0.1–1 mm); (f) Gastropoda (snails, 1–10 mm); (g) Oligochaeta (common worms, 0.5–100 mm); (h) Hydrachnellae (water mites, 0.3–1 mm); (i) Cladocera (water fleas, 0.5–1 mm); (j) Copepoda (0.5–2 mm); (k) larvae of Copepoda (nauplius larvae, 0.2–0.5 mm), (l) Ostracoda (0.5–2 mm); (m) Asellidae (aquatic sow bugs, 1–12 mm, excluding extremities), (n) larva of Chironomidae (midge flies, 4–30 mm); (o) pharate adults of Chironomidae (c. 4 mm). (Figure modified from van Lieverloo *et al.* 2002, John Wiley & Sons, reprinted with permission.)

Consumer complaints about invertebrates emerging from taps do not occur often, but persistent complaints were and still are of major concern to water supply companies. In the Netherlands, the current number of consumer complaints about invertebrates is estimated at less than 10 cases per year.

The presence of invertebrates in drinking water distribution systems has caused negative publicity in the Netherlands at the end of the 19th century (Asellidae), in 1939 (Oligochaeta), in 1964 and 1981 (Asellidae), in 1988–1989 (small invertebrates after microscopic investigation) and 1990 (Asellidae) (van Lieverloo *et al.* 2002). The latter two incidents occurred in distribution systems supplied with groundwater-derived drinking water. The incident in 1988–1989 led to questions in national parliament and to a national survey from 1993 to 1995 and subsequent investigations of the effect of control measures, the results of which have been presented during conferences (van Lieverloo *et al.* 1995, 1997, 1998).

9.2 ECOLOGY

9.2.1 Hypotheses

A review of the history of investigations of invertebrates in water supply, their significance, descriptions of the taxa found, a summary of data on abundance and a number of hypotheses explaining their presence and abundance have been described earlier (van Lieverloo *et al.* 2002). The hypotheses can be summarised as follows (illustrated in Figure 9.2):

- Invertebrates, probably in small numbers, once migrated and may still migrate into distribution systems from groundwater or surface water. Some have also colonised treatment works (filters, reservoirs), causing frequent presence of invertebrates in finished water.
- Dissolved organic and inorganic compounds enter the distribution system from the treatment work. Some of these compounds can serve as food for bacteria in the water, but also for bacteria growing in biofilms on the inner side of mains and on sediment particles. Some materials release compounds that may also contribute to biofilm formation (see Chapter 12). The biofilm in the distribution system eventually sloughs off and precipitates, adding to the sediment.
- Along with dissolved compounds, particulate matter (with both organic and inorganic compounds) enters the distribution system. It is very likely that these particles are coagulates (in case of surface water supply) and biofilm particles sloughed off from filter materials in treatment works. These particles precipitate as sediment or attach to the biofilms in the mains.
- The bacteria, free-living protozoans and particulate organic matter (POM) in the biofilm as well as in the sediments most likely serve as the primary food sources for the invertebrate community.

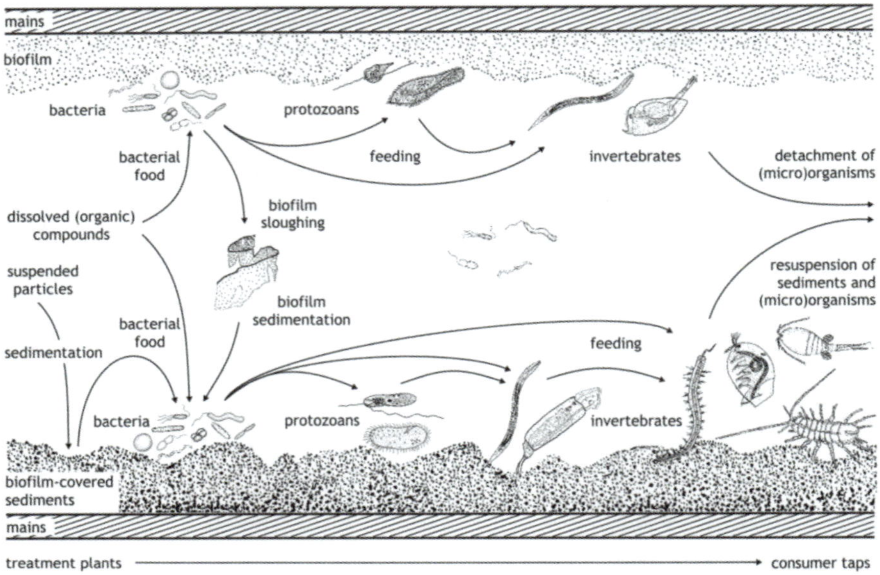

Figure 9.2 Schematic depiction of the hypothesis on the basic interactions in the food web in drinking-water distribution system biocoenoses (from: van Lieverloo *et al.* 2002, reprinted with permission from John Wiley & Sons).

9.2.2 Feeding, growth and reproduction

Availability of sufficient and suitable food most likely determines the abundance of invertebrates in distribution systems. Planktonic species (mostly Rotifera, Cladocera and Copepoda) and sessile filter feeders (Porifera (sponges), Coelenterata (polyps), Bivalvia (mussels) and Ectoprocta (bryozoans) feed on relatively high densities of algae or POM (particulate organic matter) in the water. Removal of POM from surface water by coagulation and sedimentation, especially if followed by filtration, usually suffices to limit growth of these taxa in distribution systems. A study of invertebrates in sediments in finished-water reservoirs in Germany revealed that two planktonic surface-water species of Rotifera (*Notholca caudate* and *Triarthra longiseta*) and larvae of Copepoda, decreased significantly with distance from the treatment works, but numbers of benthic (=bottom-dwelling) Rotifera slightly increased (Schreiber 1996). These observations indicate that a benthic community is more likely to succeed than a planktonic community.

On all water-exposed surfaces in the distribution systems, benthic bacteria form the basis of a biofilm, a slime layer coloured yellow, brown or black by suspended particles, for example, POM and (hydr)oxides of iron and manganese, that attach to it. Probably most of the bacteria are heterotrophic, using organic matter both as an energy source and as a source of base materials. Additionally, incomplete removal of ammonium, nitrite and methane in treatment works using anoxygenous groundwater will lead to growth of nitrifying bacteria and methane-oxidizing bacteria in distribution system biofilms as well. In these biofilms, the bacteria and protozoans form the living food for Nematoda, benthic Rotifera, Gastrotricha, Tardigrada, Turbellaria, Gastropoda, Ostracoda, benthic Copepoda, benthic Cladocera, juvenile Hydrachnellae and Chironomidae larvae. Large invertebrate species will most likely feed on the biofilms at the bottom of mains and the biofilm-covered sediment particles accumulating there. These sediments are formed by (i) precipitation of particles suspended in finished water, (ii) insoluble compounds formed in mixed water types, (iii) parts of mains materials and (iv) sloughed-off biofilms, increasing the surface area available for bacteria and protozoans to grow on. Oligochaeta burrowing in these sediments ingest whole sediment particles and Asellidae most likely predominantly scrape off biofilm from surfaces, meanwhile ingesting rust, bitumen and asbestos fibres, as shown by the composition of faecal pellets of Asellidae ubiquitously found in sediments collected from mains (van Lieverloo *et al.* 2012). Small invertebrates such as Rotifera and small Nematoda are attached to the biofilm while feeding and are likely to be ingested by biofilm grazers and sediment feeders. Planktonic Copepoda are found whirling up particles to feed on while skimming the sediment surface. Predators of many taxa may thrive on other invertebrates, provided the abundance of prey is sufficient and predators are not visually orientated.

Those species successfully competing for food in distribution systems will also depend on reproduction for establishing and maintaining a population. Though protozoans and most drinking-water invertebrates can reproduce asexually, some taxa, for example, Copepoda and Asellidae, can only reproduce sexually, causing their chances of colonizing a distribution systems to depend on the initial population density, that is, the probability of finding a mate.

9.3 ABUNDANCE IN DISTRIBUTION SYSTEMS

9.3.1 Methods, sampling sites and sampling programme

An overview of methods for quantification of invertebrates in drinking water distribution systems has been presented by Van Lieverloo *et al.* (2002). Also, methods for quantification used in the Netherlands are described by van Lieverloo *et al.* (2004, 2012). In most cases, invertebrates are filtered from tap water or from water flushed from mains. Tap water samples are filtered with a fine mesh filter (e.g. 10 μm), flushed water samples are filtered with coarse mesh filters (e.g. 500 followed by 100 μm), sometimes followed by fine mesh filters (e.g. 30 and 10 μm). An example of a flushing water sampling device used in the Netherlands is presented in Figure 9.3. Samples are conserved either chemically or by cooling. Quantification of invertebrates in samples is performed by microscopy.

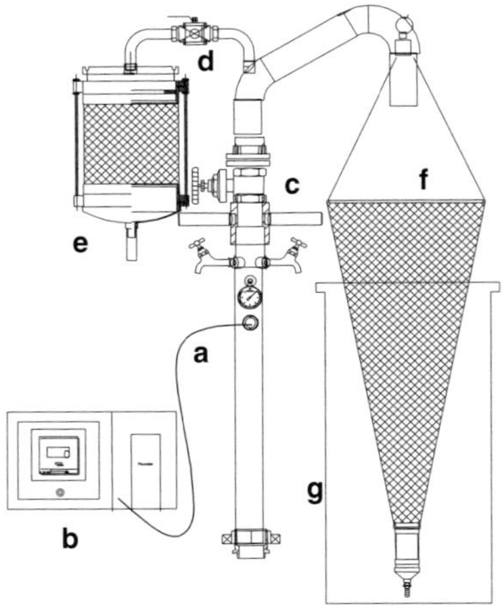

Figure 9.3 Drawing of the split top-piece and filters for flushing and filtering water from mains via underground hydrants. a = flow sensor; b = flow and volume indicator; c = slide valve; d = ball valve; e = 100 µm mesh plankton gauze filter containing a removable 500 µm mesh plankton gauze filter; f = 500 µm mesh plankton gauze net; g = filtrate vessel with overflow. Not shown: 30 µm mesh plankton gauze net (form identical to f) to be suspended under e (since 1996). (From van Lieverloo *et al.* 2004, reprinted with permission from Elsevier Publishing Ltd.)

In 1993, in the Netherlands, 1.3 billion m^3 drinking water was produced annually in a total of about 250 treatment works (18 use surface water, producing 37%, 232 used groundwater, producing 63%) (van der Kooij *et al.* 1999). The water was distributed via 110,000 km of mains with an inner diameter of over 50 mm (35% asbestos-cement, 45% unplasticized polyvinylchloride, 14% cast-iron, 6% concrete, steel, etc.). Ninety per cent of the water was distributed without a disinfectant residual and levels were low in the remaining 10% (approximately 0.05 mg $litre^{-1}$). Since 2006, all distributed drinking water gas been free of a disinfectant residual; although low doses of ClO_2 (0.1–0.2 mg $litre^{-1}$) are added at the end of a few treatment works, no detectable levels are present in finished water.

In 1992, 34 treatment works of 17 water companies were selected for a nationwide survey of invertebrates in finished water and in distribution systems. Primarily groundwater treatment works were selected in this study, because the abundance of invertebrates was already being monitored and had been for decades in distribution systems of surface water treatment works. The 31 groundwater treatment works (three of which use aerobic groundwater) were selected over the complete range of the mean permanganate ($KMnO_4$) index of finished water. This variable represents the concentration of natural organic matter (NOM) and was considered a rough index of the availability of food for bacteria and indirectly of food for invertebrates, in the absence of data on biofilm formation rate (van der Kooij *et al.* 1995a) at that time. During the selection, concentrations of assimilable organic carbon (AOC, van der Kooij *et al.* 1982), mostly <10 μg C $litre^{-1}$, were taken into account as well, as were NH_4^- concentrations in untreated water and the plate counts of aeromonads in tap water (van der Kooij *et al.* 1995b). Other criteria were the size of the treatment work (production at least 0.5 $Mm^3\ yr^{-1}$) and a largely unvarying treatment from 1991 to 1995.

The selection is not representative of water treatment works in the Netherlands, but it is representative of the range of groundwater treatment works in the Netherlands. The selected treatment works produced 18% of all drinking water (23% of all groundwater-derived drinking water). Disinfectant residuals were absent in all distribution systems.

Data from two groundwater treatment works were disregarded because data were incomplete and sampling quality proved unreliable during evaluations (van Lieverloo *et al.* 2004). Characteristics of the remaining 34 treatment works have been described by van Lieverloo *et al.* (2012).

Samples were collected from finished water (from a tap on the trunk main), from water flushed from mains (via c. five underground hydrants per distribution system) and from tap water (from taps in buildings near the hydrants). In total, samples from 176 hydrants and buildings were investigated. Nearly all mains (174) had an inner diameter of 100 to 105 mm (one site cast-iron 80 mm, one site cast-iron 125 mm). Sites were representative of all sorts of areas in distribution systems (residence time, mains lay-out plans, mains material and age), except cast-iron mains in densely populated and meshed areas such as old parts of cities and villages. It was not considered acceptable to frequently cause large numbers of consumer complaints about discoloured water while sampling in these areas. Usually one of every five sites was selected in mixing zones of two treatment works, other sites were estimated to receive water from one treatment work only for at least 95% of the time. Samples were collected every three months on all sites, starting September 1993, ending August to October 1995. Test samples were collected three months before the first samples, to allow for practice of the samplers and analysts, but also to start the survey on invertebrate communities that were used to being sampled quarterly. Capacity for sampling and analysis was distributed over these periods, equally distributing treatment works producing high and low concentrations of bacterial food in finished water over these periods, in order to prevent a bias in evaluation of seasonal effects (water temperature).

9.3.2 Abundance in drinking water

The 200 litre 10 μm filtered samples of finished water from treatment works were dominated by protozoans (present in 78% of the samples, mean 105 specimen per m^3), Rotifera (present in 76%, mean 65 m^{-3}) and Nematoda (present in 38%, mean 5.4 m^{-3}). More statistics are presented in Figure 9.4a. It is very likely that higher numbers of protozoans were present, as most flagellates and non-shelled amoebae pass the 10 μm filters. Valster *et al.* (2009), using the quantitative polymerase chain reaction technique in samples collected with 1.2 μm filters, found concentrations of up to 815 cells litre^{-1} for a single protozoan species *Hartmanella vermiformis* and a large variety of other free-living protozoans. As only larger organisms were the focus of the survey described in this chapter, no effort was taken to improve the quantification method. Although mean abundance of Gastrotricha and Copepoda and their larvae was comparable to the abundance of Nematoda, they were present only in 13%, 15% and 21% of the samples, respectively. Other invertebrates were found in less than 10% of the samples and were mostly limited to a maximum of four specimens in a 200 litre sample. The mean abundance of protozoans and invertebrates in tap water was fairly identical to that of finished water (Figure 9.4b).

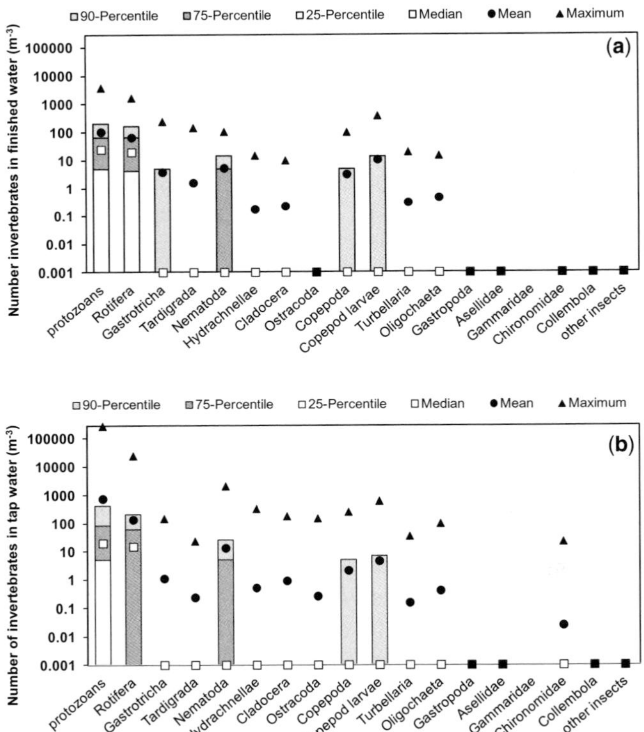

Figure 9.4 Detected abundance of protozoans and invertebrates in 34 water supplies representative of water supply in the Netherlands (1993–1995). (a) Finished water of treatment works (c. 8 samples per treatment work). (b) Tap water (176 locations, c. 8 samples per tap, 1296 in total). Abundance of invertebrates taxa that were not found is presented as 0.001 per m^3. Note that numbers of protozoans and small invertebrates are underestimated as an unknown fraction of these organisms may pass the pressurised filters (10 μm). (Adapted from: van Lieverloo *et al.* 2012. Reprinted with permission from Elsevier Publishing Ltd.)

Numbers of protozoans observed in one 200 litre sample were as high as 270 $litre^{-1}$, the sample was taken during the winter in a remote holiday resort that was fed by a 200 mm water main. As sampling sites in distribution systems were not selected in areas with increased risk of discoloured water complaints during collection of invertebrates and sediments flushed from mains (e.g. in cities with many old cast-iron mains), it is possible that higher numbers of protozoans and invertebrates can be found. Occasionally, numbers of relatively large invertebrates (Cladocera, Copepoda and Oligochaeta) in tap water were up to 0.1 $litre^{-1}$ per taxon. Most likely, these samples were taken during events of increased flow in the mains.

9.3.3 Abundance in water flushed from mains

Assuming that, due to gravity, large invertebrates may live predominantly near the bottom quarter of the mains, which was 7.8 cm in almost all sampled mains, the density of invertebrates per 100 cm^2 bottom of these mains equals the abundance per litre water flushed from mains (~12.7 cm length). This rule of thumb can be used to estimate surface densities in the mains. However, Asellidae mainly foraged on the bottom of aquarium tanks in laboratory experiments, although they were found on the vertical walls as well (unpublished observations). This indicates that many invertebrates, including the larger ones, may be foraging on biofilm on the sides and top of the mains as well.

Copepoda (present in 87% of the samples), Cladocera (69%) and Oligochaeta (71%) dominated abundance in samples filtered from water flushed from mains. Mean abundance of these taxa was 280, 350 and 135 N m^{-3} (1 m^3=10 m^2 mains bottom) respectively, but extremes could be as high as 14 specimens per litre (1.4×10^4 m^{-3}) (Figure 9.5), which is equal to approximately 7 cm^{-2}. Other abundant taxa are Asellidae (in 49%, mean 37 m^{-3} or 3.7 m^{-2}), Hydrachnellae (present in 67% of the samples, mean 55 m^{-3} or 5.5 m^{-2}), Turbellaria (in 35%, mean 10 m^{-3} or 1 m^{-2}) and Nematoda (in 32%, mean 15 m^{-3} or 1.5 m^{-2}).

Using estimates of the mean biomass, expressed as mg wet weight (ww) of specimen found in tapwater and water flushed from mains (Van Lieverloo *et al.* 2012), the mean invertebrate biomass in tapwater was estimated at 0.16 μg $litre^{-1}$ ww (maximum 40 μg $litre^{-1}$), dominated by Rotifera (27%), Oligochaeta (26%) and Copepoda (14%).

In water flushed from mains, the mean biomass is 420 μg $litre^{-1}$ ww (maximum 15.4 mg $litre^{-1}$), dominated by Asellidae (88.5%) and Oligochaeta (9.5%). In most distribution systems, *Asellus aquaticus* (L.) was found in a limited number of random identifications of Asellidae. In a few groundwater supplies, *Proasellus coxalis* (Dollfus 1892) was found. Both organisms are surface water species, indicating immigration from surface water, for example, during mains breaks (Van Lieverloo *et al.* 2002). Although Copepoda and Cladocera are much more abundant in number, the estimated contribution to the total biomass of these taxa is only 0.66% and 0.83% respectively. As only a small fraction (1%) of the Chironomidae larvae are collected by flushing mains with 1 m s^{-1} and filtering the flushed water over 100 μm filters (van Lieverloo *et al.* 2004), the actual biomass distribution in the mains most likely is larger.

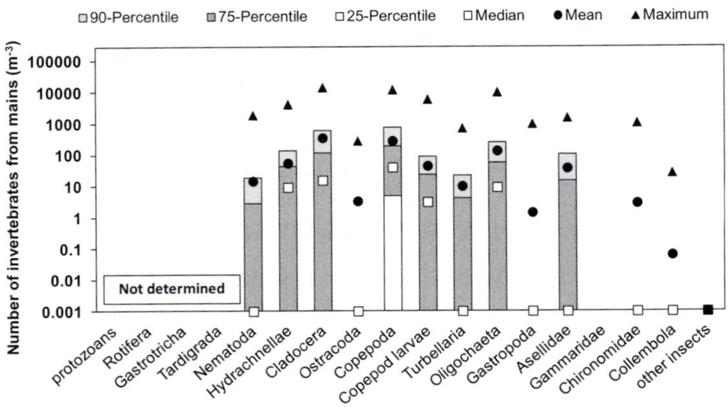

Figure 9.5 Detected abundance of invertebrates in water flushed from mains (176 locations, c. 8 samples per hydrant, 1301 in total) in 34 water supplies representative of water supply in the Netherlands (1993–1995). Abundance of invertebrates taxa that were not found is presented as 0.001 per m^3. Protozoans and small invertebrates were not determined. Note that Turbellaria, Nematoda, Oligochaeta and Copepoda larvae in mains probably are underestimated (approximately a factor of five to ten, other taxa a factor of two, Chironomidae a factor of 100, van Lieverloo *et al.* 2004). (From: van Lieverloo *et al.* 2012, reprinted with permission from Elsevier Publishing Ltd.)

A comparison of the mean results of the 1993–1995 survey in the Netherlands with means of data from 6 water supplies using surface water (1964–1995) and with means reported in international literature (Europe, United States, Asia, 1948–1996) shows that mean abundance is fairly comparable and that for some taxa the differences are quite small (Figure 9.6). Although the means of some taxa presented in literature are 10–100 times higher, for example, Nematoda in tap water and Gammaridae in water flushed from mains, generally there seems to be an optimum ratio of abundance between taxa in the drinking water distribution system biocoenosis. A comparison of means in finished water shows similar results (van Lieverloo *et al.* 2002, data not shown).

9.3.4 Variability within and between distribution systems

Although mean abundance in literature is comparable with the mean abundance in distribution systems in the Netherlands, variability between and within distribution systems can be large. Figures 9.7 (tap water) and 9.8 (water

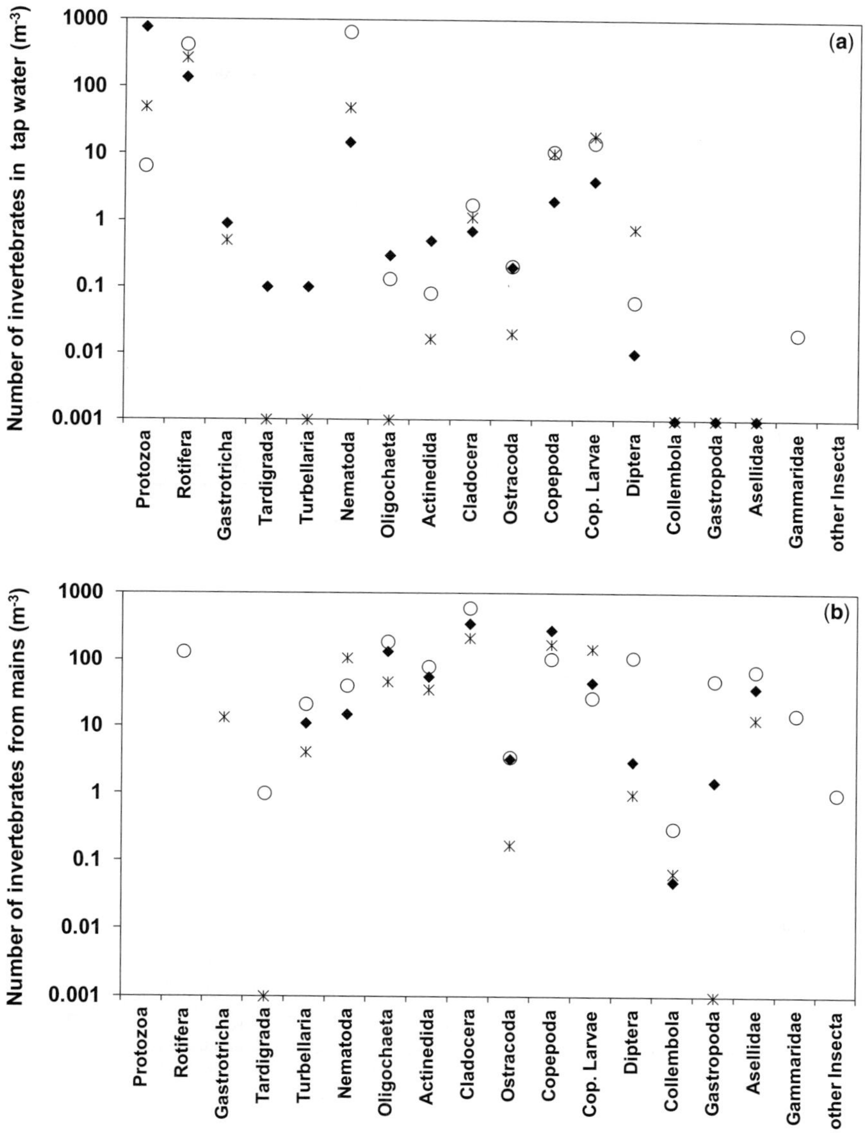

Figure 9.6 Mean abundance of protozoans and invertebrates in water supplies, mean of means per study. ○ Literature data (1948–1996); *Data from 6 surface water utilities in the Netherlands, one report per utility (1964–1995); ◆ Means from the 1993–1995 survey (31 groundwater and 3 surface water treatment plants) in the Netherlands. Missing markers indicate that numbers of these groups were not determined. The lowest *y*-axis level indicates no specimens are detected. (Adapted from van Lieverloo *et al.* 2002, reprinted with permission from John Wiley & Sons.)

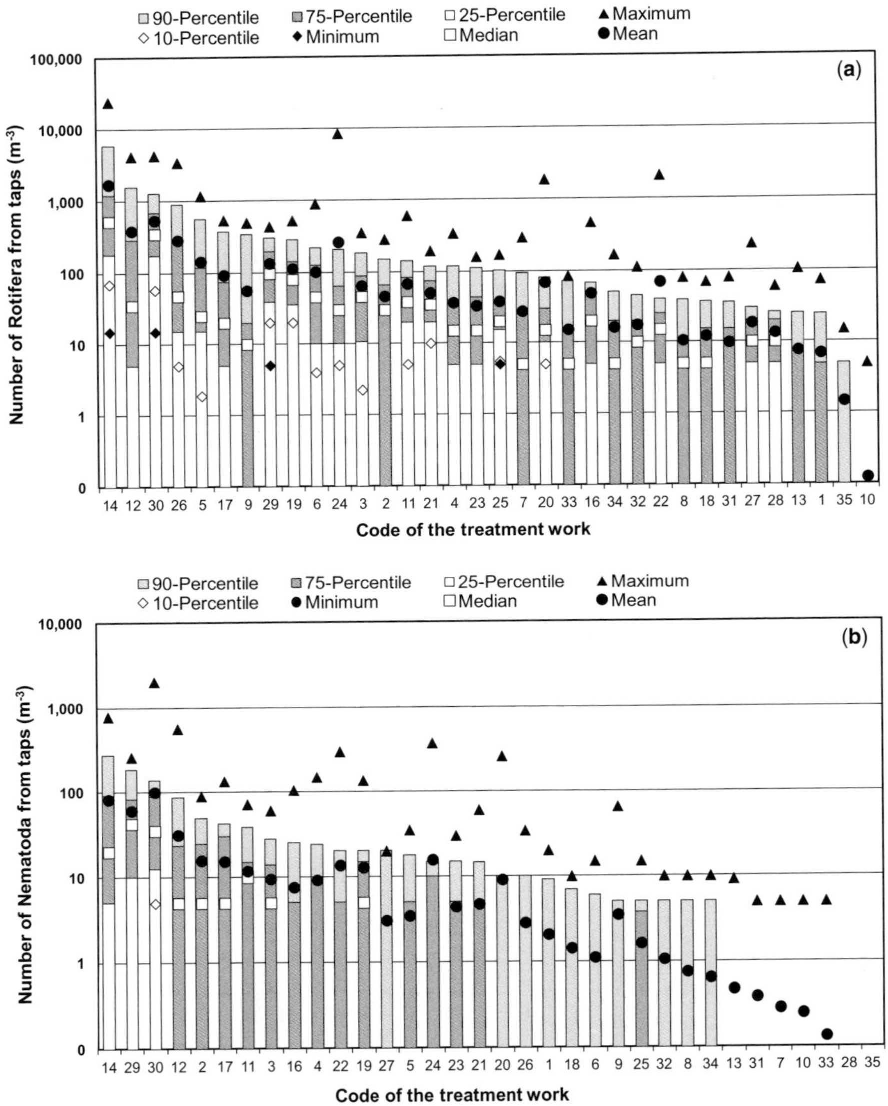

Figure 9.7 (a) Rotifera in tap water from houses and buildings, (b) Nematoda in tap water. Treatment works 1, 2 and 6 use surface water (in treatment work 1 via soil passage), treatment works 18, 22 and 35 use aerobic groundwater, other treatment works use anaerobic groundwater. The lowest *y*-axis level indicates no specimens are detected. (From: van Lieverloo *et al.* 2012, reprinted with permission from Elsevier Ltd.)

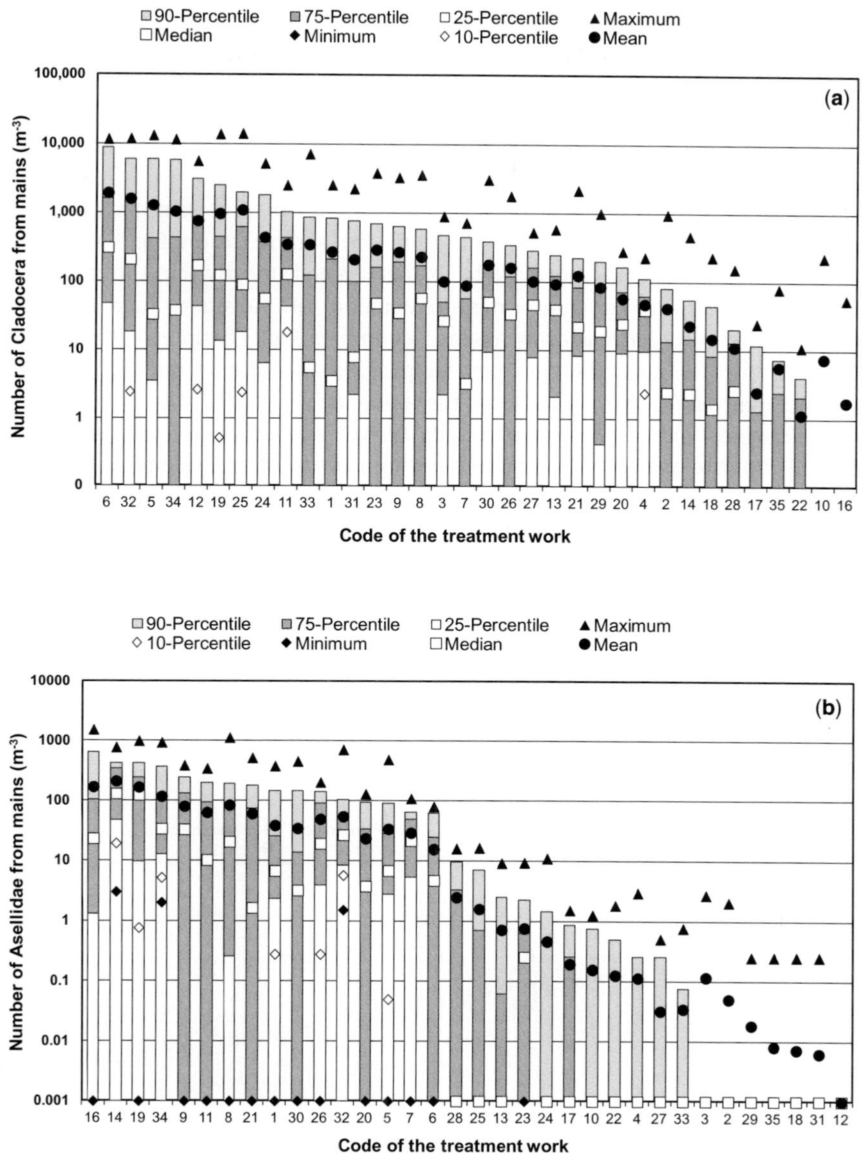

Figure 9.8 (a) Cladocera in water flushed from mains; (b) Asellidae in water flushed from mains. Treatment works numbers: see Figure 9.7. The lowest *y*-axis level indicates no specimens are detected. (From van Lieverloo *et al.* 2012, reprinted with permission from Elsevier Ltd.)

flushed from mains) show representative examples of variability in abundance of invertebrates within and between the 34 distribution systems in the 1993–1995 survey. The ratio between the 90-percentile and the 10-percentile within a distribution system is always higher than 10 and in some cases around 1000, even when the 10-percentile is higher than the detection limit. Variability within distribution systems is mostly determined by differences in abundance between sampling sites, although numbers at a sampling site may sometimes vary almost thousand-fold from one sample to another. Variability between sampling sites most likely is determined mostly by differences in hydraulics (determining transport/sedimentation of substrates and particles) and the presence of cast-iron mains, couplings and hydrants (Christensen *et al.* 2011). The total biomass per sample was estimated from the estimated average dimensions per taxon (van Lieverloo *et al.* 2012). The variability of total estimated biomass of invertebrates between distribution systems is large as well, but the variability of biomass within distribution systems is less than the variability of individual taxa (Figure 9.9). This is remarkable as 88% of the mean overall biomass is determined by Asellidae (Figure 9.8b) and 10% by Oligochaeta (van Lieverloo *et al.* 2012). It is likely that variability of different taxa is complementary, thus damping the variability of total biomass. Figure 9.10 gives the cumulative distribution functions (with percentile) of median, 90-percentile, mean and maximum values of these estimated biomasses.

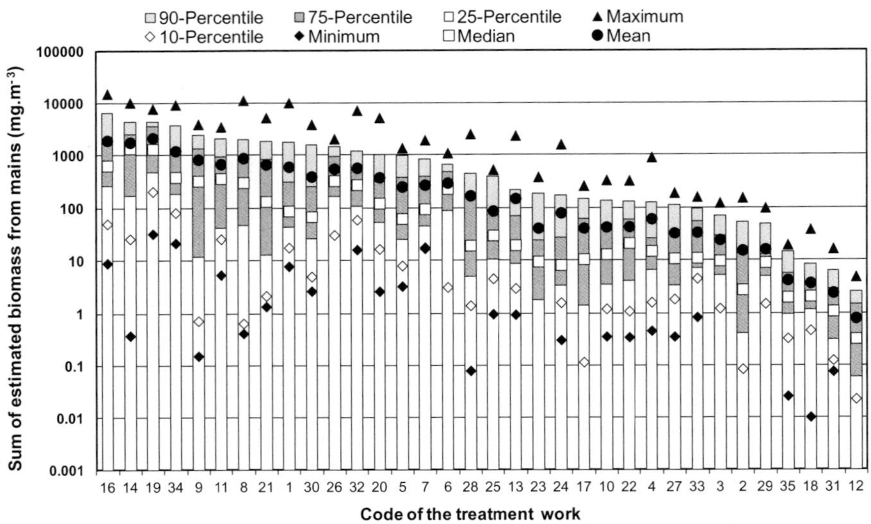

Figure 9.9 Total estimated biomass in water flushed from mains per treatment work.

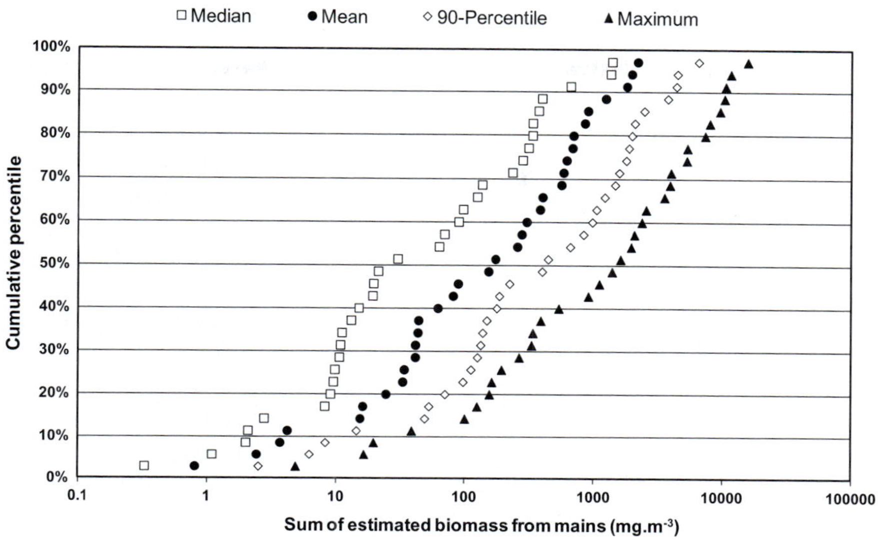

Figure 9.10 Cumulative distribution functions (cdf) of statistics per distribution system of total estimated biomass in water flushed from mains (identical to data in Figure 9.9, however cdfs are given per statistic, statistics are not linked).

9.4 CORRELATIONS WITH BIOFILM, NOM AND SEDIMENTS

9.4.1 Biofilm Formation Rate (BFR) and natural organic matter (NOM)

The biofilm formation rate (BFR) in finished water of 24 of the treatment works was determined using a biofilm monitor according to van der Kooij *et al.* (1995a). Water flows at a constant flow rate of 0.2 m s^{-1} downwards through a glass column containing a stack of glass cylinders. The concentration of adenosinetriphosphate (ATP, a measure of active biomass) per cm^2 is determined on glass rings periodically removed from the biofilm monitor. The BFR is expressed in pg ATP cm^{-2} day^{-1}. The concentration of NOM in finished water of treatment works was assessed as non-purgable organic carbon (NPOC, in mg $litre^{-1}$ C) and the permanganate index ($KMnO_4$ consumption in mg $litre^{-1}$ O_2). NPOC was determined as either total organic carbon (TOC) or dissolved organic carbon (DOC), according to methods that are now described in NEN-EN-ISO 8467 (European Committee for Standardisation 1997). As drinking water from treatment works is virtually free of particulate matter, the differences between TOC and DOC are low. The permanganate index of finished water at treatment works was determined according to a method that is described in NEN-EN-ISO 8467 (European Committee for Standardisation 1995).

9.4.2 Correlation with biostability of finished water

The quality of the finished water is determined both by the source water and by treatment. Figures 9.7 and 9.8 show that there are little or no differences in abundance of invertebrates in mains between distribution systems supplied by the three surface water treatment works (no. 1, 2 and 6) and treatment works using anaerobic groundwater (whether the water is filtered over two consecutive rapid gravity filters or over just one). This is corroborated by statistical testing of differences between medians ($p > 0.05$, Mann Whitney U tests with Bonferroni correction (MW-B)).

Distribution systems supplied by treatment works using aerobic groundwater (no. 18, 22 and 35) however clearly harbour much fewer Cladocera and Asellidae (as well as Oligochaeta and Copepoda, not shown) and therefore also less total biomass than distribution systems supplied by treatment works using anaerobic water. These differences are significant ($p < 0.01$, MW-B), differences between distribution systems from anaerobic groundwater and surface water treatment works are significant for Cladocera, Asellidae and biomass (for the latter two variables, $p < 0.01$, MW-B). When the source water is aerobic, there is no ammonium or methane present and almost all available organic material is already metabolized by microorganisms during passage of the sandy soil. Therefore this water has a high level of biological stability and BFRs of the finished waters were very low (1.1, 0.03 and 0.35 pg ATP cm^{-2} day^{-1}, respectively). For only one of the surface water treatment works (no. 6) the BFR of finished water was determined: 2.5 pg ATP cm^{-2} day^{-1}, which is also low, but biomass in the mains is not low, ~275 mg m^{-3}, possibly due to high levels of POM in the effluents of the granular activated carbon filters. The BFRs of the anaerobic groundwater treatment works range from 1.6 to 80 pg ATP cm^{-2} day^{-1}.

Figure 9.11 shows the linear regression model of the ^{10}log of the mean biomass of invertebrates in water flushed from the mains and the ^{10}log of the BFR of treated water. The percentage of variance of the mean biomass is explained for 55% ($R^2 = 0.55$) by the BFR (Table 9.1). The low BFR values of 0.03 and 0.34 pg ATP cm^{-2} day^{-1} have a high leverage, causing the R^2 to inflate. Without these values, the R^2 would be 0.45, however, in this latter model the biomass variance is no longer constant throughout the BFR value range and other biomass values have a high leverage. It is likely that the presented model is correct and the low BFR values are representative.

Invertebrate biostability in a distribution system could for example be expressed as a 95% probability of not exceeding a mean estimated invertebrate biomass of 500 mg m^{-3} water flushed from mains. This arbitrary level of biomass was achieved in two-thirds of the distribution systems in the survey (see Figure 9.10). The exceedance probability of 95% would be reached at a BFR of 2.2 pg ATP cm^{-2} day^{-1}, a permanganate index of 0.98 mg l^{-2} O_2 and an NPOC-level of 0.99

mg litre^{-1} C (including the variability of the intercept, the slope and the residuals of the log-linear models). These levels are achieved in the Netherlands in drinking water from aerobic groundwater treatment works and surface water treatment works applying slow sand filtration, but low mean biomass levels are achieved in distribution systems with less stable treatment works as well (Figure 9.11). As detailed data concerning consumer complaints are lacking, no objective acceptable level of invertebrate abundance can be suggested yet.

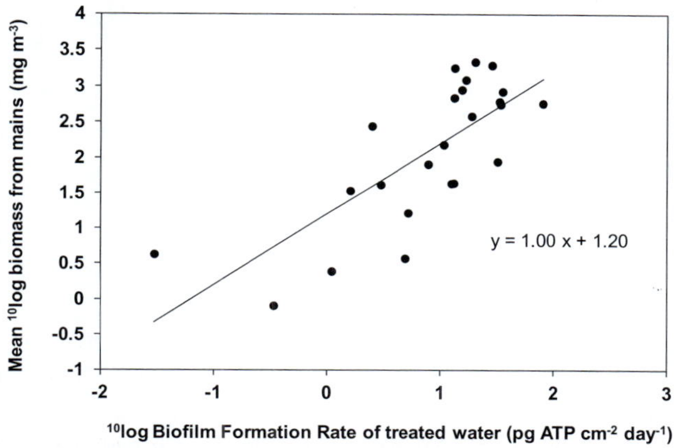

Figure 9.11 Log-linear regression model of the mean total estimated invertebrate biomass (>100 μm, wet weight) per distribution system in water flushed from mains (118 hydrants, total 933 samples) and the biofilm formation rate of the finished water of 24 supplying treatment works (the BFR of 10 treatment works was not determined). The R^2 of the ^{10}log model is 0.55; the relation is statistically significant ($p < 0.001$). (From van Lieverloo *et al.* 2012, reprinted with permission from Elsevier Ltd.)

9.4.3 Correlation with NOM and sediment

Table 9.1 shows the Pearson's correlation coefficients (R^2) of linear regression of the logarithms of mean numbers of abundant or large invertebrates in mains and the logarithms of the NPOC concentration and the permanganate ($KMnO_4$) index of finished water, and the amount of coarse sediments in water flushed from mains. Course sediment is 100–500 μm, which is only about 5% of the total sediment (Van Lieverloo *et al.* 2011). Total biomass correlates best with BFR, NPOC, $KMnO_4$ index and sediments 100–500 μm. Moreover, NPOC, $KMnO_4$ index and BFR also correlate with each other (log BFR vs. log NPOC $R^2 = 0.51$; log BFR vs. log $KMnO_4$ $R^2 = 0.67$, log NPOC vs. log $KMnO_4$ $R^2 = 0.87$; two lowest BFR and two low $KMnO_4$ values have a high leverage). Sediment volumes and BFR are not correlated (log-linear $R^2 = 0.19$), but there is some correlation with NPOC ($R^2 = 0.35$) and $KMnO_4$ ($R^2 = 0.46$).

Table 9.1 Correlations between the abundance of invertebrates in water flushed from mains (N m^{-3}), treated water quality parameters and sediments. Pearson's regression coefficients (percentage of variance accounted for, R^2) of linear regression models. Response variables are the logarithms of mean numbers or mean total biomass of invertebrates in water from mains (per distribution system). Explanatory variables are the logarithms of indexes of biostability or sediment (100–500 μm) flushed from mains (all data sets normally distributed, variance fairly constant).

Invertebrate taxon	NPOC (N = 34)	$KMnO_4$ (N = 34)	BFR (N = 24)	Sediment (100–500 μm) N = 34
Turbellaria	0.22**	0.20**	0.20*	0.22**
Oligochaeta	0.34***	0.35***	0.37***	0.24**
Hydrachnellae	0.00	0.00	0.00	0.00
Cladocera	0.15*	0.21**	0.36**	0.05
Copepoda	0.02	0.00	0.06	0.00
Copepoda larvae	0.00	0.00	0.03	0.00
Asellidae	0.26**	0.29**	0.24**	0.40***
Total biomass (mg m^{-3})	0.60***	0.63***	0.55***	0.52***

DOC = dissolved organic carbon, NPOC = non-particulate organic carbon, $KMnO_4$ = permanganate index; BFR = Biofilm Formation Rate. Significance levels (two-tailed) of the regression: *, $p \leq 0.05$;**, $p \leq 0.01$;***, $p \leq 0.001$. N = number of data sets.
Source: From van Lieverloo *et al.* (2012), reprinted with permission from Elsevier Ltd.

None of the parameters tested seems to be the best index for possible multiplication of large and/or abundant invertebrates in distribution systems. The R^2 of the correlation with BFR is somewhat lower, probably due to the lower number of BFR values available (24 vs. 34 of $KMNO_4$ and NPOC).

The BFR is more likely to have a direct causal relation as biofilm is the most likely base of the food web. Furthermore, the BFR includes possible inorganic bacterial substrates such as methane and ammonium. Looking at causal relationships, it is theoretically possible that all invertebrates are constantly entering the distribution system from the treatment works, crawling and swimming into the distribution system over the bottom of trunk mains. It is more likely, however, that (micro)organisms and POM, in the biofilm growing on the large surfaces in the mains (walls, sediment particles) and in the sediment in the distribution system, supply the majority of the base of the food web. In any case, the BFR of finished water most likely is correlated with the density of biofilm in the treatment works, where similar water flows through the treatment systems. Separation of all causal effects would require a far more detailed study, including laborious sampling of invertebrates crawling into trunk mains and quantification of biofilm densities and POM inside mains.

Christensen *et al.* (2011) found that the abundance of Asellidae is higher in cast-iron mains in Denmark than in PVC mains (52 samples from 31 hydrants). In the Netherlands, cast-iron mains are predominantly found in the centres of old cities, where flow characteristics are likely to be different as well.

9.5 CONCLUSIONS

Invertebrates may cause consumer complaints and although the probability of complaints is low, some complaints have led to negative publicity, in the Netherlands and elsewhere. There are no indications however of any direct health effect of the presence of invertebrates in distribution systems and in drinking water. The results of the 1993–1995 national survey show that:

- Protozoans and small invertebrates such as Rotifera and Nematoda are common in finished water from treatment works and in tap water.
- Cladocera and Copepoda dominate the numbers of larger organisms (in water flushed from mains), whereas Asellidae and Oligochaeta are dominant in biomass. Together, Asellidae (88%) and Oligochaeta (10%) form 98% of the mean biomass in distribution systems.
- Mean abundance in the Netherlands is fairly comparable to mean abundance reported in literature. There are however large differences within sampling points, distribution systems and between distribution systems, most likely due to differences in flow and the presence of cast-iron materials.
- Most likely, the differences between distribution systems are largely determined by biostability (measured as BFR) or organic matter (measured as NPOC, permanganate index) of the finished water of treatment works.

Information on benchmark values and control strategies can be found in a previous publication (van Lieverloo *et al.* 2002). Conclusions regarding these topics are:

- Monitoring invertebrate abundance and comparing data to benchmark values is necessary to efficiently aim at control measures, but also to be able to adequately inform authorities and if necessary customers and media on water quality. In distribution systems where large invertebrates such as Asellidae and Oligochaeta are abundant, probabilities of consumer complaints most likely are relatively high as well, increasing the incentive for control measures.
- Control of invertebrate abundance probably is best attained by distributing biostable water (e.g. with a low BFR and POM) and by using inert materials. As invertebrate populations quickly re-establish after cleaning mains, only intensive and frequent cleaning is likely to have effect in distribution systems where the level of biostability is insufficient.

Water companies in the Netherlands have been and are still improving treatment works to further reduce the microbial growth potential of the finished drinking water, especially when using anaerobic groundwater, and thus increase their control of invertebrate abundance. As improvement of biostability may take large investments and longer periods, research into the causes of the large differences of abundance of especially Asellidae within distribution systems and the causes of occasional migration into in-house plumbing may be of interest. In the meantime, it might be good to inform consumers about the presence of invertebrates in distribution systems, very much like in clean, pristine mountain brooks people have drunk safely from since prehistoric times. A chapter of the Reading Book for Primary Schools in The Netherlands by Wijma (1899) may be of help: "Just a glass of water! You see a glass in front of you, filled with pure water from a water supply. It is without taste and odour; it is clear; there is nothing remarkable in sight. Nonetheless, this glass contains a world of wonders (…). Although your eye presumably sees nothing but pure water, it is the residence of a large number of animals that are invisible to the naked eye, to which the glass is a large pool, a vast lake, a sea (…). However, drink it without concern: the little creatures are completely harmless and will, taken in by your body, soon have lost their lives."

Acknowledgements

This research was funded by the Joint Research Programme (Bedrijfstakonderzoek [BTO]) of the Netherlands drinking water companies. This chapter was composed from parts of articles published in Water Research, by Elsevier Publishing Ltd. (Van Lieverloo *et al.* 2004, 2012) and a chapter in the Encyclopedia of Environmental Microbiology, published by John Wiley & Sons, New York, USA (van Lieverloo *et al.* 2002).

9.6 REFERENCES

Anonymous (1996). Problem Organisms in Water: Identification and Treatment. Manual M7. 2nd edn. American Water Works Association, Denver Co, The United States of America, 153.

Bichai F., Payment P. and Barbeau B. (2008). Protection of waterborne pathogens by higher organisms in drinking-water. *A Review. Can. J. Microbiol.*, **54**, 509–524.

Bichai F., Barbeau B. and Payment P. (2009). Protection against UV disinfection of E. coli bacteria and B. subtilis spores ingested by C. elegans nematodes. *Wat. Res.*, **43**, 3397–3406.

Christensen S. C. B., Nissen E., Arvin E. and Albrechtsen H. J. (2011). Distribution of Asellus aquaticus and microinvertebrates in a non-chlorinated drinking water supply system – Effects of pipe material and sedimentation, *Wat. Res.*, **45**, 3215–3224.

Evins C. (2004). Small animals in drinking water distribution systems. In: Safe Piped Water: Managing Microbial Water Quality in Piped Distribution Systems. World Health Organization, IWA Publishing, London, The United Kingdom, pp. 101–120.

European Committee for Standardisation (1995). Water Quality – Determination of Permanganate Index (ISO 8467:1993). EN ISO 8467:1995, Brussels, Belgium.

European Committee for Standardisation (1997). Water Analysis – Guidelines for the Determination of Total Organic Carbon (TOC) and Dissolved Organic Carbon (DOC). EN 1484:1997, Brussels, Belgium.

Schreiber H. (1996). *Invertebrates in Water Supplies: A Survey from a Hygienic View-Point* (in German: Invertebraten in Wasserversorgungsanlagen: eine Bestandsaufnahme unter hygienischen Aspekten). PhD Thesis. Rheinische Friedrich–Wilhems–Universität Bonn, Germany, 154 p.

Valster R. M., Wullings B. A., Bakker G., Smidt H. and Van der Kooij D. (2009). Free-living protozoa in two unchlorinated drinking water supplies, identified by phylogenic analysis of 18S rRNA gene sequences, *Appl. Environ. Microbiol.*, **75**, 4736–4746.

van der Kooij D., Visser A. and Hijnen W. A. M. (1982). Determining the concentration of easily assimilable organic carbon in drinking water. *J. Am. Wat. Works Assoc.*, **74**, 540–545.

van der Kooij D., Veenendaal H. R., Baars-Lorist C., Van der Klift D. W. and Drost Y. C. (1995a). Biofilm formation on surfaces of glass and Teflon exposed to treated water. *Wat. Res*, **29**, 1655–1662.

van der Kooij D., Vrouwenvelder J. S., Veenendaal H. R. and Van Raalte-Drewes M. J. C. (1995b). Multiplication of Aeromonads in ground-water supplies in relation with the biofilm formation characteristics of drinking water. *Proc. Am. Wat. Works Assoc. Wat. Qual. Technol. Conf., San Francisco*, **1994**, 1349–1363.

van der Kooij D., Van Lieverloo J. H. M., Schellart J. and Hiemstra P. (1999). Maintaining quality without a disinfectant residual. *J. Am. Wat. Works Assoc.*, **91**, 55–64.

van Lieverloo J. H. M. and Van der Kooij D. (1995). National survey of invertebrates in drinking water distribution systems in the Netherlands. *Proc. Am. Wat. Works Assoc. Wat. Qual. Technol. Conf., San Francisco*, **1994**, 2065–2081.

van Lieverloo J. H. M., Van Buuren R., Veenendaal G. and Van der Kooij D. (1997). How to control invertebrates in distribution systems: by starvation or by flushing? *Proc. Am. Wat. Works Assoc. Wat. Qual. Technol. Conf.*, Denver, 1997. Paper 4C–2, 26 p.

van Lieverloo J. H. M., Van Buuren R., Veenendaal G. and Van der Kooij D. (1998). Controlling invertebrates in distribution systems with zero or low disinfectant residual. *Wat. Supply*, **16**(3–4), 199–204.

van Lieverloo J. H. M., Van der Kooij D. and Hoogenboezem W. (2002). Invertebrates and protozoans (free-living). In: Drinking Water Distribution Systems. Encyclopedia of Environmental Microbiology, G. Bitton (ed.), J. Wiley & Sons, Inc., New York, The United States of America, pp. 1718–1733.

van Lieverloo J. H. M., Bosboom D. W., Bakker G. L., Brouwer A. J., Voogt R. and De Roos J. E. M. (2004). Sampling and quantification of invertebrates from drinking water distribution systems. *Wat. Res*, **38**, 1101–1112.

van Lieverloo J. H. M., Drevijn A., Schaap P. and Hoogenboezem W. (2011). *Biological Activity and Sediment in The Supply Zone of PWN 2000–2010.* (In Dutch: Biologische activiteit en sediment in het voorzieningsgebied van PWN 2000–2010). Report 2010.03, Viaeterna, Rosmalen, The Netherlands, 81 p.

van Lieverloo J. H. M., Hoogenboezem W. and Van der Kooij D. (2012). Variability of invertebrate abundance in drinking water distribution systems in the Netherlands in relation to biostability and sediment volumes. *Wat. Res.*, **46**, 4918–4932.

Wijma J. (1899). Only a glass of water! (or 'A glass of water, but!', in Dutch: Een glas water maar!), Chapter 29, Leesboek voor de Lagere Scholen, Joh. Ykema. 's-Gravenhage, The Netherlands.

Chapter 10

Emerging issues of biological stability in drinking water distribution systems

M. Prévost, M. C. Besner, P. Laurent and P. Servais

10.1 INTRODUCTION

Distribution systems (DSs) have long been considered as inert vehicles for transporting drinking water. From this perspective, the focus was placed on the quantitative aspects of these systems and their hydraulic effectiveness in meeting water demand at any point of the network at any time. Water quality was achieved by a combination of conventional treatments and disinfection processes at the plant, and often extended into the distribution system, which is generally referred to as the multiple barrier concept. However, it has also been well known that DSs are susceptible to accidental contamination and that chemical and biological contamination do occur within these systems. For example, corrosion is a major internal source of water quality deterioration, and utilities are aware of the importance of implementing corrosion control to maintain water quality and to prevent customer complaints. Water quality can dramatically decrease in a DS, mainly due to excessive microbial growth. Bacterial proliferation, biofilm development and its associated trophic food web can be the source of various types of problems for drinking water producers and consumers. These problems include regulatory violations, aesthetic and organoleptic nuisances and health threats.

The degradation of the microbiological quality of water during distribution has been widely documented. Indeed, all essential elements for the growth of microorganisms are present in DSs. Furthermore, the residence time of the water in the pipes and reservoirs is more than sufficient for chemical and biological reactions to occur. As the residence time increases, various changes in water quality may occur such as the disappearance of the disinfectant residual (generally chlorine or chloramines), the consumption of biodegradable organic matter (BOM), the growth of bacteria (so-called regrowth or aftergrowth) and the repair of injured bacteria. Fixed and suspended bacterial biomass constitute the starting point for complex trophic chains that include rotifers and small invertebrates (Levy *et al.* 1986; Block *et al.* 1997) that can be controlled by reducing the BOM levels (van Lieverloo *et al.* 1998). This phenomenon represents a water quality challenge to the drinking water industry that strives to limit bacterial regrowth during water distribution.

BOM has been identified as the key factor determining biostability, often referred to as the absence of significant bacterial regrowth in a distribution system. A thorough critical review of BOM occurrence, removal by treatment impact on distribution and water quality was presented by Prévost *et al.* (2005). Work on BOM and methods to measure nutrient levels in drinking waters started at the beginning of the eighties and resulted in several approaches and methodologies to measure BOM. In the late eighties, BOM was first demonstrated to be the cause of regrowth in some DSs stirring the development of BOM removal treatment processes and bringing further development to biological filtration. It was then anticipated that its removal would be the major and perhaps even the only way to control bacterial regrowth and the occurrence of indicator bacteria such as coliform bacteria in full-scale systems (Laurent *et al.* 2005) Several studies have indeed shown that removing carbon nutrients such as biodegradable dissolved organic carbon (BDOC) decreases the amount of biofilm bacteria in areas of the distribution system without disinfectant residual (Figure 10.1) (Servais *et al.* 1995b; Servais *et al.* 2004).

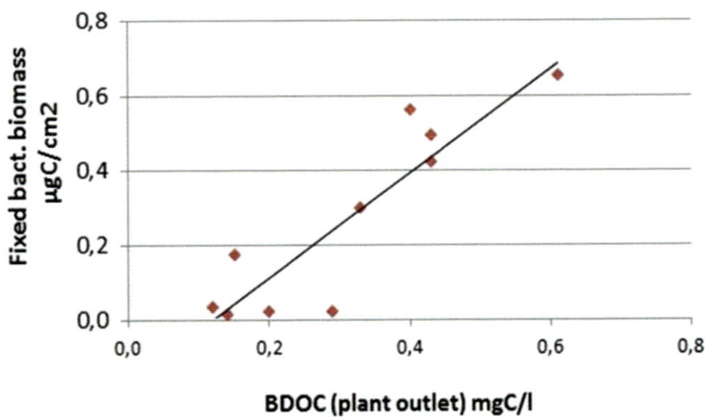

Figure 10.1 Average fixed bacterial biomass (expressed in µg C/cm^2) measured on cast-irons coupons plotted against the biodegradable dissolved organic carbon (BDOC) at the plant outlet for different sampling campaigns in the Parisian suburbs networks (situations with an average free chlorine residual in the DS < 0.1 mg Cl_2/l) adapted from Servais *et al.* (2004). In this study, bacterial biomass was estimated by the potential exoproteolytic activity method described by Laurent and Servais (1995).

Table 10.1 lists the various threshold concentrations of biodegradable organic matter below which water can be considered biologically stable in the DS. These thresholds refer to different methodologies for the measurement of BOM such as assimilable organic carbon (AOC) (van der Kooij *et al.* 1982), $BDOC_{SUSP}$ using a suspended mixed indigenous inoculums (Servais *et al.* 1987), or $BDOC_{SAND}$ using a fixed mixed indigenous inoculums (Joret and Lévi 1986). These thresholds are also adjusted to the reference organism chosen to define biostability such as total bacterial biomass, culturable bacteria heterotrophic plate counts (HPCs) or total coliforms. Van der Kooij (1992) proposed an AOC level of 10 μg/L to maintain the geometric mean of the colony counts of heterotrophic below the Dutch guideline value of 100 CFU HPC/ml incubated on agar medium at 22°C. LeChevallier *et al.* (1991, 1996) and Volk and LeChevallier (2000) later proposed AOC concentrations that should not be exceeded to limit the occurrence of total coliforms in the DS. Using a large full-scale DS database, Volk and Joret (1994) also defined a value of 0.15 mg C/L ($BDOC_{SAND}$) as a one of 4 critical thresholds to limit total coliform occurrence along with other parameters. Another threshold value of $BDOC_{SUSP}$ was defined by Servais *et al.* (1995a), it is based on the relationship between the consumption of BDOC observed in the DS and the BDOC content in the treated water of various DSs. Using a similar approach on Belgian DSs, Niquette *et al.* (2001) determined a higher value of 0.25 mg C/L ($BDOC_{SUSP}$) as the biostability threshold value. Volk (1994) defined a value of 0.15 mg C/L ($BDOC_{SAND}$) at 20°C and of 0.30 mg C/L ($BDOC_{SAND}$) at 15°C as thresholds under which no BDOC consumption will occur, based on a large experimental database obtained on a DS located in suburban Paris. Finally, using a deterministic model of bacterial dynamics in pipes, Laurent *et al.* (1997) showed that bacterial growth in the DS is minimal below 0.15 mg C/L of $BDOC_{SUSP}$, even in the absence of chlorine. Apparent discrepancies between these recommended thresholds reflect the stringencies of the biostability goals, the organisms used to measure biostability, the affinity for carbon substrates of the BOM test organisms and the practice of maintaining chlorine residual (Laurent *et al.* 2005).

In the nineties, however, further work showed that bacterial regrowth in full-scale distribution systems was driven by a combination of various factors, and that, in many cases, the removal of BOM alone was not feasible or sufficient to ensure the maintenance of water quality. Pipe materials, the accumulation of corrosion deposits and the excess ammonia present in chloraminated systems are examples of other drivers that can surpass the impact of BOM (Jiang *et al.* 2012). Finally, the development of several models and their validation in full-scale systems (Volk and Joret 1994; Gatel *et al.* 1995; Dukan *et al.* 1996; Laurent *et al.* 1997; Piriou *et al.* 1998), allowed utilities to develop a holistic approach that includes treatment and distribution operations to reach their biostability goals.

Table 10.1 BOM thresholds of biostability adapted from Laurent *et al.* (2005).

Parameter monitored in the DS	Threshold values	BOM measurement techniques	References
Total coliform occurrence	<0.05 mg ac-C eq. C L	AOC	LeChevallier *et al.*1991
	<0.10 mg ac-C eq. CL	AOC	LeChevallier *et al.* 1996
	≤0.15 mg C L	$BDOC_{SAND}$	Volk and LeChevallier 2000
			Volk and Joret 1994
HPCs increase	<0.01 mg ac-C eq. C L	AOC	van der Kooij 1992
No BDOC decrease	≤0.15 mg C L at 20°C	$BDOC_{SAND}$	Volk and Joret 1994
	≤0.30 mg C L at 15°C	$BDOC_{SAND}$	Volk and Joret 1994
	≤0.15 mg C L	$BDOC_{SUSP}$	Servais *et al.* 1995c
	≤0.25 mg C L	$BDOC_{SUSP}$	Niquette *et al.* 2001
No BDOC decrease	≤0.15 mg C L	Modelling approach $BDOC_{SUSP}$	Laurent *et al.* 1997

While utilities and regulators engaged in controlling bacterial regrowth in distribution systems, questions were raised on the actual health impacts of these biofilms. This questioning addressed directly the health relevance of biostability. Early epidemiological studies on the health impact of the consumption of drinking water were conducted with the objective of determining the relative contribution of treatment breakthrough and distribution on the incidence of waterborne highly credible gastro-intestinal (ICGI) illnesses (Payment *et al.* 1991a; Payment *et al.* 1991b; Payment *et al.* 1993; Payment *et al.* 1997). Results of these household intervention studies showed a significant contribution of distribution, more so than treatment breakthrough, while no clear effects of water quality parameters, including BOM, HPCs or total viable bacteria, were observed. These early studies were followed by several epidemiological and observational studies which provided conflicting conclusions regarding the impact of treatment breakthrough and contamination during distribution, especially during low pressure events (Hunter *et al.* 2005; Calderon and Craun 2006). These studies pointed to the potential intrusion of contaminated water during distribution and stirred considerable research in the last 10 years to quantify the occurrence and impacts of ingress of contaminated water during transient negative pressure events and pipe repairs. This recent focus has shifted the focus from biostability to intrusion and raised the question of colonization and amplification of pathogens resulting from ingress of contaminated water.

More recently, two recent developments that may impact biostability in distribution systems have raised concerns for the water distributors:

- The presence and role of viable but non culturable (VBNC) bacteria in distribution systems was documented as early as the nineties (Byrd *et al.* 1991; Singh *et al.* 1990). It is now well understood that the understanding of regrowth and our ability to predict it need to be based on the estimation of total active bacteria in the biofilm and the bulk water. However, recent advances in molecular methods have enhanced our ability to detect low concentrations of viable but non culturable bacterial indicators and pathogens in distribution system biofilm and water. For example, Baudart *et al.* (2002) demonstrated the presence of viable but non culturable *Enterobacteriacea* in drinking water using fluorescent *in situ* hybridization (FISH). Their application has revealed the presence of pathogens in drinking water biofilms challenging the established knowledge of their occurrence and raising concern over their potential health effects. The conditions favourable to their establishment and amplification and proliferation in drinking water biofilms are not well known and need to be determined.
- The European community has promulgated new and lower lead standards at the tap (10 μg/L) that come into effect at the latest in 2013, following the footsteps of the Lead and Cooper Rule in the USA in 1981. A main consequence of these regulations is the widespread use of phosphate based corrosion inhibitors, as shown in the UK. This addition of excess bioavailable orthophosphates raises the issue of their potential impact on biostability.

This chapter proposes a critical review of emerging issues of biostability in drinking water DSs that have been the focus of recent concern and research namely:

(1) the biodiversity of the biofilm, including occurrence and growth of bacterial pathogens;
(2) the importance of materials on biostability, including the contribution of materials to nutrients and the addition of phosphorous based corrosion inhibitors; and
(3) the impact of operations of distribution systems on biostability, namely the management of residence time and deposits.

10.2 BIODIVERSITY OF BIOFILMS

10.2.1 Bacterial abundance

Bacteria are the microorganisms found in greatest abundance in distributed water (Amblard *et al.* 1996; Niquette *et al.* 2001) ranging from 10^5 to 10^9 bacteria cells per litre of water, of which less than 0.1 to 10% can be detected by plate counts (Block 1992; Rompré *et al.* 1995; Prévost *et al.* 1998; Schwartz *et al.* 1998). Bacteria found in the DS can originate from the source water (surface and groundwater), from intrusion of contaminated water, soil and air (deterioration of the DS, cross-connections, pipe breakage, repairs, etc.) (LeChevallier *et al.* 1996; Block *et al.* 1997; Karim *et al.* 2003; Besner *et al.* 2008), or from the proliferation followed by detachment occurring in certain treatment processes such as biofilters (Maul *et al.* 1991; Servais *et al.* 1991; Servais and Menon 1991; Morin and Camper 1997; Beaudet *et al.* 2000; Camper and Jones 2000).

Several studies have quantified the abundance of fixed bacteria in drinking water biofilm, with reported abundances in the range of 10^6 to 10^7 cells per cm^2 in full-scale and pilot DSs (Mathieu *et al.* 1992; Coallier *et al.* 1994; Niquette *et al.* 2001; Servais *et al.* 2004). Some studies have compared the relative importance of the abundance of suspended and biofilm bacteria in DS pipes (Servais *et al.* 1995a; Boe-Hansen *et al.* 2002) and reported a large predominance of fixed bacteria over suspended bacteria. Several studies have shown that the production of bacterial biomass in a DS takes place primarily at the inner surface of the pipes as fixed bacterial biomass, these cells being more protected against chlorine than those in the adjacent bulk water (van der Wende *et al.* 1989; Block *et al.* 1993).

Most of these bacteria entering DSs are autochthonous, that is perfectly well adapted to an oligotrophic environment, and thus able to grow in the water phase and on the surface of distribution pipes where they can be detached by the flow of circulating water. The generic term "biofilm" is used to describe the presence of microorganisms on the surface of materials in contact with water. The literature dealing with the mechanisms by which biofilms are formed is extensive (Marshall 1984; Characklis and Wilderer 1989; Characklis and Marshall 1990; Olson *et al.* 1991; Melo *et al.* 1992; Geldreich 1996; Keevil *et al.* 1999; Bryers 2000; Evans 2000). However, the term "biofilm" does not necessarily imply the existence of a continuous biological structure of regular thickness, either in time or in space. A drinking water biofilm can be defined as an accumulation on materials of aggregates of microorganisms (dead or alive) that are fairly irregularly dispersed and of organic polymers (Pedersen 1990). This structure becomes more complex in a DS in operation for several months or years due to the presence of organic and inorganic deposits (flocculated organic matter, carbonate precipitates, metallic hydroxides caused by corrosion, ions retained by exo-cellular polymers, etc.) (Allen *et al.* 1980; Sly *et al.* 1988; Delanoue *et al.* 1999; Straub *et al.* 2002). Corrosion reactions that can be amplified by bacteria lead to the formation of incrustations and tubercles (Allen *et al.* 1980; Tuovinen *et al.* 1980; Ridgway and Olson 1981; LeChevallier *et al.* 1987; Flemming and Geesey 1990), considerably modifying the surface characteristics of the distribution pipes and creating new accumulation zones on the surface and within the corrosion tubercles, that are available to bacteria. In fact, although BOM levels have a significant impact, the presence of corrosion by-products is the most important factor determining the density of heterotrophic biofilms in some situations (Camper *et al.* 1996; Rompré *et al.* 1998).

Culture-based methods are inadequate to study the phylogeny of bacteria in DS as only a very small fraction of bacteria present in DS in suspension or in the biofilm are able to grow in typical culture media. With the development of molecular biology, culture-independent techniques are now available to study biodiversity. FISH using oligonucleotide probes designed specifically for 16S rRNA of targeted bacteria can be used to determine the relative abundance of different genera of bacteria present in biofilm and bulk samples (Schwartz *et al.* 2003; Mathieu *et al.* 2009). Polymerase chain reaction (PCR) of the DNA coding for 16S rRNA coupled with denaturating gel gradient electrophoresis (DGGE) provides a fingerprint of the bacterial community (Eichler *et al.* 2006) as DNA from the bands of DGGE gel can be sequenced to identify the most probable bacterial strains.

The use of molecular techniques has shown that the bacterial community is often dominated by Proteobacteria (α, β and γ Proteobacteria) in the biofilm (Emtiazi *et al.* 2004; Bai *et al.* 2010) and in the water phase (Kormas *et al.* 2010) of full-scale DSs. It should be noted that the age of the biofilm influences the composition of the bacterial community. 16S rRNA gene sequence analysis revealed that biofilm species richness was comparable to the species richness in the bulk water during the initial stages of biofilm formation (Martiny *et al.* 2003). However, these authors also observed that a longer period (>500 days) was needed before a stable biofilm community composition was reached, questioning the validity of findings reported on young biofilms. These findings suggest that the results obtained with sort term biofilm studies conducted at bench and pilot scales may not be representative of those in a full-scale DS.

Numerous studies in DSs have investigated the impact of disinfection and materials on the bacterial community composition. Several studies using molecular methods now converge to show that the microbial community diversity will be impacted by the disinfection strategy. In a pilot study, the application of free chlorine and chlorine dioxide resulted in a considerable shift of bacterial community composition of the biofilm (Roeder *et al.* 2010). A clear impact of discontinuous chlorination on the relative abundance of three subclasses of Proteobacteria (α, β and γ) in a pilot loop DS was shown using the FISH technique (Mathieu *et al.* 2009). The β and γ-Proteobacteria were favoured by the presence of a free chlorine residual, but the observed shifts in the Proteobacteria populations were reversible. Yu *et al.* (2010) compared the DGGE profiles of 16S rRNA genes of bacteria from biofilm developed on different materials (copper, stainless steel and plastics) in laboratory reactors. They showed that the pipe materials affected not only the biofilm formation potential but also the microbial diversity of the formed biofilm.

More recently, a high throughput sequencing method (454 pyrosequencing) has been successfully applied to samples from DS biofilm (P.-Y. Hong *et al.* 2010).

This approach maximizes the amount of 16S rRNA gene sequences (several thousand sequences) providing thus a more complete picture on the community composition of the bacteria contained in a given sample. The use of this new molecular technique will provide new insights into microbial colonization of DSs.

10.2.2 Bacterial pathogens and faecal indicators in drinking water biofilms

The presence of a heterotrophic biofilm leading to elevated concentrations of HPCs in a full-scale drinking water DS has not been considered of health significance by the drinking water industry. Prospective epidemiological studies were conducted to investigate the relationship between heterotrophic plate counts exposure from drinking water and human health effects in the USA (Calderon 1991) and Canada (Payment *et al.* 1991a; Payment *et al.* 1991c; Payment *et al.* 1993; Payment *et al.* 1997). All these studies concluded that, in the absence of faecal contamination, the presence of high levels of HPC in tap water does not contribute to an increase in gastrointestinal illnesses in the general population (non immune-compromised population) Indeed, a World Health Organization (WHO) consensus was reached on the absence of an association between any opportunistic pathogens commonly found among the heterotrophic bacteria in drinking water (*Pseudomonas aeruginosa*, *Acinetobacter sp.*, *Aeromonas sp.*) and gastrointestinal infections in the general population (Bartram *et al.* 2003).

The fundamental goal of drinking water suppliers is to minimize or eliminate pathogens through water treatment and to properly manage the distribution system. Pathogens may enter the DS following unintentional ingress, plant breakthrough or intentional contamination, such as resulting from a bioterrorist attack. Their ability to survive and amplify in full-scale distribution systems and premise plumbing has recently received growing attention. Indeed, the presence of bacterial pathogens in water and biofilm samples from distribution systems has been confirmed and species isolated include: *Legionella sp.*, *Aeromonas sp.*, *Mycobacteria sp.*, *P. aeruginosa*, *Burkholderia cepacia*, and so on (van der Kooij 1977; Kuhn *et al.* 1997; Falkinham III *et al.* 2001; Park *et al.* 2001; Le Dantec *et al.* 2002; Emtiazi *et al.* 2004; Sen and Rodgers 2004). Even if HPCs are not considered to be a health concern to the general population, heterotrophic opportunistic pathogens, including some that are not detected by HPC measurements, such as *Legionella* and non tuberculous mycobacteria, may regrow in water and biofilms. Moreover, their incorporation into biofilms can increase their resistance to disinfection. In fact, the resistance of *Legionella spp.* to free chlorine is well documented The incorporation of *Acinetobacter calcoaceticus*, *Burkholderia cepacia*, *Methylobacterium spp.*, *Mycobacterium mucogenicum*, and *Sphingomonas capsulata* isolated from the distribution system into multi-species biofilm increases their resistance to sodium hypochlorite (Simões *et al.* 2010).

The most evident case of a health significant amplification in engineered water systems is *Legionella pneumophila*. Legionellosis outbreaks have been widely attributed to inhalation of aerosols from contaminated hot water systems, cooking towers or components of water DSs (Lin *et al.* 1998). *Legionella pneumophila* is now recognized as the single most common cause of drinking water associated outbreak in the US (Brunkard *et al.* 2011). *Legionella pneumophila* are heterotrophic bacteria able to out-compete other organisms because of their ability to grow at elevated temperatures. Evidence now clearly indicates that amoeba and other protozoa are natural hosts for *Legionella* (Barbaree *et al.* 1986; Fields 1993; Berk *et al.* 1998; Lin *et al.* 1998; Atlas 1999; Valster *et al.* 2009). Many *Legionella* species in bulk drinking water and biofilms can at present not be cultured from drinking water as strongly suggested by PCR detection (Wullings and van der Kooij 2006; Wullings *et al.* 2011). In addition, VBNC forms of *L. pneumophila* might also be present in water as suggested by the presence of bright signals after *in situ* hybridization of *L. pneumophila* in drinking water biofilms using a peptide nucleic acid probe (Giao *et al.* 2009a; Giao *et al.* 2009b). VBNC of *L. pneumophila* can be resuscitated after contact with amoebic trophozoites (García *et al.* 2007). However, *L. pneumophila*, the main causative agent of Legionnaires' disease, represents less than 1% of the *Legionella* isolated from drinking water systems operated at low water temperatures (<12°C) (Wullings and van der Kooij 2006; Wullings *et al.* 2011).

At low levels of nutrients typical of tap water and in the presence of higher nutrient levels from the leaching of plasticised polyvinyl chloride (pPVC), *Legionella pneumophila* is not able to replicate unless incorporated in a biofilm and in the presence of parasitizing amoebal species (Steinert *et al.* 1997; Declerck *et al.* 2009; Giao *et al.* 2009b). This suggests that depressing biofilm and protozoan densities by limiting nutrient levels would be an indirect option to control *Legionella pneumophila*. However, field observations have revealed significant levels of *Legionella* estimated by PCR in 97 drinking water supplies with a wide range of nutrient concentrations (<0.5–7.6 mg/L DOC) (Wullings and van der Kooij 2006; Wullings *et al.* 2011). Moreover, total cells of chemostat grown *L. pneumophila* detected by peptide nucleic acid (PNA) probes in a drinking water biofilm were not significantly influenced by carbon concentrations (low tap water and 8.8 mg-C/L) while constituting 10–21% of the total cells present (Giao *et al.* 2009b). These findings appear to question the validity of a carbon based nutrient threshold for the control of *Legionella*. As for the efficacy of disinfection, total native *L. pneumophila* in a young biofilm estimated from PNA probes were highly resistant to chlorine disinfection (Giao *et al.* 2009a). On the contrary, while *Legionella* remained ubiquitous, *L. pneumophila* were no longer detectable in biofilm devices installed in full-scale distribution systems after long-term (6 months) disinfection using UV or ClO_2(Obst and Schwartz 2007). Once released from the biofilm, the resistance to chlorine of *Legionella* in the bulk water phase was similar to that of other Gram-negative bacteria (Saby *et al.* 2003).

Mycobacterium avium complex (MAC) bacteria are a major cause of disseminated opportunistic infections in immune-compromised patients and were the second most common cause of death in HIV-positive patients. Non-tuberculous mycobacteria are prevalent in hospital water systems (Shin *et al.* 2007) and a known cause of nosocomial infections in developed countries (von Reyn *et al.* 1994). MAC bacteria have repeatedly been isolated from full-scale DSs (du Moulin *et al.* 1988; Glover *et al.* 1994), but also other potentially pathogenic mycobacteria have been isolated from full-scale DSs (Le Dantec *et al.* 2002). In a field study of eight full-scale DSs, culturable *M. avium* were isolated from biofilm on water meters at a low frequency (5%) and in low numbers (<0.3 CFU/cm^2); the observed levels of slowly growing *Mycobacterium* were significantly correlated with AOC and BDOC levels but not those of *M. avium* (Falkinham III *et al.* 2001). At pilot scale, the elimination of competing heterotrophic bacteria resulted in a quasi-dominance of *M. avium* that can grow even at relatively low nutrient levels (AOC > 53 µg/L and BDOC > 0.17 mg/L) (Norton *et al.* 2004). These authors conclude that the control of *M. avium* by limiting BDOC or AOC would require substantial reduction of nutrient levels in treated water. When using qPCR, high regional occurrence of *M. avium* subspecies *paratuberculosis* (MAP) was recently reported in drinking water with 88% of samples at first draw and 76% of biofilm from taps swabs positive (Beumer *et al.* 2010). Hussein *et al.* (2009) have observed that culturable non-tuberculous mycobacteria were highly prevalent in cold (43%) and warm (73%) hospital taps. Conflicting results have been presented on the efficacy of chlorination of *M. avium*. Norton *et al.*(2004) observed a significant reduction of fixed *M. avium* following the application of a high free chlorine dose (5 mg/L). However, the presence of a waxy cell envelope enhances their resistance to chlorine (Le Dantec *et al.* 2002).

The presence of enteric and non-enteric coliform bacteria in DSs has long been used to trace faecal contamination and monitor failures in the process of disinfection. Both total and faecal coliforms, and more specifically *E. coli*, are able to persist in drinking water biofilms (Camper *et al.* 1991; Rice *et al.* 1991; Momba *et al.* 1999), using nutrients present in the bulk liquid or produced by heterotrophic biofilm bacteria (Szewzyk *et al.* 1995). In fact, *E. coli* can survive and multiply in the oligotrophic conditions prevailing in many DSs, suggesting that this organism has a high affinity for low substrate concentrations (Botzenhart and Kufferath 1976). The presence of an indigenous heterotrophic biofilm does not prevent the establishment of coliform bacteria (LeChevallier *et al.* 1987; Szewzyk *et al.* 1995; Parent *et al.* 1996). More recently, verocytotoxic *E. coli* O157 has emerged as an important public health concern associated with faecal contamination of drinking water in several countries (Licence *et al.* 2001; Hrudey *et al.* 2003; Hunter, 2003) and information on its ability to survive and develop in biofilms is limited. Porous media colonized with drinking water biofilms are able to retain significantly more *E. coli*

O157:H7 cells, than clean porous media not previously colonized by biofilm (Bauman *et al.* 2009).

The relative impact of factors affecting the development of heterotrophic biofilm in distribution systems, such as the level of BOM, the type of materials, temperature and sheer stress, has been studied extensively and to a great extent established (Laurent *et al.* 2005; Jiang *et al.* 2012). However, the conditions prevailing in premise plumbing and hot water systems such as flow patterns, residence time, materials, temperature and surface to volume ratio, differ considerably from those found in main distribution systems. Overall, they are likely to be more favourable to biofilm formation, raising the question of the need of more stringent criteria to control regrowth in premise plumbing. Relatively short overnight stagnations in short residential service lines have been shown to cause significant increases in total bacterial counts and HPCs that were related to the rapid consumption of biodegradable organic carbon (Prévost *et al.* 1997). Finally, international drinking water standards such as the World Health Organization (WHO) Guidelines for Drinking Water Quality set water quality levels that can be considered safe for ingestion, contact and aerosol inhalation. However, these standards may not be suitable for all individuals, particularly for people suffering from certain specific immune disorders or with weakened immunity (Bartram *et al.* 2003).

Therefore, it appears important to determine the conditions leading to the amplification of opportunistic pathogens in premise plumbing and hot water system in order to define best practice guidance. This is especially needed for large buildings serving susceptible populations such as long term care facilities and hospitals, where waterborne transmission of pathogens is now a recognized concern (Anaissie 2002; Anaissie *et al.* 2002; Reuter *et al.* 2002).

10.3 EFFECTS OF PIPE MATERIAL AND CORROSION CONTROL

The impact of materials on biofilms has been long recognized but recent developments support the importance of this factor and have led to efforts in the standardization of material testing for biostability. Materials commonly used in DSs for transmission/distribution mains and service lines are of three main types: metals (cast iron, ductile iron, steel, stainless steel, lead, copper, and galvanized iron), polymers (polyvinyl chloride (PVC), polyethylene (PE), fibreglass, and polybutylene), and cement (asbestos cement, reinforced and pre-stressed concrete) Additionally, the control of corrosion of lead containing materials will be intensified as more stringent regulations on lead in drinking water are enforced (Official Journal of the European Communities 1998). The addition of orthophosphates is recognized as an efficient and cost effective control option in many situations and should be reviewed in terms of its impact on biostability.

10.3.1 Comparison of biofilm support potential

Different tests are available to determine the growth-promoting properties of materials in contact with drinking water (i) the mean dissolved oxygen difference method (MDOD) (UK) (British Standard Institution 1988); (ii) the W270 test (Germany) (Deutscher Verein des Gas- und Wasserfaches-DVGW-Technologiezentrum Wasser 2007); (iii) the biomass production potential (BPP) (Netherlands) (van der Kooij and Veenendaal 2001); and (iv) the testing of biofilm formation potential of pipes in contact with drinking water (Austria) (Austrian Standards Institute ÖNORM 2002). Because these tests are conducted under diverse operating conditions, with different biomass measurement parameters and threshold compliance values they may lead to conflicting results for a same material. Harmonization efforts will probably result in the selection of the BPP test, the MDOD test and the W270 test for incorporation into the European Acceptance Scheme for construction products in contact with drinking water (van der Kooij *et al.* 2003).

Biofilm densities on various pipe materials can result from the release of substrates from the materials (as indicated by the tests described above) and from the attachment of bacteria and/or organic matter to the surface of the pipe material. The surface properties of the material and environmental conditions determine the potential for attachment but the local release of nutrients enhances it. Biofilm densities generally vary from 10^1 to 10^7 CFU/cm^2 based on heterotrophic plate counts and from 10^1 to 10^4 pg adenosine triphosphate - ATP/cm^2 when the biofilm formation potential or BPP methods are used. Ranking of pipe materials generally show the following trend in terms of decreasing levels of attached bacteria densities (Neden *et al.* 1992; Rogers *et al.* 1994a; Kerr *et al.* 1999; Niquette *et al.* 2000; Norton and LeChevallier 2000; Camper *et al.* 2003):

$$\textit{Unlined iron} > \textit{plastic},\ \textit{cement} > \textit{stainless steel},\ \textit{copper}$$

The highest levels of bacterial attachment have been found on unlined iron pipes. These pipes are subject to corrosion, which decreases the efficiency of disinfection and allows bacterial attachment to persist (LeChevallier *et al.* 1990). Corroded iron pipes (typically containing 0.03–0.2% phosphorus by weight) could also release levels of phosphorus high enough to sustain biofilm growth (Morton *et al.* 2005). Plastic pipes and cement-lined pipes sustain less attached biomass than iron-based materials. However, PE pipes can in some cases sustain more biomass density than PVC due to the release of organic compounds in the water (Brocca *et al.* 2002). For cement and plastic surfaces, biofilm densities are usually slightly lower on cement than on plastic (Hallam *et al.* 2001; Momba and Makala 2004).

Stainless steel and copper are the materials that generally sustain lower densities of attached bacteria. Antimicrobial properties have been associated with these two materials (Domek *et al.* 1984; Percival 1999). The antimicrobial nature of copper may be reflected by less microorganisms diversity (Rogers *et al.* 1994b) and the selection of microorganisms with resistance mechanisms (Critchley *et al.* 2002). Slower biofilm formation was observed in copper than in PE pipes, but after 200 days, there was no difference in microbial numbers between the two surfaces (Lehtola *et al.* 2004b). This indicates that long-term studies may be needed to accurately assess the effect of pipe materials as other authors have also observed a disappearance of the dependency of biofilm formation on pipe material with time (Obst and Schwartz 2007).

The different types of materials used for water distribution not only have an effect on fixed biomass densities, but they also influence the rate of biofilm development (Norton and LeChevallier 2000), the composition and diversity of attached microbial communities (Jang *et al.* 2011), and may also stimulate the growth of opportunistic pathogens, often in a viable but non culturable state (Moritz *et al.* 2010). Using model viruses, Helmi *et al.* (2010) showed that the viral adsorption rate although very low (below 1% under the tested conditions), was higher on biofilm developed on cast iron than on cement or PVC, probably because of asperities available on pipe surfaces. The impact of these lower adsorption rates on the transport of virus particles remains to be determined.

10.3.2 Lining, coating, gasket and lubricant materials

Linings and coatings, generally used to cover metallic surfaces for corrosion protection, can also act as substrata for bacterial growth (Schoenen, 1986). Common materials used for lining are cement-mortar and epoxy resin. Careful application of linings is important as excessive leaching of organics from epoxy resins may occur and result in higher bacterial densities in epoxy-lined tanks (Bae *et al.* 2002). Other materials may also provide a source of degradable organic compounds to be used by microorganisms (Burman and Colbourne, 1977) such as petroleum-based grease coatings (Kirmeyer *et al.* 1999), rubber-coated valves (Kilb *et al.* 2003), and some types of lubricating oils (White and LeChevallier 1993).

10.3.3 Corrosion and the addition of phosphate-based inhibitors

The presence of corrosion has been shown to increase the potential for biofilm through (i) the provision of attachment sites for bacteria on metallic surfaces and corrosion deposits (Rompré *et al.* 2000; Camper *et al.* 2003); (ii) the creation of a substantial number of corrosion particles that can accumulate into the DS and that can be significantly colonized by heterotrophic bacteria (Gauthier *et al.* 1999), that are less susceptible to chlorination than suspended bacteria (Morin *et al.* 1999); (iii) the reduction of chlorine/chloramines by iron oxides, limiting the ability of the oxidant to act on the biofilm (Rompré *et al.* 1998); and (iv) the adsorption of organic compounds by iron oxides, which can increase the availability of complex and/or slowly biodegradable organic compounds to support biofilm growth (Camper 2004). The relative importance of these various mechanisms may vary widely within a DS due to decreasing pipe diameters, increasing water residence times, and presence of dead-ends.

The easiest and most widely used corrective actions to control iron corrosion are to adjust water chemistry or to apply corrosion inhibitors. However, orthophosphates are bacterial nutrients and thus may stimulate the growth of fixed and suspended bacteria if phosphorus is the limiting nutrient (Miettinen *et al.* 1997b; Lehtola *et al.* 2004a). Typical phosphate dosages for corrosion control (1 to 5 mg-P/L) greatly exceed the nutritional requirement of bacteria and could potentially have an effect on bacterial physiology (enhancing storage, enhancing disinfectant resistance, etc.) The addition of phosphate inhibitors resulted in a notable accumulation of phosphorus in the biofilm but did not influence its overall density (HPCs or epifluorescence counts), activity or composition (Batté *et al.* 2003a). However, the addition of phosphate to the treated water to control corrosion caused a shift of bacterial populations enhancing the proportion of γ-Proteobacteria (Batté *et al.* 2003b). The addition of phosphates has been shown to increase the bacterial growth of culturable indicators in some DSs with very low phosphorus content (Miettinen *et al.* 1997a, 1997b; Sathasivan and Ohgaki 1999; Lehtola 2002; Fang *et al.* 2009). Basing their evaluation on HPCs, Park and Hu (2010) recommend that biodegradable organic carbon levels should be controlled in systems where orthophosphate-based anticorrosion treatments are used. A recent study by Fang *et al.* (2009) provides some insights into the complex effect of phosphorus on biofilm formation. These authors show that, in the absence of chlorine, although phosphorus addition led to increased levels of biofilm growth, (i) it was inhibitory to exopolymeric substances (EPS) production which could make the biofilm less resistant to disinfectants, (ii) it induced thicker and less homogeneous biofilms that could result in a less disinfectant-resistant structure, and (iii) it dramatically increased the biofilm cell metabolic potential which may result in overproduction of bacteria in drinking water once the carbon source level increases. A subsequent study by Fang *et al.* (2010) showed that the addition of phosphorus could further increase the disinfection efficacy, especially with monochloramine. Finally recent evidence also points to a role of phosphorus in limiting biofouling of reverse osmosis membranes in very low nutrient treated water (Vrouwenvelder *et al.* 2010).

Conversely, several pilot studies and general industrial experience show that adding phosphates will not significantly impact heterotrophic bacterial biomass in carbon-limited distribution systems (Lyons *et al.* 1995; Batté *et al.* 2000; Rompré *et al.* 2000; Appenzeller *et al.* 2001; Butterfield *et al.* 2002; Batté *et al.* 2003b; Hong *et al.* 2008). Utility experience also shows that using phosphate-based inhibitors for corrosion control contributes to improving the compliance records for coliforms, which may be the result of better maintenance of the chlorine residual along with the limitation of corrosion deposits (LeChevallier *et al.* 1996).

These apparently contradictory results most likely reflect differences in water quality (C:N:P ratio), the bacterial measurement techniques used, the experiment scale factor, the characteristics of the biofilms, and the synergistic effects of disinfectant combined with the impact of the water matrix.

10.4 OPERATION OF THE DISTRIBUTION SYSTEM

DSs have been traditionally designed to meet the following criteria: (i) firefighting flow needs; (ii) seasonal demands; and (iii) anticipated future population growth. Such design criteria usually result in DSs that are too large to meet daily demands, with low water velocities in pipes common under average conditions. High water age (or retention time) in a system can potentially result in various changes in water quality.

10.4.1 Water velocity and pipe biofilms

Water velocity in distribution mains will affect biofilm growth by influencing nutrient and disinfectant transport and shear stress at the pipe wall. It could be assumed that high water velocities would be beneficial in limiting biofilm accumulation because of highly turbulent conditions resulting in higher shear stresses promoting detachment of bacteria, and a better diffusion of disinfectant into the biofilm. However, the situation is more complex as increased velocities can also lead to a better diffusion of nutrients into the biofilm (Lehtola *et al.* 2006), and higher velocities have been associated with higher biofilm densities (Manuel *et al.* 2007).

The fraction of biofilm detachment from erosion (single cells or small aggregate) is usually larger than from sloughing (large cell aggregates) (Telgmann *et al.* 2004). However, abrupt changes in flow conditions may result in detachment of cells from biofilms and lead to significant increases in effluent-suspended bacteria concentration (Lehtola *et al.* 2006). A direct quantitative correlation between increase in shear stress in a test loop facility and DAPI (4′,6′-diamidino-2-phénylindole) cell counts was obtained by Sharpe *et al.* (2010).

10.4.2 Low water movement and/or stagnation in the distribution system

In a DS, water storage facilities are locations that are susceptible to low water velocity or water stagnation. Design considerations, mixing characteristics inside the tank, its operation, and its location in the system may lead to long water retention times. Tank operation may result in significant loss of disinfectant residuals, coliform occurrences, and nitrification episodes in DSs as well as increased biomass and abundance of microorganisms inside reservoirs (Amblard *et al.* 1996; Gauthier *et al.* 2000). However, several actions are possible to control water residence time in storage facilities such as adjusting pump schedules, reducing the operational top water level, and eliminating dead zones. Installation of booster disinfection can also be used to maintain water quality in the DS.

Long water residence times also characterize dead-end mains. Although changes in water quality have been observed in dead-ends, microbial levels are not always systematically higher at these locations (Carter *et al.* 1997; Lu *et al.* 1997). Settling of bacteria during stagnant or low-flow conditions in the water main and nutrient limitation are possible hypotheses for such observation. Some practices may be set up to help control water residence time in DSs, including altering the valving within the network, installing time-varying valves, performing manual/automated flushing, downsizing mains, and looping dead-end mains (Brandt *et al.* 2004).

10.4.3 Sediment accumulation and pipe flushing

The low-flow conditions found in some DS areas are also favourable to the sedimentation of suspended particles to form loose deposits. The particulate matter found in water pipes may originate from various sources, including incomplete removal of particles from raw water, release of fines or particles from filters, precipitation of metal oxides or calcium carbonate, external contamination introduced during repairs in pipes and reservoirs, post-flocculation, biological growth and corrosion.

Sediments can be a cause for concern as they can:

- serve as a nutrient source for microorganisms and support microbial growth. Although deposits are usually mostly mineral in nature, they may include variable fractions of organic matter (Gauthier *et al.* 1999; Carrière *et al.* 2005), and they may be heavily colonized by micro- and macroorganisms (Torvinen *et al.* 2004; Christensen *et al.* 2011).
- shield microorganisms from disinfectants (Morin *et al.* 1999) and exert a high disinfectant demand.
- create aesthetic problems if rapid changes in velocity and flow direction take place (Vreeburg and Boxall 2007).

Flushing programs may be undertaken to remove sediments from water pipes and different flushing techniques are available including unidirectional flushing (UDF) (water velocity is maximized through mains closure), conventional flushing (water velocity is not maximized as water flows from several mains), and continuous blowoff (bleeding of water, typically in dead-ends). Conventional flushing, on an intermittent or "spot" basis, is widely used by water utilities to resolve either acute or chronic water quality problems in parts of the DS.

However, only short-term benefits to water quality improvements are usually observed (Barbeau *et al.* 2005). On the other hand, because of the use of higher water velocities and cleaner water, water quality benefits such as improved chlorine residual, decreased HPC concentrations and coliform occurrences, and decreased turbidity and customer complaints have been associated with UDF.

It is difficult to provide any specific guidelines regarding the frequency at which unidirectional flushing should be conducted. Sediment accumulation and composition vary from one DS to another and will be influenced by factors such as the accumulation period prior to flushing, the pipe material and the treated water characteristics. The supplied water quality has recently been identified as a key factor governing regeneration (Husband and Boxall 2011; Vreeburg and Tankerville 2011). Guidelines for flushing velocities of at least 1.5 m/s have been provided to promote scouring, remove biofilm/loose deposits, and reduce disinfectant demand. In the Netherlands, this velocity criterion is used along with a 2–3 pipe turnovers and clear water front (Vreeburg and Boxall 2007). In contrast, Cloete *et al.* (2003) suggested that a pipe velocity of 3–4 m/s would be ideal to achieve reduced biofilm growth.

The concept of self-cleaning networks was developed in the Netherlands with design guidelines based on higher velocities (a once-a-day flow velocity of 0.4 m/s) and unidirectional flow in pipes to prevent settling and accumulation of particles (van den Boomen *et al.* 2004). The main advantages of this concept include no stagnant water, short residence times, improved water quality, and reduction of material costs. Field measurements have shown that this concept is feasible in real DSs and recent investigations indicate that a self-cleaning velocity of 0.2–0.25 m/s would be adequate for pipes to remain clean (Blokker *et al.* 2010).

10.5 CONCLUSION

Recent developments in analytical tools have allowed progress to be made in our understanding of the biodiversity of the biofilms and their ability to harbour and amplify bacterial pathogens. Although the relative impact of factors affecting the development of heterotrophic biofilms in distribution systems has been studied extensively and is to a great extent established, the conditions prevailing in premise plumbing and hot water systems differ considerably from those found in main distribution systems. Overall, they are likely to be more favourable to biofilm formation and they warrant a reassessment of our understanding of the factors affecting biostability. Therefore, it appears important to better determine the conditions leading to the amplification of opportunistic pathogens in premise plumbing and hot water system in order to define best practice guidance. Results of this work will demonstrate the importance of maintaining high levels of biostability to avoid the amplification of pathogens in this last frontier of the distribution system.

10.6 REFERENCES

Allen M. J., Taylor R. H. and Geldreich E. E. (1980). The occurrence of microorganisms in water main encrustations. *J. Am. Water Works Assoc.*, **72**(11), 614–625.

Amblard C., Bourdier G., Carrias J.-F., Maurin N. and Quiblier C. (1996). Seasonal evolution of microbial community structure in a drinking water reservoir. *Water Res.*, **30**(3), 613–624.

Anaissie E. J. (2002). Dangers on tap in hospitals. *Pulm. Re.*, **7**(9), 1–5.

Anaissie E. J., Penzak S. R. and Dignani M. C. (2002). The hospital water supply as a source of nosocomial infections: a plea for action. *Arch. Intern. Med.*, **162**(13), 1483–1492.

Appenzeller B. M. R., Batté M., Mathieu L., Block J.-C., Lahoussine V., Cavard J. and Gatel D. (2001). Effect of adding phosphate to drinking water on bacterial growth in slightly and highly corroded pipes. *Water Res.*, **35**(4), 1100–1105.

Atlas R. M. (1999). *Legionella*: from environmental habitats to disease pathology, detection and control. *Environ. Microbiol.*, **1**(4), 283–293.

Bae B., Jeong J. H. and Lee S. J. (2002). The quantification and characterization of endocrine disruptor bisphenol-A leaching from epoxy resin. *Water Sci. Tech.*, **46**(11–12), 381–387.

Bai X., Wu F., Zhou B. and Zhi X. (2010). Biofilm bacterial communities and abundance in a full-scale drinking water distribution system in Shanghai. *J. Water Health*, **8**(3), 593–600.

Barbaree J. M., Fields B. S., Feeley J. C., Gorman G. W. and Martin W. T. (1986). Isolation of protozoa from water associated with a Legionellosis outbreak and demonstration of intracellular multiplication of *Legionella pneumophila*. *Appl. Environ. Microbiol.*, **51**(2), 422–424.

Barbeau B., Gauthier V., Julienne K. and Carrière A. (2005). Dead-end flushing of a distribution system: short and long-term effects on water quality. *J. Water Supply Res. T.*, **54**(6), 371–383.

Bartram J., Cotruvo J., Exner M., Fricker C. and Glasmacher A. (2003). Heterotrophic plate counts and drinking-water safety. World Health Organization (WHO) and International Water Association (IWA) Publishing, London, United Kingdom.

Batté M., Koudjonou B., Laurent P., Coallier J. and Prévost M. (2000). Does phosphate-based anticorrosion treatment enhance biofilm resistance to disinfection? Paper presented at the American Water Works Association-Water Quality Technology Conference Salt Lake City, Utah, USA.

Batté M., Koudjonou B., Laurent P., Mathieu L., Coallier J. and Prévost M. (2003a). Biofilm responses to ageing and to a high phosphate load in a bench-scale drinking water system. *Water Res.*, **37**(6), 1351–1361.

Batté M., Mathieu L., Laurent P. and Prévost M. (2003b). Influence of phosphate and disinfection on the composition of biofilms produced from drinking water, as measured by fluorescence in situ hybridization. *Can. J. Microbiol.*, **49**(2), 741–753.

Baudart J., Coallier J., Laurent P. and Prévost M. (2002). Rapid and sensitive enumeration of viable diluted cells of members of the family *Enterobacteriaceae* in freshwater and drinking water. *Appl. Environ. Microbiol.*, **68**(10), 5057–5063.

Bauman W. J., Nocker A., Jones W. L. and Camper A. K. (2009). Retention of a model pathogen in a porous media biofilm. *Biofouling*, **25**(3), 229–240.

Beaudet J.-F., Prévost M., Niquette P., Clément B., Coallier J. and Arcouette N. (2000). Simple methods for evaluating the population of naidids in drinking water treatment plants. *Can. J. Civil Eng.*, **27**(4), 601–609.

Berk S. G., Ting R. S., Turner G. W. and Ashburn R. J. (1998). Production of respirable vesicles containing live *Legionella pneumophila* cells by two *Acanthamoeba* spp. *Appl. Environ. Microbiol.*, **64**(1), 279–286.

Besner M. C., Lavoie J., Morissette C., Payment P. and Prévost M. (2008). Effect of water main repairs on water quality. *J. Am. Water Works Assoc.*, **100**(7), 95–109.

Beumer A., King D., Donohue M., Mistry J., Covert T. and Pfaller S. (2010). Detection of *Mycobacterium avium* subsp. *paratuberculosis* in drinking water and biofilms by quantitative PCR. *Appl. Environ. Microbiol.*, **76**(21), 7367–7370.

Block J. C. (1992). Biofilms in drinking water distribution system. Paper presented at the Biofilms-Science and Technology, Nato Advanced-Study Institute, Portugal.

Block J. C., Haudidier K., Paquin J. L., Miazga J. and Lévi Y. (1993). Biofilm accumulation in drinking water distribution systems. *Biofouling*, **6**(4), 333–343.

Block J. C., Sibille I., Gatel D., Reasoner D. J., Lykins B. and Clark R. M. (1997). Biodiversity in drinking water distribution system: a brief review. In: The Microbiological Quality of Water, D. W. Sutcliffe (ed.), Freshwater Biological Association London, United Kingdom, pp. 63–70.

Blokker E., Vreeburg J., Schaap P. and van Dijk J. (2010). The self-cleaning velocity in practice. Paper presented at the Water Distribution System Analysis Conference, Tucson, Arizona, USA.

Boe-Hansen R., Albrechtsen H.-J., Arvin E. and Jorgensen C. (2002). Bulk water phase and biofilm growth in drinking water at low nutrient conditions. *Water Res.*, **36**(18), 4477–4486.

Botzenhart K. and Kufferath R. (1976). On the growth of various *Enterobacteriaceae*, *Pseudomonas aeruginosa* and *Alkaligenes* spec. in distilled water, de-ionized water, tap water, and mineral salt solution. *Zbl. Bakt. Hyg.*, **163**(5–6), 470–485.

Brandt M., Clement J., Powell J., Casey R., Holt D., Harris N. and Ta C. T. (2004). Managing Distribution Retention Time to Improve Water Quality – Phase 1, Report, American Water Works Association Research Foundation Denver, Colorado, USA.

British Standard Institution (1988). Suitability of Non-Metallic Products for Use in Contact with Water Intended for Human Consumption with Regard to their Effect on the Quality of Water, Report BS 6920–1:1988.

Brocca D., Arvin E. and Mosbaek H. (2002). Identification of organic compounds migrating from polyethylene pipelines into drinking water. *Water Res.*, **36**(15), 3675–3680.

Brunkard J. M., Ailes E., Roberts V. A., Hill V., Hilborn E. D., Craun G. F., Rajasingham A., Kahler A., Garrison L., Hicks L., Carpenter J., Wade T. J., Beach M. J. and Yoder J. S. (2011). Surveillance for waterborne disease outbreaks associated with drinking water - United States, 2007–2008. *Morbidity and Mortality Weekly Report (MMWR)*, **60**(ss12), 38–68.

Bryers J. D. (2000). Biofilms II: process analysis and applications. Wiley-Liss, New York.

Burman N. P. and Colbourne J. S. (1977). Techniques for the assessment of growth of micro-organisms on plumbing materials used in contact with potable water supplies. *J. Appl. Bacteriol.*, **43**(1), 137–144.

Butterfield P. W., Camper A. K., Biederman J. A. and Bargmeyer A. M. (2002). Minimizing biofilm in the presence of iron oxides and humic substances. *Water Res.*, **36**(15), 3898–3910.

Byrd J. J., Xu H. S. and Colwell R. R. (1991). Viable but nonculturable bacteria in drinking water. *Appl. Environ. Microbiol.*, **57**, 875–878.

Calderon R. L. (1991). Bacteria Colonizing Point-of-Entry Granular Activated Carbon Filters and their Relationship to Human Health, Report CR-813978-01-0, United States Environmental Protection Agency.

Calderon R. L. and Craun G. F. (2006). Estimates of endemic waterborne risks from community-intervention studies. *J. Water Health.*, **4**(S2), 89–99.

Camper A. K. (2004). Involvement of humic substances in regrowth. *Int. J. Food Microbiol.*, **92**(3), 355–364.

Camper A. K. and Jones W. L. (2000). Factors Affecting Microbial Growth in Model Distribution Systems, Report 90785, American Water Works Association Research Foundation and American Water Works Association, Denver, Colorado, USA.

Camper A. K., McFeters G. A., Characklis W. G. and Jones W. L. (1991). Growth kinetics of coliform bacteria under conditions relevant to drinking water distribution systems. *Appl. Environ. Microbiol.*, **57**(8), 2233–2339.

Camper A. K., Jones W. L. and Hayes J. T. (1996). Effect of growth conditions and substratum composition on the persistence of coliforms in mixed-population biofilms. *Appl. Environ. Microbiol.*, **62**(11), 4014–4018.

Camper A. K., Brastrup K., Sandvig A., Clement J., Spencer C. and Capuzzi A. J. (2003). Effect of distribution system materials on bacterial regrowth. *J. Am. Water Works Assoc.*, **95**(7), 107–121.

Carrière A., Gauthier V., Desjardins R. and Barbeau B. (2005). Evaluation of loose deposits in distribution systems through unidirectional flushing. *J. Am. Water Works Assoc.*, **97**(9), 82–92.

Carter J. T., Lee Y., Buchberger S. G., Rossman L. A. and Rice E. W. (1997). Correlations between travel time and water quality in a deadend loop. Paper presented at the American Water Works Association-Water Quality Technology Conference Denver, Colorado, USA.

Characklis W. G. and Marshall K. C. (1990). Biofilms: a basis for an interdisciplinary approach (Chapter 1) In: Biofilms, John Wiley and Sons, New York, USA, pp. 3–15.

Characklis W. G. and Wilderer P. A. (1989). Structure and function of biofilms. John Wiley and Sons, New York, USA.

Christensen S. C. B., Nissen E., Arvin E. and Albrechtsen H.-J. (2011). Distribution of *Asellus aquaticus* and microinvertebrates in a non-chlorinated drinking water supply system - Effects of pipe material and sedimentation. *Water Res.*, **45**(10), 3215–3224.

Cloete T. E., Westaard D. and van Vuuren S. J. (2003). Dynamic response of biofilm to pipe surface and fluid velocity. *Water Sci. Technol.*, **47**(5), 57–59.

Coallier J. M., Prévost M., Rompré A. and Duchesne D. (1994). The optimization and application of two direct viable count methods for bacteria in distributed drinking water. *Can. J. Microbiol.*, **40**(10), 830–836.

Critchley M. M., Cromar N. J., McClure N. and Fallowfield H. J. (2002). Biofilms in copper plumbing systems: sensitivity to copper and chlorine and implications for corrosion. *Water Sci. Technol. Water Supply*, **2**(4), 81–87.

Declerck P., Behets J., Margineanu A., van Hoef V., De Keersmaecker B. and Ollevier F. (2009). Replication of *Legionella pneumophila* in biofilms of water distribution pipes. *Microbiol. Res.*, **164**(6), 593–603.

Delanoue A., Holt D. M., Anderson H. A., McMath S. M., Smith S. E., Woodward C. A., Fraser A. R. and Roe M. (1999). Analysis of material (biofilm) present on the internal surfaces of a drinking water distribution system. In: Biofilms in the aquatic environment, C. W. Keevil, A. Godfree, D. Holt and C. Dow (eds), The Royal Society of Chemistry, pp. 202–208.

Domek M. J., LeChevallier M. W., Cameron S. C. and McFeters G. A. (1984). Evidence for the role of copper in the injury process of coliform bacteria in drinking water. *Appl. Environ. Microbiol.*, **48**(2), 289–293.

Dukan S., Levi Y., Piriou P., Guyon F. and Villon P. (1996). Dynamic modelling of bacterial growth in drinking water networks. *Water Res.*, **30**(9), 1991–2002.

du Moulin G. C., Stottmeier K. D., Pelletier P. A., Tsang A. Y. and Hedley-Whyte J. (1988). Concentration of *Mycobacterium avium* by hospital hot water systems. *J. Am. Med. Assoc.*, **260**(11), 1599–1601.

DVGW-Technologiezentrum Wasser (2007). Microbial enhancement on materials to come into contact with drinking water - Testing and assessment, W 270 Technical Standard: Deutscher Verein des Gas- und Wasserfaches e.V.

Eichler S., Christen R., Holtje C., Westphal P., Botel J., Brettar I., Mehling A. and Hofle M. G. (2006). Composition and dynamics of bacterial communities of a drinking water supply system as assessed by RNA- and DNA-based 16S rRNA gene fingerprinting. *Appl. Environ. Microbiol.*, **72**(3), 1858–1872.

Emtiazi F., Schwartz T., Marten S. M., Krolla-Sidenstein P. and Obst U. (2004). Investigation of natural biofilms formed during the production of drinking water from surface water embankment filtration. *Water Res.*, **38**(5), 1197–1206.

Evans L. V. (2000). Biofilms: recent advances in their study and control. Harwood Academic Publishers, Amsterdam, The Netherlands.

Falkinham J. O., III, Norton C. D. and LeChevallier M. W. (2001). Factors influencing numbers of *Mycobacterium avium*, *Mycobacterium intracellulare*, and other Mycobacteria in drinking water distribution systems. *Appl. Environ. Microbiol.*, **67**(3), 1225–1231.

Fang W., Hu J. Y. and Ong S. L. (2009). Influence of phosphorus on biofilm formation in model drinking water distribution systems. *J. Appl. Microbiol.*, **106**(4), 1328–1335.

Fang W., Hu J. Y. and Ong S. L. (2010). Effects of phosphorus on biofilm disinfections in model drinking water distribution systems. *J. Water Health*, **8**(3), 446–454.

Fields B. S. (1993). *Legionella* and protozoa: interaction of a pathogen and its natural host (Chapter 5). In: Legionella: current status and emerging perspectives, J. M. Barbaree, R. F. Breiman and A. P. Dufour (eds), American Society for Microbiology, Washington, DC, USA.

Flemming H. C. and Geesey G. G. (1990). Biofouling and biocorrosion in industrial water systems. Springer-Verlag Publishers, Berlin, RFA.

García M. T., Jones S., Pelaz C., Millar R. D. and Abu Kwaik Y. (2007). *Acanthamoeba polyphaga* resuscitates viable non-culturable *Legionella pneumophila* after disinfection. *Environ. Microbiol.*, **9**(5), 1267–1277.

Gatel D., Fahmy T., Parent E., Goldman J. C., Joret J. C. and Cavard J. (1995). Forecasting the number of coliform positive samples. Paper presented at the American Water Works Association-Water Quality Technology.

Gauthier V., Gérard B., Portal J.-M., Block J.-C. and Gatel D. (1999). Organic matter as loose deposits in a drinking water distribution system. *Water Res.*, **33**(4), 1014–1026.

Gauthier V., Besner M. C., Barbeau B., Millette R. and Prévost M. (2000). Storage tank management to improve drinking water quality: case study. *J. Water Res. Plan. Manage.*, **126**(4), 221–228.

Geldreich E. E. (1996). Microbial quality of water supply in distribution systems. CRC Lewis Publishers, Boca Raton, Florida, USA.

Giao M. S., Wilks S., Azevedo N. F., Vieira M. J. and Keevil C. W. (2009a). Incorporation of natural uncultivable *Legionella pneumophila* into potable water biofilms provides a protective niche against chlorination stress. *Biofouling*, **25**(4), 345–351.

Giao M. S., Wilks S. A., Azevedo N. F., Vieira M. J. and Keevil C. W. (2009b). Comparison between standard culture and peptide nucleic acid 16S rRNA hybridization quantification to study the influence of physico-chemical parameters on *Legionella pneumophila* survival in drinking water biofilms. *Biofouling*, **25**(4), 335–343.

Glover N., Holtzman A., Aronson T., Froman S., Berlin O. G. W., Dominguez P., Kunkel K. A., Overturf G., Stelma G. Jr. Smith C. and Yakrus M. (1994). The isolation and identification of *Mycobacterium avium* complex (MAC) recovered from Los Angeles potable water, a possible source of infection in AIDS patients. *Int. J. Environ. Health Res.*, **4**(2), 63–72.

Hallam N. B., West J. R., Forster C. F. and Simms J. (2001). The potential for biofilm growth in water distribution systems. *Water Res.*, **35**(17), 4063–4071.

Helmi K., Menard-Szczebara F., Lenes D., Jacob P., Jossent J., Barbot C., Delabre K. and Arnal C. (2010). Adenovirus, MS2 and ϕX174 interactions with drinking water biofilms developed on PVC, cement and cast iron. *Water Sci. Technol.*, **61**, 3198–3207.

Hong H. C., Mazumder A., Wong M. H. and Liang Y. (2008). Yield of trihalomethanes and haloacetic acids upon chlorinating algal cells, and its prediction via algal cellular biochemical composition. *Water Res.*, **42**(20), 4941–4948.

Hong P.-Y., Hwang C., Ling F., Andersen G. L., LeChevallier M. W. and Liu W.-T. (2010). Pyrosequencing analysis of bacterial biofilm communities in water meters of a drinking water distribution system. *Appl. Environ. Microbiol.*, **76**(16), 5631–5635.

Hrudey S. E., Payment P., Huck P. M., Gillham R. W. and Hrudey E. J. (2003). A fatal waterborne disease epidemic in Walkerton, Ontario: comparison with other waterborne outbreaks in the developed world. *Water Sci. Technol.*, **47**(3), 7–14.

Hunter P. R. (2003). Climate change and waterborne and vector-borne disease. *J. Appl. Microbiol.*, **94**, 37S–46S.

Hunter P. R., Chalmers R. M., Hughes S. and Syed Q. (2005). Self-reported diarrhea in a control group: a strong association with reporting of low-pressure events in tap water. *Clin. Infect. Dis.*, **40**(4), e32–e34.

Husband P. S. and Boxall J. B. (2011). Asset deterioration and discolouration in water distribution systems. *Water Res.*, **45**(1), 113–124.

Hussein Z., Landt O., Wirths B. and Wellinghausen N. (2009). Detection of non-tuberculous mycobacteria in hospital water by culture and molecular methods. *Int. J. Med. Microbiol.*, **299**(4), 281–290.

Jang H.-J., Choi Y.-J. and Ka J.-O. (2011). Effects of diverse water pipe materials on bacterial communities and water quality in the annular reactor. *J. Microbiol. Biotechnol.*, **21**(2), 115–123.

Jiang J., Pang S. Y., Ma J. and Liu H. (2012). Oxidation of phenolic endocrine disrupting chemicals by potassium permanganate in synthetic and real waters. *Environ. Sci. Technol.*, **46**(3), 1774–1781.

Joret J. C. and Lévi Y. (1986). Méthode rapide d'évaluation du carbone éliminable des eaux par voies biologiques. *La Tribune du CEBEDEAU*, **510**(39), 3–9.

Karim M. R., Abbaszadegan M. and LeChevallier M. (2003). Potential for pathogen intrusion during pressure transients. *J. Am. Water Works Assoc.*, **95**(5), 134–146.

Keevil C. W., Godfree A., Holt D. and Dow C. (1999). Biofilms in the aquatic environment. The Royal Society of Chemistry, Cambridge, United Kingdom.

Kerr C. J., Osborn K. S., Robson G. D. and Handley P. S. (1999). The relationship between pipe material and biofilm formation in a laboratory model system. *J. Appl. Microbiol.*, **85**, 29S–38S.

Kilb B., Lange B., Schaule G., Flemming H.-C. and Wingender J. (2003). Contamination of drinking water by coliforms from biofilms grown on rubber-coated valves. *Int. J. Hyg. Environ. Health*, **206**(6), 563–573.

Kirmeyer G. J., Kirby L., Murphy B. M., Noran P. F., Martel K. D., Lund T. W., Anderson J. L. and Medhurst R. (1999). Maintaining Water Quality in Finished Water Storage Facilities, Report 90763, American Water Works Association Research Foundation, American Water Works Association and United Kingdom Water Industry Research Limited, Denver, Colorado, USA.

Kormas K. A., Neofitou C., Pachiadaki M. and Koufostathi E. (2010). Changes of the bacterial assemblages throughout an urban drinking water distribution system. *Environ. Monit. Assess.*, **165**(1–4), 27–38.

Kuhn I., Allestam G., Huys G., Janssen P., Kersters K., Krovacek K. and Stenstrom T. (1997). Diversity, persistence, and virulence of *Aeromonas* strains isolated from drinking water distribution systems in Sweden. *Appl. Environ. Microbiol.*, **63**(7), 2708–2715.

Laurent P. and Servais P. (1995). Fixed bacterial biomass estimated by potential exoproteolytic activity. *Can. J. Microbiol.*, **41**(8), 749–752.

Laurent P., Servais P., Prévost M., Gatel D. and Clément B. (1997). Testing the SANCHO model on distribution systems. *J. Am. Water Works Assoc.*, **89**(7), 92–103.

Laurent P., Servais P., Gauthier V., Prévost M., Joret J.-C. and Block J. C. (2005). Biodegradable organic matter and bacteria in drinking water distribution systems (Chapter 4). In: Biodegradable organic matter in drinking water treatment and distribution, M. Prévost, P. Laurent, P. Servais and J. C. Joret (eds), 1st edn., American Water Works Association, Denver, Colorado, USA, pp. 147–204.

LeChevallier M. W., Babcock T. M. and Lee R. G. (1987). Examination and characterization of distribution system biofilms. *Appl. Environ. Microbiol.*, **53**(12), 2714–2724.

LeChevallier M. W., Lowry C. D. and Lee R. G. (1990). Disinfecting biofilms in a model distribution system. *J. Am. Water Works Assoc.*, **82**(7), 87–99.

LeChevallier M. W., Schulz W. and Lee R. G. (1991). Bacterial nutrients in drinking water. *Appl. Environ. Microbiol.*, **57**, 857–862.

LeChevallier M. W., Welch N. J. and Smith D. B. (1996). Full-scale studies of factors related to coliform regrowth in drinking water. *Appl. Environ. Microbiol.*, **62**(7), 2201–2211.

Le Dantec C., Duguet J.-P., Montiel A., Dumoutier N., Dubrou S. and Vincent V. (2002). Occurrence of mycobacteria in water treatment lines and in water distribution systems. *Appl. Environ. Microbiol.*, **68**(11), 5318–5325.

Lehtola M. J. (2002). Microbially available phosphorus in drinking water (pp. 77), National Public Health Institute.

Lehtola M. J., Juhna T., Miettinen I. T., Vartiainen T. and Martikainen P. J. (2004a). Formation of biofilms in drinking water distribution networks, a case study in two cities in Finland and Latvia. *J. Ind. Microbiol. Biotechnol.*, **31**(11), 489–494.

Lehtola M. J., Miettinen I. T., Keinanen M. M., Kekki T. K., Laîné O., Hirvonen A., Vartiainen T. and Martikainen P. J. (2004b). Microbiology, chemistry and biofilm development in a pilot drinking water distribution system with copper and plastic pipes. *Water Res.*, **38**(17), 3769–3779.

Lehtola M. J., Laxander M., Miettinen I. T., Hirvonen A., Vartiainen T. and Martikainen P. J. (2006). The effects of changing water flow velocity on the formation of biofilms and water quality in pilot distribution system consisting of copper or polyethylene pipes. *Water Res.*, **40**(11), 2151–2160.

Levy R. V., Hart F. L. and Cheetham R. D. (1986). Occurrence and public health significance of invertebrates in drinking water systems. *J. Am. Water Works Assoc.*, **78**(9), 105–110.

Licence K., Oates K. R., Synge B. A. and Reid T. M. (2001). An outbreak of *E. coli* O157 infection with evidence of spread from animals to man through contamination of a private water supply. *Epidemiol. Infect.*, **126**(1), 135–138.

Lin Y. E., Vidic R. D., Stout J. E. and Yu V. L. (1998). *Legionella* in water distribution system. *J. Am. Water Works Assoc.*, **90**(9), 112–121.

Lu L., Fang M., Wang X., Patni N., Ashendorff A. and Principe M. (1997). In-situ water main biological study in New-York City. Paper presented at the American Water Works Association-Water Quality Technology Conference Denver, Colorado, USA.

Lyons J. J., Pontes J. and Karalekas P. C. (1995). Optimizing corrosion for lead and copper using phosphoric acid and sodium hydroxide. Paper presented at the American Water Works Association-Water Quality Technology Conference New Orleans, Louisiana, USA.

Manuel C. M., Nunes O. C. and Melo L. F. (2007). Dynamics of drinking water biofilm in flow/non-flow conditions. *Water Res.*, **41**(3), 551–562.

Marshall K. C. (1984). Microbial Adhesion and Aggregation. Springer-Verlag, Berlin, RFA.

Martiny A. C., Jorgensen T. M., Albrechtsen H.-J., Arvin E. and Molin S. (2003). Long-term succession of structure and diversity of a biofilm formed in a model drinking water distribution system. *Appl. Environ. Microbiol.*, **69**(11), 6899–6907.

Mathieu L., Paquin J. L., Block J.-C., Randon G., Maillard J. and Reasoner D. (1992). Paramètres gouvernant la prolifération bactérienne dans les réseaux de distribution. *Rev. Sci. de l'Eau*, **5**(Spécial), 91–112.

Mathieu L., Bouteleux C., Fass S., Angel E. and Block J. C. (2009). Reversible shift in the alpha-, beta- and gamma-proteobacteria populations of drinking water biofilms during discontinuous chlorination. *Water Res.*, **43**(14), 3375–3386.

Maul A., Block J.-C. and Vagost D. (1991). Microbiological Analysis in Water Distribution Networks: Sampling Strategies, Methods and Computer Programs. Ellis Horwood Publishers, New York, USA.

Melo L., Bott T. R., Fletcher M. and Capdeville B. (1992). Biofilms - Science and Technology. Springer.

Miettinen I. T., Vartiainen T. and Martikainen P. J. (1997a). Changes in water microbial quality during bank filtration of lake water. *Can. J. Microbiol.*, **43**(12), 1126–1232.

Miettinen I. T., Vartiainen T. and Martikainen P. J. (1997b). Phosphorus and bacterial growth in drinking water. *Appl. Environ. Microbiol.*, **63**(8), 3242–3245.

Momba M. N. B. and Makala N. (2004). Comparing the effect of various pipe materials on biofilm formation in chlorinated and combined chlorine-chloraminated water systems. *Water SA*, **30**(2), 175–182.

Momba M. N. B., Cloete T. E., Venter S. N. and Kfir R. (1999). Examination of the behaviour of *Escherichia coli* in biofilms established in laboratory-scale units receiving chlorinated and chloraminated water. *Water Res.*, **33**(13), 2937–2940.

Morin P. and Camper A. K. (1997). Attachment and fate of carbon fines in simulated drinking water distribution system biofilms. *Water Res.*, **31**(3), 399–410.

Morin P., Gauthier V., Saby S. and Block J.-C. (1999). Bacterial resistance to chlorine through attachment to particles and pipes surfaces in drinking water distribution systems. In: Biofilms in the aquatic environment, C. W. Keevil, A. F. Godfree, D. M. Holt and C. S. Dow (eds), Vol. 242, Royal Society of Chemistry, Cambridge, United Kingdom, pp. 171–190.

Moritz M. M., Flemming H. C. and Wingender J. (2010). Integration of *Pseudomonas aeruginosa* and *Legionella pneumophila* in drinking water biofilms grown on domestic plumbing materials. *Int. J. Hyg. Environ. Health*, **213**(3), 190–197.

Morton S. C., Zhang Y. and Edwards M. A. (2005). Implications of nutrient release from iron metal for microbial regrowth in water distribution systems. *Water Res.*, **39**, 2883–2892.

Neden D. G., Jones R. J., Smith J. R., Kirmeyer G. J. and Foust G. W. (1992). Comparing chlorination and chloramination for controlling bacterial regrowth. *Am. Water Works Assoc.*, **84**(7), 80–88.

Niquette P., Servais P. and Savoir R. (2000). Impacts of pipe materials on densities of fixed bacterial biomass in a drinking water distribution system. *Water Res.*, **34**(6), 1952–1956.

Niquette P., Servais P. and Savoir R. (2001). Bacterial dynamics in the drinking water distribution system of Brussels. *Water Res.*, **35**(3), 675–682.

Norton C. D. and LeChevallier M. W. (2000). A pilot study of bacteriological population changes through potable water treatment and distribution. *Appl. Environ. Microbiol.*, **66**(1), 268–276.

Norton C. D., LeChevallier M. W. and Falkinham J. O. (2004). Survival of *Mycobacterium avium* in a model distribution system. *Water Res.*, **38**(6), 1457–1466.

Obst U. and Schwartz T. (2007). Microbial characteristics of water distribution: compiled investigations in a German drinking water distribution system. *Pract. Period. Hazard. Tox. Radioact. Waste Manag.*, **11**(2), 78–82.

Official Journal of the European Communities. (1998). Council Directive 98/83/EC of 3 November 1998 on the Quality of Water Intended for Human Consumption (the Drinking Water Directive), Report L330/32, The Council of the European Union.

Olson B. H., McCleary R. and Meeker J. (1991). Background and models for bacterial biofilm formation and function in water distribution systems (Chapter 12) In: Modelling the environmental fate of microorganisms, American Society for Microbiology, Washington, DC, pp. 255–295.

ÖNORM. (2002). Testing of biofilm formation potential of pipes in contact with drinking water. (Vol. B 5018), Austrian Standards Institute.

Parent A., Fass S., Dincher M. L., Reasoner D., Gatel D. and Block J.-C. (1996). Control of coliform growth in drinking water distribution systems. *J. Chart. Inst. Water Environ. Manage.*, **10**(6), 442–445.

Park S. K. and Hu J. Y. (2010). Interaction between phosphorus and biodegradable organic carbon on drinking water biofilm subject to chlorination. *J. Appl. Microbiol.*, **108**(6), 2077–2087.

Park S. R., Mackay W. G. and Reid D. C. (2001). Helicobacter *sp.* recovered from drinking water biofilm sampled from a water distribution system. *Water Res.*, **35**(6), 1624–1626.

Payment P., Franco E., Richardson L. and Siemiatycki J. (1991a). Gastrointestinal health effects associated with the consumption of drinking water produced by point-of-use domestic reverse-osmosis filtration units. *Appl. Environ. Microbiol.*, **57** (4), 945–948.

Payment P., Richardson L., Edwardes M., Franco E. and Siemiatycki J. (1991b). A prospective epidemiological study of drinking water related gastrointestinal illnesses. *Water Sci. Technol.*, **24**(2), 27–28.

Payment P., Richardson L., Siemiatycki J., Dewar R., Edwardes M. and Franco E. (1991c). A randomized trial to evaluate the risk of gastrointestinal disease due to consumption of drinking water meeting current microbiological standards. *Am. J. Public Health*, **81** (6), 703–708.

Payment P., Franco E. and Siemiatycki J. (1993). Absence of relationship between health effects due to tap water consumption and drinking water quality parameters. *Water Sci. Technol.*, **27**(3–4), 137–143.

Payment P., Siemiatycki J., Richardson L., Renaud G., Franco E. and Prévost M. (1997). A prospective epidemiological study of gastrointestinal health effects due to the consumption of drinking water. *Int. J. Environ. Health Res.*, **7**(1), 5–31.

Pedersen K. (1990). Biofilm development of stainless steel and PVC surfaces in drinking water. *Water Res.*, **24**(2), 239–243.

Percival S. L. (1999). The effect of molybdenum on biofilm development. *J. Ind. Microbiol. Biotechnol.*, **23**(2), 112–117.

Piriou P., Dukan S. and Kiene L. (1998). Modelling bacteriological water quality in drinking water distribution systems. *Water Sci. Technol.*, **38**(8–9), 299–307.

Prévost M., Rompré A., Baribeau H., Coallier J. and Lafrance P. (1997). Service lines: their effect on microbiological quality. *J. Am. Water Works Assoc.*, **89**(7), 78–91.

Prévost M., Rompré A., Coallier J., Servais P., Laurent P., Clément B. and Lafrance P. (1998). Suspended bacterial biomass and activity in full-scale drinking water distribution systems: impact of water treatment. *Water Res.*, **32**(5), 1393–1406.

Prévost M., Laurent P., Servais P. and Joret J.-C. (2005). Biodegradable Organic Matter in Drinking Water Treatment and Distribution (First Edition) American Water Works Association, Denver, Colorado, USA.

Reuter S., Sigge A., Wiedeck H. and Trautmann M. (2002). Analysis of transmission pathways of *Pseudomonas aeruginosa* between patients and tap water outlets. *Crit. Care Med.*, **30**(10), 2222–2228.

Rice E. W., Scarpino P. V., Reasoner D. J., Logsdon G. S. and Wild D. K. (1991). Correlation of coliform growth response with other water quality parameters. *J. Am. Water Works Assoc.*, **83**(7), 98–102.

Ridgway H. F. and Olson B. H. (1981). Scanning electron microscope evidence for bacterial colonization of a drinking water distribution system. *Appl. Environ. Microbiol.*, **41**(1), 274–287.

Roeder R. S., Lenz J., Tarne P., Gebel J., Exner M. and Szewzyk U. (2010). Long-term effects of disinfectants on the community composition of drinking water biofilms. *Int. J. Hyg. Environ. Health*, **213**(3), 183–189.

Rogers J., Dowsett A. B., Dennis P. J., Lee J. V. and Keevil C. W. (1994a). Influence of plumbing materials on biofilm formation and growth of *Legionella pneumophila* in potable water systems. *Appl. Environ. Microbiol.*, **60**(6), 1842–1851.

Rogers J., Dowsett A. B., Dennis P. J., Lee J. V. and Keevil C. W. (1994b). Influence of temperature and plumbing materials selection on biofilm formation and growth of *Legionella pneumophila* in model potable water systems containing complex microbial flora. *Appl. Environ. Microbiol.*, **60**(5), 1585–1592.

Rompré A., Prévost M., Coallier J., Gauthier C., Lafrance P., Serkedjieva R. and Laurent P. (1995). Impact of biodegradable organic matter on fixed and suspended biomass: results from full-scale distribution systems and annular reactors studies. Paper presented at the American Water Works Association-Water Quality Technology Conference New Orleans, Louisiana, USA.

Rompré A., Prévost M., Coallier J. and Servais P. (1998). Impact of reducing biodegradable organic matter on fixed and suspended biomass in distribution systems (Chapter 4). In: Microbial impact of biological filtration, American Water Works Association Research Foundation and American Water Works Association, Denver, Colorado, USA, pp. 57–120.

Rompré A., Prévost M., Coallier J., Brisebois P. and Lavoie J. (2000). Impacts of implementing a corrosion control strategy on biofilm growth. *Water Sci. Technol.*, **41**(4–5), 287–294.

Saby S., Vidal A., Gilouppe S., Tournieux D. and Suty H. (2003). The resistance of Legionella to disinfection explained by experiments on a hot water distribution system. Paper presented at the American Water Works Association-Water Quality Technology Conference Philadelphia, Pennsylvania, USA.

Sathasivan A. and Ohgaki S. (1999). Application of new bacterial regrowth potential method for water distribution system - a clear evidence of phosphorus limitation. *Water Res.*, **33**(1), 137–144.

Schoenen D. (1986). Microbial growth due to materials used in drinking water systems (Chapter 20) In: Biotechnology, H. J. Rehm and G. Reed (eds), VCH Publishers, Weinheim, pp. 627–647.

Schwartz T., Kalmbach S., Hoffmann S., Szewzyk U. and Obst U. (1998). PCR-based detection of mycobacteria in biofilms from a drinking water distribution system. *J. Microbiol. Meth.*, **34**(2), 113–123.

Schwartz T., Hoffmann S. and Obst U. (2003). Formation of natural biofilms during chlorine dioxide and U.V. disinfection in a public drinking water distribution system. *J. Appl. Microbiol.*, **95**(3), 591–601.

Sen K. and Rodgers M. (2004). Distribution of six virulence factors in *Aeromonas* species isolated from US drinking water utilities: a PCR identification. *J. Appl. Microbiol.*, **97**(5), 1077–1086.

Servais P. and Menon P. (1991). Fate of autochthonous and fecal bacteria in marine ecosystems. *Kieler Meeresforsch*, **8**, 290–296.

Servais P., Billen G. and Hascoet M. C. (1987). Determination of the biodegradable fraction of dissolved organic matter in water. *Water Res.*, **21**(4), 445–500.

Servais P., Billen G., Ventresque C. and Bablon G. P. (1991). Microbial activity in GAC filters at the Choisy-le-Roi treatment plant. *J. Am. Water Works Assoc.*, **83**(2), 62–68.

Servais P., Laurent P., Billen G. and Gatel D. (1995a). Développement d'un modèle décrivant les variations de CODB et de biomasse bactérienne dans les réseaux de distribution. *Rev. Sci. de l'Eau*, **8**(4), 427–462.

Servais P., Laurent P. and Gatel D. (1995b). Characterization of dissolved organic matter biodegradability in waters: impact of water treatment and bacterial regrowth in distribution systems. Paper presented at the American Water Works Association-Water Quality Technology Conference, New Orleans, Louisiana, USA.

Servais P., Laurent P. and Randon G. (1995c). Comparison of the bacterial dynamics in various french distribution systems. *Water Supply: Research and Technology–Aqua*, **4**, 10–17.

Servais P., Anzil A., Gatel D. and Cavard J. (2004). Biofilm in the Parisian suburbs drinking water distribution system. *Water Supply Res. T.*, **53**(5), 313–324.

Sharpe R. L., Smith C. J., Biggs C. A. and Boxall J. B. (2010). Pilot scale laboratory investigations into the impact of steady state conditioning hydraulics on potable water discolouration. Paper presented at the 11th Annual International Symposium on Water Distribution Systems Analysis, Tucson, Arizona, USA.

Shin J. H., Lee E. J., Lee H. R., Ryu S. M., Kim H. R., Chang C. L., Kim Y. J. and Lee J. N. (2007). Prevalence of non-tuberculous mycobacteria in a hospital environment. *J. Hosp. Infect.*, **65**(2), 143–148.

Simões L. C., Simões M. and Vieira M. J. (2010). Influence of the diversity of bacterial isolates from drinking water on resistance of biofilms to disinfection. *Appl. Environ. Microbiol.*, **76**(19), 6673–6679.

Singh A., Yu F. P. and McFeters G. A. (1990). Rapid detection of chlorine-induced bacterial injury by the direct viable count method using image analysis. *Appl. Environ. Microbiol.*, **56**(2), 389–394.

Sly L. I., Hodgkinson M. C. and Arunpairojana V. (1988). Effect of water velocity on the early development of manganese-depositing biofilm in a drinking-water distribution system. *FEMS Microbiol. Ecology*, **53**(3–4), 175–186.

Steinert M., Emody L., Amann R. and Hacker J. (1997). Resuscitation of viable but nonculturable *Legionella pneumophila* Philadelphia JR32 by *Acanthamoeba castellanii*. *Appl. Environ. Microbiol.*, **63**(5), 2047–2053.

Straub T. M., Daly D. S., Wunshel S., Rochelle P. A., DeLeon R. and Chandler D. P. (2002). Genotyping *Cryptosporidium parvum* with an hsp70 single-nucleotide polymorphism microarray. *Appl. Environ. Microbiol.*, **68**(4), 1817–1826.

Szewzyk U., Manz W., Amann R., Scheilfer K.-H. and Stenström T. H. (1995). Growth and in situ detection of a pathogenic *Escherichia coli* in biofilms of a heterotrophic water-bacterium by use of 16S- and 23S-rRNA directed fluorescent oligonucleotide probes. *FEMS Microbiol. Ecology*, **13**(3), 169–176.

Telgmann U., Horn H. and Morgenroth E. (2004). Influence of growth history on sloughing and erosion from biofilms. *Water Res.*, **38**(17), 3671–3684.

Torvinen E., Suomalainen S., Lehtola M. J., Miettinen I. T., Zacheus O., Paulin L., Katila M.-L. and Martikainen P. J. (2004). Mycobacteria in water and loose deposits of drinking water distribution systems in Finland. *Appl. Environ. Microbiol.*, **70**(4), 1973–1981.

Tuovinen O. H., Button K. S., Vuorinen A., Carlson L., Mair D. M. and Yut L. A. (1980). Bacterial, chemical, and mineralogical characteristics of tubercles in distribution pipelines. *J. Am. Water Works Assoc.*, **72**(11), 626–635.

Valster R. M., Wullings B. A., Bakker G., Smidt H. and van der Kooij D. (2009). Free-living protozoa in two unchlorinated drinking water supplies, identified by phylogenic analysis of 18s rRNA gene sequences. *Appl. Environ. Microbiol.*, **75**(14), 4736–4746.

van den Boomen M., van Mazijk A. and Beuken R. H. S. (2004). First evaluation of new design concepts for self-cleaning distribution networks. *Water Supply Res. T.*, **53**(1), 43–50.

van der Kooij D. (1977). The occurrence of *Pseudomonas* spp. in surface water and in tap water as determined on citrate media. *Anton. Leeuw.*, **43**(2), 187–197.

van der Kooij D. (1992). Assimilable organic carbon as an indicator of bacterial regrowth. *J. Am. Water Works Assoc.*, **84**(2), 57–65.

van der Kooij D. and Veenendaal H. R. (2001). Biomass production potential of materials in contact with drinking water: method and practical importance. *Water Sci. Technol. Water Supply*, **1**(3), 39–45.

van der Kooij D., Visser A. and Hijnen W. A. M. (1982). Determining the concentration of easily assimilable organic carbon in drinking water. *J. Am. Water Works Assoc.*, **74**(10), 540–545.

van der Kooij D., Albrechtsenz H.-J., Corfitzen C. B., Ashworth J., Parry I., Enkiri F., Hambsch B., Hametner C., Kloiber R., Veenendaal H. R., Verhamme D. and Hoekstra E. J. (2003). Assessment of the Microbial Growth Support Potential of Products in Contact with Drinking Water (CPDW), Report EVK1-CT2000-00052, European Commission.

van der Wende E., Characklis W. G. and Smith D. B. (1989). Biofilms and bacterial drinking water quality. *Water Res.*, **23**(10), 1313–1322.

van Lieverloo J. H. M., van Buuren R., Veenendaal G. and van der Kooij D. (1998). Controlling invertebrates in distribution systems with zero or low disinfectant residual. *Water Supply*, **16**(3–4), 199–204.

Volk C. (1994). Contrôle de la matière organique biodégradable au cours des traitements d'oxydation et de la distribution des eaux d'alimentation. Doctorate (Ph.D.), École Nationale des Ponts et Chaussées. Université de Paris, Paris, France. Retrieved from http:hal.archives-ouvertes.frdocs00529964PDF1994TH_VOLK_C_ NS17722.pdf

Volk C. and Joret J. C. (1994). Paramètres prédictifs de l'apparition des coliformes dans les réseaux de distribution d'eau d'alimentation. *Rev. Sci. de l'Eau*, **7**(2), 131–152.

Volk C. J. and LeChevallier M. W. (2000). Assessing biodegradable organic matter. *J. Am. Water Works Assoc.*, **92**(5), 64–76.

von Reyn C. F., Maslow J. N., Barber T. W., Falkinham J. O., III. and Arbeit R. D. (1994). Persistent colonisation of potable water as source of *Mycobacterium avium* infection in AIDS. *The Lancet*, **343**(8906), 1137–1141.

Vreeburg J. H. G. and Boxall J. B. (2007). Discolouration in potable water distribution systems: a review. *Water Res.*, **41**(3), 519–529.

Vreeburg J. H. G. and Tankerville M. Z. (2011). Build-up of discolouration potential in networks as function of treatment performance. Paper presented at the World Environmental and Water Resources Congress Bearing Knowledge for Sustainability, Palm Springs, California, USA.

Vrouwenvelder J. S., Beyer F., Dahmani K., Hasan N., Galjaard G., Kruithof J. C. and Van Loosdrecht M. C. M. (2010). Phosphate limitation to control biofouling. *Water Res.*, **44**(11), 3454–3466.

White D. R. and LeChevallier M. W. (1993). AOC associated with oils for lubricating well pumps. *J. Am. Water Works Assoc.*, **85**(8), 112–114.

Wullings B. A. and van der Kooij D. (2006). Occurrence and genetic diversity of uncultured *Legionella* spp. in drinking water treated at temperatures below 15°C. *Appl. Environ. Microbiol.*, **72**(1), 157–166.

Wullings B. A., Bakker G. and van der Kooij D. (2011). Concentration and diversity of uncultured *Legionella* spp. in two unchlorinated drinking water supplies with different concentrations of natural organic matter. *Appl. Environ. Microbiol.*, **77**(2), 634–641.

Yu J., Kim D. and Lee T. (2010). Microbial diversity in biofilms on water distribution pipes of different materials. *Water Sci. Technol.*, **61**(1), 163–171.

Chapter 11

Regrowth problems and biological stability assessment in the Netherlands

D. van der Kooij and H. R. Veenendaal

11.1 INTRODUCTION

Drinking water in the Netherlands is distributed without a disinfectant residual. This chapter describes (i) microbial growth in distribution systems in the Netherlands, (ii) the microbial growth potential of treated water assessed with methods developed in the Netherlands and (iii) biostability classes and the adenosine triphosphate (ATP)-based biostability assessment (ABBA) approach.

11.1.1 Water treatment and distribution systems

About two thirds of the volume of drinking water in the Netherlands, with a total annual production of about 1.1×10^9 m^3, is produced from groundwater and one third from surface water (Vewin 2010). Treatment of anaerobic groundwater (50% of the total production) in most cases includes aeration followed by one or two stages of rapid sand filtration (Figure 11.1). Hardness reduction is applied in more than 50% of the plants and ion exchange has been introduced at a few locations for the reduction of the colour of the water (Schippers and Sjoerdsma 2007). Drinking water prepared from aerobic groundwater comprises about 15% of the total production. This water contains a low concentration of natural organic matter (NOM; < 0.5 mg C/litre) and is either distributed without any treatment or after limestone filtration for CO_2 removal. In surface-water treatment, dune passage, river-bank passage or storage in open reservoirs, is applied as pre-treatment. Further treatment includes physico-chemical processes in combination with one or more biological filtration processes, such as rapid sand filtration, multimedia filtration, granular activated carbon (GAC) filtration and slow sand filtration (cf. Figure 11.1). These (combinations of) processes serve as barriers against microorganisms of faecal origin and chemical contaminants and also remove natural organic matter (NOM). Ozonation is used for disinfection and the removal (by oxidation) of organic contaminants and

NOM (colour). However, this treatment increases the concentration of biodegradable organic carbon, including a variety of easily assimilable low-molecular-weight compounds, from refractory humic and fulvic acids (Kuo and Chang 1977; van der Kooij and Hijnen 1984; Gilbert 1988; Gagnon *et al.* 1997; Richardson *et al.* 1999). Biological filtration is applied to reduce the concentration of these and other biodegradable compounds (Figure 11.1; van der Kooij 1987; van der Kooij *et al.* 1989; van der Hoek *et al.* 2000). Treated water is distributed without addition of a disinfectant, but at a few surface-water supplies a low dosage of ClO_2 (<0.1 mg/litre; no residual in the distribution system) is used for heterotrophic plate count (HPC) reduction in GAC filtrates.

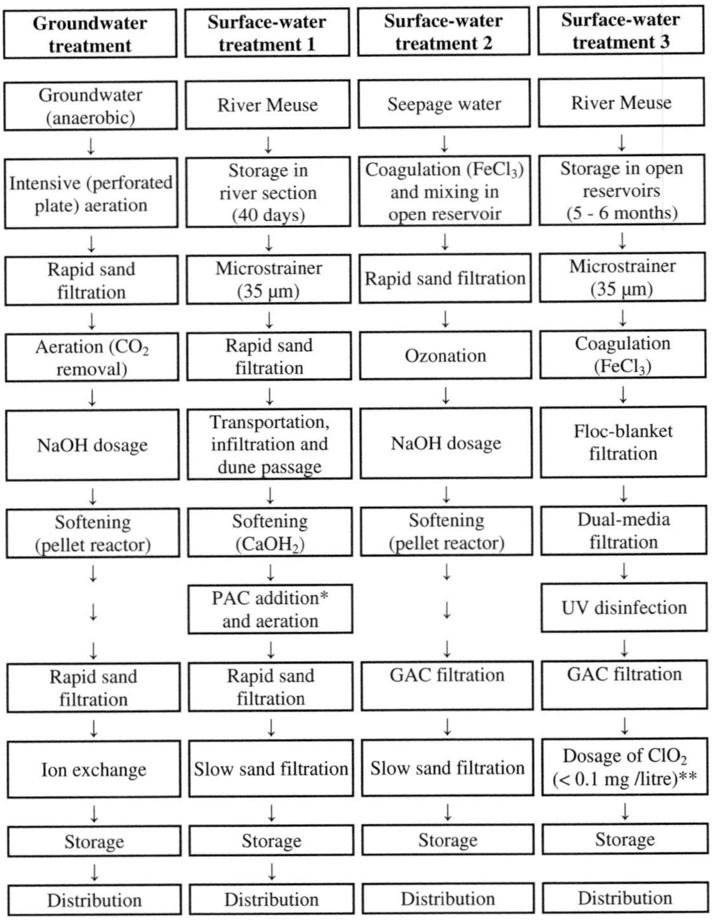

Figure 11.1 Examples of water treatment schemes in the Netherlands. *PAC, powdered activated carbon; **ClO_2 residual <0.05 mg/litre after storage at the treatment plant.

Treated water is distributed in systems with a total pipe length of about 116,000 km (in 2008) (Vewin 2010). Asbestos cement (AC) was the main pipe material from 1965–1990, but since 2010 unplasticized polyvinyl chloride (PVC-U) has constituted more than 50% of the distribution mains (Figure 11.2). The percentage of cast iron has strongly declined, and also the total length of this material, which is removed to reduce corrosion problems affecting water quality and leakages. The total length of AC is also decreasing. Certain synthetic pipe materials can play a role in regrowth (see Section 11.3). Copper is a major pipe material in premise-plumbing systems but synthetic materials, for example chlorinated PVC, cross-linked polyethylene, polypropylene, and polybutylene, are used in about 50% of recently constructed installations (Scheffer 2012).

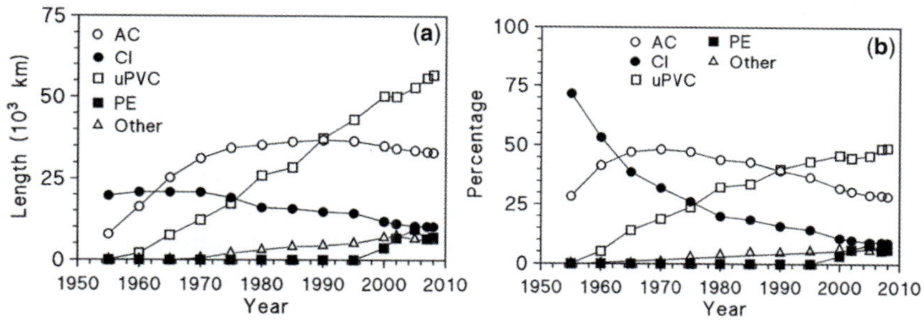

Figure 11.2 Pipe materials in distribution systems (diameter > 60 mm). (a) Total length. (b) Percentage of total length. AC: asbestos cement; CI, cast iron (usually cement lined); PVC-U, unplasticized polyvinylchloride; PE, polyethylene; Other: other materials, for example steel, concrete, glass-fibre reinforced plastic (Vewin 2010).

The regulatory standards for the microbiological quality of drinking water are shown in Table 11.1. These standards are based on European legislation (European Union [EU] 1998), with a few extensions, that is aeromonads and *Legionella*.

Table 11.1 Regulatory standards for the microbiological quality of drinking water (Ministerie van Infrastructuur en Milieu IM 2011a, b).

Parameter	*Treated water*	*Distribution system*
Escherichia coli (CFU)*	<1/100 ml	<1/100 ml
Faecal streptococci (CFU)	<1/100 ml	<1/100 ml
*Clostridium perfringens*** (CFU)	<1/100 ml	<1/100 ml
Total coliforms (CFU)	<1/100 ml	<1/100 ml
Heterotrophic plate count# (CFU)	ND##	<100/ml$

(*Continued*)

Table 11.1 Regulatory standards for the microbiological quality of drinking water (Ministerie van Infrastructuur en Milieu IM 2011a, b) (*Continued*).

Parameter	*Treated water*	*Distribution system*
Aeromonads 30°C (CFU)	<1000/100 ml	<1000/100 ml
Legionella (CFU)	<100/litre	<100/litre[$$]

*CFU, colony-forming units; **including spores; #pour plate of glucose yeast extract agar incubated at 22°C during 3 days; ##ND, not defined, $yearly geometric mean, $$also in premise-plumbing systems.

11.1.2 NOM and bacteria in treated water

The concentration of NOM in drinking water, measured as total organic carbon (TOC), ranges from less than 0.5 to about 6 mg/litre (Figure 11.3). The lowest concentrations are observed in aerobic-groundwater supplies and the highest in supplies with anaerobic groundwater as source. In most surface-water supplies, the TOC concentration ranges from about 1.5 to 2.0 mg/litre. The UV absorbance (254 nm) showed a maximum value of about 15 m^{-1} and correlates with the TOC concentration (Figure 11.3b). The quotient of UV absorbance and TOC, a measure of aromatic carbon content (Trains *et al.* 1990), varies from about 1 to 4.

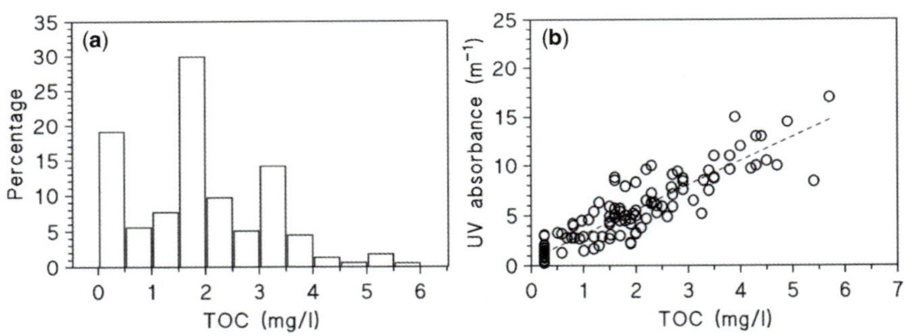

Figure 11.3 Total organic carbon (TOC) in finished water (186 facilities) in the Netherlands in 2011; (a) Relative frequency of concentration ranges; (b) UV absorption (254 nm) in relation to TOC concentration [UVabs = 2.46 (±0.10) TOC + 0.64 (±0.23) $R^2 = 0.82$)]. Data from Rewab (2011).

The concentration of bacteria in treated water entering the distribution system has been measured in a survey using different media and methods (Figure 11.4). HPCs, which depend on medium composition, incubation temperature and incubation period, were only a small fraction (≪1%) of the total direct cell count (TDC), determined with epifluorescence microscopy after acridine-orange staining (Hobbie *et al.* 1977). The lowest values of TDC (<10^4 cells/ml) and ATP (<1 ng/litre) were observed in treated aerobic groundwater. The ATP to cell ratio ranged from 0.02 to 0.3 fg ATP/cell with a median (P50) value of 0.03 fg ATP/cell and P90 of 0.078 fg ATP/cell. The percentile values of the various parameters, which can serve as reference, are listed in Table 11.2.

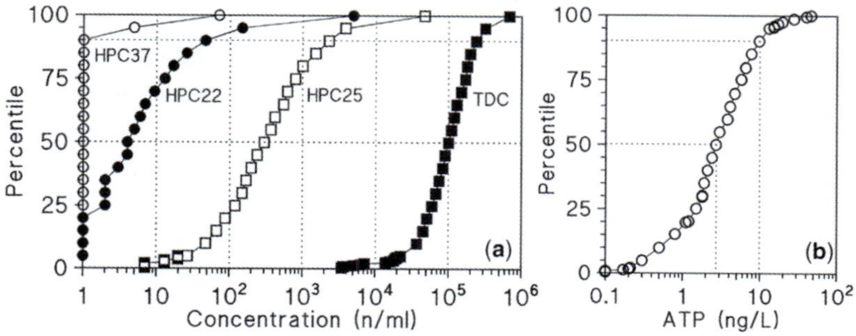

Figure 11.4 Concentration of microorganisms in finished water at 243 treatment plants in the Netherlands in 1996; (a) Cumulative percentages of the heterotrophic plate count (HPC) and total direct cell count (TDC); (b) Cumulative percentage of ATP concentration. Abbr.: HPC37, HPC (CFU/ml) after 24 h incubation at 37°C; HPC22, HPC (CFU/ml) after 72 h incubation at 22°C; HPC25, HPC (CFU/ml) after 10 days incubation at 25°C on R2A medium (spread plate method); TDC, total direct (microscopic) cell count.

Table 11.2 Ranges and percentile values of concentrations of bacteria and ATP in treated water in the Netherlands (243 treatment facilities) in 1996.

Percentile	***Parameter***				
	HPC 37°C (CFU/ml)	**HPC 22°C (CFU/ml)**	**HPC 25°C (CFU/ml)**	**TDC (n/ml)**	**ATP (ng /litre)**
Range	<1–72	<1–5000	7–4.7×10^4	3.5×10^3–6.8×10^5	<0.5–46
P50	<1	4	300	1.0×10^5	2.7
P90	1	50	2280	2.4×10^5	9.6

11.2 REGROWTH

11.2.1 Compliance with standards

In 2010, drinking water during distribution in all supplies complied with the regulatory standard for HPC22 (Table 11.1) (Versteegh and Dik 2011). Coliforms were observed in 4 (repeat) samples (4 supplies) of a total of about 50,000 samples collected from distribution systems in 2010 (Versteegh and Dik 2011). In 2.0% of all samples (21 supplies) aeromonads were observed at levels above the standard of 1000 CFU/100 ml. *Legionella* was detected with the culture method in 4.8% of the samples (27 supplies) routinely collected from distribution systems.

The cultured bacteria were identified as *L. anisa* (Versteegh and Dik 2011). Hence, non-compliance with microbiological drinking water quality standards related to regrowth is mainly caused by *Legionella* and *Aeromonas.*

11.2.2 Aeromonads and coliforms

Bacteria of the genus *Aeromonas* have been detected in drinking water in many countries, indicating that certain species are ubiquitous in water supplies (Leclerc and Buttiaux 1962; Schubert 1976; LeChevallier *et al.* 1982; Burke *et al.* 1984). Aeromonads, cultured on ampicillin-dextrin agar at 30°C (Havelaar *et al.* 1987), are included in drinking water regulations in the Netherlands as a regrowth parameter. *Aeromonas* counts generally represent $< 10\%$ of HPC22 in distributed drinking water, and do not correlate with HPC22 (Havelaar *et al.* 1990). In summer, their numbers are clearly higher than in winter.

Coliforms isolated from drinking water in the Netherlands usually are identified as *Enterobacter, Citrobacter* and *Klebsiellla* spp. (van der Kooij 1997). In 1961–1963, 1.5 to 3.0% coliform-positive samples were observed in a surface-water supply with treatment consisting of slow sand filtration in parallel with break-point chlorination followed by coagulation/sedimentation and 0.2 mg/litre free-chlorine residual in finished water (Rook 1970). The currently observed low percentage of coliform-positive samples (0.01%) demonstrates that changes in water treatment in this and also other supplies effectively have reduced the regrowth of coliforms in water supplies in the Netherlands.

11.2.3 Legionella

Cases of Legionnaires' disease (LD) in the Netherlands, with a nosocomial origin and related to potable-water installations, have been observed in the early 1980s (van Ketel *et al.* 1983; Meenhorst *et al.* 1985). LD became a notifiable disease in 1987, following the advice of the Health Council (Health Council 1986). In 1999, an LD outbreak caused 31 dead, 188 diseased and a much larger number of infected persons (Den Boer *et al.* 2002; Nagelkerke *et al.* 2003). Whirlpools on display at a flower show, containing water without chlorine residual, were identified as the source of *L. pneumophila*. Regulations aiming at preventing growth of *Legionella* in warm tap water installations were introduced by the end of 1999, adapted in 2004 and revised in June 2011 (IM 2011a, b). In these regulations, risk analysis, based on water temperature, stagnancy, dead-end pipes etc., and preventive measures based on temperature control, as well as periodic monitoring are prescribed for potable-water installations with aerosol-generating devices in hospitals, other healthcare institutions, hotels, jails, asylum-seeker centres, swimming facilities, camping sites, truck-stop facilities, and a few other facilities. A standard for *Legionella* on Buffered Charcoal Yeast Extract (BCYE) agar of 100 CFU/litre is defined for the involved installations. This standard applies to finished and distributed drinking water as well, combined with a low monitoring frequency.

The notification rate of LD has not declined in the period 2000–2010 despite all regulations and measures applied since 1999 (Figure 11.5). The increased notification rate after 1999 is attributed to the introduction of the urine-antigen test for the diagnosis of LD and improvements in reporting and surveillance (Versteegh *et al.* 2007). About 50% of the reported cases were associated with travel abroad. The years with an elevated notification rate (2002, 2006 and 2010) had a warm and wet summer. The weekly incidence of LD in 2006 correlated with the weekly averages of temperature, humidity and precipitation intensity (Karagiannis *et al.* 2009). These correlations may suggest that infection is associated with outside sources, for example cooling towers. In a recent study, *L. pneumophila* was detected with the quantitative polymerase chain reaction (qPCR) technique in 32% of 50 investigated rooftop cooling towers in the Netherlands (Oesterholt *et al.* 2012), but proof that cooling towers are implicated as a major source of infection is lacking (Euser *et al.* 2012).

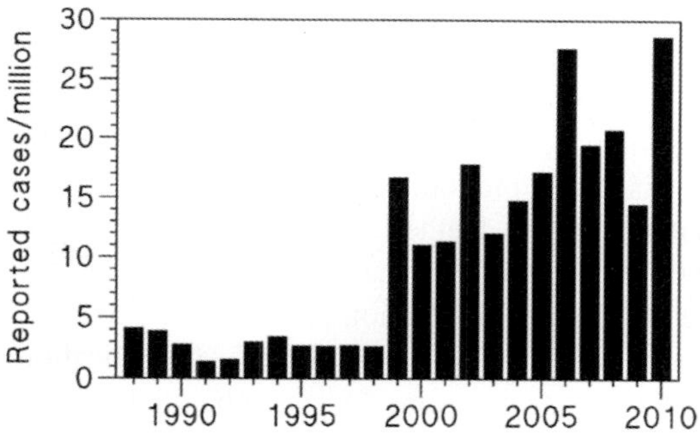

Figure 11.5 Notification rate of LD in the Netherlands (based on Bijkerk *et al.* 2011). 1999: outbreak "Bovenkarspel"; 2002, 2006 and 2010, years with a warm and wet summer.

A survey of finished water of 96 treatment plants, at a temperature below 15°C, yielded no *Legionella*-positive results on BCYE-medium incubated at 37°C. However, by using a quantitative PCR (qPCR) method, *Legionella* spp. were detected at all investigated plants at concentrations ranging from about 10^3 to 10^6 cells/litre (Wullings *et al.* 2006). Cloning and partial 16S rRNA sequencing yielded sequences which all clustered within the genus *Legionella*, but only one third of these sequences was most closely (>97%) related to defined *Legionella* spp., including Legionella-like amoebal pathogens. A more detailed study was conducted at two groundwater supplies, that is supply A with a low TOC concentration (<0.5 mg C/litre) and a low ATP concentration (<1 ng/litre) and supply B with a TOC concentration of 8 mg/litre and an ATP concentration of ~10 ng/litre (Wullings and van der Kooij 2011). Also here, legionellae were not detected by culture but qPCR revealed that *Legionella* spp. were present in all samples of finished water (10°C), distributed water and biofilms scraped from the inner surface of pipe segments (PVC-U, Ø 10 cm). The abundance of *Legionella* in distributed water, determined with qPCR, was clearly higher in supply B (avg. 2.5 (±1.9) × 10^3cells/litre) than in supply A (avg. 2.9(±1.6) × 10^2 cells/litre). However, the average total species richness estimated for supply A (153 operational taxonomic units, OTUs) was clearly higher than in the supply B (58 OTUs). The total of 161 *Legionella*-related OTUs, most of which belonging to yet-uncultivated species, collected in these two studies, demonstrates the presence of a high diversity of *Legionella* spp. in water supplies in a temperate climate. This conclusion is consistent with the detection of *Legionella* spp. in 87% of water samples from cold-water taps in 250 public buildings in the Netherlands (Diederen *et al.* 2007). In this study 2.2% of the samples were culture positive.

Legionella was detected with the culture method in 4% of the drinking water installations in a total of 400 family homes in the distribution systems of two surface-water supplies and two groundwater supplies (Oesterholt and van der Kooij 2003). These organisms were all identified as *L. anisa*. *Legionella* was also cultured from 18.5% of 11,500 samples collected from warm tap water installations in buildings (van der Kooij *et al.* 2007). *L. pneumophila* and *L. anisa* were observed in 3.1% and in 13.4% of these samples, respectively. *L. anisa* obviously is the predominating culturable *Legionella* species in potable-water systems in the Netherlands. This organism has a low virulence (Fields *et al.* 1990) and is only very rarely (≤0.2%) observed in clinical samples (European Centre for Disease Prevention and Control [ECDC] 2012). Hence, the colony counts on BCYE medium, used for the quantification of *Legionella* in water, are not an adequate measure for the potential of exposure to virulent *Legionella* spp., that is *L. pneumophila,* and identification of typical colonies is needed for health-risk evaluation.

The ubiquitous presence of *Legionella* spp. in all investigated water supplies in the Netherlands indicates that treated water has a potential for growth of these organisms, even at a low NOM concentration. This growth potential may lead to the presence of *L. pneumophila* when water temperature exceeds 25°C, as was demonstrated in a study with a model potable-water system. This study showed that warm (36°C) tap water (dissolved organic carbon [DOC] 1.9 ppm; assimilable organic carbon [AOC] 5 μg ac-C eq C/litre), promoted growth of *Legionella pneumophila* with P90 values of the colony counts in water of 2.5 × 10^3 CFU/litre (copper pipes), 8.2 × 10^3 CFU/litre (stainless steel pipes) and 1.5 × 10^4 CFU/litre (cross-linked polyethylene pipes) (van der Kooij *et al.* 2005).

11.2.4 Free-living protozoa (FLP)

FLP are part of the microbial communities in water supply systems. Their activities have a large impact on the bacterial community because of (i) consumption of bacteria in water, biofilms, and sediments; (ii) excretion of inorganic and organic nutrients and (iii) harbouring protozoan-resistant microorganisms. Grazing reduces the amount of attached and suspended biomass, but preferences of FLP for certain bacteria also strongly affect the composition of the bacterial community (Simek *et al.* 1992; Weekers *et al.* 1993; Pernthaler *et al.* 2001; Sherr and Sherr 2002). Bacteriovorous amoeba, flagellates and ciliates excrete considerable amounts of inorganic nutrients (ammonia and phosphate) and amino acids during grazing (Anderson *et al.* 1985; Weekers *et al.* 1993; Nagata and Kirchman 1991; Dolan 1997; Ferrier-Pagès *et al.* 1998; Sherr and Sherr 2002). These compounds are the favourite growth substrates of fluorescent pseudomonads and *Aeromonas* spp. (van der Kooij *et al.* 1982a; van der Kooij and Hijnen 1988). The survival and intracellular growth of protozoan-resistant bacteria have attracted much attention since the discovery of the role of protozoa as host for *Legionella* (Rowbotham 1980; Wadowsky *et al.* 1988) in the aquatic environment. A large variety of amoebae-resisting bacteria, including *Legionella spp.* and mycobacteria, has been identified in free-living amoebae cultured from finished water, distributed drinking water and water in a hospital water network (Thomas *et al.* 2006; Thomas *et al.* 2008; Thomas *et al.* 2010; Thomas and Ashbolt 2011).

Traditionally, cultivation procedures and microscopy were used for detection of FLP in drinking water supplies and their identification (Michel *et al.* 1995; Thomas *et al.* 2006). However, these techniques are time consuming, complicated and incomplete, because many protists are not cultivable and/or difficult to identify. For elucidating the presence and role of FLP in drinking water and in related biofilms, in particular in relation with the presence of *Legionella* spp. in water supplies and in premise-plumbing systems, molecular methods for detection and identification were used in studies conducted in the Netherlands. Observations on biofilms grown at a high concentration ($3.8 \times 10^4 \pm 2.7 \times 10^3$ pg ATP/cm^2) on plasticized PVC (PVC-P) showed that the amoeba *Hartmannella vermiformis* served as host for *L. pneumophila* in the biofilm (Kuiper *et al.* 2004). When the inoculum was membrane-filtered (3 μm) for FLP removal, no growth of *H. vermiformis* and *L. pneumophila* was observed, despite a similar level of biofilm formation and the presence of *L. pneumophila* introduced with the inoculum. In a survey using a specific qPCR for the detection of *H. vermiformis,* the organism was detected in 21 of 28 samples (200 ml) of a variety of surface-water types in the Netherlands (Kuiper *et al.* 2006).

The two drinking water supplies mentioned above, supply A with a NOM concentration of <0.5 mg C/litre and supply B with a NOM concentration of 8 mg C/litre, were also investigated for the presence of FLP, by using molecular methods. In these supplies, a total of 127 OTUs of FLP were observed (Valster *et al.* 2009). The protists in the biofilms in supply A were predominated by Cercozoa (flagellates) whereas Ciliophora (ciliates) predominated in the biofilms in supply B. *H. vermiformis* was not detected in the samples of drinking water collected from the distribution system nor in biofilms of supply A, but was observed in all samples of distributed drinking water collected in summer from supply B. These observations are consistent with those of Michel *et al.* (1995), who cultured *H. vermiformis* from treated groundwater. An investigation of three Caribbean water supplies yielded a total of 59 OTUs of FLP, with *H. vermiformis* as the predominating protozoan (Valster *et al.* 2011). A significant correlation was observed between *L. pneumophila* and the concentration of *H. vermiformis* in water sampled from the distribution system of the supply with the highest concentrations of *L. pneumophila*. Ciliates predominated in the protozoan community in the supply with elevated, sediment-related, turbidities and ATP concentrations. In the supply with a low ATP concentration (<1 ng/litre) in most samples, flagellates predominated and *H. vermiformis* was not observed. *H. vermiformis* was the most commonly observed host for *L. pneumophila* growing at 37°C in biofilm-batch tests with samples collected from a variety of engineered water systems (Valster *et al.* 2010). In these tests, a number of candidate hosts were observed as well, that is *Diphylleia rotans*, *Echinamoeba thermum* and *Neopara* sp., which are not (yet) identified as hosts in laboratory tests with pure cultures.

Overall, the presence of a large diversity of FLP in water supplies is consistent with the ubiquitous presence of *Legionella* spp. in drinking water at temperatures <20°C. Hence, the inability of *L. pneumophila* to grow in distributed drinking water in the Netherlands is not due to the absence of host protozoa, but related to the relatively low water temperature at which other *Legionella* species apparently predominate.

11.2.5 Other (micro)organisms in drinking water

Fluorescent pseudomonads, which are regarded as nuisance organisms, comprised a small proportion (usually < 1%) of the HPCs in drinking water (van der Kooij 1977). A total of 48 biotypes were identified, based on physiological tests, demonstrating large diversities among representatives of the species *P. fluorescens* and *P. putida* in drinking water (van der Kooij 1979). A recent study, using sequence analysis of cloned 16S rRNA specific PCR fragments, retrieved from finished water and distribution-system biofilms of the earlier mentioned water supplies A and B, confirmed the high bacterial diversity in microbial communities. These diversities differed strongly between and within distribution systems. Average values of the Chao estimate of species richness for the predominating bacteria ranged from 353 OTUs in supply B (TOC of 8 ppm) to 515 OTUs in supply A (TOC < 0.5 ppm), with 39–45% belonging to the Proteobacteria and 20–45% of the bacteria belonging to yet-unclassified phyla (Heijnen *et al.* 2014). These observations are consistent with other reports showing large diversities of bacteria in drinking water systems (Hong *et al.* 2010; Henne *et al.* 2012). However, the interpretation of these data is hampered by the absence of information about the physiological properties of most identified and all unidentified bacteria.

Nitrifying bacteria and archaea, which do not contribute to HPCs, are also present in drinking water during distribution (van der Wielen and van der Kooij. 2009). Fungi represented about 10% of the partial 18S rRNA gene sequences (27 OTUs) of the eukaryotic clones retrieved from two water supplies in the Netherlands (Valster *et al.* 2009), but constituted a larger proportion (33% of clones; 83 OTUs) in three supplies in a tropical region (Valster *et al.* 2011). A recent study with qPCR methods, aiming at detecting opportunistic pathogens, revealed that fungi were present in all water samples collected from six supplies (van der Wielen and van der Kooij 2013).

A national survey, conducted in 1993–1995, showed that a large diversity of invertebrates was present in all 34 investigated supplies, with abundance correlating with the concentration of NOM, biofilm formation and presence of sediments (van Lieverloo *et al.* 2012). The arthropod *Asellus aquaticus* represented the major proportion (88%) of the biomass of collected invertebrates in flush water (van Lieverloo *et al.* 2002).

The total concentration of bacteria in distributed drinking water has been assessed with ATP and TDC analysis. Surveys revealed that typical ATP concentrations range from <1 to 10 ng/litre, with values up to about 30 ng/litre in the peripheral parts in some of the investigated distribution systems (van der Kooij 1992; van der Wielen and van der Kooij 2010). ATP concentrations in the distribution systems differ between water types and are affected by the distance to the treatment plant and by season. TDC values in distributed drinking water ranged from about 5×10^4 to 7×10^5 cells/ml. A significant correlation was observed between TDCs and ATP in water collected from distribution systems in summer, with an average ATP amount of 0.018 ($\pm$ 0.0095) fg per cell (van der Wielen and van der Kooij 2010).

11.3 ASSESSMENT OF REGROWTH POTENTIAL

At the beginning of the 20th century, Heymann (1928) developed a method for determining the fraction of biodegradable organic matter removed by a test filter with a volume of 10 litre, containing sand from a ripened slow sand filter (van der Kooij 1990). At a contact time of one day, a considerable reduction (20–40%) of the $KMnO_4$ demand, a measure for the NOM concentration, was observed even with water treated by two stages of rapid sand filtration. This method was not used for optimization of water treatment, which, at that time mainly focused on drinking water safety. Elevated HPCs associated with the introduction of ozonation for the removal of persistent organic contaminants in the 1970s and the discovery of the formation of undesired by-products of chlorination (Rook 1974) renewed interest in regrowth problems in the Netherlands. As a result, the need for a method to determine the microbial growth potential of treated water emerged.

11.3.1 Assimilable organic carbon (AOC)

11.3.1.1 Method

A few micrograms per litre of easily biodegradable compounds can promote microbial growth (Chapter 1). Therefore, a method was developed for determining the concentration of easily assimilable organic carbon (AOC) in treated water. This method is based on measuring the growth of two bacterial strains in 600 ml of pasteurized (60°C) water contained in thoroughly cleaned glass-stoppered pyrex Erlenmeyer flasks with a volume of 1 litre (van der Kooij *et al.* 1982c; van der Kooij 1992). The organisms used are *P. fluorescens* strain P17 and *Spirillum* sp. strain NOX. The nutritional versatility of these strains, their growth kinetics and growth yields at concentrations of a few μg per litre have been reported in detail (van der Kooij *et al.* 1982a; van der Kooij and Hijnen 1984). Table 11.3 shows the number of tested compounds serving as sole source of carbon and energy for these organisms at a concentration of 1 mg C/litre as well as the yields for acetate and oxalate.

Table 11.3 Characteristics of the test strains used in the AOC method.

Characteristic	***Organism***	
	Strain P17	**Strain NOX**
Utilization of amino acids	19/19*	2/19
Utilization of carboxylic acids	11/17	14/17
Utilization of carbohydrates	10/17	0/17
Utilization of aromatic acids	7/11	0/11
Yield (CFU/μg C) on acetate	4.1×10^6	1.2×10^7
Yield (CFU/μg C) on oxalate	No growth	2.9×10^6

*number of compounds utilized and number of compounds tested at 1 mg C/litre.

The difference between the yield values of strain NOX for acetate and oxalate (Table 11.3) is consistent with the difference in Gibbs energy dissipation associated with the production of biomass from these compounds (Heijnen and van Dijken 1992). The AOC concentration, expressed as acetate-C eq/litre, therefore may differ from the concentration of organic carbon used for growth, but enables comparison between water types based on biomass production potential. However, the two test strains mentioned are not able to utilize all biodegradable compounds that may be present in drinking water. Therefore, strains of bacteria which can utilize other (groups of) compounds, for example polysaccharides and proteins, have been selected for use in AOC tests as well (van der Kooij and Hijnen 1985; Sack *et al.* 2010, 2011). Growth tests with specific bacteria, for example coliforms, *E. coli*, *Aeromonas*, *P. aeruginosa,* can be used to assess the potential ability of a water type to support growth of these organisms (van der Kooij and Hijnen 1982b; van der Kooij *et al.* 1988; Rice *et al.* 1991; van der Kooij 1997, 2002; Vital *et al.* 2010). However, this growth potential does not reflect the level of growth of these organisms in drinking water in the presence of the indigenous microbial community, because of competition for substrates and temperature effects, as has been shown for *P. aeruginosa* (van der Kooij *et al.* 1982b) and *Aeromonas* (van der Kooij 1991).

AOC tests have been applied in the Netherlands in finished water, distributed water and water at various treatment stages and in pilot plants (van der Kooij *et al.* 1989; Hijnen and van der Kooij 1992a, b; van der Kooij 1992; van der Hoek *et al.* 2000). A few typical examples of the growth curves of strain P17 and strain NOX, grown simultaneously in duplicate water samples, are shown in Figure 11.6A. The AOC concentration of finished water ranged from < 1 to ~ 60 µg ac-C eq/litre (van der Kooij *et al.* 1989; van der Kooij 1992; van der Kooij *et al.* 1999). The fraction utilized by strain P17 usually is below the fraction utilized by strain NOX and generally accounts for less than 1 µg of ac-C eq/litre in treated water. The AOC concentration of drinking water declined during distribution, depending on the distance to the treatment plant (van der Kooij *et al.* 1989; van der Kooij 1992, 1999). Figure 11.6b shows that the observed maximum reduction (dAOC) is a function of the AOC concentration of the water leaving the treatment plant (AOC_T). The AOC reduction was very limited at AOC_T concentrations below 10 µg C/litre and AOC concentrations did not decline to zero during distribution. Elevated HPCs were observed at AOC_T concentrations >40 µg ac-C eq/litre (van der Kooij *et al.* 1989), but HPC22 remained below 100 CFU/ml at AOC_T < 10 µg ac-C eq/litre (van der Kooij 1992). From these observations, an AOC concentration of 10 µg ac-C eq/litre was derived as guideline value of biologically stable drinking water.

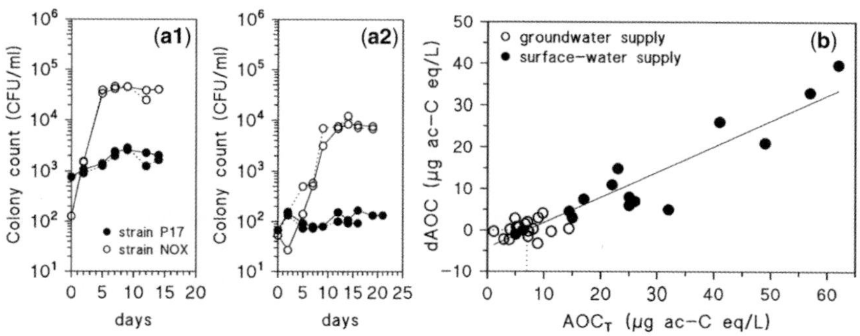

Figure 11.6 (a) Growth curves (in duplicate) of AOC tests. (a1): slow–sand filtrate (AOC = 4.5 µg ac-C eq/litre); (a2): treated aerobic groundwater (AOC = 0.9 µg ac-C eq/litre); (b) Observed maximum AOC reduction (dAOC) in drinking water during distribution without disinfectant residual; AOC_T = AOC concentration at treatment plant. Reprinted from J. Am. Wat. Works Assoc. **91**(1), 55–64 by permission. Copyright © 1999 the American Water Works Association.

11.3.1.2 AOC-reduction tests

To simulate the effects of distribution and to assess the utilization of AOC by the indigenous bacterial community, AOC concentrations were determined in water on the day of sampling (day zero) and in samples incubated unpasteurized for several weeks at 15°C. Typical results of these AOC-reduction tests are depicted in Figure 11.7. A rapid decline of the AOC concentration to less than 10 µg ac-C eq/litre was observed in water collected from open storage reservoirs (Figure 11.1, SW supply 1). The compounds utilized by strain P17, for example amino acids, comprised the main AOC fraction in this water. Also in ozonated water, a concentration of about 10 µg ac-C eq/litre was reached after 7 days, but further decline was slow.

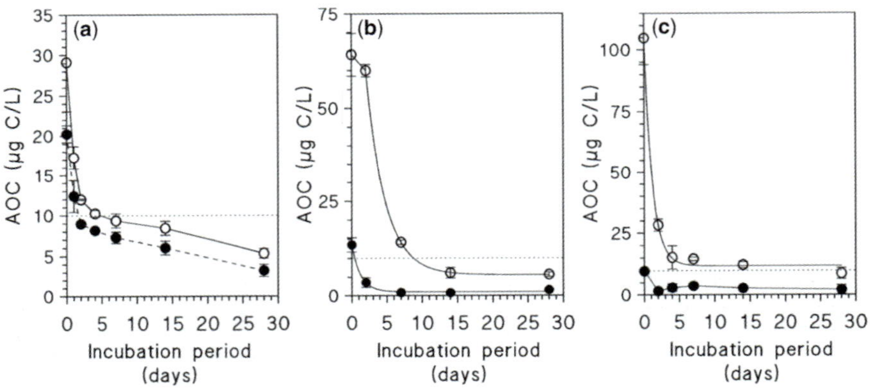

Figure 11.7 Effect of incubation at 15°C on the AOC concentration in unpasteurized water. (a) water from open storage reservoirs; (b, c) water after ozonation at two different treatment plants. Symbols: ● AOC-P17 in the presence of strain NOX; ○ AOC-P17 + NOX. Standard deviations (duplicate flasks) are indicated; the dotted line represents the AOC level of 10 µg ac-C eq/litre.

AOC-reduction tests with a variety of finished water types and water from treatment stages revealed a clear relationship between the initial AOC concentration (day zero) and the reduction of the AOC concentration after 7 days of growth of the indigenous bacteria at 15°C (Figure 11.8). The linear relationship obtained for water with an initial AOC concentration of > 10 µg ac-C eq/litre suggests that no AOC decline occurs within 7 days of incubation at AOC concentrations of <7.3 µg ac-C eq/litre (Table 11.4). From the distribution system data, a value of 4.7 µg ac-C eq/litre was derived, which is close to the value observed for the entire dataset. Figure 11.8B shows that a further reduction of the AOC concentration also occurred in treated water at an initial AOC concentration < 10 µg ac-C eq/litre. Hence, the relationship most likely is nonlinear at very low concentrations, with no uptake at AOC = 0. The low utilization rate at AOC < 10 µg ac-C eq/litre supports the guideline value of AOC for biological stability of tested water.

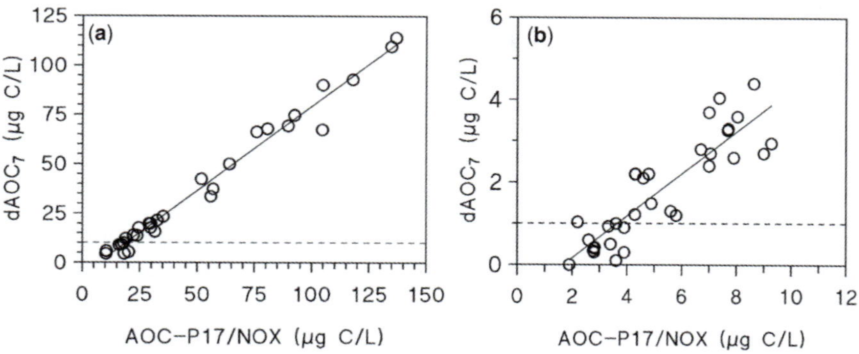

Figure 11.8 Relationship between the AOC decrease in water after 7 days of incubation at 15°C ($dAOC_7$) and the AOC concentration at day 0. (a) initial AOC concentration >10 µg ac-C eq/litre; (b) initial AOC concentration < 10 µg ac-C eq/litre.

Table 11.4 Relationships between the initial AOC concentration (AOC_0) and the AOC concentration after incubation (7 days at 15°C) or distribution.

Water type	Relationship*	AOC_0 (µg C/ litre) for $dAOC_7 = 0$
Water types with AOC > 10 µg C /litre	$dAOC_7 = 0.85\ AOC_0 - 6.2$ ($R^2 = 0.98$)	7.3
Water types with AOC < 10 µg C /litre	$dAOC_7 = 0.51\ AOC_0 - 0.85$ ($R^2 = 0.77$)	1.7
All water types tested	$dAOC_7 = 0.82\ AOC_0 - 4.1$ ($R^2 = 0.98$)	5.0
Water from distribution systems	$dAOC = 0.6\ AOC_T - 4.1$ ($R^2 = 0.88$)	4.7

*AOC_o, initial AOC concentration; $dAOC_7$, decrease of AOC concentration after 7 days of incubation at 15°C. AOC_T, AOC concentration at treatment plant.

11.3.1.3 Effects of water treatment

Ozonation of water generally increases the AOC concentration. A linear relationship has been observed between the reduction of UV absorbance (254 nm), caused by increasing concentrations of ozone, and the increase of the AOC concentration

(Figure 11.9a). The AOC increase was 26.1 ± 1.3 μg ac-C eq/litre for the reduction of the UV absorbance with one unit. This observation is consistent with reports that showed that transformation of the aromatic structures of humic acid is the major cause of the AOC formation in water by ozonation (Kuo and Chang 1977; Gilbert 1988). Post treatment following ozonation is applied to reduce the AOC concentration to an acceptable level for regrowth limitation. For this purpose biological filtration processes, that is dual-media filtration (DMF), rapid sand filtration (RSF), granular activated carbon (GAC) filtration and slow sand filtration (SSF) are used (Figure 11.1; van der Hoek *et al.* 2000). These processes clearly reduce the AOC concentration, but the removal efficacy depends on the initial AOC concentration (Figure 11.9b), that is a maximum reduction of about 84% and no apparent uptake at AOC-P17 < 7 μg ac-C eq/litre.

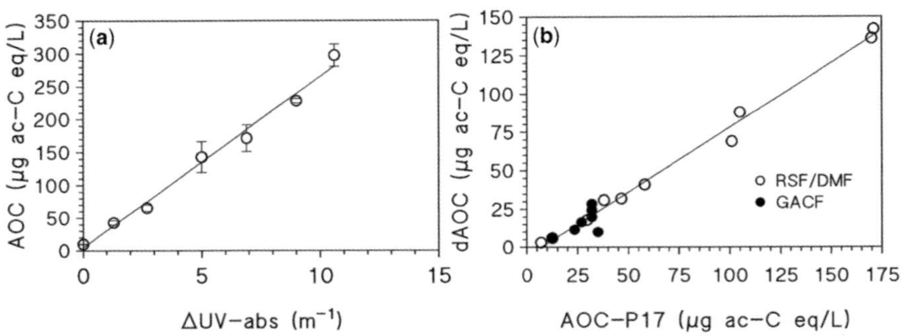

Figure 11.9 (a) Relationship between the AOC concentration and the decrease of UV absorbance (254 nm) of water (DOC 6.4 mg/litre) after ozonation (adapted from: van der Kooij *et al.* 1989). The relationship between AOC and ΔUV is given by: AOC = 26.1 ΔUV + 4.5 ($r^2 = 0.99$); (b) Relationship between the removal of AOC-P17 and the AOC concentration in the influent during biological filtration in water treatment. RSF: rapid sand filtration; DMF, dual-media filtration; GACF, granular activated carbon filtration. The relationship is given by: dAOC = 0.84 AOC–6.0 ($R^2 = 0.98$). Source: van der Kooij 1984.

AOC-reduction tests with water collected from different water treatment stages also enable an evaluation of the effect of the applied treatment processes. GAC filtration at a surface water treatment plant reduced the AOC concentration of ozonated water to a value below the concentration remaining after 7 days of incubation at 15°C (Figure 11.10). The average rate of AOC uptake during GACF and DMF, with an effective contact time of approximately 10 min, obviously is about 1000 times higher than in water incubated at 15°C during 7 days. SSF further reduced the concentration to below 10 μg ac-C eq/litre and even below the level attained after 14 days of incubation at 15°C. In another treatment plant, however, DMF failed to reduce the AOC concentration of ozonated water to the residual concentration observed after 7 days of incubation (Figure 11.10b). Hence, the AOC concentration observed after 7 days of incubation at 15°C is attained either by applying DMF followed by GAC filtration or GAC filtration followed by SSF. The presented examples do not give a complete picture of the effects of biological filtration, because also temperature and contact time affect AOC removal (Zhang and Huck 1996; Gagnon *et al.* 1997).

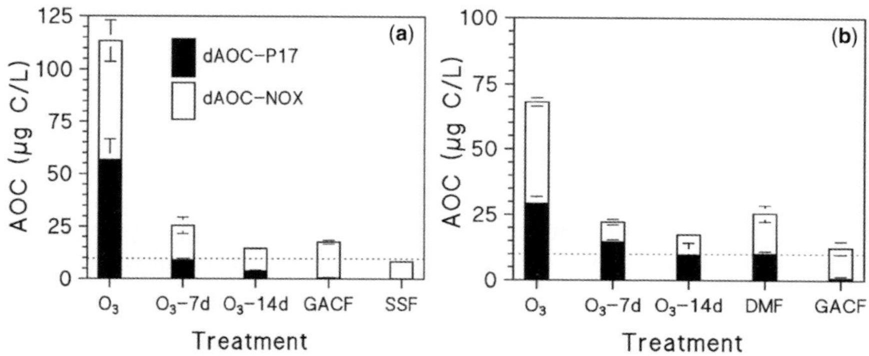

Figure 11.10 AOC decline in ozonated water during incubation at 15°C compared with the effects of post treatment at two surface-water treatment plants (a and b). O_3, ozonated water; O_3-7d, ozonated water after 7 days of incubation at 15°C; O_3-14d, after 14 days of incubation; GACF, granular activated carbon filtration; DMF, dual-media filtration; SSF slow sand filtration. Temperature ~ 10°C.

Membrane filtration is used in water treatment for the removal of particles, microorganisms, chemical contaminants or NOM and for hardness reduction. Ultrafiltration (Molecular Weight Cut Off 100 kD) did not reduce the concentrations of TOC and AOC, which is consistent with the molecular weight of dissolved compounds in treated water (Amy *et al.* 1992). Dosage experiments showed that nanofiltration (NF) membranes (MWCO 200 D) removed more than 90% of most low-molecular weight compounds tested at pH 8, but the removal of ethanol (MW 46) and glycol (MW 62) was less than 30% (Figure 11.11A). Benzoate (MW 144) showed only 80% removal at pH 5. By feeding this NF membrane with water after ozonation and biological filtration in three different water treatment plants, more than 95% reduction of the AOC concentration was observed (Figure 11.11b). However, reducing elevated AOC concentrations with spiral-wound NF membranes may lead to operational problems as a result of biofouling of the membranes (see below).

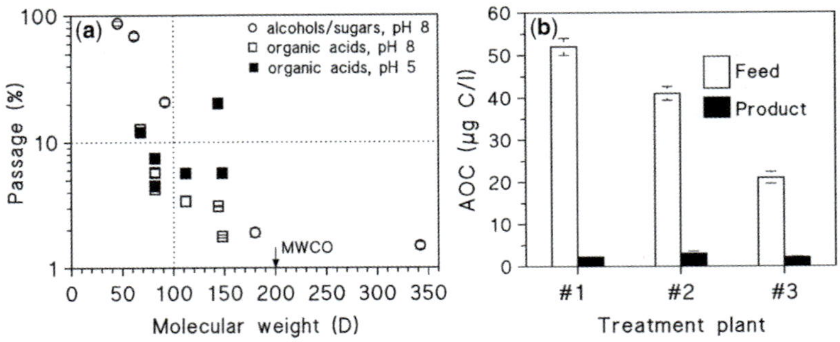

Figure 11.11 AOC removal by nanofiltration; (a) Effect of molecular weight on passage of selected compounds through NF membrane (MWCO 200 D) at pH 8 and pH 5; (b) AOC removal by NF filtration of water after ozonation and dual-media filtration (feed) at three treatment plants.

11.3.1.4 Clogging of filter beds, injection wells and spiral-wound membranes

Biological activity in water treatment processes may cause clogging problems, for example in injection wells, filter beds or the feed channel of spiral-wound membranes. The effect of the AOC concentration on biological clogging of filter beds (biofouling) has been studied by (i) dosage experiments using acetate and (ii) observations of the effect of the AOC concentration *in situ* (Hijnen and van der Kooij 1992a). Dosage experiments revealed that acetate concentrations as low as 10 μg C/litre caused a pressure drop increase in experimental sand filters (H = 20 cm; D = 9 cm) at a flow rate of 1.2 m/h. Observations on water injection wells, used for infiltration of pre-treated river water into the dune area, also revealed that clogging occurred at AOC concentrations >10 μg ac-C eq/litre (Hijnen *et al.* 1992b). In spiral-wound NF membranes, acetate concentrations as low as a few μg ac-C/litre induced a pressure drop increase in the feed channel and from these observations a threshold value of 1 μg ac-C/litre was derived (Hijnen *et al.* 2009). Furthermore, a significant correlation was observed between the biomass concentration in the feed channel, measured as carbohydrate-carbon (CHC) and ATP, respectively (Hijnen *et al.* 2011). A pressure-drop increase of 100% corresponded with an active biomass concentration of 3.7 ng ATP/cm^2 and a total carbohydrate concentration of 8.1 μg CHC/cm^2 in the feed channel. Hence, a concentration of a few μg of C/litre of an easily biodegradable organic compound, for example acetate, can cause biofouling leading to clogging problems with granular materials (e.g. filter beds) and spiral-wound membranes.

11.3.2 Biofilm formation rate (BFR)

AOC concentrations in groundwater supplies in the Netherlands generally are <10 μg C/litre (Figure 11.6), but in a number of these supplies regrowth of

Aeromonas has been observed (Havelaar *et al.* 1990). Most microorganisms in the distribution system grow in biofilms and sediments. Biofouling of the distribution system most likely is related with the biofilm formation potential of treated water. Therefore, the biofilm formation characteristics of treated water were included in the assessment of the microbial growth potential. For this purpose, the biofilm monitor (BFM) was developed (van der Kooij *et al.* 1992; van der Kooij *et al.* 1995a). A BFM consists of a glass column (i.d. of 2.5 cm) with a length of 60 cm, containing cylinders ('rings') of borosilicate glass each with an external surface of 17.5 cm^2, on top of each other. Water is supplied in top down mode to the column at a flow of 275 litre/h, to achieve a flow rate (empty column) of 0.2 m/sec, corresponding with an average residence time of less than 3 seconds. The attached-biomass concentration on the cylinders is determined with ATP analysis. For this purpose, cylinders are periodically collected in duplicate from the column and the attached biomass is removed by using ultrasonic treatment. Subsequently, the active-biomass concentration of the obtained suspension is determined with ATP analysis and the biofilm concentration is calculated. Also TDC, HPC, community analysis using molecular techniques and chemical analysis (e.g. iron, manganese) can be used for biofilm characterization. The biofilm formation rate (BFR, pg ATP/($cm^2 \cdot d$)) is obtained by calculating the linear slope of the increase of the ATP concentration vs. time. Examples of biofilm formation are presented in Figure 11.12, which shows that the maximum level of biofilm formation (biofilm formation potential, BFP) of treated water is typically reached after about 100 days. For the calibration of the BFM, a low concentration of acetate or benzoate was added to the water (Figure 11.12c). A linear relationship was observed between BFR and the applied concentration of acetate-C, ranging from 1 to 100 μg C/litre, with a slope of 35 pg ATP/($cm^2 \cdot d$) per μg ac-C/litre (van der Kooij *et al.* 1995b). With this conversion factor, also BFR values can be expressed as ac-C eq/litre.

The BFM has been applied at a large number of water supplies, including water in different treatment stages, finished water and water in distribution systems (van der Kooij *et al.* 1995c; Saftic *et al.* 1997; van der Kooij *et al.* 1997). BFR values of treated water in the Netherlands generally range from values of < 1 pg ATP/($cm^2 \cdot d$) to about 100 pg ATP/($cm^2 \cdot d$) (Figure 11.13a), with a median of 7.4 pg ATP/($cm^2 \cdot d$). BFRs of < 1 pg ATP/($cm^2 \cdot d$) were observed with drinking water prepared from aerobic groundwater and in slow sand filtrates. BFRs of >35 pg ATP/($cm^2 \cdot$ day) were observed in a few supplies with drinking water prepared from anaerobic groundwater and in treated surface water after ClO_2 dosage. BFRs of finished water of less than 35 pg ATP/($cm^2 \cdot d$) correspond with less than 1 μg of ac-C eq/litre. This concentration is clearly below the AOC concentration in treated water. Furthermore, no correlation was observed between BFR and AOC (van der Kooij *et al.* 1995c). The discrepancy between the AOC concentration (μg ac-C eq/litre) derived from the BFR and the AOC concentration measured in the batch test, indicates that compounds which are as easily biodegradable as acetate generally constitute only a small fraction of the AOC concentration of finished water. Hence, uptake and utilization in the biofilm of most of the biodegradable compounds contributing to the AOC concentration is much slower than for easily biodegradable low-molecular-weight compounds such as acetate or benzoate.

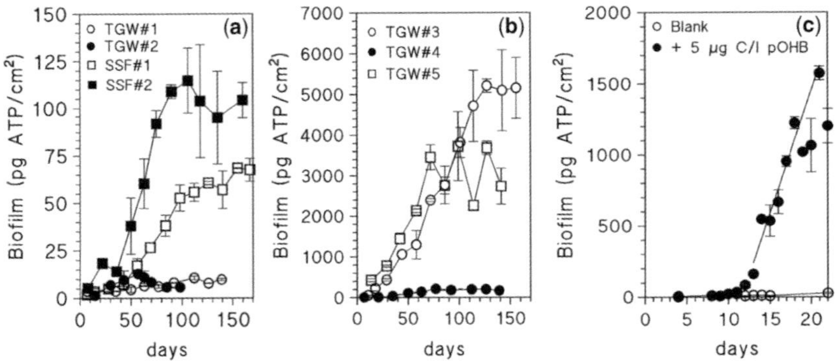

Figure 11.12 Biofilm formation in the biofilm monitor (BFM) supplied with treated water; (a) TGW#1 and TGW#2, drinking water prepared from aerobic groundwater; SSF#1, SSF#2, drinking water produced from surface water with slow sand filtrate as final treatment process; (b) drinking water prepared from anaerobic groundwater; (c) drinking water supplemented with 5 µg of 4-hydroxybenzoate-C/litre (BFR = 173 ± 11 pg ATP/(cm^2 · d); $R^2 = 0.95$). The BFR of the blank (drinking water) was 1.5 ± 0.3 pg ATP/(cm^2 · d).

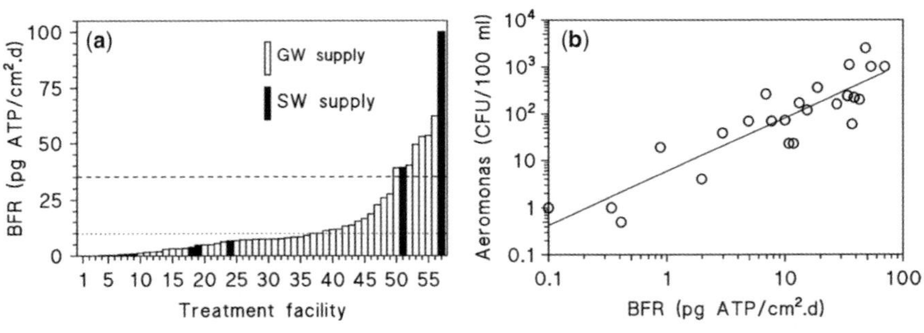

Figure 11.13 (a) Biofilm formation rate (BFR) of drinking water at 57 water treatment plants (c. 25% of total number). Period of investigation: 1992–2008. Open bars: treated groundwater (GW); black bars: treated surface water (SW). The broken line indicates the BFR corresponding with 1 µg acetate-C/litre and the dotted line indicates BFR = 10 pg ATP/(cm^2 · d) (b) relationship between BFR and the P90 values (one year) of *Aeromonas* in groundwater supplies. The relationship is given by: log CFU (*Aeromonas*) = 1.14 log BFR + 0.76 ($R^2 = 0.80$; $p < 0.001$). Reprinted from Journal Am. Wat. Works Assoc. **91**(1), 55–64, by permission. Copyright © 1999 the American Water Works Association.

BFRs of finished water of a number of drinking water types have been measured to investigate the relationship of the parameter with regrowth of aeromonads in the connected distribution system. A significant relationship was observed between the 90 percentiles (P90) of the colony counts of *Aeromonas* in groundwater supplies and BFR (Figure 11.13b). A guideline value for BFR of 10 pg ATP/($cm^2 \cdot d$) was derived from this relationship (van der Kooij *et al.* 1997). This value corresponds with a 20% risk of exceeding a P90 value of 200 CFU/100 ml.

At three groundwater supplies the biofilm concentration on the inner surface of PVC-U pipes (Ø 110 mm) in the distribution system was determined by using ATP analysis (van der Kooij *et al.* 1995c). The concentrations ranged from less than 100 pg ATP/cm^2 in a water supply distributing drinking water prepared from aerobic groundwater (TOC $<$ 0.5 mg/litre) with a BFR of $<$ 1 pg ATP/($cm^2 \cdot d$) to about 6000 pg ATP/cm^2 in a groundwater supply distributing drinking water with a high TOC concentration (8 mg/litre) and a BFR of $\sim$ 50 pg ATP/($cm^2 \cdot d$). Hence, the level of biofilm formation of finished water in the biofilm monitor corresponds with the biofilm concentration on the water-exposed surfaces of the PVC-U pipes in these supplies.

11.3.3 Biomass production potential (BPP) of materials in contact with drinking water

Materials in contact with treated water can enhance biofilm formation by releasing biodegradable compounds into the water. Test methods for determining the enhancement of microbial growth have been developed and standardized in the UK (British Standards Institute [BSI] 2000) and in Germany (Deutscher Verein des Gas- und Wasserfaches [DVGW] 2007). In the Netherlands, a batch-test method was developed based on measuring microbial growth on pieces of material incubated in drinking water in a batch test (van der Kooij and Veenendaal 1993, 2001). The amount of biomass attached to the surface of the exposed material is measured using ATP analysis, which enables the quantification of active biomass per unit of surface area at concentrations as low as 10 pg ATP/cm^2. Initially, samples of materials were incubated at 25°C in drinking water (600 ml) without water replacement and a surface (S) to volume (V) ratio of 0.167 cm^2/cm^3. Material samples were collected in duplicate from the flasks after 8, 12 and 16 weeks of incubation (van der Kooij *et al.* 2001). Directly after sampling, the water volume was reduced to maintain the S/V value at the initial level.

PVC-U is the main material in distribution systems in the Netherlands (Figure 11.2). As mentioned above, drinking water in the Netherlands is distributed without a disinfectant residual and HPC values comply with the regulated water quality standard. Therefore, it was anticipated that the BFP value of PVC-U would be very low. Incubation of PVC-U in the BPP test, both without and with water replacement, confirmed that the biofilm formation potential of the material was low (<100 pg ATP/cm^2), although growth was clearly enhanced directly after incubation (Figure 11.14a). This phenomenon is attributed to the presence of easily-biodegradable compounds on the surface of the new material, which obviously were not removed with the standard cleaning procedure of washing with water only. Typical results of BPP tests with other materials, most of which with clearly higher BPP values than PVC-U, are shown in Figure 11.14b, c. The BPP method, with weekly water replacement, together with the Mean Dissolved Oxygen Difference (MDOD) method and the W270 method, is included in the draft European Committee for Standardization (CEN) Standard "Enhancement of microbial growth (EMG)" (CEN/TC164 2012). Details of these methods are described in Chapter 12.

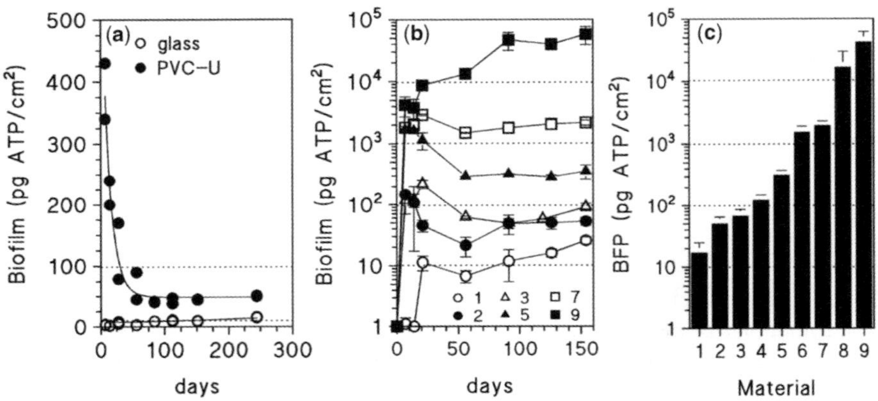

Figure 11.14 (a) Biofilm concentration on PVC-U and glass in the BPP test incubated at 25°C without water replacement; (b) Biofilm concentration on materials incubated at 37°C in tap water with weekly water replacement; (c) Biofilm Formation Potential (pg ATP/cm^2) of the materials; 1 Glass (negative control), 2 Stainless steel, 3 Chlorinated polyvinyl chloride PVC-C, 4 Copper, 5 Silicone rubber, 6 Polyethylene PE80, 7 Cross-linked polyethylene PE-Xa, 8 Plasticized PVC, 9 Natural rubber. Products tested may not represent products of similar material.

Biofilm formation on materials in contact with water may promote multiplication of *Legionella* (Schoenen *et al.* 1988; Rogers *et al.* 1994; van der Kooij *et al.* 2002). The effect of biofilm formation on PVC-U on the growth of *L. pneumophila* was assessed in the BPP test at 30°C, without water replacement. The test water was inoculated with *L. pneumophila* and associated microbiota from a tap-water installation. Within a few days, a maximum biofilm concentration of about 2×10^3 pg ATP/cm^2 was observed on PVC-U and *L. pneumophila* attained a level of about 2×10^5 CFU/cm^2 after 10 days (Figure 11.15a). Cleaning of PVC-U with a detergent, prior to the standard procedure of washing with water, reduced the biofilm formation by 90% and the growth of *L. pneumophila* by 99%.

Multiplication of *L. pneumophila* was also limited on glass, with a biofilm concentration below 100 pg ATP/cm^2 (Figure 11.15b).

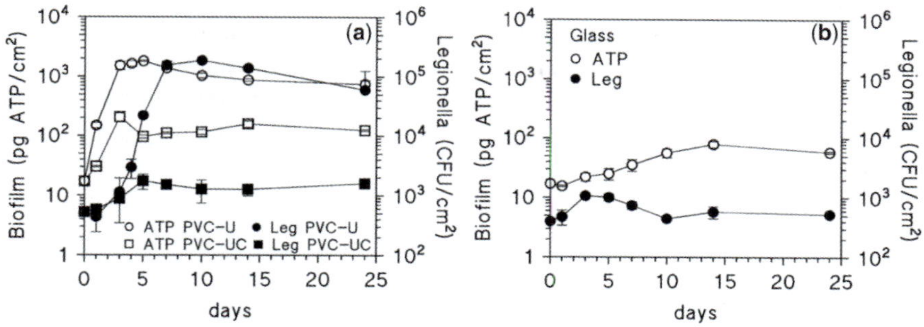

Figure 11.15 Biofilm formation and growth of *Legionella* in the biofilm on PVC-U and glass incubated in tap water at 30°C without water replacement; (a) PVC-U is washed with water (standard procedure); PVC-UC is PVC-U cleaned with detergent; (b) glass (washed with water).

11.4 IMPROVED METHODS FOR ASSESSING THE MICROBIAL GROWTH POTENTIAL OF WATER

The limitations of the AOC test with two pure cultures, the long duration of the test with the BFM and growth of *Legionella* in relation to the biofilm concentration initiated studies for further improvement of methods for determining the biological (in)stability of drinking water. The following additional methods have been developed:

- the biomass production potential (BPP) test of (drinking) water;
- the continuous biofouling monitor (CBM);
- the boiler-biofilm monitor (BBM).

The principle and significance of these methods are briefly described below.

11.4.1 Biomass production potential (BPP) of water

The BPP test of water is based on collecting water samples (600 ml) in thoroughly cleaned glass-stoppered Erlenmeyer flasks of 1 litre (AOC procedure), followed by incubation, without any treatment, at 25°C in the dark without shaking. Also, 1 ml of membrane-filtered (1.2 µm) river water is added to ensure the presence of a large diversity of bacteria. The ATP concentration in the water (in duplicate flasks) is measured periodically during an incubation period of 14 days or longer. This BPP test is a modification of the AOC test described by Stanfield and Jago (1987).

Typical results obtained with the BPP test with different types of drinking water (Figure 11.16a) show that: (i) the ATP concentration in treated water during incubation in most cases did not differ much from the initial value and (ii) the ATP level differs between water types. The lowest level (<1 ng ATP/litre) was observed with drinking water produced from aerobic groundwater with a low concentration of TOC (<0.5 mg/litre) and AOC (1.4 ± 0.6 μg ac-C eq/litre). A level of >10 ng ATP/litre was attained in treated anaerobic groundwater with a TOC concentration of 8 mg/litre and an AOC concentration of 9.1 ± 0.6 μg ac-C eq/litre.

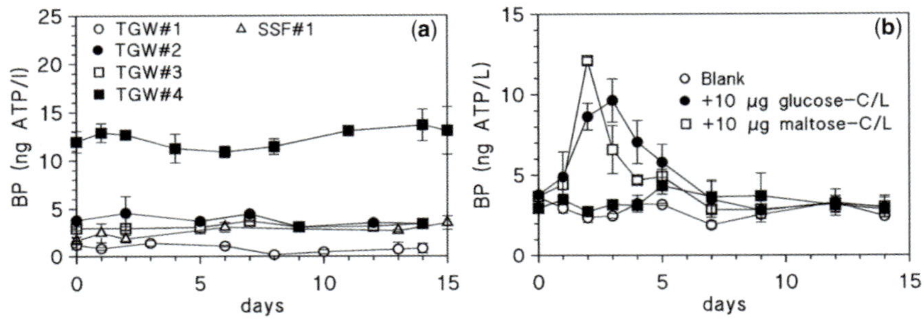

Figure 11.16 ATP concentrations in drinking water incubated in duplicate Erlenmeyer flasks at 25°C without sample treatment and inoculated with 1 ml of membrane-filtered river water. BP, biomass production; (a) TGW, treated groundwater from different treatment plants; SSF, slow-sand filtrate; (b) treated groundwater supplemented with substrates.

Addition of a low concentration of an easily biodegradable compound resulted in a clear increase of the ATP concentration within a few days (Figure 11.16b). The observed maximum ATP concentrations are consistent with a yield of 1 ng ATP/μg of C derived from the linear relationship between the maximum ATP concentration and the concentration of added acetate-C (van der Kooij 1999; van der Kooij *et al.* 2010). The observations presented in Figure 11.16 demonstrate that the maximum ATP concentration attained within 7 days and the ATP level maintained during prolonged incubation are the relevant characteristics of the growth potential of (treated) water in the BPP test.

11.4.2 Continuous biofouling monitor (CBM)

Observations with the BFM revealed clear differences between water types (Figure 11.12). However, the long operational period and the relatively large water volume needed limited the application of the BFM in practice and in research. Therefore, a monitor system was developed to enable a more rapid assessment of the BFR, with a lower water demand and on a continuous basis.

The principle of the CBM is based on using four glass cylinders in parallel, containing glass beads (2 mm). Each two weeks two of these cylinders are removed, the amount of biomass (ATP) on the glass beads is analyzed and the Biomass Accumulation Rate (BAR) (pg ATP/($cm^2 \cdot d$)) is calculated.

Figure 11.17 shows that the BAR value obtained at groundwater treatment plant A (TGW-A) was close to the BFR value obtained with the BFM, but the BAR at TGW-B initially was clearly higher than the BFR and showed more variation. The difference is attributed to the accumulation of particulate biomass and iron in the CBM at TGW-B. These observations demonstrate that the CBM gives additional information as compared to the BFM. In combination with the limited water usage and its continuous use, the CBM is attractive for monitoring of (partly) treated water as well as drinking water in the distribution system. Analysis of other microbiological and chemical parameters, for example, iron, manganese, enables the detection of variations in water quality on a semi-continuous basis. The utility of this approach will increase with further extension of the database.

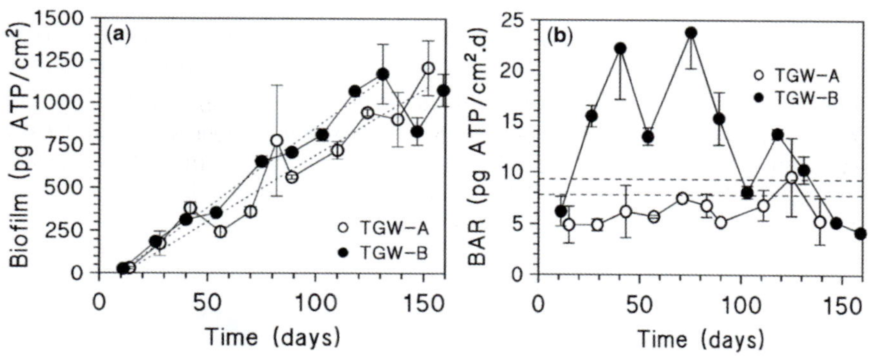

Figure 11.17 Biofilm formation characteristics of finished water of two groundwater supplies (TGW-A and TGW-B) in the biofilm monitor (a) and in the continuous biofouling monitor (b). Biofilm formation rate (BFR) of TGW-A = 7.8 ± 0.65 pg ATP/($cm^2 \cdot d$); BFR of TGW-B = 9.4 ± 0.4 pg ATP/($cm^2 \cdot d$). These BFR values are indicated by the broken lines in Figure 11.17b.

11.4.3. Boiler-biofilm monitor (BBM)

To elucidate the effects of drinking water composition on the multiplication of *L. pneumophila*, a boiler-biofilm monitor (BBM) was developed. In this system, water is heated to 70°C in an electric boiler and mixed with cold water to obtain a temperature of 38 ± 1°C. This warm water was supplied to a biofilm monitor (with glass cylinders) at a discontinuous flow, that is, each 20 minutes a volume of two litres of warm water was supplied to the glass column. The column was placed in a closed box in which the warm-water temperature was maintained. After one to two weeks of operation for enabling the establishment of a biofilm, the system was inoculated with *L. pneumophila* cultured with associated microbiota on pieces of silicone tubing as reported elsewhere (Kuiper *et al.* 2004).

Figure 11.18 shows that a biofilm concentration of about 500 pg ATP/cm^2 (Figure 11.18a) promoted the growth of *L. pneumophila* to a level of $\sim 10^3$ CFU/cm^2. The Legionella Growth Potential (LGP) of this water, which is defined as the 90 percentile of the colony counts of *Legionella* for the period in which growth of the organism is observed, was 10^3 CFU/cm^2. Similarly, the biofilm formation potential (BFP, 90 percentile of biofilm concentrations) was 600 pg ATP/cm^2. Water type B gave a BFP of 2625 pg ATP/cm^2 in the BBM, with concentrations of *L. pneumophila* exceeding 10^5 CFU/cm^2 (LGP 1.4×10^5 CFU/cm^2). This LGP in relation to the biofilm concentration is consistent with the level of growth of *L. pneumophila* on PVC-U in a batch test (Figure 11.15). The colony counts of *L. pneumophila* showed more variation than the biofilm concentration (Figure 11.18). The changes in these colony counts may be attributed to changes in the biofilm composition, affecting protozoan activity. These observations indicate that the BBM, which simulates a worst case situation in a potable-water installation, can be used to (i) differentiate between water types on the basis of BFP and LGP, (ii) determine the relationship between the biofilm concentration and growth of *L. pneumophila* in the biofilm. The BBM can also be used to determine the effect of the composition of the microbial biofilm community on the growth of *L. pneumophila.*

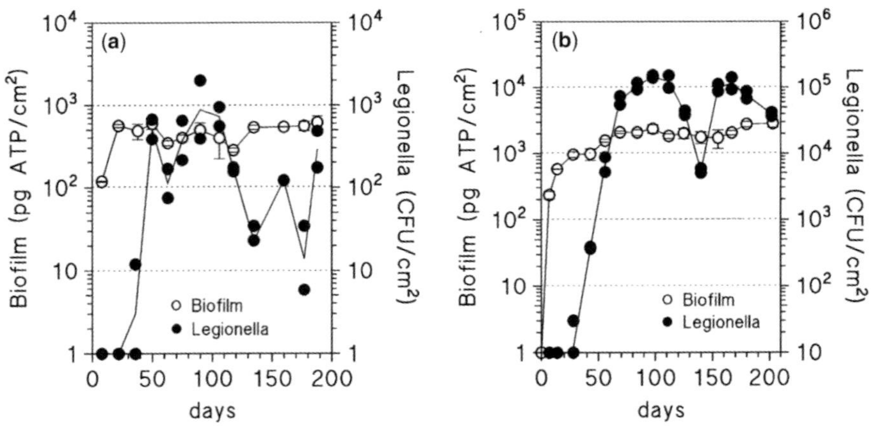

Figure 11.18 Biofilm formation and multiplication of *L. pneumophila* in the biofilm in the boiler-biofilm monitor (BBM) supplied with two types of drinking water; (a) drinking water (DOC = 0.9 mg/litre) prepared from river water by dune passage and post treatment including slow sand filtration; (b) drinking water (DOC = 3.2 mg/litre) prepared from seepage water (cf. Figure 11.1).

11.5 EVALUATION

11.5.1 Regrowth assessment

The concentrations of ATP and TDC and the presence *of Aeromonas*, *Pseudomonas*, *Legionella,* nitrifying bacteria and archaea, fungi, FLP and invertebrates in drinking water, biofilms and sediments in distribution systems in the Netherlands demonstrate that (micro)biological growth occurs in these environments, despite compliance with the regulated standard for HPC22 (geometric mean < 100 CFU/ml). A survey of HPC22 data, collected in the period 2004 to 2007, revealed that the geometric mean of HPC22 was <10 CFU/ml in 90% of the supplies, that is, far below the regulatory standard (van der Wielen and van der Kooij 2011). HPC22, and also improved cultivation methods, for example, R2A agar, only represent a small fraction of the concentration of the bacteria present in drinking water (Figure 11.4) and therefore are not suited to monitor the level of regrowth or microbiological activity in distribution systems.

Elevated colony counts of *Aeromonas spp.* in distribution systems are observed in summer (LeChevallier *et al.* 1982; Burke *et al.* 1984; Havelaar *et al.* 1990), despite the ability of these organisms to multiply at low temperatures (Palumbo *et al.* 1985; Holmes *et al.* 1996). A strong seasonal effect was also observed in lake water, with the highest *Aeromonas* counts at a water temperature above 16° C (Rippey and Cabelli 1980). These counts correlated with the trophic state (amount of biomass) of the water, indicating that growth of *Aeromonas* is biomass-associated. Such an association is also indicated by the relationship between *Aeromonas* counts and the biofilm formation rate of treated water (Figure 11.13), a measure of the biofouling potential of the distribution system (van der Kooij *et al.* 1995c).

Members of the genus *Aeromonas* are specialized in the utilization of amino acids at the microgram per litre level (van der Kooij and Hijnen 1988). These compounds, as well as ammonia and phosphate, are excreted by FLP grazing on bacteria in biofilms and sediments (Section 11.2.3). The digestion rate of flagellates and ciliates increases 2 to 2.8 times at a temperature increase of 10°C (Sherr *et al.* 1988). Therefore, it is hypothesized that the increased growth of aeromonads in distribution systems in summer is associated with the activity of FLP grazing on biofilms and sediments, that is, the result of the effect of temperature on both *Aeromonas* and FLP. These FLP do not serve as host for *Aeromonas,* as is reported for *Legionella* (Section 11.2.4). Establishment of a relationship between *Aeromonas,* FLP and biomass would increase the significance of *Aeromonas* as regrowth parameter.

Multiplication of coliforms in distribution systems is not observed in water supplies in the Netherlands with <0.01% positive samples. These organisms occur more frequently in supplies in the US (Smith *et al.* 1990; LeChevallier *et al.* 1996). Mono- and oligo-saccharides are the preferred substrates for coliforms, but these organisms require much higher concentrations for growth on these compounds than *Aeromonas* (van der Kooij 1997). Elevated concentrations of coliforms have been observed on corroding surfaces (LeChevallier *et al.* 1987; Camper *et al.* 1996; LeChevallier *et al.* 1996) and in sediments (Zacheus *et al.* 2001). Biodegradation of accumulated organic compounds at a low oxygen concentration might enhance coliform growth, but quantitative data about these factors do not seem available.

As is shown for aeromonads and coliforms, explaining the growth of specific types of heterotrophic microorganisms in distribution systems is very complicated, even with information about the physiological properties of the organisms. Complicating factors include: (i) the large diversity of physico-chemical conditions (e.g. oxygen, pH, temperature, disinfectant residual, corrosion products, nutrients and hydraulics) in microenvironments (microniches), (ii) differences in properties between genera and within species affecting the behaviour in the aquatic environment, and (iii) interactions with other microorganisms, including FLP. The existence of microniches is demonstrated by the large diversities of microorganisms in drinking water, biofilms and sediments (e.g. Martiny *et al.* 2003; Hong *et al.* 2010; Henne *et al.* 2012), almost all with unknown physiological characteristics. Consequently, regrowth studies should focus on selected microorganisms, for example, opportunistic pathogens and microorganisms with specific physiological properties, for example, nitrifying bacteria and archaea, iron-oxidizing bacteria, amoebae, for example, *H. vermiformis* and ciliates, as indicators for biofilms and sediments, respectively, and so on. Cultivation of most of these organisms is not possible or difficult and therefore, further development and standardization of culture-independent methods, for example qPCR, is needed.

Quantification of the total concentration of microorganisms in water is a more practical approach. The range of ATP concentrations in finished water (Figure 11.4) confirms the utility of this parameter for monitoring the concentration of active microbial biomass in water supplies, as has been shown earlier (van der Kooij 1992; Deiniger and Lee 2001; Delahaye *et al.* 2003; van der Kooij 2003; van der Wielen and van der Kooij 2010). Observations of periodically elevated ATP concentration in finished water and/or in water during distribution in comparison with the database for this parameter, may initiate further measurements or control measures. As compared to ATP, the use of TDC seems to have less potential as a parameter for active microbial biomass, because of differences in cell size and viability/activity. However, TDCs in combination with ATP give information about the level of activity of the bacteria present and can be used to elucidate microbial processes in research projects (Magic-Knezev and van der Kooij 2004).

11.5.2 Biostability assessment

The microbial growth potential of water is defined by the concentration of organic and inorganic energy sources available for the production of microbial biomass.

Prediction of the level of microbial growth in drinking water distribution systems and potable-water installations, is difficult because of the (micro)biological interactions and the variety of environmental conditions affecting regrowth processes. Most of these conditions, for example, temperature, residence time and hydraulics, cannot be changed easily in existing distribution systems. Therefore, research aiming at defining biostability was mainly focused on establishing relationships between regrowth phenomena, as described above, and the microbial growth potential of finished water.

A large variety and still increasing number of methods for quantifying the microbial growth potential of finished water is available. A distinction can be made between static (batch-test) methods and dynamic (flow-through) methods, but also slowly degraded colloidal organic compounds, for example, biomass components and (locally) accumulating in sediments, can enhance regrowth. Static methods are based on the incubation of a representative sample of the involved water under defined conditions, followed by either measuring microbial growth (AOC-related tests) or the decrease of the concentration of dissolved organic carbon (Biodegradable Dissolved Organic Carbon [BDOC] tests). Modifications of the original AOC test (see Section 11.3.1) include: (i) ATP analysis for determining the growth of the test strains (LeChevallier *et al.* 1993) and (ii) the use of bioluminescent strains of P17 and NOX (Weinreich *et al.* 2009).

The AOC concentrations of finished water in the Netherlands (Figure 11.6) are about 10 times lower than those observed in the US (LeChevallier *et al.* 1996; Volk and LeChevallier 2000). The difference is consistent with the difference in coliform-positive samples between the Netherlands and the US (see above). The low AOC concentrations in the Netherlands may be attributed to the extended water treatment (Figure 11.1), for example, (i) the application of soil passage, including dune passage and river-bank filtration, (ii) one or more stages of biological filtration, (iii) absence of a disinfectant residual in treated water and to differences between the applied AOC methods as well. In Finland, AOC concentrations in finished water at most investigated supplies were above 100 μg ac-C eq/litre, at DOC concentrations of about 3 mg/litre (Miettinen *et al.* 1999). Treatment generally consisted of coagulation/sedimentation with or without sand filtration, and post disinfection.

Measuring the growth of the indigenous microbial population in the static test is an alternative for the AOC method. This approach was already used about a century ago, but unreliable results were obtained due to insufficiently cleaned and closed containers and the use of plate counts (Heymann 1928). With ATP analysis all active bacteria in growth tests can be quantified (Stanfield and Jago 1987). Also TDC measurements are used to quantify the growth of the indigenous population to assess the AOC concentration (Hammes and Egli 2005). The BPP test which uses ATP as the parameter for active biomass, appears to have the potential of a simple, rapid, quantitative and sensitive test. Figure 11.16 shows that information about the growth potential due to easily biodegradable compounds is obtained within a few days. Furthermore, the ATP concentration that is maintained during continuing incubation gives information about the availability of slowly-degraded compounds. No reduction of the DOC concentration was observed in the BPP tests with drinking water. Obviously, the reduction was less than the difference ($\pm$ s.d.) between the DOC concentrations measured before and after incubation.

The BDOC concentration in finished water is always higher than the AOC concentration (Joret and Lévi 1986; Volk and LeChevallier 2000). The difference is caused by the use of the indigenous microbial community in the BDOC test, either attached to sand (100 g in 300 ml of water sample) from a biological filter bed in intensive contact with the water (by aeration) (Joret and Lévi 1986), or present in suspension in combination with an incubation period of 30 days (Servais *et al.* 1987). Uptake of BDOC during distribution generally is rather slow (Matthieu *et al.* 1995; Capellier *et al.* 1992), but increases in pipes with a small diameter at a low disinfectant residual and an elevated temperature. BDOC levels of 0.43–0.61 mg C/litre corresponded with biofilm concentrations of 0.48–0.65 μg C/cm^2, that is, 2.0–2.6 $\times$ 10^7 bacteria/cm^2 in a surface-water supply (Servais *et al.* 1992). The biofilm concentration on surfaces exposed to treated groundwater, with BDOC ranging from 0.23-0.47 mg C/litre, and containing 0.08 mg Cl_2/litre, was less than 0.017 mg C/litre (Ninquette *et al.* 2001). However, the BDOC value of treated water is not equivalent to the microbial growth potential, because the bacterial growth efficiency for NOM may differ between water types (0.4–20%) (Tranvik *et al.* 1987; Eiler *et al.* 2003).

Dynamic testing for assessing the microbial growth potential of (finished) water can be applied in-line in a semi-continuous mode. An in-line test has been developed for measuring the BDOC concentration of treated water after passage of a column containing sintered-glass beads with an attached and adapted microbial community (Lucena *et al.* 1990; Ribas *et al.* 1991). This approach resembles the use of small sand filters for determining the removal of NOM (Heymann 1928). The long time (up to six months) needed for adaptation of the column with sintered-glass beads (Kaplan and Newbold 1995), the impact of a disinfectant and difficulties in measuring a small decrease of the DOC concentration may have hampered its general application. Quantification of the concentration of attached biomass on the exposed surface with ATP analysis enables the assessment of the biofilm formation characteristics of water types, even at a low level of biodegradable compounds. With the biofilm monitor, a wide range of BRF values (Figure 11.13) has been observed in finished water in the Netherlands, which corresponds with the biofilm concentration in PVC-U pipes in the distribution system (van der Kooij *et al.* 1995c). The BFR value therefore may represent the biofouling potential of finished water related to attached microbial growth, because biofilm sloughing leads to sediment accumulation. Fluctuations in time in the accumulation rate of biomass, and also of iron and manganese, can be detected with the in-line operating CBM, with glass-beads containing cylinders resembling those used by Halam *et al.* (2001). The CBM, which provides quantitative information each 14 days or more frequently if desired, facilitates the elucidation of the effects of (operational and seasonal) changes in water treatment as well as the effects of distribution. Such data will be helpful for defining measures to improve biostability.

The observations on ATP and TDC in finished water revealed that the concentration of bacterial biomass in water entering the distribution system ranges over about two orders of magnitude (Figure 11.4). Finished water also contains living and dead biomass of algae (surface-water supplies), free-living protozoa, fungi and invertebrates (Valster *et al.* 2009; van Lieverloo *et al.* 2012). Slowly-degraded biomass (components), accumulating locally in the distribution system, for example, as a result of coagulation in the presence of iron, and sedimentation depending on hydraulics and retention time, may contribute to the formation of regrowth-stimulating deposits (Gauthier *et al.* 1999; Zacheus *et al.* 2001). However, it is yet unclear which fraction of the biomass (components) present in finished water is retained within the distribution system.

11.5.3 Defining biostability

11.5.3.1 AOC, BDOC and BFR

Distribution of water at AOC concentrations greater than 40 μg of ac-C eq/litre yielded elevated HPCs (van der Kooij *et al.* 1989), but HPC22 complies with legislation at AOC concentrations below 10 μg of ac-C eq/litre. AOC reduction during distribution is very limited at this concentration and no decrease of the TOC concentration was observed (van der Kooij 1992). BDOC concentrations of about 0.15 to 0.25 mg C/litre did not (further) decline during distribution (Servais *et al.* 1992; Ninquette *et al.* 2001). This value is clearly higher than the AOC value of 10 μg ac-C eq/litre, but the bacterial growth yield on NOM constituents may resemble the yield for oxalate, which is about four times lower than for acetate-C (Table 11.3). Taking also into account the differences in methodology between AOC and BDOC tests (see previous section), an AOC level of 10 μg of ac-C eq/litre may not be much more stringent than a BDOC of 0.15 mg C/litre. Furthermore, elevated AOC levels, as observed in the US showed less difference with BDOC concentrations (Volk and LeChevallier 2000). However, data for comparing AOC and BDOC at low levels are lacking, partly because of the difficulty in measuring low BDOC concentrations.

Figure 11.8 shows that AOC concentrations <10 μg of ac-C eq/litre further decline during incubation of a water sample for 7 days. Utilization of 1 μg of C/litre of an easily-biodegradable compound can result in an HPC increase of 3×10^3 CFU/ml to 2×10^4 CFU/ml, depending on the nature of the compounds and the cell size of the involved microorganisms (Table 11.3). These values exceed the standard for HPC22 of 100 CFU/ml (Table 11.1). In distribution systems however, most substrates are utilized by the bacteria present in biofilms and sediments, most of which do not contribute to HPC22. Endogenous metabolism in these environments as well as grazing of protozoa reduces the bacterial growth yield. Furthermore, the AOC-reduction test showed that the AOC uptake is slow at concentrations <10 μg of ac-C eq/litre (Figures 11.7 and 11.8), which is consistent with the low BFR values of finished water (Figure 11.13). Consequently, at low AOC concentrations, the utilization of biodegradable compounds per km of pipe length is low and AOC removal remains limited within the residence time of the water in the distribution system (van der Kooij 1999).

As is shown with acetate (van der Kooij *et al.* 1995b) and *p*-OH-benzoate (Figure 11.12), a low concentration of an easily biodegradable compound gives a much higher BFR than treated water with an equivalent AOC concentration. Therefore, a certain BFR induced by rapidly utilized organic compounds, will decrease rapidly during distribution. A similar BFR caused by compounds with less favourable uptake kinetics, that is, a low rate of uptake at a relatively high concentration, will decrease more slowly and therefore produce more biomass in the distribution system. The utility of BFR for defining biostability is supported by the relationship between BFR and *Aeromonas* (Figure 11.13) and between BFR and the amount of invertebrate biomass in water flushed from mains (van Lieverloo *et al.* 2012). The use of the CBM for continuously measuring the biomass accumulation rate (BAR) offers the advantage of observing water quality changes.

The presence of both rapidly and slowly degraded compounds in finished water is clearly demonstrated in the BPP test (Figure 11.16). The database of ATP concentrations in finished water indicates that a concentration below a few ng ATP/litre in the BPP test is indicative for a high level of biostability. Hence, biostability assessment should be based on measuring the microbial growth potential (AOC, BPP, BDOC) in a batch test, as well as the rate of utilization in a dynamic test. The effect of elevated concentrations of biomass (components) in finished water on regrowth is not yet resolved, but this parameter may represent another dimension of the microbial growth potential.

11.5.3.2 Biological-stability classification

An AOC concentration of about 6 μg ac-C eq/litre induced biofilm formation (2–3 × 10^6 bacteria/cm^2) on stainless steel in a model distribution system (Boe-Hansen *et al.* 2003). In a model plumbing system, supplied with drinking water (36°C) with a similar AOC concentration, growth of *L. pneumophila* was observed (van der Kooij *et al.* 2005). Furthermore, fungi and non-tuberculous mycobacteria are present in distribution systems at a low AOC concentration (van der Wielen and van der Kooij 2013). Clearly, the definition of biostability becomes more complicated when the presence (or compliance to standards) of specific microorganisms is taken into account. Therefore the question arises: what level of biostability is needed to prevent growth of undesirable microorganisms?

The highest level of biostability in the Netherlands has been observed in drinking water prepared from aerobic groundwater, that is, water from aquifers of sand or chalk, which represents about 15% of the drinking water produced in the Netherlands. This water is characterized by: DOC < 0.5 mg/litre, ATP < 1 ng/litre (Figure 11.4b), BPP < 1 ng ATP/litre; AOC < 1–2 μg ac-C eq/litre (Figure 11.6a); BFR ≪1 pg ATP/($cm^2 \cdot d$) (Figure 11.12a). Counts of *Aeromonas* and HPC22 in these supplies are several orders of magnitude below the regulatory standard. Also concentrations of uncultivated *Legionella* spp. and free-living protozoa were low in an aerobic groundwater supply, as compared to concentrations in a supply with elevated concentrations of DOC and ATP (Valster *et al.* 2009; Wullings *et al.* 2011). BFR < 1 pg ATP/($cm^2 \cdot d$) corresponds with a low biofilm concentration (<100 pg ATP/cm^2) in the distribution system (van der Kooij *et al.* 1995c; Valster *et al.* 2009). Growth of *L. pneumophila* was limited at a biofilm concentration of about 100 pg ATP/cm^2 on glass and PVC-U (Figure 11.15), which may be attributed to the inability of amoebae to proliferate at a biofilm concentration below a threshold level (Pickup *et al.* 2007). Using materials that do not enhance biofilm formation therefore would imply a BFP of less than 100 pg ATP/cm^2. An example of such a material is PVC-U, the major pipe material in the distribution systems in the Netherlands (Figure 11.2). Other materials which do not increase the ATP concentration in the BPP test include: PVC-C and stainless steel (Figure 11.14; van der Kooij and Veenendaal 2001).

Based on the observations of regrowth and the microbial growth potential in water supplies in the Netherlands, it is proposed to distinguish between the following biostability classes (BSC):

BSC I: water with the highest observed level of biostability, that is, ATP < 1 ng/litre (finished water and during distribution); AOC < 1 μg acetate-C eq/litre; BPP < 1 ng ATP/litre; BFR <1 pg ATP/($cm^2 \cdot d$), corresponding with a biofilm formation potential of <100 pg ATP/cm^2 and a biofilm concentration <100 pg ATP/cm^2 in distribution system pipes (PVC-U). The TOC concentration is <0.5 mg/litre. This water complies with the standards for HPC22, *Aeromonas* and coliforms, and has no or a low potential for growth of legionellae.

BSC II: water characterized by: ATP < 10 ng/litre (finished water and during distribution), AOC < 10 μg ac-C eq/litre; BPP < 10 ng ATP/litre; and BFR < 10 pg ATP/($cm^2 \cdot d$), corresponding with a biofilm formation potential of < 1000 pg ATP/cm^2. The TOC concentration is <5 mg/litre in most cases. This drinking water complies with the standards for HPC22, *Aeromonas* and coliforms, but has a distinct potential for growth of legionellae. Also fungi and nontuberculous mycobacteria have been observed in this type of drinking water (Chapter 7; van der Wielen and van der Kooij 2013).

The proposed description of the BSCs is incomplete because seasonal effects (temperature), residence time, hydraulic conditions and materials (e.g. corroding iron), complicate the impact of the growth potential of the water. Furthermore, data on biomass concentrations, total direct cell counts and BDOC values should be added for completion and to make the classification more generally applicable.

Finished water that periodically or continuously exceeds the concentrations of ATP, AOC, BPP or BFR indicated for BSC II, most likely will experience regrowth problems in the distribution system (e.g. elevated HPCs, *Aeromonas*, invertebrates, coliforms). These problems depend on the actual levels of the microbial growth potential, seasonal effects and distribution system conditions as well. Regrowth control may require maintenance of a disinfectant residual in the distribution system and/or improvement of water treatment to attain biostability.

11.5.3.3 ATP-based biostability assessment (ABBA)

As described above, the biological (in)stability of finished water can be assessed with a combination of methods, including both static (batch test) and dynamic (flow-through) tests. In the approach developed in the Netherlands, ATP is used for determining microbial growth in water and in biofilms in growth tests, as well as in samples collected from water supplies and plumbing systems. This approach, the ***ATP-based biostability assessment (ABBA)***, provides a framework that previously was designated as the Unified Biofilm Approach (van der Kooij *et al.* 2003). In the ABBA approach, which is schematically depicted in Figure 11.19, the use of ATP reaches further than biofilm analysis and is the key parameter for:

- the concentration of active microbial biomass in finished water, in drinking water in the distribution system (regrowth), in flush water and in sediments;
- the level of growth in the BPP test with drinking water;
- the biomass concentration in the BFM and the CBM for assessment of the biofilm formation rate of (finished) water;
- the biofilm concentration on construction materials in contact with drinking water in the BPP test;
- the concentration of attached, active microbial biomass (biofilm) in distribution system pipes, in plumbing systems and in experimental installations.

Applications of ATP analysis in water treatment include: (i) quantification of the amount of biomass in biological filters, for example GAC filters, rapid sand filters (Magic-Knezev *et al.* 2004), and (ii) assessment of the level of biofouling in spiral-wound membranes (Hijnen *et al.* 2009; Hijnen *et al.* 2011).

The definition of the biostability classes of treated water (without disinfectant residual), based on the ABBA approach, can be improved by extending the database of ATP concentrations in relation with (i) results of routine monitoring of microbial regrowth (coliforms, HPC, *Aeromonas*) and (ii) observations of the presence and concentration of undesirable (micro)organisms in water, biofilms and sediments. Establishment of a relationship between the biofilm concentration and the growth of *L. pneumophila* and other undesirable micro-organisms is useful for defining a biofilm-based standard for construction materials in contact with drinking water.

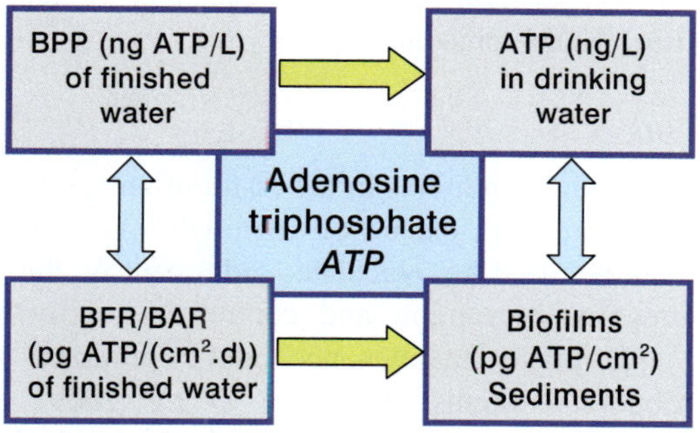

Figure 11.19 ATP-based biostability assessment (ABBA). Left: monitoring methods; right, distribution system.

11.5.4 Regrowth control

11.5.4.1 Treatment

The microbial growth potential of treated water depends on the composition of the raw water and the effects of water treatment processes on the concentration of growth-promoting compounds and biomass. In the Netherlands, many data of effects of treatment processes on AOC and BFR are available for this purpose (van der Kooij 1987; van der Hoek *et al.* 2000; Bonné *et al.* 2002; van der Veer *et al.* 2005). A high degree of biostability can be expected for aerobic groundwater, which is derived from soil without peat layers, for example, sand or chalk aquifers. Also membrane filtration (NF), or reverse osmosis (RO) for desalination, can lead to drinking water with a high degree of biostability. Biostability monitoring of these water types can be restricted to measuring ATP. Anaerobic groundwater may contain a considerable concentration of NOM and also ammonia, methane, iron and manganese. Aeration for oxygen supply and removal of CH_4 and CO_2, followed by rapid sand filtration to remove iron, manganese and ammonia, is used to achieve biostability (Figure 11.1), but slowly-degraded NOM is not removed. Coagulation/sedimentation is the first step in surface-water treatment and in a number of cases is followed by oxidation. In the Netherlands two stages of biological filtration are generally applied after ozonation to achieve biological stability. A systematic analysis of the data about the effects of individual treatment processes and combinations of processes can lead to the development of a model for the quantitative assessment of the biostability of finished water based on data about source water quality and treatment. Subsequently, biostability assessment can be restricted to monitoring of the ATP concentration and periodically measuring the Biomass Production Potential of finished water. However, even with extended biological treatment, achievement of Biostability Class 1 is not feasible in most water supplies, because of seasonal effects and/or a limited removal of slowly-degraded compounds. Physical methods, for example adsorption (GAC), ion-exchange or membrane filtration, are required for further NOM removal.

11.5.4.2 Distribution

A number of environmental conditions in the distribution system and in plumbing systems affect the nature and extent of regrowth. Application of the tools described above can be used to elucidate and quantify the effects of these conditions. Measures for prevention and control of regrowth in distribution systems include appropriate maintenance and optimized design, that is self-cleaning distribution systems (Vreeburg *et al.* 2009) and the use of construction materials that do not promote microbial growth. Premise-plumbing systems, which are beyond the jurisdiction of the water supplies, will benefit from improvement of the biostability of drinking water, but in most cases additional control measures are needed to prevent growth of undesirable microorganisms, that is *Legionella pneumophila* (Chapter 6).

11.6 SUMMARY AND CONCLUSIONS

11.6.1 Regrowth

- Drinking water in the Netherlands, with NOM concentration ranging from <0.5 to 6 mg C/litre, is distributed without a disinfectant residual. Growth of coliforms is not observed in the supplies, which all comply with the regulatory standard for HPC22 (yearly geometric mean < 100 CFU/ml).
- In samples collected from a few distribution systems *Aeromonas* counts exceed the standard of 1000 CFU/100 ml and counts of *Legionella* (identified as *L. anisa*) exceed the standard of 100 CFU/litre.
- In all water supplies microbial activity is observed with ATP analysis and total direct cell counts. Furthermore, large diversities of bacteria, including *Aeromonas*, fluorescent pseudomonads, uncultured *Legionella* spp., non-tuberculous mycobacteria, fungi and invertebrates have been observed in drinking water and biofilms.
- HPC22, in combination with the standard of 100 CFU/ml, is not suited for assessing the level of microbial growth in distribution systems.
- ATP is a suited parameter for monitoring the concentration of active microbial biomass and regrowth in drinking water supplies.

- The effect of regrowth on drinking water quality can be assessed with a selection of culture methods and qPCR techniques for the detection of undesirable microorganisms.
- *L. pneumophila* multiplies in warm water in premise-plumbing systems. The organism can proliferate at very low biofilm concentrations and may be the most critical parameter for biostability evaluation of water and construction materials.

11.6.2 Biostability assessment

- A combination of a batch test (BPP, AOC, BDOC), for determining the concentration of growth substrates, and a dynamic test (BFM, CBM) for determining the rate of uptake, is needed to assess the microbial growth potential of (finished) water.
- The ATP-based biostability assessment (ABBA) provides a framework for the evaluation of both the level of regrowth and the microbial growth potential of treated water and construction products in contact with drinking water.
- Further research is needed to (i) substantiate the ABBA approach, (ii) assess the effects of dead and living biomass entering the distribution system, (iii) assess the effects of environmental conditions in the distribution system on regrowth and (iv) improve the definition of biological stability (classes).

Acknowledgement

This review, which includes results of recent experimental work, has been prepared as part of the Joint Research Programme (Bedrijfstakonderzoek [BTO]) of the water supply companies in the Netherlands. The cooperation of the water supply companies in the described investigations is gratefully acknowledged.

11.7 REFERENCES

Amy G. L., Sierca R. A., Bedessen J., Price D. and Tan L. (1992). Molecular size distributions of dissolved organic matter. *J. Am. Wat. Works Assoc.*, **6**, 67–75.

Anderson A., Lee C., Azam F. and Hagstrom A. (1985). Release of amino acids and inorganic nutrients by heterotrophic marine flagellates. *Mar. Ecol. Prog. Ser.*, **23**, 99–106.

Bijkerk P., van Lier E. A., Swaan C. M. and Kretzschmar M. E. E. (2011). Staat van infectieziekten in Nederland 2010. Rapport 21022110072011 Rijksinstituut voor Volksgezondheid en Milieu Bilthoven, the Netherlands.

Boe-Hansen R., Albrechtsen H. J., Arvin E. and Jorgensen C. (2002). Bulk water phase and biofilm growth in drinking water at low nutrient conditions. *Wat. Res.*, **36**, 4477–4486.

Bonné P. A. C., Hofman J. A. M. H. and van der Hoek J. P. (2002). Long term capacity of biological activated carbon filtration for organics removal. *Wat. Sci. Technol. Wat. Supply*, **2**(1), 139–146.

BSI (2000). BS 6920 – Suitability of non-metallic products for use in contact with water intended for human consumption with regard to their effect on the quality of the water. Part 1 (Specification) and Part 2.4 (Growth of aquatic microorganisms test). British Standards Institute.

Burke V., Robinson J., Gracey M., Peterson D. and Partridge K. (1984). Isolation of *Aeromonas hydrophila* from a metropolitan water supply: seasonal correlation with clinical isolates. *Appl. Environ. Microbiol.*, **48**, 361–366.

Camper A. K., Jones W. L. and Hayes J. T. (1996). Effect of growth conditions and substratum composition on the persistence of coliforms in mixed-population biofilms. *Appl. Environ. Microbiol.*, **62**(11), 4014–4018.

Cappelier M., Pichoce C. and Deguin A. (1992). Devenir de CBOD dans les reseaux de distribution; etude de 2 cas. *Rev. Sci. de L'eaux*, **5**, 51–67.

CEN/TC164 (2012). Influence of materials on water for human consumption-Enhancement of microbial growth (EMG), prEN 16421.

Deiniger R. A. and Lee J. Y. (2001). Rapid determination of bacteria in drinking water using an ATP assay. *Field Anal. Chem. Technol.*, **5**, 185–189.

Delahaye E., Welte B., Levy Y., Leblon G. and Montiel A. (2003). An ATP-based method for monitoring the microbiological water quality in a distribution network. *Wat. Res.*, **37**, 3689–3696.

Den Boer J. W., Yzerman E. P. F., Schellekens J., Lettinga K. D., Boshuizen H. C., van Steenbergen J. E., Bosman A., van den Hof S., van Vliet H. A., Peeters M. F., van Ketel R. J., Speelman P., Kool J. L. and Conyn-van Spaerdonk M. A. E. (2002). A large outbreak of Legionnaires' disease at a flower show, the Netherlands, 1999. *Emerg. Infect. Dis.*, **8**(1), 37–43.

Diederen B. M. W., de Jong C. M. A., Arrts I., Peeters M. F. and de Vries A. (2007). Molecular evidence for the ubiquitous presence of *Legionella* species in Dutch tap water installations. *J. Wat. Health*, **5**(3), 375–383.

Dolan J. R. (1997). Phosphorus and ammonia excretion by planktonic protists. *Mar. Geol.*, **139**, 109–122.

DVGW (2007). Vermehrung von Mikroorganismen auf Werkstoffen für den Trinkwasserbereich–Prüfung und Bewertung –. Technische Regel Arbeitsblatt W270. Bonn, Germany.

ECDC (2012). Legionnaires' disease in Europe, 2010. ECDC Surveillance report, European Centre for Disease Prevention and Control, Stockholm.

Eiler A., Langenheder S., Bertilsson S. and Tranvik L. J. (2003). Heterotrophic bacterial growth efficiency and community structure at different natural organic carbon concentrations. *Appl. Environ. Microbiol.*, **69**(7), 3701–3709.

European Union (1998). Council Directive 98/83/EC of 3 November 1998 on the quality of water intended for human consumption. *Off. J. Eur. Commun.*, **L330**, 32–53.

Euser S. M., Bruim J. P., van der Hoek W., Schop W. A. and den Boer J. W. (2012). Wellness centres, an important but overlooked source of Legionnaires' disease. Eight years of source investigation in the Netherlands, 1 August 2002–1 August 2010. *Euro Surveill.*, **17**(8), 1–6.

Ferrier-Pagès C., Karner M. and Rassoulzadegan F. (1998). Release of dissolved amini acids by flagellates and ciliates grazing on bacteria. *Oceanol. Acta*, **21**, 485–494.

Fields B., Barbaree J., Sanden G. and Morill W. (1990). Virulence of a *Legionella anisa* strain associated with Pontiac Fever: an evaluation using protozoan, cell culture, and guinea pig models. *Infect. Immun.*, **58**(9), 3139–3142.

Gagnon G. A., Booth S. D. J., Peldzus S., Mutti D., Smith F. and Huck P. M. (1997). Carboxylic acids; formation and removal in full-scale plants. *J. AWWA*, **89**(8), 88–97.

Gauthier V., Gérard B., Portal J. M., Block J. C. and Gatel D. (1999). Organic matter as loose deposits in a drinking water distribution system. *Wat. Res.*, **33**(4), 1014–1026.

Gilbert E. (1988). Biodegradability of ozonation products as a function of COD, and DOC elimination by the example of humic acids. *Wat. Res.*, **22**(1), 123–126.

Hallam N. B., West J. R., Forster C. F. and Simms J. (2001). The potential for biofilm growth in water distribution systems. *Wat. Res.*, **35**(17), 4063–4071.

Hammes F. and Egli T. (2005). A new method for assimilable organic carbon determination and biodegradable dissolved organic carbon (BDOC). *Env. Sci. Technol.*, **39**, 3289–3294.

Havelaar A. H., During M. and Versteegh J. F. (1987). Ampicillin-dextrin agar medium for the enumeration of *Aeromonas* species in water by membrane filtration. *J. Appl. Bacteriol.*, **62**, 279–287.

Havelaar A. H., Versteegh J. F. and During M. (1990). The presence of *Aeromonas* in drinking water supplies in the Netherlands. *Zentralbl. Hyg. Umweltmed.*, **190**, 236–256.

Health Council (1986). Prevention of legionellosis. Publication no. 200312, Health Council of the Netherlands, The Hague.

Heijnen J. J. and van Dijken J. P. (1992). In search of a thermodynamic description of biomass yields for the chemoheterotrophic growth of microorganisms. *Biotechnol. Bioeng.*, **39**, 833–858.

Heijnen L., Bakker G. and van der Kooij D. (2014). Biodiversity and population dynamics of the predominant bacteria in the communities of water and biofilms at two unchlorinated groundwater supplies with different concentrations of natural organic matter. Submitted for publication.

Henne K., Kahlisch L., Brettar I. and Hofle M. (2012). Analysis of structure and composition of bacterial core communities in mature drinking water biofilms and bulk water of a citywide network in Germany. *Appl. Environ. Microbiol.*, **78**(10), 3530–3538.

Heymann J. A. (1928). De organische stof in het waterleidingbedrijf. *Water Gas*, **12**, 61–65, 69–72.

Hijnen W. A. M. and van der Kooij D. (1992a). The effect of low concentrations of assimilable organic carbon in water on biological clogging of sand filters. *Wat. Res.*, **26**(7), 963–972.

Hijnen W. A. M. and van der Kooij D. (1992b). Biologische Kolmation von Schluckbrunnen unter dem Einfluss des AOC-gehalts des Wassers. *Wasser-Abwasser*, **113**(3), 148–153.

Hijnen W. A. M., Biraud D., Cornelissen E. and van der Kooij D. (2009). Threshold concentration of easily assimilable organic carbon in feed water for biofouling of spiral-wound membranes. *Environ. Sci. Technol.*, **43**, 4890–4895.

Hijnen W. A. M., Cornelissen E. R. and van der Kooij D. (2011). Threshold concentrations of biomass and iron for pressure drop increase in spiral-wound membrane elements. *Wat. Res.*, **45**, 1607–1616.

Hobbie J. E., Daley R. J. and Jasper S. (1977). Use of nucleopore filter for counting bacteria by fluorescence microscopy. *Appl. Environ. Microbiol.*, **33**(5), 1225–1228.

Holmes P., Niccolls L. M. and Sartory D. (1996). The ecology of mesophilic *Aeromonas* in the aquatic environment. In: The Genus Aeromonas, B. Austin, M. Altwegg, P. J. Gosling and S. W. Joseph (eds), John Wiley and Sons, New York, pp.127–150.

Hong P. Y., Hwang C., Ling F., Andersen G. L., LeChevallier M. W. and Liu W. T. (2010). Pyrosequencing analysis of bacterial biofilm communities in water meters of a drinking water distribution system. *Appl. Environ. Microbiol.*, **76**(16), 5631–5635.

IM (2011a). Besluit van 23 mei 2011, houdende bepaling inzake de productie en distributie van drinkwater en de organisatie van de openbare drinkwatervoorziening (drinkwaterbesluit). Staatsblad 293:1-90. Ministerie van Infrastructuur en Milieu (IM), Den Haag, Nederland.

IM (2011b). Regeling van de Staatssecretaris van Infrastructuur en Milieu van 27 juni 2011, nr. BJZ2011046957, houdende nadere regels met betrekking tot enige onderwerpen inzake legionellapreventie in drinkwater en warm tapwater (Regeling legionellapreventie in drinkwater en warm tapwater). Staatscourant 10828

Joret J. C. and Levi Y. (1986). Méthode rapide d'évaluation du carbone éliminable des eaux par voie biologique. *Trib. Cebedeau*, **510**(39), 3–9.

Kaplan L. A. and Newbold J. D. (1995). Measurement of streamwater biodegradable dissolved organic carbon with a plug-flow bioreactor. *Wat. Res.*, **29**(12), 2696–2706.

Karagiannis I., Brandsema P. and van der Sande M. (2009). Warm wet weather associated with increased Legionnaires' disease incidence in the Netherlands. *Epidemiol. Infect.*, **137**, 272–277.

Kuiper M. W., Wullings B. A., Akkermans A. D. L., Beumer R. R. and van der Kooij D. (2004). Intracellular proliferation of *Legionella pneumophila* in *Hartmannella vermiformis* in aquatic biofilms grown on plasticized polyvinyl chloride. *Appl. Environ. Microbiol.*, **70**(11), 6826–6833.

Kuiper M. W., Valster R. M., Wullings B. A., Boonstra H., Smidt H. and van der Kooij D. (2006). Quantitative detection of the free-living amoeba *Hartmannella vermiformis* in surface water by using real-time PCR. *Appl. Environ. Microbiol.*, **72**(9), 5750–5765.

Kuo P. P. K., Chiang E. S. K. and Chang B. J. (1977). Identification of end products resulting from ozonation and chlorination of organic compounds found in water. *Environ. Sci. Technol.*, **11**(13), 1177–1181.

LeChevallier M. W., Evans T. M., Seidler R. J., Daily O. P., Morrell B. R., Rollins D. M. and Joseph S. W. (1982). *Aeromonas sobria* in chlorinated drinking water supplies. *Microb. Ecol.*, **8**, 325–333.

LeChevallier M. W., Babcock T. M. and Lee R. G. (1987). Examination and characterization of distribution system biofilms. *Appl. Environ. Microbiol.*, **53**(12), 2714–2724.

LeChevallier M. W., Shaw N. E., Kaplan L. A. and Bott T. L. (1993). Development of a rapid assimilable organic carbon method for water. *Appl. Environ. Microbiol.*, **59**(5), 1526–1531.

LeChevallier M. W., Welch N. J. and Smith D. B. (1996). Full-scale studies of factors related to coliform regrowth in drinking water. *Appl. Environ. Microbiol.*, **62**(7), 2201–2211.

Leclerc H. and Buttiaux R. (1962). Fréquence des *Aeromonas* dans les eaux d'alimentation. *Ann. Inst. Pasteur Lille*, **103**, 97–100.

Lucena F., Frais J. and Ribas F. (1990). A new dynamic approach to the determination of biodegradable dissolved organic carbon in water. *Environ. Technol.*, **12**, 343–347.

Magic-Knezev A. and van der Kooij D. (2004). Optimization and significance of ATP analysis for measuring active biomass in granular activated carbon filters used in water treatment. *Wat. Res.*, **38**, 3971–3979.

Martiny A. C., Jørgensen T. M., Albrechtsen H. J., Arvin E. and Molin S. (2003). Long-term succession and diversity of a biofilm formed in a model drinking water distribution system. *Appl. Environ. Microbiol.*, **69**(11), 6899–6907.

Matthieu L., Block J. C., Prévost M., Maul A. and De Bischop R. (1995). Biological stability of drinking water in the city of Metz distribution system. *J. Wat. SRT Aqua*, **44**(5), 230–239.

Meenhorst P. L., Reingold A. L., Groothuis D. G., Gorman G. W., Wilkinson H. W., McKinney R. M., Feeley J. C., Brenner D. J. and van Furth R. (1985). Water-related nosocomial pneumonia caused by *Legionella pneumophila* serogroups 1 and 10. *J. Infect. Dis.*, **132**(2), 356–364.

Michel R., Hoffmann R., Giese A. and Műller K.-D. (1995). Untersuchung von drei Grundwasserwerken auf Vorkommen von Acanthamoeben, Naeglerien and anderen freilebenden Amöben. *Acta Hydrochim. Hydrobiol.*, **23**(5), 202–211.

Miettinen I. T., Vartinainen T. and Martikainen P. J. (1999). Determination of assimilable organic carbon in humus-rich drinking waters. *Wat. Res.*, **33**(10), 2277–2282.

Nagata T. and Kirchman D. L. (1991). Release of dissolved free and combined amino acids by bacteriovorous marine flagellates. *Limnol. Oceanogr.*, **36**(3), 433–443.

Nagelkerke N. J. D., Boshuizen H. C., de Melker H. E., Schellekens J. F. P., Peeters M. F. and Conyn-van Spaendonck M. (2003). Estimating the incidence of subclinical infections with *Legionella pneumophila* using data augmentation: analysis of an outbreak in the Netherlands. *Statist. Med.*, **22**, 3713–3724.

Niquette P., Servais P. and Savoir R. (2001). Bacterial dynamics in the drinking water distribution system of Brussels. *Wat. Res.*, **35**(3), 675–682.

Oesterholt F. and van der Kooij D. (2003). *Legionella* in woninginstallaties. *H_2O*, **38**(4), 19–20.

Oesterholt F., van der Kooij D. and van der Wielen P. W. J. J. (2012). *Legionella pneumophila* finds niche in roof-top cooling towers. *H_2O*, **47**(24), 33–35.

Palumbo S. A., Morgan D. R. and Buchanan R. L. (1985). Influence of temperature, NaCl, and pH on the growth of *Aeromonas hydrophila*. *J. Food Sci.*, **50**, 1417–1421.

Pernthaler J., Posch T., Simek K., Vrba J., Pernthaler A., Glockner F. O., Nubel U., Psenner R. and Amann R. (2001). Predator specific enrichment of Actinobacteria from a cosmopolitan freshwater clade in mixed continuous culture. *Appl. Environ. Microbiol.*, **67**(5), 2145–2155.

Pickup Z. L., Pickup R. and Parry J. D. (2007). Effects of bacterial prey species and their concentration on growth of the amoeba *Acanthamoeba castellani* and *Hartmannella vermiformis*. *Appl. Environ. Microbiol.*, **73**(8), 2631–2634.

Rewab (2011). Registration of drinking water quality data in the Netherlands, Vewin, Rijswijk, the Netherlands.

Ribas F., Frias J. and Lucena F. (1991). A new dynamic method for the rapid determination of the biodegradable dissolved organic carbon (BDOC) in drinking water. *J. Appl. Bacteriol.*, **71**, 371–378.

Rice E. W., Scarpine P. V., Reasoner D. J., Logsdon G. S. and Wild D. K. (1991). Correlation of coliform growth response with other water quality parameters. *J. Am. Wat. Works Assoc.*, **83**(7), 98–102.

Richardson S. D., Thruston A. D., Caughran T. V., Chen P. H., Collette T. W., Floyd T. L., Schenck K. M., Lykins B. W., Sun G. R. and Majetich G. (1999). Identification of new ozone disinfection byproducts in drinking water. *Environ. Sci. Technol.*, **33**, 3368–3377.

Rippey S. R. and Cabelli V. J. (1980). Occurrence of *Aeromonas hydrophila* in limnetic environments: relationship of the organism to trophic state. *Microb. Ecol.*, **6**, 45–54.

Rogers J. R., Dossett A. D., Dennis P. J., Lee J. V. and Keevil C. W. (1994). Influence of temperature and plumbing material selection on biofilm formation and growth of *Legionella pneumophila* in a model potable water system containing complex microbial flora. *Appl. Environ. Microbiol.*, **60**(3), 1585–1592.

Rook J. J. (1970). Über das Verhalten desinfizierter Wässer in Rohrleitungen. *SchrReihe Ver. Wass.-Boden- Lufthyg. Berlin Dahlem*, **31**, 137–150.

Rook J. (1974). The formation of haloforms during chlorination of natural waters. *Water Treat. Exam.*, **23**, 234–243.

Rowbotham T. J. (1980). Preliminary report on the pathogenicity of *Legionella pneumophila* for freshwater and soil amoeba. *J. Clin. Pathol.*, **33**, 1179–1183.

Sack E. L. W., van der Wielen P. W. J. J. and van der Kooij D. (2010). Utilization of oligo-and polysaccharides at microgram-per-litre levels in freshwater by *Flavobacterium johnsoniae*. *J. Appl. Microbiol.*, **108**, 1430–1440.

Sack E. L. W., van der Wielen P. W. J. J. and van der Kooij D. (2011). *Flavobacterium johnsoniae* as a model organism for characterizing biopolymer utilization in oligotrophic freshwater environments. *Appl. Environ. Microbiol.*, **77**(19), 6931–6938.

Saftic S., Vrouwenvelder H., Heller O., Schubert J., van der Kooij D. and Rohns H. P. (1997). Biofilmuntersuchungen im Trinkwasserversorgungssystemen der Düsseldorfer Wasserwerke. *Wasser-Abwasser*, **138**(10), 526–530.

Scheffer W. J. H. (2012). Personal communication.

Schippers D. and Sjoerdsma P. (2007). Kwaliteitsverbetering op meerdere fronten door ontkleuring via ionenwisseling. *H_2O*, **20**, 38–40.

Schoenen D., Schulze-Robbecke R. and Schirdewahn N. (1988). Mikrobielle Kontamination von Wasser durch Rohr und Schlauchmaterialien. 2. Mitteilung: Wachstum von *Legionella pneumophila*. *Zbl. Bakt. Hyg.*, **B186**, 326–332.

Schubert R. H. W. (1976). Der Nachweis von Aeromonaden der "Hydrophila-Punctata Gruppe" in Rahmen der hygienischen Trinkwasser-Beurteilung. *Zb. Bakt Hyg.*, **B 161**, 482–497.

Servais P., Billen G. and Hascoët M. C. (1987). Determination of the biodegradable fraction of dissolved organic matter in waters. *Wat. Res.*, **21**(4), 445–450.

Servais P., Laurent P., Billen G. and Levi Y. (1992). Etude de la colonisation bacterienne des reseaux de distribution. *T.S.M. de L'Eau*, **87**(6), 321–326.

Sherr E. B. and Sherr B. F. (2002). Significance of predation by protists in aquatic microbial food webs. *Antonie van Leeuwenhoek*, **81**, 293–308.

Sherr B. F., Sherr E. B. and Rassoulzadegan F. (1988). Rates of digestions of bacteria by marine phagotrophic protozoa: temperature dependence. *Appl. Environ. Microbiol.*, **54**(5), 1091–1095.

Simek K. and Chranowski T. H. (1992). Direct and indirect evidence of size selective grazing on pelagic bacteria by freshwater nanoflagellates. *Appl. Environ. Microbiol.*, **58**(11), 3715–3720.

Smith D. B, Hess A. B. and Hubbs S. A. (1990). Survey of distribution system coliform occurrence in the United States. In: Proceedings Water Quality Technology Conference, American Water Works Association, Denver, CO, pp. 1103–1117.

Stanfield G. and Jago P. H. (1987). The development and use of a method for measuring the concentration of assimilable organic carbon in water. *WRc Report* PRU 1628–M, Medmenham, UK.

Thomas J. M. and Ashbolt N. J. (2011). Do free-living amoebae in treated drinking water systems present an emerging health risk? *Environ. Sci. Technol.*, **45**, 860–869.

Thomas V., Herrera-Rimann K., Blanc D. S. and Greub G. (2006). Biodiversity of amoebae and amoeba-resistant bacteria in a hospital water network. *Appl. Environ. Microbiol.*, **72**(4), 2428–2438.

Thomas V., Loret J. J., Jousset M. and Greub G. (2008). Biodiversity of amoebae and amoebae-resisting bacteria in a drinking water treatment plant. *Environ. Microbiol.*, **10**(10), 2728–2745.

Thomas V., McDonnell G., Denyer S. P. and Maillard J. Y. (2010). Free-living amoebae and their intracellular pathogenic micro-organisms: risks for water quality. *FEMS Microbiol. Rev.*, **34**, 231–259.

Traina S. J., Novak J. and Smeck N. E. (1990). An ultraviolet absorbance method of estimating the percent aromatic carbon content of humic acids. *J. Environ. Qual.*, **19**, 151–153.

Tranvik L. J. and Höfle M. G. (1987). Bacterial growth in mixed cultures on dissolved organic carbon from humic and clear waters. *Appl. Environ. Microbiol.*, **53**(3), 482–488.

Valster R. M., Wullings B. A., Bakker G., Smidt H. and van der Kooij D. (2009). Free-living protozoa in two unchlorinated drinking water supplies, identified by phylogenic analysis of the 18S rRNA gene sequences. *Appl. Environ. Microbiol.*, **75**(14), 4736–4746.

Valster R. M., Wullings B. A. and van der Kooij D. (2010). Detection of protozoan hosts for Legionella pneumophila in engineered water systems by using a biofilm batch test. *Appl. Environ. Microbiol.*, **76**(21), 7144–7153.

Valster R. M., Wullings B. A., van den Berg R. and van der Kooij D. (2011). Free-living protozoa and *Legionella* in three Caribbean water supplies. *Appl. Environ. Microbiol.*, **77**(20), 7321–7328.

van der Hoek J. P., Hofman J. A. M. H. and Graveland A. (2000). Benefits of ozone-activated carbon filtration in integrated treatment processes, including membrane systems. *J. Wat. SRT Aqua*, **49**(6), 341–356.

van der Kooij D. (1977). The occurrence of *Pseudomonas* spp. in surface water and in tap water as determined on citrate media. *Antonie van Leeuwenhoek J. Microbiol.*, **43**, 187–197.

van der Kooij D. (1979). Characterization and classification of fluorescent pseudomonads isolated from tap water and surface water. *Antonie van Leeuwenhoek J. Microbiol.*, **45**, 225–240.

van der Kooij D. (1984). The growth of bacteria on organic compounds in drinking water. Dissertation, University of Wageningen, the Netherlands.

van der Kooij D. (1987). The effect of treatment on assimilable organic carbon in drinking water. In: Treatment of drinking water for organic contaminants, P. M. Huck and P. Toft (eds), Proceedings Second National Conference on Drinking Water, Edmonton Canada, 7–8 April, 1986. Pergamon Press, New York, pp. 317–328.

van der Kooij D. (1990). Assimilable organic carbon (AOC) in drinking water. In: Drinking water microbiology, progress and recent developments, G. A. McFeters (ed.), Springer Verlag, New York, pp. 57–87.

van der Kooij D. (1991). Nutritional requirements of aeromonads and their multiplication in drinking water. *Experientia*, **47**, 444–446.

van der Kooij D. (1992). Assimilable organic carbon as an indicator of bacterial regrowth. *J. Am. Wat. Works Assoc.*, **84**, 57–65.

van der Kooij D. (1997). Multiplication of coliforms at very low concentrations of substrates in tap water. In: Coliforms and *E. coli*: problem or solution? D. Kay and C. Fricker (eds), The Royal Society of Chemistry, Cambridge, UK, pp. 195–207.

van der Kooij D. (1999). Potential for biofilm development in drinking water distribution systems. *J. Appl. Microbiol. Symp. Suppl.*, **85**, 39S–44S.

van der Kooij D. (2002). Assimilable organic carbon (AOC) in treated water: determination and significance. In: Encyclopedia of Environmental Microbiology, G. Bitton (ed.), John Wiley & Sons, New York, pp. 312–327.

van der Kooij D. (2003). Managing regrowth in drinking water distribution systems. In: Heterotrophic plate counts and drinking water safety, J. Bartram *et al.* (ed.), IWA Publishing, London, pp. 199–232.

van der Kooij D. and Hijnen W. A. M. (1984). Substrate utilization by an oxalate-consuming *Spirillum* species in relation to its growth in ozonated water. *Appl. Environ. Microbiol.*, **47**, 551–559.

van der Kooij D. and Hijnen W. A. M. (1985). Determination of the concentration of maltose and starch-like compounds in drinking water by growth measurements with a well-defined strain of a *Flavobacterium* species. *Appl. Environ. Microbiol.*, **49**(7), 65–71.

van der Kooij D. and Hijnen W. A. M. (1988). Nutritional versatility and growth kinetics of an *Aeromonas hydrophila* strain isolated from drinking water. *Appl. Environ. Microbiol.*, **54**, 2842–2851.

van der Kooij D. and Veenendaal H. R. (1992). Assessment of the biofilm formation characteristics of drinking water. In: Proceedings American Water Works Association Water Quality Technology Conference, 15–19 November, American Water Works Association, Denver, CO, pp. 1099–1110.

van der Kooij D. and Veenendaal H. R. (1993). Assessment of the biofilm formation potential of synthetic materials in contact with drinking water during distribution. In: Proceedings American Water Works Association Water Quality Technology Conference, 7–11 November, American Water Works Association, Denver, CO, pp. 1395–1407.

van der Kooij D. and Veenendaal H. R. (2001). Biomass production potential of materials in contact with drinking water: method and practical importance. *Wat. Sci. Technol. Wat. Supply*, **1**(3), 39–45.

van der Kooij D., Hijnen W. A. M. and Kruithof J. C. (1989). The effects of ozonation, biological filtration and distribution on the concentration of easily assimilable organic carbon (AOC) in drinking water. *Ozone Sci. Eng.*, **11**, 297–311.

van der Kooij D., Visser A. and Oranje J. P. (1982a). Multiplication of fluorescent pseudomonads at low substrate concentrations in tap water. *Antonie van Leeuwenhoek J. Microbiol.*, **48**, 229–243.

van der Kooij D., Oranje J. P. and Hijnen W. A. M. (1982b). Growth of *Pseudomonas aeruginosa* in tap water in relation to utilization of substrates at concentrations of a few micrograms per liter. *Appl. Environ. Microbiol.*, **44**(5), 1086–1095.

van der Kooij D., Visser A. and Hijnen W. A. M. (1982c). Determining the concentration of easily assimilable organic carbon in drinking water. *J. Am. Wat. Works Assoc.*, **74**, 540–545.

van der Kooij D., Veenendaal H. R., Baars-Lorist C., van der Klift H. W. and Drost Y. C. (1995a). Biofilm formation on surfaces of glass and Teflon exposed to treated water. *Wat. Res.*, **29**, 1655–1662.

van der Kooij D., Vrouwenvelder J. S. and Veenendaal H. R. (1995b). Kinetic aspects of biofilm formation on surfaces exposed to drinking water. *Wat. Sci. Technol.*, **32**, 61–65.

van der Kooij D., Vrouwenvelder J. S. and Veenendaal H. R. (1995c). Density and composition of biofilms in drinking water distribution systems in the Netherlands. In: Proceedings American Water Works Association Water Quality Technology Conference, American Water Works Association, Denver, CO, pp. 1055–1062.

van der Kooij D., Vrouwenvelder J. S. and Veenendaal H. R. (1997). Assessment and significance of the biofilm formation characteristics of finished water (in Dutch). *H_2O*, **30**(25), 767–771.

van der Kooij D., van Lieverloo J. H. M., Schellart J. and Hiemstra P. (1999). Maintaining quality without a disinfectant residual. *J. Am. Wat. Works Assoc.*, **91**(1), 55–64.

van der Kooij D., Veenendaal H. R., Slaats N. P. G. and Vonk D. (2002). Biofilm formation and multiplication of *Legionella* on synthetic pipe materials in contact with treated water under static and dynamic conditions. In: Legionella, R. Marre *et al.* (eds), ASM Press, Washington DC, pp. 176–180.

van der Kooij D., Veenendaal H. R. and Scheffer W. J. H. (2005). Biofilm formation and multiplication of *Legionella* in a model warm water system with pipes of copper, stainless steel and cross-linked polyethylene. *Wat. Res.*, **39**, 2789–2798.

van der Kooij D., Wubbels G. and van der Kooij D. (2007). Legionellabacteriën in leidingwaterinstallaties behoren meestal tot de ongevaarlijke soort *Legionella anisa*. *H_2O*, **40**(5), 33–35.

van der Kooij D., Hijnen W. A. M. and Cornelissen E. R. (2010). Biofouling of spiral-wound membranes in water treatment. *Report Water Research Foundation, Denver, CO.*

van der Veer A. J., Hijnen W. A. M., Groenendijk M. and van der Kooij D. (2005). Pilot plant studies on AOC and biofilm formation rates of ozonated water after filtration. In *Proceedings 17th World Ozone Congress*, pp. IV.4.11.1–11.8.

van der Wielen P. W. J. J. and van der Kooij D. (2009). Ammonia-oxidizing bacteria and archae in groundwater treatment and drinking water distribution systems. *Appl. Environ. Microbiol.*, **108**(14), 1430–1440.

van der Wielen P. W. J. J. and van der Kooij D. (2010). Effect of water composition, distance and season on the adenosine triphosphate concentration of unchlorinated drinking water in the Netherlands. *Wat. Res.*, **44**, 4860–4867.

van der Wielen P. W. J. J. and van der Kooij D. (2011). Survey of Aeromonas and Colony count 22°C in drinking water and relationship with physical/chemical parameters *(in*

Dutch). BTO Report 2011.001. KWR Watercycle Research Institute, Nieuwegein, the Netherlands.

van der Wielen P. W. J. J. and van der Kooij D. (2013). Nontuberculous mycobacteria, fungi, and opportunistic pathogens in unchlorinated drinking water in the Netherlands. *Appl. Environ. Microbiol.*, **79**(3), 825–834.

van Ketel R. J., Rietra P. J. G. M., Zanen-Lim O. G. L and Zanen H. C. (1983). Een epidemie van pneumonie door *Legionella pneumophila* in een Nederlands ziekenhuis. *Ned. Tijdschr. Geneeskd.*, **127**, 324–327.

van Lieverloo J. H. M., van der Kooij D. and Hoogenboezem W. (2002). Invertebrates and protozoa (free–living) in drinking water distribution systems. In: Encyclopedia of Environmental Microbiology, G. Bitton (ed.), John Wiley & Sons, New York, pp. 1718–1733.

van Lieverloo J. H. M., Hoogenboezem W., Veenendaal G. and van der Kooij D. (2012). Variability of invertebrate abundance in drinking water distribution systems in the Netherlands in relation to biostability and sediment volumes. *Wat. Res.*, **46**, 4918–4932.

Versteegh J. F. M. and Dik H. H. J. (2011). The quality of drinking water in the Netherlands in 2010, RIVM report nr. 7037190812011(in Dutch). Ministerie van Infrastructuur en Milieu, The Hague, the Netherlands.

Versteegh J. F. M., Brandsema P. S., van der Aa N. G. F. M., Dik H. H. J. and de Groot G. M. (2007). Evaluation of legionella prevention. Report 7037190202007 (in Dutch), RIVM Bilthoven, the Netherlands.

Vewin (2010). Drinking water statistics 2012 (in Dutch), Vewin, Rijswijk, the Netherlands.

Vital M., Stucki D., Egli T. and Hammes F. (2010). Evaluating the growth potential of pathogenic bacteria in water. *Appl. Environ. Microbiol.*, **76**(19), 6477–6484.

Volk C. J. and LeChevallier M. W. (2000). Assessing biodegradable organic matter. *J. Am. Wat. Works Assoc.*, **92**(5), 1198–1204.

Vreeburg J. H. G., Blokker E. J. M., Horst P. and van Dijk J. C. (2009). Velocity-based self-cleaning residential drinking water distribution systems. *Wat. Sci. Technol. Wat. Supply*, **9**(6), 635–641.

Wadowsky R. M., Butler L. J., Cook M. K., Verma S. M., Paul M. A., Fields B. S., Keleti G., Sykora J. L. and Yee R. B. (1988). Growth-supporting activity for *Legionella pneumophila* in tap water cultures and implication of Hartmannellid amoeba as growth factors. *Appl. Environ. Microbiol.*, **54**(11), 2677–2682.

Weekers P. H. H., Bodelier P. L., Wijen J. P. H. and Vogels G. D. (1993). Effects of grazing by the free-living soil amoebae *Acanthamoeba castellanii*, *Acanthamoeba polyphaga*, and *Hartmannella vermiformis* on various bacteria. *Appl. Environ. Microbiol.*, **59**(7), 2317–2319.

Weinrich L. A., Giraldo E. and LeChevallier M. W. (2009). Development and application of a bioluminescence-based test for assimilable organic carbon in reclaimed waters. *Appl. Environ. Microbiol.*, **75**(23), 7385–7390.

Wullings B. A. and van der Kooij D. (2006). Occurrence and genetic diversity of uncultured *Legionella* spp. in drinking water treated at temperatures below 15°C. *Appl. Environ. Microbiol.*, **72**(1), 157–166.

Wullings B. A., Bakker G. and van der Kooij D. (2011). Concentration and diversity of uncultured *Legionella* spp. in two unchlorinated drinking water supplies with different concentrations of natural organic matter. *Appl. Environ. Microbiol.*, **77**(3), 634–641.

Zacheus O. M., Lehtola M. J., Korhonen L. K. and Martikainen P. J. (2001). Soft deposits, the key site for microbial growth in drinking water distribution networks. *Wat. Res.*, **35**(7), 1757–1765.

Zhang S. and Huck P. M. (1996). Removal of AOC in biological water treatment processes: a kinetic model. *Wat. Res.*, **30**(5), 1195–1207.

Chapter 12

Enhancement of microbial growth by materials in contact with drinking water: problems and test methods

B. Hambsch, J. Ashworth and D. van der Kooij

12.1 INTRODUCTION

Water intended for human consumption (drinking water) comes into contact with a variety of materials (construction products) on its way from the source to the consumer's tap. Such materials are used for storage reservoirs (tanks), pipes for transportation and distribution, pipes in drinking water installations in homes and buildings (in-premise plumbing systems), in pipe joints, valves, fittings, taps, gaskets, pumps, and so on, and also include products such as kits, lubricants and greases. The degree of the contact between a material and drinking water depends on the surface to volume (S/V) ratio and the contact time. In plumbing systems, with pipe diameters of about 1 to 2 cm, the S/V ratio is 10 to 100 times higher than in distribution systems and contact times can be relatively long, for example, at weekends and during holidays.

Physico–chemical and microbiological processes at the surface of materials can adversely affect drinking water quality. To prevent a deterioration of the quality of drinking water during storage and distribution, the drinking water directive in the European Community (EC) (EU 1998), requires that "member states shall take all measures to ensure that no substances or materials for new installations used in the preparation or distribution of water intended for human consumption or impurities associated with such substances or materials for new installations remain in water intended for human consumption in concentrations higher than is necessary for the purpose of their use and do not, either directly or indirectely, reduce the protection of human health provided for in this Directive". The member states of the EC have the obligation to incorporate this requirement in their national legislation.

The main processes at the interface between water and materials affecting water quality include (i) corrosion of metal surfaces resulting in the release of inorganic compounds into the water, for example, iron, copper or lead, (ii) release of organic compounds from natural or synthetic organic materials, and (iii) attachment, growth and release of microorganisms. Release of metals can impair the colour or the flavour of the water and/or result in noncompliance with quality criteria for metals. Organic compounds released into drinking water may also (i) affect colour, turbidity, odour and flavour, (ii) include compounds with toxic properties and/or (iii) include compounds which enhance the growth of microorganisms on the material (biofilm formation) and/or in the water. Methods for testing these products for potential effects on drinking water quality have been developed and standardized in several countries. In Europe, harmonisation and standardisation of test methods, related to materials used in contact with drinking water, including those described in the Construction Products Directive (EU 1989) and the replacement EU Regulation (EU 2011), are on-going activities. These activities aim at protecting drinking water quality and the reduction of trade barriers between member states.

Enhancement of microbial growth by materials in contact with drinking water was already reported in the 19th century (de Vries 1890) and at the beginning of the 20th century (Schoenen and Schöler 1983). These problems were caused by natural materials such as wood, natural rubber, jute, hemp yarn, and leather, but from the 1950s onwards it became clear that synthetic materials, which replaced most natural materials, also could enhance microbial growth (Colbourne 1985). Synthetic materials are mostly composed of a polymer with additives for functionality or protection against physical and chemical degradation, viz. plasticizers, stabilizers, anti-oxidants, fillers, antistatics, and so on. The number of types of synthetic materials, including plastic polymers, elastomers, composite materials, varieties of polyvinyl chloride (PVC), polyethylene (PE), polypropylene (PP), polybutylene (PB), rubbers, and so on, is increasing steadily.

This chapter briefly describes (i) the main microbial problems induced by materials in contact with drinking water and (ii) methods used in Europe for assessment of the potential for enhancement of microbial growth.

12.2 MICROBIAL WATER QUALITY PROBLEMS

12.2.1 Microbial growth

Microorganisms attach to surfaces in contact with water and subsequently multiply at a rate depending on the availability of growth-promoting compounds and other environmental conditions, for example, temperature, flow rate (variation), disinfectant. Release of biodegradable compounds from materials can enhance planktonic (suspended) growth and lead to the formation of biofilms and accumulation of sediments. Microbial growth may have adverse effects on drinking water quality. Problems related to the enhancement of microbial growth by materials have been reviewed in the 1980s (Schoenen and Schöler 1983; Colbourne 1985; Schoenen 1989). In this chapter, these problems therefore are described briefly, with a focus on recent developments.

12.2.2 Problems

The following types of problems can be distinguished: (i) elevated heterotrophic plate counts; (ii) presence of coliform bacteria; (iii) aesthetic problems (colour, odour, flavour, visible biofilms or flocs) and (iv) multiplication of opportunistic pathogens.

Elevated heterotrophic plate counts (HPCs) is the most commonly reported microbial problem related to materials in contact with drinking water during storage and distribution (Schoenen and Schöler 1983; Colbourne 1985). The release of solvents by coatings still is a major cause of this problem (Frensch *et al.* 1987; Bernhardt and Liesen 1988). Recently, the DVGW-Technologiezentrum Wasser (TZW) in Germany observed such a problem in a newly installed closed vessel for an activated carbon filter. This vessel was coated with an epoxy resin, which was applied on-site; curing should have been carried out at 60°C to 80°C for 24 h. As the filter vessels were installed in the water works, curing was not possible at these high temperatures; instead, curing was done at 15–22°C for 6 weeks, before placing the vessels into service. Subsequently, the HPC increased to more than 500 CFU/ml. After a chlorination step the HPCs in the granular activated carbon effluent fell to below 100 CFU/ml. After about 6 months, the HPCs in the effluent had returned to the normal level below 5 CFU/ml and chlorination was terminated. Another case occurred at a water utility, where water from a deep groundwater source, distributed after treatment without addition of chlorine, was stored in two renovated reservoirs with plasticized (soft) PVC (PVC-P) linings. As a consequence, HPCs of $>$100 CFU/ml were observed in the effluent of these reservoirs. Moreover, a visible biofilm developed on the reservoir wall at the water/air interface. After the addition of chlorine to the influent of the reservoirs, the effluent showed HPCs below 10 CFU/ml, but biofilm formation could not be stopped by this measure. Additionally, cleaning was done every 3 months to minimize the visible biofilm. After 15 years, the chlorine dosage could be stopped. Obviously, the release of organic compounds had decreased to such a low level that elevated HPCs no longer occurred. Elevated HPCs caused by the release of solvents from an epoxy-based coating in the pipes at a water treatment plant in the Netherlands led to the decision to replace the coating (KWR, unpublished observation).

Multiplication of coliforms, resulting into non-compliance with drinking water quality criteria, has been observed in the presence of natural materials such as wood, jute, leather, and hemp (Seidler *et al.* 1977; Schoenen and Schöler 1983; Colbourne 1985) and also with certain coatings (Ellgas and Lee 1980). Coliforms also multiplied in the presence of materials during testing (Burman and Colbourne

1977). In the Netherlands, coliform growth has been observed on certain lubricants, for example, soft soap, used in pipe joints (Van der Kooij 1987) and on certain rubber-coated valves, which has also been reported in Germany (Kilb *et al.* 2003).

Aesthetic problems can be the result of visible accumulation of biomass or the formation of flavour- and odour-affecting compounds. Microbial growth visible to the naked eye in most cases is caused by specific materials, promoting a local accumulation of biomass on the surface (e.g. Figure 12.1).

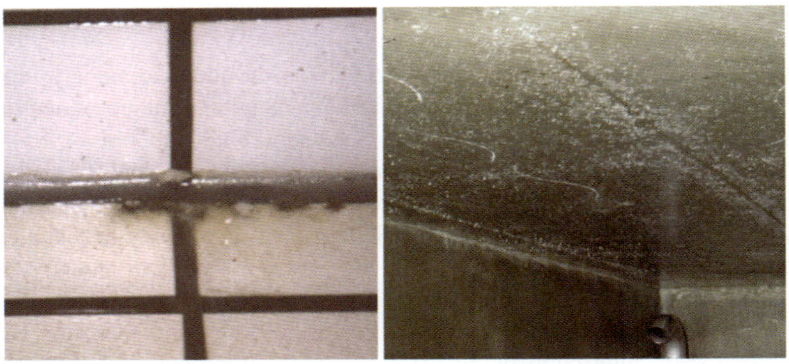

Figure 12.1 Left: Microbial growth on a tile joint of silicone in a drinking water reservoir. Right: Fungi at the ceiling of a small concrete reservoir.

Odour and flavour complaints have been attributed to growth of fungi on materials (Burman 1979; Ashworth and Colbourne 1987). A recent problem was observed in Germany in a small reservoir that had been built in a new treatment plant. This reservoir was made of concrete, and no organic compounds were included in the materials used in its construction. Nevertheless, after some time growth of fungi was observed at the ceiling (Figure 12.1). At the same time, mouldy odours were detected in the water. To prevent consumer complaints, measures were taken to get rid of these fungi. Mechanical cleaning was done every two weeks to eliminate the fungi. After eight cleaning regimes no further growth occurred. The most probable explanation for the growth of fungi was the use of lining boards for the ceiling that were water proofed by a phenolic resin. A diffusion of these organics into the concrete at the ceiling seemed to be the explanation for this fungal growth. For the walls no such lining boards were used and no fungi were growing there.

Opportunistic pathogens mainly cause disease in immunocompromised persons, but can also infect healthy individuals. Enhancement of the growth of such organisms by materials in contact with drinking water has been reported for *Legionella pneumophila* and less frequently for *Pseudomonas aeruginosa* and *Mycobacterium* spp. (Colbourne 1985; Schulze-Röbbecke and Fischeder 1989; Bressler *et al.* 2009). *L. pneumophila* is the etiological agent for Legionnaires' disease, a life-threatening pneumonia, which can be the result of exposure to aerosols containing *L. pneumophila* (Fields *et al.* 2002). This organism multiplies in biofilms where certain amoebae serve as its host (Wadowsky *et al.* 1988). Multiplication of *L. pneumophila* in a building plumbing system was reported for the first time by Tobin *et al.* (1980). Subsequently, it was shown that certain rubber components within shower heads and taps can support growth of this organism (Colbourne *et al.* 1984; Niedeveld *et al.* 1986; van der Kooij and Hijnen 1984; Schofield and Locci 1985). Further studies have shown that a large variety of synthetic materials in contact with drinking water can enhance growth of *L. pneumophila* (Schoenen *et al.* 1988; Rogers *et al.* 1994; van der Kooij *et al.* 2001), in parallel with enhanced biofilm formation. Recently, *P. aeruginosa* growth in drinking water in hospitals was identified as the cause of infections and fatalities in neonatal and high dependency units. In some cases there was an indication that the use of certain grades of flexible EPDM rubber hoses may have contributed to these instances (Department of Health 2012a, b; RQIA 2012).

Hence, materials can affect the microbiological quality of water intended for human consumption and used for personal hygiene, resulting in noncompliance with microbiological water quality criteria, customer complaints and exposure to potentially pathogenic micro-organisms. Prevention of these problems is needed to protect public health, comply with water quality criteria and maintain the confidence of the customers in the safety of water supplies, including installations in houses and buildings.

12.3 TEST METHODS

12.3.1 Need for test methods

Microbial growth on materials and/or in the water in distribution systems and in tap water installations will occur when (i) biodegradable organic compounds are available, (ii) temperature or inorganic elements like N or P do not limit growth and (iii) growth-inhibiting substances (e.g. biocides/disinfectants) are absent. Maintaining a disinfectant residual, for example, free chlorine, in the distributed drinking water prevents microbial growth in the water phase and limits biofilm formation, but does not eliminate the growth potential. Furthermore, maintaining a residual at the tap requires high disinfectant dosages, which adversely affect drinking water quality, and in some cases, adversely affect polymeric materials in contact with the water.

The use of a disinfectant residual in drinking water during distribution therefore is not common practice in Germany, the Netherlands and in a number of other European countries and is even considered undesirable. A requirement for materials in contact with drinking water therefore should be that these materials do not enhance microbial growth. Consequently, assessment of such materials for enhancement of microbial growth (EMG), by using reliable, standardized methods, is needed.

12.3.2 Available test methods

The microbial problems related to materials stimulated the development of test methods in several European countries. In the UK a method for EMG assessment was first published in 1957 (Metropolitan Water Board 1958) based on material samples completely immersed in 900 ml of drinking water plus 100 ml of river water and incubated at 20°C for a minimum of 12 weeks. Water was replaced twice a week and the microbiological quality of the water was assessed at weekly intervals using culture methods for the detection of heterotrophic bacteria, coliforms, fungi, actinomycetes and *P. aeruginosa*, plus a visual examination for turbidity of the test water (planktonic growth or pieces of biofilm) and microscopic examination of any biofilm present on the surface of the material. The performance of the test method was subsequently subject to improvements, including increasing the incubation temperature to 30°C and reducing the test period (Burman and Colbourne 1977). Adaptation of this method, including the assessment of overall growth in the test system by measuring dissolved oxygen uptake from the test water by growing organisms (Colbourne and Brown 1979), was introduced in 1982 (BSI 1982). After some time of running the two alternative methods for assessing overall microbial growth in parallel, counts of heterotrophic bacteria were excluded in the 1988 version of the test method BS6920-2.4 (BSI 1988) in favour of the more reliable and repeatable measurement of dissolved oxygen. This latter method has been subject to minor modification and was last revised in 2000 (BSI 2000). This mean dissolved oxygen difference (MDOD) method is described in more detail in Paragraph 12.4.3.1.

Developments aiming at assessing the growth-promoting properties of materials in contact with drinking water in Germany date back to the first half of the 20th century (Schoenen and Schöler 1983). Also here, initially, culture methods were used for the assessment of the microbiological quality of water in contact with the material under batch test conditions. Subsequently, test plates exposed to drinking water in reservoirs to simulate conditions in practice were investigated by using contact-culture methods and microbiological testing of the water dripping from the exposed surface when removed from the system (Schoenen and Schöler 1983). These observations resulted in a method in which biomass, present on the surface of test pieces (800 cm^2, on plates of 20 cm × 20 cm) exposed to slowly flowing test water, was collected after six months by scraping and quantified by measuring the collected volume (referred to as the biomass volume) after sedimentation for 24 hrs. This method was published as *Code of Practice* W270 in Germany (DVGW 1990). This *Code of Practice* W270 has been modified and optimized based on investigations conducted in Germany (Gerhardy 2008). In the present method W270 (DVGW 2007), surfaces of plates or pipes are exposed to test water for defined periods of time. The biofilm is scraped from the exposed surface and after centrifugation its volume is measured as parameter for microbial growth enhancement of the involved material. Details of this method are described in Paragraph 12.4.3.2.

Research in the Netherlands revealed that adenosinetriphosphate (ATP) is a rapid and sensitive parameter for measuring the biomass concentration of active bacteria in (drinking) water and in biofilms (van der Kooij 1992; van der Kooij and Veenendaal 1992). Consequently, a batch test procedure, using ATP as biomass parameter was developed for determining the EMG by materials (van der Kooij and Veenendaal 1993; van der Kooij and Veenendaal 2001). The principle of this method was based on incubation (at 25°C) of material samples during 16 weeks in tap water at a surface area (S) to volume (V) ratio (S/V) of 0.15 cm^{-1}, and measuring the concentration of active biomass (ATP) in water and on sample pieces (biofilm) after 8, 12 and 16 weeks. The S/V-ratio was kept constant by discharging a proportional part of the test water after collection of material samples. The method has been standardized in the Netherlands (NEN 2004). Further details of the standardized biomass production potential (BPP) method are given in Paragraph 12.4.3.3.

In Austria a method has been standardized for testing the EMG of plastic pipes. This method is based on measuring the HPC of water contained in vertically-placed pipes after contact times of one, two and three months at room temperature. The biofilm concentration is assessed with ATP measurements (Önorm 2002a, b).

Effects of materials on microbial growth have also been reported in the US (Ellgas and Lee 1980; Talbot and Seidler 1979) and the use of the MDOD method (BS 6920-2.4) was studied as well (Bellen *et al.* 1993). However, testing of materials for enhancement of microbial growth is not prescribed in the US. The UK BS 6920 MDOD method has been incorporated in the national standards of Australia, New Zealand and Singapore.

Recently, a method has been described, based on determining the concentration of assimilable organic carbon (AOC) in migration water that had been in contact with the material at 60°C (Bucheli-Witschel *et al.* 2012).

12.4 STANDARDISATION OF TEST METHODS IN EUROPE

12.4.1 Research related to the European Acceptance Scheme

In 1999/2000, a European Acceptance Scheme for materials used in contact with drinking water was planned to be established in connection to Article 10 of the Drinking Water Directive (EU 1998) and Article 3 of the Construction Products Directive (EU 1989). A Regulators Group on Construction Products Drinking Water (RG-CPDW) was formed, which identified the research needed to achieve the establishment of the European Acceptance Scheme (Fielding *et al.* 2001). These research needs included the development or harmonisation of methods for testing materials for (i) enhancement of microbial growth, (ii) cytotoxicity, (iii) analysis non-target compounds by gass-chromatopgraphy and mass-spectrometry of non-target compounds and (iv) the disinfection by-products (DBP)–formation potential. Subsequently, a European Research Project "Development of harmonised tests to be used in the European Approval Scheme concerning CPDW" was initiated. Work Package 1 of this project concerned the enhancement of microbial growth by materials. Research institutes in six European countries particpated in the investigations. The results of these investigations, which have been reported in detail (van der Kooij *et al.* 2003), are summarized below.

On the basis of existing methods, an improved batch-test method was developed. This method determines the Biomass Production Potential (BPP) of materials using ATP as the parameter for the concentration of active biomass in the test water and on the material. In brief, product samples with an external surface of 50 cm^2, which had been cleaned by washing with water, were incubated in tap water (900 ml, contained in wide-neck flasks of 1000 ml) at a surface to volume ratio of 0.166 cm^{-1} at 30°C for a period of 16 weeks. The tap water, supplemented with nitrate and phosphate to prevent growth limitation by N and P and with a 1% river-water inoculum, was replaced weekly to simulate a stagnation period of one week. Samples of water and material were collected after 8, 12 and 16 weeks for measuring the concentration of planktonic biomass (PB) and attached biomass (biofilm) with ATP analysis. Attached biomass (AB) was removed from the sample pieces using ultrasonic treatment and the ATP concentration of the obtained biomass suspension was determined. Subsequently, the concentration of attached biomass (AB, pg ATP/cm^2) was calculated and the biomass production (BP, pg ATP/cm^2) was obtained by using: $BP = AB + PB/0.166$. The Biomass Production Potential (BPP, pg ATP/cm^2) of a material is the mean of the BP values determined after 8, 12 and 16 weeks of incubation, corrected for the mean BP of the blank (glass).

The BPP value of six selected materials (stainless steel, chlorinated poly vinyl chloride (PVC-C), silicone rubber, high density polyethylene (HDPE), nitrile rubber and plasticized poly vinyl chloride (PVC-P), after correction for the BPP of the blank (glass), ranged from less than 50 pg ATP/cm^2 (PVC-C) to more than 40,000 pg ATP/cm^2 (PVC-P). In this first application of the revised test, the coefficient of variation of the reproducibility (COV) for the six participating laboratories ranged from 34% (silicone rubber) to 107% (nitrile rubber) for the different materials tested at 25°C and from 20 to 34% for three materials tested at 30°C (Figure 12.2). This reproducibility was affected by differences between laboratories in (i) the procedure for ATP analysis, (ii) the quality of the test water and (iii) procedures for biomass detachment. Most BPP values at 30°C were lower than those at 10 and 25°C, which was attributed to elevated endogenous respiration. The number of samples tested in the MDOD test and the W270 test was too limited to establish quantitative relationships between the results of these tests and the revised BPP test.

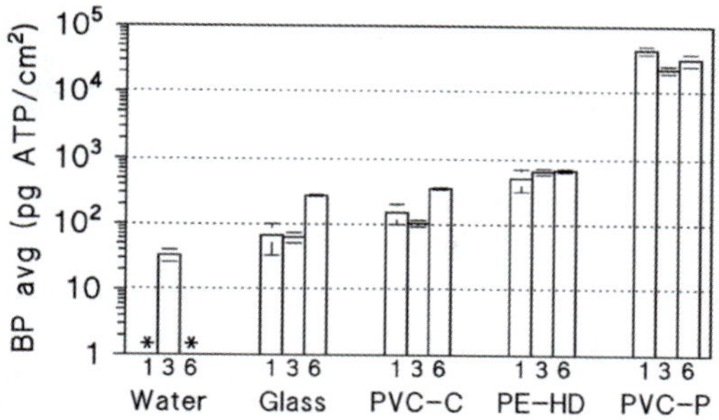

Figure 12.2 Average BP values with standard deviation of selected materials, tested by three participants in the project. Incubation at 30°C. BP avg. is the mean of the BP values after 8, 12 and 16 weeks of incubation, without correction for the effect of the blank (glass). BP of water is calculated using: PB/0.166. Numbers 1, 3 and 6 represent the involved laboratories. Abbreviations: PVC-C, chlorinated PVC; PE-HD, high-density polyethylene; PVC-P, plasticised (soft) PVC (positive control) and glass, negative control. Water: test water without material; *, water without material not tested.

From these results, it was concluded that standardisation of the adapted BPP test required further investigations aiming at:

- optimization and standardization of the ATP analysis;
- optimization of the procedure for the removal of attached biomass from the materials;
- a procedure and criteria for selection of suitable test water;
- establishment of the effect of temperature and selection of the most appropriate temperature for incubation.

12.4.2 Optimisation of the BPP test

The proposed research was conducted as part of the EC Proposal "Inter-laboratory testing for enhancement of microbial growth (EMG)". Research institutes in Denmark, France and Germany, the Netherlands and the United Kingdom participated in the investigations. The project addressed the issues mentioned above, but not the effect of test temperature, for which 30°C was selected by the participants. The results, which have been reported in detail (van der Kooij *et al.* 2006), are summarized below.

Procedures for ATP analysis were evaluated in a series of three round robin inter-laboratory tests with samples containing either freely dissolved ATP or active biomass grown on materials in contact with tap water. In the third round robin test, the coefficient of variation (CV_R) values of the ATP measurements were below 20% (mostly $<15\%$), when using the average value of the slopes of the calibration curves in each laboratory with ATP solutions prepared in buffer. The use of an average slope value reduced both the variations in the ATP measurements and the workload in the laboratory. The composition of tap water, for example, the concentration of calcium, has an impact on the ATP analysis. Therefore, a buffer solution was used for the calibration curves, to reduce the variation in the measurements in the provided ATP samples. However, measurement of the ATP concentration in the BPP test requires calibration in the tap water used in the test to exclude matrix effects.

The selection of the tap water used for the test is another critical aspect of the BPP test. The impact of test water composition on biomass production during the test should be limited in order to obtain a low level of detection (greatest sensitivity) and to improve the reliability of the result. The main quality aspects of test water are: (i) the growth-promoting properties of the water itself, (ii) the presence of a disinfectant residual and (iii) the possibility of growth limitation by inorganic trace elements. The growth-promoting properties of the tap water selected for use in the tests in each of the participants' laboratories were assessed in a batch test using ATP analysis for biomass quantification. Tap water in these laboratories gave maximum ATP concentrations (BPP) below 5 ng/litre in most cases. Addition of acetate at concentrations of 1 and 5 mg C/litre, with and without the addition of trace elements, demonstrated that acetate carbon was growth limiting in the selected tap waters. Trace elements neither promoted nor delayed growth. A procedure for determining the suitability of water for use in materials testing has been defined, which includes assessment of its growth-promoting properties and the effect of trace elements on the growth in the presence of added acetate.

Accurate determination of the growth-promoting properties of materials requires complete detachment of biomass growing on the material. Detachment techniques tested included low energy sonication, high energy sonication and swabbing. The applied technique(s) used repeatedly and/or combined, should be capable of removing >90% of the attached biomass. The suitability of a procedure can be assessed by determining the effect of each detachment technique.

After optimisation of the various test procedures, an inter-laboratory test was conducted using a total of seven materials. Typical test results, shown in Figure 12.3, demonstrate that the average PB concentration is proportionally related to the BPP (not corrected for growth in glass control), except for concentrations at the level of the blank. PB values remained at the level of the blank (water without material) in the presence of glass, PVC-C, and SS, indicating that these materials did not enhance microbial growth in the test system. The no-effect level of PB in the tests at laboratory A was below the no-effect level at laboratory B. The average BP values at these levels were <200 pg ATP/cm^2. PE-Xc showed some effect at laboratory A, but this effect was less pronounced at laboratory B. All other materials gave clearly elevated PB values. The average of the BPP values, reported by the five participating laboratories, ranged from 57 $\pm$ 46 pg ATP/cm^2 for glass (negative control for measuring the effect of water quality) to 31,600 $\pm$ 14,800 pg ATP/cm^2 for PVC-P. These results demonstrate that the BPP test enables differentiation of the growth-promoting properties over 2 to 3 orders of magnitude. The fraction of planktonic biomass (PB) comprised less than 15% for the materials with clearly elevated growth. COV values for these materials ranged from 31 to 47%. Two participants had used tap water with growth-promoting properties exceeding the level of 10 ng ATP/litre, leading to deviating results. Lower COV values can be expected with optimised biomass removal techniques, appropriate selection of test water and by following the procedure defined for ATP measurement. Water purified in the laboratory, and subsequently supplemented with inorganic nutrients and so on, may be an alternative for tap water.

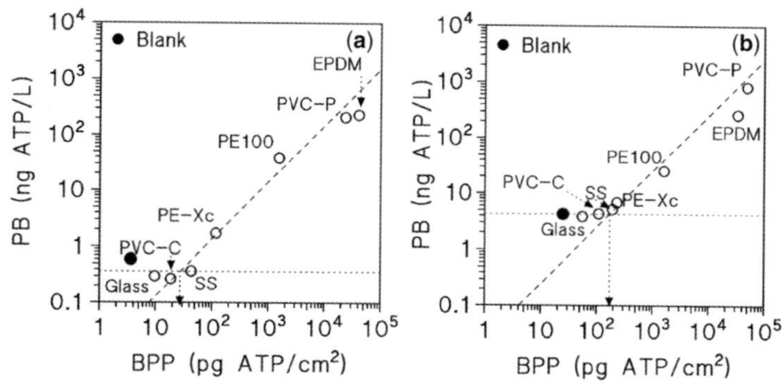

Figure 12.3 Relationship between the average concentration of planktonic biomass (PB, ng ATP/litre) and the Biomass Production Potential (BPP, pg ATP/cm^2) (not corrected for effect of water) of a selection of materials in the optimized BPP test at 30°C at two institutes (A, B). The dotted horizontal line represents the level of no-effect on PB. Abbreviations: SS. stainless steel; PVC-C, chlorinated polyvinylchloride; PVC-P, plasticized PVC (positive control); PE100, polyethylene; PE-Xc, polyethylene structurally cross-linked by radiation; EPDM rubber, ethylene propylene diene monomer; glass, negative control.

Statistical evaluation of the test results of all laboratories revealed that the average BPP values of the materials (not corrected for blank) differed significantly from the BPP value of glass. Two of the participating laboratories were unable to differentiate between the BPP of SS and PVC-C and the BPP of glass. In each laboratory, the BPP concentrations observed with PE100, EPDM and PVC-P differed significantly from the glass control and from the other materials. The following BPP–based ranking of the materials was obtained: Glass < PVC-C < SS < PE-Xc < PE100 < EPDM < PVC-P. Two laboratories had a different ranking for some of the materials with low BPP values.

The results of the inter-laboratory test showed that the overall performance of the BPP test at all laboratories was satisfactory, despite differences in procedures and test conditions, for example, water quality. Still, the performance of the BPP test can be further improved by standardising the procedures for (i) ATP analysis, (ii) selection of water type and (iii) biomass detachment. The performance criteria

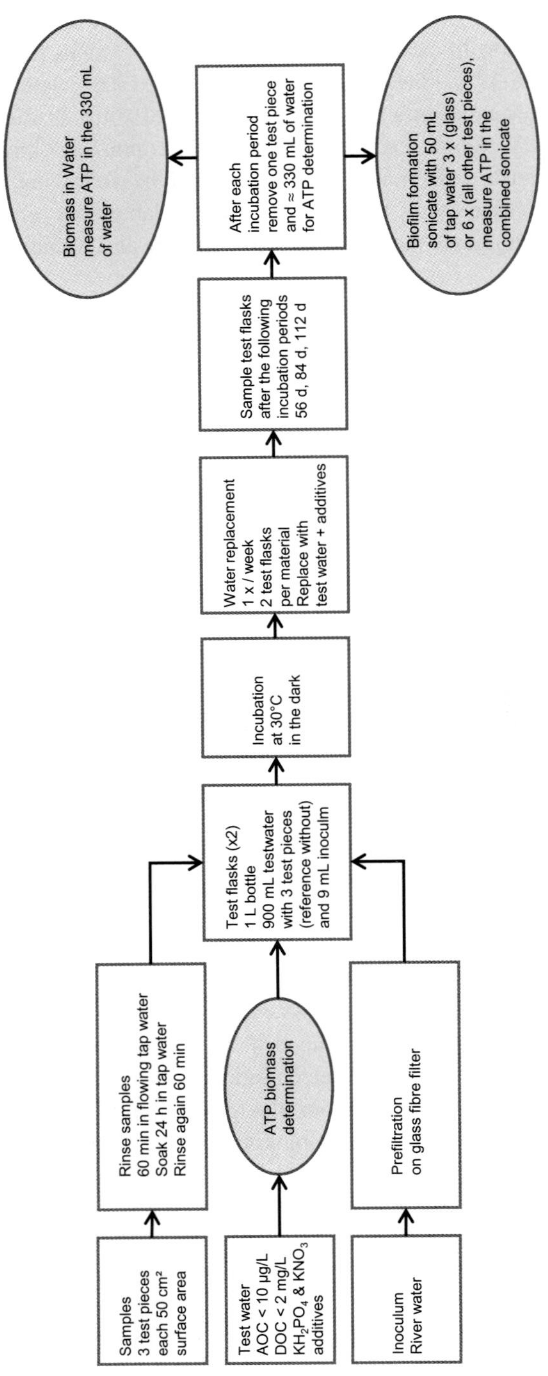

Figure 12.4 Scheme of the revised BPP-test.

required for the BPP test, viz. the lower level of detection and the COV value in inter–laboratory testing are critical at low BPP levels and therefore depend on the establishment of pass–fail limit (PFL) values.

The revised biofilm production potential (BPP) test is presented schematically in Figure 12.4. The test water should be drinking water complying with the EC drinking water directive (EU, 1998) with a number of additional conditions, for example, no free chlorine present, dissolved oxygen of ≥6.5 mg/litre, maximum ATP concentration <10 ng/litre in test water incubated under test conditions or AOC <10 µg/litre. Addition of P (phosphate) and N (nitrate) to the test water is required and testing for the need of other inorganic nutrients (trace elements) is needed. Inoculation should be done with 1% of membrane filtered (1,2 µm) river water. For the test, two containers are used for each material, each containing 3 sample pieces with an external surface area of 50 cm^2 each. Examples of sample pieces and test containers are shown in Figure 12.5.

Figure 12.5 Material sample pieces and test container for the BPP test. Stainless steel wire is attached to sample pieces to prevent them floating.

12.4.3 Three standardized test methods

12.4.3.1 Mean dissolved oxygen difference (MDOD, BS 6920)

The current MDOD test method is the latest version of the British Standard Method BS 6920–2.4 (BSI, 2000). In brief: the test is conducted in airtight jars of 1000 ml containing tap water without a disinfectant residual. The normal test S/V ratio is $0.15 \pm 0.2\ cm^{-1}$. A river water inoculum is added and the containers are incubated at $30 \pm 1°C$ for a period of 7 weeks. Twice a week the water is replaced and the oxygen concentration is measured at the time of the 4-day water change during weeks 5 to 7 (inclusive) of the test. The MDOD is the arithmic mean of the concentrations of dissolved oxygen measured in weeks five, six and seven in the test sample container, after subtraction of the mean of the test water control. The lower limit of detection for the MDOD method is 0.1 ± 0.1 mg O_2/litre and the glass reference (negative control) should have an MDOD value of < 0.1 mg O_2/litre. The method is used for materials approval with an established pass–fail limit (PFL) value of $\leq$2.39 mg O_2/litre, published in BS 6920-1 (Specification). This PFL value is based on observations of visible growth on the surface of the test pieces at MDOD values >2.1 mg O_2/litre and a 10% surface area tolerance factor (Colbourne and Brown 1979; Colbourne 1985). A large variety of materials has been tested with this method. Once work started in CEN to develop a European test method, no major changes in the British Standard were permitted (due to CEN standstill regulations) and therefore no further method development and optimisation work was done on the MDOD method.

12.4.3.2 Biomass volume (DVGW-W270)

The *Code of Practice* W270 method (DVGW 1990) uses the volume of the biofilm (including matrix) as parameter for microbial growth enhancement. The test is conducted in a flow-through system (reservoir of 100 L) using tap water without a disinfectant at ambient temperature and a continuous flow of 20 L/h. Test sample plates of 800 cm^2 surface each are completely immersed. After three and six months of exposure (original procedure), the sample plates are removed for examination, viz. biomass is scraped off and its volume is determined in ml after storage during 24 hrs at 6 ± 2°C. This original procedure has been modified and optimized based on investigations conducted in Germany (Gerhardy 2008). In the revised method (DVGW 2007a), sample plates are examined after one, two and three months of exposure and the volume of the biofilm is measured after centrifugation. The measurement of the biomass volume is evaluated after centrifugation of the collected biomass in a graduated tube with a calibrated scale between 0.01 and 0.4 ml. Centrifugation reduced the time needed for quantification of the biomass volume, and also reduced the biomass volume as compared to the original procedure of sedimentation. The test pieces are then washed with test water and placed once again in the test container. A further modification of W270 enables testing of pipe materials. Vertically-placed samples of pipes with a surface area 800 cm^2 are supplied with drinking water, from bottom to top (flow rate: 0.4 to 3 m/h). Biomass is removed with a conical stopper made of silicone or PVC-U. Figure 12.6 shows the test container with the plates (left) and the test module for pipes and hoses (right). The rubber scraper (for plates), the stopper for removing the biomass from the surfaces inside pipe samples and the centrifugation vial are shown in Figure 12.7. For the original Code of Practice W270, a pass–fail limit value of 0.1 ml of biomass/800 cm^2 after an exposition period of 6 months had been defined in Germany (DVGW 1990). This criterion was similar to the lower detection limit of the method, thus representing a zero tolerance limit for growth on materials under the test conditions. For the optimised W270 method (DVGW 2007b), using centrifugation

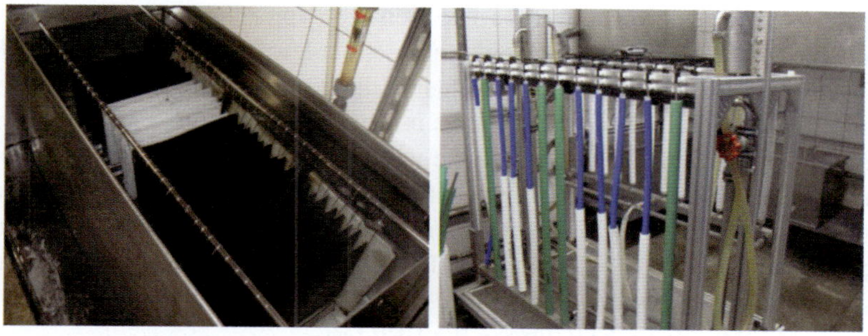

Figure 12.6 Test container and test module for DVGW-W270.

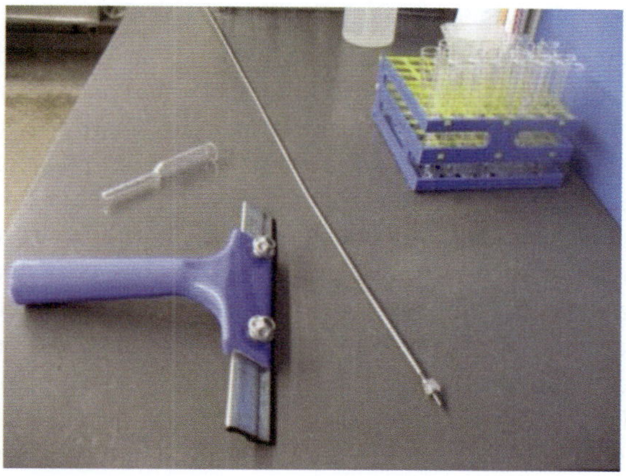

Figure 12.7 Rubber scraper and stopper for removing the biofilm in the W270-method.

to reduce the collected biomass volume, a PFL value for general use of 0.05 ($\pm$0.02) ml/800 cm² after each of the test periods has been defined (Gerhardy 2008). This level is similar to the previously defined test criterion, because centrifugation gives about 50% smaller biomass volumes than sedimentation.

12.4.3.3 Biomass production potential (BPP) based on ATP

The BPP test, modified on the basis of the inter-laboratory testing, has been described above and is shown in Figure 12.4. The lower level of detection of planktonic growth in this test is 0.5 ng ATP/litre, corresponding with a contribution of $0.5 \times S/V = 3$ pg/cm^2 of BPP. A PB value of 10 ng ATP/litre of the test water, measured in the absence of a material, corresponds with a contribution of 60 pg ATP/cm^2. The detection level for attached growth is lower, for example, treatment of a sample of 50 cm^2 for biomass removal in a total of

40 ml of water gives a lower level of detection of <1 pg ATP/cm^2. Hence, the BPP test enables the assessment of BPP at the no-effect level which is less than 100 pg ATP/cm^2, but the detection limit depends on the growth potential of the test water (Figure 12.3). Overall, the BPP test gives information about the effect of the tested material on the ATP concentration in the water (PB) and the biofilm formation on the material. A database of ATP concentrations in drinking water is available for comparison (van der Kooij 2003; van der Wielen and van der Kooij 2010). In the Netherlands a PFL value of 1000 pg ATP/cm^2 has been proposed for materials in contact with drinking water. Figure 12.3 shows that this value corresponds with a BP value of 20 to 30 ng ATP/litre in the BPP test.

12.4.4 Comparison of test characteristics

The characteristics of the three standardized methods are presented in Table 12.1. Natural microbial communities present in treated water and in surface water are used in these methods to assess the enhancement of growth by materials. These organisms are adapted to oligotrophic environments as prevailing in drinking water distribution systems. Culture techniques are not suited for measuring the level of planktonic microbial growth and attached growth in the tests. Therefore, the overall amount of microorganisms or microbial activity is assessed by using alternative measurements viz. ATP, biomass volume and oxygen consumption. As explained above, ATP represents the concentration of active biomass, and both the impact of the test material on the concentration of planktonic biomass and attached biomass is measured in the BPP test.

The biomass volume measured in the W270 method represents the amount of attached microbial biomass (biofilm), alive and dead, including the extracellular polymeric substances (EPS). The effect of the material on planktonic biomass is not tested in the W270 method. In the MDOD method, oxygen consumption is used as a parameter for active biomass. The utilisation of oxygen is a measure for the respiration by the microorganisms to obtain energy from the available food sources originating from the material under test. The oxygen utilisation measured represents the sum of the microbial activity in the water (planktonic) and on the material (biofilm).

The incubation temperature of the W270 test differs from the test temperature for the BPP and MDOD tests. A temperature of 30°C represents a worst case situation for example, in building plumbing systems and in distribution systems in mediterranean areas. Many studies have shown that temperature affects the composition of the aquatic microbial community, for example, enhanced growth of opportunistic pathogens at elevated temperature. In the European projects it was shown that the biofilm formation was lower at higher temperature (see 12.4.1).

In all tests, drinking water without disinfectant residual is used as test water. The use of water without a disinfectant residual and with a low microbial growth

Table 12.1 Characteristics of the three standardized test methods.

Characteristic	Method		
	MDOD (BS 6920)	**Biomass volume (W 270)**	**BPP**
Microbial biomass and/or activity	O_2-consumption	Biomass volume	ATP
Temperature	30±1°C	12±5°C	30±1°C
Test water	Drinking water without disinfectant	Drinking water without disinfectant	Drinking water without disinfectant
Testwater replacement	Twice a week	Continuously	Weekly
Test container	Sealed jars	Tanks	Sealed jars
Duration	7–9 weeks	12–16 weeks	16 weeks
Water volume	1000 cm^3	100,000 cm^3	900 cm^3
Surface of sample pieces	150 cm^2	800 cm^2	3 x 50 cm^2
Surface to volume ratio (S/V) (cm^{-1})	0.15	n.a.	0.166

potential is essential to prevent bias in the test by the effect of water on the growth-promoting potential of the material to be tested. The incubation conditions in the BPP test represent stagnation of one week at an S/V ratio, similar to a pipe diameter of 24 cm; the MDOD test simulates a stagnation of 3 to 4 days at a pipe diameter of 27 cm. These S/V ratio's are larger than those in distribution system pipes with a diameter >27 cm, but more than 10 times smaller than those of in-premise plumbing with pipe diameters of 1 to 2 cm. Hence, the impact of a pipe material on water quality in a water installation may be much larger than in the BPP and MDOD tests.

The different characteristics of the three methods, as shown in Table 12.1, allow their use for specific materials or product types but also have limitations. For example, multi-layer pipes cannot be tested easily with either the BPP-or MDOD methods, and testing of greases or lubricants is not easily possible or would require adaptations of the BPP test and the W270 method.

12.4.5 Comparison of test results

A comparative study using 11 commercially available PVC materials in the original W270 method and the MDOD test revealed that 9 of these materials exceeded the pass-fail limit (PFL) values in both tests (Schoenen and Colbourne, 1987). These materials, therefore, were considered unsuitable for use in contact with drinking and one was considered suitable by both tests. The average MDOD value of PVC-P in the comparative study was 4.9 $\pm$ 1.2 mg O_2/litre and the average biomass volume was 5.8 $\pm$ 1.2 ml/800 cm^2. These observations suggest that a PFL of 2.39 mg O_2/litre in the MDOD test corresponds with 2.39 $\times$ 5.8/4.9 =2.8 ml/800 cm^2 in the original W270 test, that is 28 times higher than the PFL of 0.1 ml/800 cm^2 in the original W270 test. This estimation may also apply for the optimized W270 method with a PFL of 0.05 ml/800 cm^2, because a reduction (relative to sedimentation) of 50% of the biomass volume has been observed for PVC-P (DVGW 2007). Production of planktonic biomass is not measured in the W270 method and therefore, the ratio will be larger than the estimated value of 28.

Stainless steel, PVC-C and glass did not affect the concentration of planktonic biomass (PB) and had BPP values of $<$100 pg ATP/cm^2 (Figure 12.3). Such a low BPP value has also been observed with PVC-U (van der Kooij and Veenendaal 1993; van der Kooij & Veenendaal 2001). This material comprises more than 50% of the pipe length in distribution systems in the Netherlands and does not affect the microbiological quality of drinking water distributed without a disinfectant residual. PVC-P gave a BPP of about 20,000 to 40,000 pg ATP/cm^2. Hence, the no-effect level of materials in the BPP test is about 400 times lower than the BPP of PVC-P. Applying this ratio to the observations on PVC-P in the MDOD and W270 methods, suggests that the no-effect values for oxygen consumption and biomass production are below the current PFL values and detection limits for these tests. Also, the no-effect level of 100 pg ATP/cm^2 is below the PFL value of 1000 pg ATP/cm^2 as proposed in the Netherlands. Elucidation of the relationships between the PFL values would require further investigation.

12.4.6 Further Developments Related to Standardization in Europe

Until 2006, it was intended to regulate the assessment and approval of materials for use in contact with water intended for human consumption in the European Community via the European Acceptance Scheme (EAS). Due to significant legal, regulatory issues and financial concerns, including the differing requirements of Article 10 of the Drinking Water Directive (covering all materials and products used between the source of the water and the consumer's tap) and Article 3 of the Construction Products Directive (which could only cover materials and products used in contact with water after the last point of treatment), the Commission stopped further work of the Regulatory Group – Construction Products Drinking Water. As a result of this decision, the National Regulators of the member states (MS) France, Germany, the Netherlands and the United Kingdom agreed to work together in the development of optimised and standardised test methods for approval of materials for use in contact with water intended for human consumption in accordance with the Drinking Water Directive, that is, from source to consumer's tap. This "4MS Group" has been working since 2007 towards defining and adopting common practices for these materials and products. The 4MS initiative was formalized in January 2011 and it was decided to work together to decide upon the use of standardised test methods. Meanwhile, the CEN group TC164/WG3/AHG3 had developed a test method standard (prEN 16421) for Enhancement of Microbial Growth (EMG) which includes the regulatory methods (UK and Germany) and the revised BPP method, which is the standard method in the Netherlands (CEN/TC164 2012). In the introduction to this draft standard the purpose and the methods are summarized as follows: 'The purpose of this standard is to describe three European test methods that can be applied to determine the ability of non-metallic materials to enhance microbial growth in drinking water'. The different procedures of the test methods included in the proposed standard allow selection of the best method(s) for a specific material/product type. The method(s) for use with specific product types will be specified in published harmonised product standards (hEN).

PFL values are needed for the evaluation of the result obtained for materials in the standardised method. Scientific and practical considerations are the basis for such values, aiming at maintaining compliance with water quality regulations as required in Article 10 of the Drinking water Directive (EU 1998). The establishment of PFL values for materials is complicated by differences not only between the standardized test procedures, but also by different conditions of water distribution in the various member states (for instance differences in the degree of contact between a material and water and the presence or absence of a disinfectant residual).

12.5 CONCLUSIONS

The following conclusions can be drawn from the information presented:

- Materials in contact with drinking water can have adverse effects on the microbiological quality of water intended for human consumption. The increasing variety of synthetic materials used in contact with drinking water requires the assessment of materials on the basis of enhancement of microbial growth to protect public health and maintain customers' confidence in and appreciation of the water quality.
- Three standardized test methods for measuring the EMG are included in a draft European standard prEN 16421, accepted by CEN/TC164. These methods differ in degree of exposure, duration, temperature and parameters for measuring microbial growth.
- The test waters used in the test methods must have a low microbial growth potential to enable proper assessment of the enhancement of microbial growth attributable to materials in contact with drinking water.

- The impact of materials on the microbiological water quality depends on the conditions for application in practice, for example, S/V ratio, contact time, temperature and also the presence of a disinfectant residual. These conditions may affect the definition of effective pass–fail criteria.
- The most stringent PFL values should be applied to materials with a high degree of contact with potable water, viz. pipes of distribution systems in contact with water without disinfectant residual and materials in tap-water installations in contact with water with or without disinfectant residual.

12.6 REFERENCES

Ashworth J. and Colbourne J. S. (1987). The testing of non-metallic materials for use in contact with potable water and the interrelationships with in service use. *Industrial Microbiological Testing*, 151–169.

Bellen G. E., Abrishami S. H., Colucci P. M. and Tremel C. J. (1993). Methods for Assessing the Biological Growth Support Potential of Water Contact Materials Report AWWA Research Foudation, American Water Works Association, Denver CO, USA.

Bernhardt H. and Liesen H. U. (1988). Trinkwasserverkeimungen in Verteilungsnetzen durch Korrosionschutz auf Bitumenbasis. *GWF Wasser–Abwasser*, **129**(1), 28–32.

Bressler D., Balzer M., Dannehl A., Flemming H.-C. and Wingender J. (2009). Persistence of *Pseudomonas aeruginosa* in drinking water biofilms on elastomeric material. *Water Science & Technol., Water Supply*, **9**, 81–87.

BSI (1982) Specification of requirements for suitability of materials for use in contact with water for human consumption with regard to their effect on the quality of the water. DD82 (now withdrawn and superceded)

BSI (1988) Suitability of non-metallic products for use in contact with water intended for human consumption with regard to their effect on the quality of the water. Part 2: Methods of test-Section 2.4: Growth of aquatic microorganisms. BS 6920-2.4. (revised version issued in 2000)

BSI (2000) BS 6920 – Suitability of non-metallic products for use in contact with water intended for human consumption with regard to their effect on the quality of the water. Part 1 (Specification) and Part 2.4 (Growth of aquatic microorganisms test).

Bucheli-Witschel M., Kötzsch S., Darr S., Widler R. and Egli T. (2012). A new method to assess the influence of migration from polymeric materials on the biostability of drinking water. *Wat. Res.*, **46**, 4246–4260.

Burman N. P. and Colbourne J. S. (1977). Techniques for the assessment of growth of micro-organisms on plumbing materials used in contact with potable water supplies. *J. Appl. Bacteriol.*, **43**, 137–144.

Burman N. P. (1979). Effect of non-metallic materials on water quality. *J. Inst. Water Eng. Sci.*, **33**, 11–18.

CEN/TC164 (2012) Influence of Materials on Water for Human Consumption – Enhancement of microbial growth (EMG)" prEN 16421.

Colbourne J. (1985). Materials usage and their effects on the microbiological quality of water. *J. Appl. Bacteriol. Symp. Suppl.* 47S–59S.

Colbourne J. S. and Brown D. A. (1979). Dissolved Oxygen utilization as an indicator of total microbial activity on non-metallic materials in contact with potable water. *J. Appl. Bacteriol.*, **47**, 223–231.

Colbourne J. S., Pratt D. J., Smith M. G., Fisher–Hoch S. P. and Harper D. (1984). Water fittings as sources of *Legionnella pneumophila* in a hospital plumbing system. *The Lancet,* January, **323**, 210–213.

Department of Health (2012a). Report on the Review of Evidence Regarding the Contamination of Wash-hand Basin Water Taps Within Augmented Care Units with Pseudomonads; www.dhgov.uk/health/files/2012/Pseudomonas-report-290612.pdf (accessed date: publication date is 29 June 2012, downloaded: 17 August 2012). Department of Health, London, UK.

Department of Health (2012b). Water Sources and Potential *Pseudomonas aeruginosa* Contamination of Taps and Water Systems. Advice for augmented care units. Department of Health, London, UK

De Vries H. (1890). Die Pflanzen und Thiere in den dunklen Räumen der Rotterdamer Wasserleitung. Bericht der biologischen Untersuchungen der Crenothrix-Commission zu Rotterdam vom Jahre 1887. Gustav Fischer, Jena.

DVGW (1990) Vermehrung von Mikroorganismen auf Materialien für den Trinkwasserbereich-Prüfung und Bewertung-. Technische Regeln Arbeitsblatt W270, Eschborn, Gernany.

DVGW (2007a) Vermehrung von Mikroorganismen auf Werkstoffen für den Trinkwasserbereich-Prüfung und Bewertung-. Technische Regel Arbeitsblatt W270. Bonn, Germany.

DVGW (2007b) Research into the refinement of the DVGW Code of practice W 270. "Surface biomass of micro-organisms on materials used in drinking water supply". DVGW Report W6/03/03, CEN/TC164/WG3 N885, Bonn, Germany.

Ellgas W. M. and Lee R. (1980). Reservoir coatings can support bacterial growth. *J. Am. Water Works Assoc.*, **72**, 693–695.

EN 12873-1 (2003) Influence of Materials on Water Intended for Human Consumption-Influence Due to Migration – Part 1: Test Method for Non-metallic and Noncementitious Factory Made Products.

EU (1989) Council Directive 89/106/EEC of 21 December 1988 on the Administration of Laws, Regulations and Administrative Provisions of the Member States Relating to Construction Products (89/196/EEC).

EU (1998). Council Directive 98/83/EC of 3 November 1998 on the quality of water intended for human consumption. *Official J. European Communities*, **L330**, 32–53.

EU (2011) Regulation (EU) No 305/2011 of the European Parliament and of the Council of 9 March 2011 Laying Down Harmonized Conditions for The Marketing of Construction Products and Repealing Council Directive 89/106/EEC.

Fielding M., Rogers H. R. and Wilson I. M. (2001) Effects of materials on water quality. Inventory of European Approval Schemes for Products for Use in Contact with drinking water, WRc-NSF Ltd.

Fields B. S., Benson R. F. and Besser R. E. (2002). *Legionella* and Legionnaires' disease: 25 years of investigation. *Clin. Microbiol. Rev.*, **15**, 506–526.

Frensch K., Hahn J. U., Levsen K., Niessen J., Schöler H. F. and Schoenen D. (1987). Koloniezahlerhöhungen in einem Trinkwasserbehälter verursacht durch Lösemittel des Anstrichmaterials. *Zbl. Bakt. Hyg.*, **B. 184**, 556–559.

Gerhardy K. (2008). Vermehrung von Mikroorganismen auf Werkstoffen für den Trinkwasserbereich – Prüfung und Bewertung. *Energie/Wasser-Praxis*, **3**, 31–32.

Kilb B., Lange B., Schaule G., Flemming H.-C. and Wingender J. (2003). Contamination of drinking water by coliforms from biofilms grown on rubber-coated valves. *Int. J. Hyg. Environ. Health*, **206**, 563–573.

Metropolitan Water Board (1958) The examination of New Materials and Equipment for Water Fittings. In: Thirty Eight Report on the Results of the Bacteriological, Chemical and Biological Examination of the London Waters for the Years 1957–1958, E. Windle Taylor (eds), Metropolitan Water Board, London, UK, pp. 79–80.

NEN (2004). Bepaling van de biomassaproductiepotentie (BPP) van kunststof materialen metalen en coatings in contact met (leiding)water. Ontwerp NVN 1225, Normcommissie 349 164 'Drinkwatervoorziening'.

Niedeveld C. J., Pet F. M. and Meenhorst P. L. (1986). Effect of rubbers and their constituents on proliferation of *Legionella pneumophila* in naturally contaminated hot water. *The Lancet*, July 26, **328**, 180–184.

Önorm B 5018-1 (2002a) Prüfung der Verkeimungsneigung von Trinkwasserrohren. Teil 1: Prüfverfahren. Österreichisches Normungsinstitut, 1020 Wien.

Önorm B 5018-2 (2002b) Prüfung der Verkeimungsneigung von Trinkwasserrohren. Teil 1: Bewertung. Österreichisches Normungsinstitut, 1020 Wien.

Rogers J., Dowsett A. B., Dennis P. J., Lee J. V. and Keevil C. W. (1994). Influence of plumbing materials on biofilm formation and growth of *Legionella pneumophila* in potable water systems. *Appl. Environ. Microbiol.*, **60**, 1842–1851.

RQIA (2012) Independent review of incidents of *Pseudomonas aeruginosa* infection in neonatal units in Northern Ireland. Final report. The Regulation and Quality Improvement Authority; *www.rqia.org.uk*

Seidler R. J., Morrow J. E. and Bagley S. T. (1977). Klebsielleae in drinking water emanating from redwood tanks. *Appl. Environ. Microbiol.*, **33**, 893–900.

Schoenen D. (1989). Influence of materials on the microbiological colonization of drinking water. *Aqua*, **38**, 101–113.

Schoenen D. and Colbourne J. S. (1987). Microbiological evaluation of drinking water construction materials. *Zbl. Bakt. Hyg.*, **B. 183**, 505–510.

Schoenen D. and Schöler H. F. (1983). Trinkwasser und Werkstoffe: Praxisbeobachtungen und Untersuchungsverfahren; Praxisbeobachtungen und Untersuchungsverfahren. DVGW-Schriftenreihe, Wasser Nr. 37. Gustav Fischer Verlag, Stuttgart.

Schoenen D., Schulze-Röbbecke R. and Schirdewahn N. (1988). Mikrobielle Kontamination des Wassers durch Rohr-und Schlauchmaterialien. 2. Mitteilung: Wachstum von *Legionella pneumophila. Zb. Bakt. Hyg.*, **B. 186**, 326–332.

Schofield G. M. and Locci R. (1985). Colonization of components of a model hot water system by *Legionella pneumophila. J. Appl. Bacteriol.*, **58**, 151–162.

Schulze-Robbecke R. and Fischeder R. (1989). Mycobacteria in biofilms. *Zentrblt. Hyg. Umweltmed.*, **188**, 385–390.

Talbot H. W., Morrow J. E. and Seidler R. J. (1979). Control of coliform bacteria in finished drinking water stored in redwood tanks. *J. Am. Wat. Works Assoc.*, **71**, 349–353.

Tobin J. O. H., Beare J. and Dunell M. S. (1980). Legionnaires' disease in a transplant unit: isolation of the causative organism from shower baths. *Lancet*, 118–121.

van der Kooij D. (1987). Actuele bacteriologische problemen bij opslag, transport en distributie van drinkwater. *H_2O*, **20**(1), 6–11.

van der Kooij D. (1992), Assimilable organic carbon as an indicator of bacterial regrowth. *J. Am. Wat. Works. Assoc.*, **84**(2), 57–65.

van der Kooij D. (2003). Managing regrowth in drinking-water distribution systems. In: Heterotrophic plate counts and drinking-water safety, J. Bartram, J. Cotruvo, M. Exner, C. Fricker and A. Glasmacher (eds), *IWA Publishing*, pp. 199–232.

van der Kooij D. and Hijnen W. A. M. (1984). Presence and control of *Legionella pneumophila*, the causative agent of Legionnaires' disease, in hot water installations (in dutch). *H_2O*, **17**(18), 387–391.

van der Kooij D. and Veenendaal H. R. (1992). Assessment of the biofilm formation characteristics of drinking water. Proc. Wat. Qual. Technol. Conf. Am. Water Works Assoc. Denver, CO, pp. 1099–1110.

van der Kooij D. and Veenendaal H. R. (1993). Assessment of the biofilm formation potential of synthetic materials in contact with drinking water during distribution. Proc. Water Qual. Technol. Conf, Am. Water Works Assoc Denver CO, pp. 1395–1407.

van der Kooij D. and Veenendaal H. R. (2001). Biomass production potential of materials in contact with drinking water: method and practical importance. *Water Sci. Technol. Water Supply*, **1**(3), 39–45.

van der Kooij D., Veenendaal H. R., Slaats N. P. G. and Vonk D. (2002). Biofilm formation and multiplication of *Legionella* on synthetic pipe materials in contact with treated water under static and dynamic conditions, *In*: Legionella, R. Marre (ed.), ASM Press Washington D.C, pp. 176–180.

van der Kooij D., Albrechtsen H. J., Corfitzen C. B., Ashworth J., Parry I., Enkiri F., Hambsch B., Hametner C., Kloiber R., Veenendaal H. R., Verhamme D. and Hoekstra E. J. (2003). CPDW Project Assessment of the Microbial-Growth Support Potential of Products in Contact with Drinking Water. Report EUR 20832 EN. (Contract no. EVK1-CT2000-00052).

van der Kooij D., Veenendaal H. R. and Scheffer W. J. H. (2005). Biofilm formation and multiplication of *Legiionella* in a Model Warm Water Systems with Pipes of Copper, Stainless steel and Cross-Linked Polyethylene. *Water Res.*, **39**, 2789–2798.

van der Kooij D., Baggelaar P. K., Veenendaal H. R., Moulin L., Corfitzen C. B., Albrechtsen H. J., Holt D. and Hambsch B. (2006). Standardising the Biomass Production Potential Method for Determining the Enhancement of Microbial Growth by Construction Products in Contact with Drinking Water. Report EC Grant Agreement nr SI2. 403889.

van der Wielen P. W. J. J and van der Kooij D. (2010). Effect of water composition, distance and season on the adenosine triphosphate concentration in unchlorinated drinking water in the Netherlands. *Wat. Res.*, **44**, 4860–4867.

Wadowsky R. M., Butler L. J., Cook M. K., Verma S. M., Paul M. A., Fields B. S., Keleti G., Sykora J. L. and Yee R. B. (1988). Growth supporting activity for *Legionella pneumophila* in tap water cultures and implication of hartmannelid amoebae as growth factors. *Appl. Environ. Microbiol.*, **54**, 2677–2682.

Chapter 13

Biological filtration for diverse applications: towards the development of a unified conceptual design approach*

P. M. Huck, B. K. Siembida-Lösch and M. M. Sozański

13.1 INTRODUCTION

Biological processes have been used for a long time in drinking water treatment, for instance in slow sand filtration and bank filtration. In recent decades, biological rapid filtration processes have come into relatively widespread use. Seminal work in this regard was the development of the Mülheim process (Sontheimer *et al.* 1978). Other early applications of biological filtration are described by Rice and Robson (1982). More recently, a comprehensive review has been provided by Prévost *et al.* (2005) and conference proceedings on the topic have been edited by Gimbel *et al.* (2006). Although a filter will operate biologically whenever there is no disinfectant residual throughout the depth of the filter, a major impetus to biological filtration was given by the increased use of ozonation, which increases the concentration of easily biodegradable carbon.

The first modern quantification of biodegradable carbon in drinking water treatment was provided by van der Kooij (e.g. van der Kooij *et al.* 1982), who presented a method for what he termed Assimilable Organic Carbon (AOC). Another commonly used method measures Biodegradable Dissolved Organic Carbon (BDOC) (e.g. Servais *et al.* 1987). A discussion of various measurement methods is provided by Huck (1990). In addition to reducing concentrations of easily biodegradable organic carbon implicated in bacterial regrowth in the distribution system, biological filtration has also been shown to reduce levels of

*This chapter is based on the previously published paper – Huck P. M. and Sozański M. M. (2008) Biological filtration for membrane pre-treatment and other applications: towards the development of a practically-oriented performance parameter. *J. Water Supply Res. Tech. – AQUA*, **57**(4), 203–224.

chlorine demand (e.g. Prévost *et al.* 1991) and chlorination byproduct formation potential (e.g. Niquette *et al.* 1999). Although biological filtration in drinking water treatment can also be used for other applications such as nitrogen removal (e.g. Rittmann and Huck 1989) the most common use is for the removal of organic carbon.

The uses of biological filtration mentioned just above can be considered as the classical applications, contributing to improved water quality within the distribution system, the importance of which is discussed by Sozański and Olańczuk–Neyman (2002). In addition to these, newer applications include the potential removal of trace contaminants such as pharmaceuticals and endocrine disrupting compounds that are biodegradable, the removal of odorous substances such as geosmin, and the use of biological filtration as a membrane pre-treatment (e.g. Hallé *et al.* 2009; Farr and Stampone 2007).

Over the years various modeling approaches have been applied to biological filtration (normally for its classical applications) and the major models are reviewed briefly in this paper. However, no overall design framework for biological filtration has yet emerged. In part this is because specific generally accepted quantitative goals and therefore design criteria for biofiltration have not been defined (a notable exception being the Netherlands). In addition, the models are generally complicated and require a number of parameters that are not easily measured in practice.

In summary, there is as yet no paradigm or approach for biological filtration comparable to those that exist for other processes: for example the CT (concentration × time) concept for disinfection, the use of G (velocity gradient) and GT (t is time) in flocculation, and the use of the L/d ratio (L is filter depth and d is media diameter) in granular filtration design. The objective of this paper is to develop a unifying practically-oriented conceptual approach to biofiltration based on modeling. This approach is based on a previously developed for contact time, dimensionless contact time or X*, that incorporates actual contact time and other factors important for biofiltration performance. The approach is discussed in terms of the various applications of biofiltration in relation to organic carbon – the classical one of the removal of easily biodegradable material, the removal of trace contaminants and, in somewhat more detail, the relatively new application of biological filtration as a membrane pre-treatment. For the latter application the approach needs further development. To relate X* directly to practice, the last section of the paper proposes a new design-oriented parameter, the Biofiltration Factor (BF).

In principle both X* and the BF can be further developed for other uses of biofiltration in drinking water treatment, such as nitrogen removal and the removal of iron and manganese. Similarly, the X* and BF parameters could potentially be developed for the analysis of other traditional biological processes in drinking water such as slow sand or bank filtration and subsurface passage.

13.2 BACKGROUND

13.2.1 Factors affecting the performance of biofiltration

Various authors (e.g. Urfer *et al.* 1997) have reviewed the factors affecting the removal of biodegradable organic matter (BOM) in rapid biological filtration. The factors that are important are:

- Contact time
- Surface area
- Media
- Temperature
- Nature and concentration of the influent BOM, and
- Biofilm disruption (due to backwashing)

For a given bed depth, the surface area is inversely related to media diameter. The concentration of easily biodegradable organic matter, often measured as assimilable organic carbon (AOC), entering a biofilter will be increased if there is upstream ozonation (e.g. van der Kooij *et al.* 1982).

From a practical point of view, the level of control that can be exerted over the various factors is important. Table 13.1 (Huck *et al.* 2000) provides a qualitative assessment of both the degree of control and effect of the major parameters. For example the type of media can be specified and it has a moderate influence on process performance. On the other hand temperature has a major impact on performance and essentially cannot be controlled. The high effect shown for chlorination is because of its impact on biomass. (This figure was developed fairly early in the study and in fact the effect of chlorine at full-scale was found to be less than expected.) The impact of the backwashing method may be lower than shown, except when biomass levels in the filter are low. The impact of salinity on the removal of micropollutants by biological carbon filters has been investigated by Sozański (1995).

Table 13.1 Level of control over parameters that influence (rapid) biological filtration.

	Degree of control (✓)		Or effect (♦)	
Parameter	**None**	**Low**	**Moderate**	**High**
Media type			♦	✓
Chlorination			✓	♦
Filtration rate (EBCT)			✓♦	
Backwashing method			♦	✓
BOM loading		✓	♦	
Temperature	✓			♦
Time since startup	✓			♦

Source: Huck *et al.* (2000). Reprinted with permission.

Although the extent of biodegradation is related to the amount of biomass, the latter cannot be independently controlled because it is related to biodegradation through the yield coefficient. Thus the factors listed above that control the extent of biodegradation also essentially control the amount of biomass. Although various methods of measuring biomass have been used in biofiltration investigations, a number of investigators have estimated biomass by measuring phospholipid, based on a method by Findlay *et al.* (1989). In addition, Kijowska *et al.* (2001) have reported on the use of a metabolic activity test to investigate biodegradation in biological filters. Although other methods have been used, a detailed review of biomass measurement methods is outside the scope of this article.

In European practice a number of applications of biofiltration have been second-stage filters (typically with GAC as media). A number of North American installations have implemented biofiltration in first-stage biofilters (i.e. where particle removal is also an objective). The approaches developed in this chapter are applicable to both first- and second-stage biofiltration.

Results from an American Water Works Association Research Foundation (AwwaRF) study (Huck *et al.* 2000) provide a useful context for a practically-based approach to quantify biofiltration performance. In contrast to many investigations that are conducted at pilot-scale, a substantial portion of this AwwaRF study was conducted on demonstration scale and full-scale biological filters. The objective was to optimize the performance of biological filters that were also intended for particle removal. Data were available from three locations and included a wide range of temperature conditions.

Figure 13.1 summarizes results from this study for the removal of BDOC above 10°C. There are two important messages in this figure:

- Removals of BDOC do increase with increasing EBCT, but not proportionally.
- The removal of BDOC in these mature filters was essentially the same for both anthracite/sand and GAC/sand media. This was despite the fact that the biomass (as measured by the phospholipid method) was higher on the GAC media. GAC was shown to be better for BDOC removal at temperatures below 5°C and there were indications that it might also be better in the range from 5 to 10°C. GAC also did show an advantage above 10°C for the removal of a specific easily biodegradable compound (oxalate).

Additional results from the study showed that the use of air scour in backwashing (under collapse pulsing conditions – Amirtharajah *et al.* 1991) did not affect BOM removals (oxalate, AOC and DOC), compared with water-only backwashing. These results were obtained using a GAC filter at temperatures above 10°C without chlorine in the backwash water. These data supported results of modeling conducted as part of the study (Hozalski and Bouwer 2001a, b, discussed later in this paper) that indicated backwashing would have to achieve biomass removals greater than 60% to have an effect on performance.

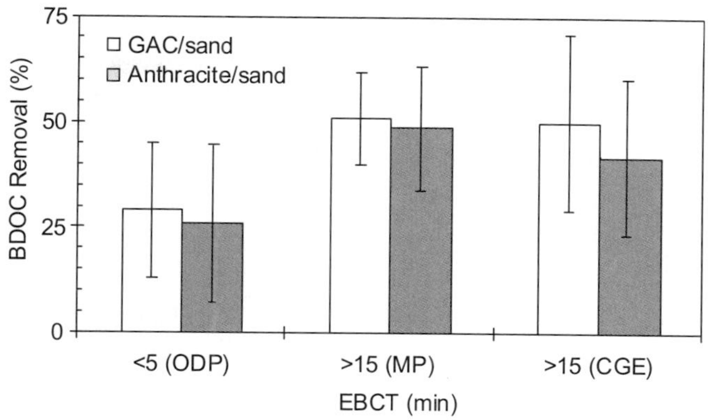

Figure 13.1 Impact of EBCT on removal of BDOC (above 10°C).
Source: Huck *et al.* (2000). Reprinted with permission. (ODP = Oxidation Demonstration Plant, MP = Mannheim Plant and CGE = Compagnie Générale des Eaux).

Additional investigations showed that the BOM removal capability of the full-scale filters was essentially unaffected by free chlorine in the backwash water. This was true whether air scour was present or absent. In contrast, the performance of bench-scale filters backwashed using air scour was substantially impaired by free chlorine, although not by combined chlorine. From a practical perspective, it is encouraging that full-scale filters appear to be more robust. The reasons for the differences between bench- and full-scale performance could not be determined within the study. However, it is possible that the older, full-scale biofilms had become more resilient.

Winter data from a site (low temperature, not shown) where lower biomass levels were observed suggested that the biomass level as measured by the phospholipid method only becomes important for performance if it drops below a certain level. The lack of a direct relationship between phospholipid biomass and removals was also noted by Urfer *et al.* (1997) in analyzing data published by Wang *et al.* (1995). Earlier, Hubele (1985) had concluded that the total biomass in a biofilter was only linked in a small way to the degradation of easily biodegradable TOC. The lack of a direct relationship between biomass and removals by biofiltration was also shown by Elhadi (2004) for geosmin and methylisoborneol (MIB).

In summary, important points for biofiltration modeling from this demonstration and full-scale study are:

- BOM removals increase with increasing EBCT, but in a less than proportional way,
- Backwashing likely does not have a measurable effect on BOM removal, and
- Biomass (as measured by the phospholipid method) is not directly quantitatively related to BOM removal.

13.2.2 Review of modeling approaches

Before considering the various available models, it is useful to place them in a contextual framework. Three important reasons for which modeling may be undertaken are:

- Design
- Operation
- Research

Table 13.2 shows some key attributes of models, in relation to these three goals. The number of diamonds indicates qualitatively the importance of each attribute, depending on the intended use of the model. For example, a model to be used for operation should be relatively easily understood by operations staff and based on easily-measured parameters, whereas a very important aspect of a model used for research is its comprehensiveness. Models to be used for design would generally fall between these two categories in terms of their desired attributes. Most of the available models have been developed by researchers and their complexity, along with the general lack of specific quantitative goals for biofiltration, has hindered the use of modeling in practice. It is important to note that several of the models were developed in terms of a single rate-limiting substrate, whereas drinking water organic matter is complex, typically consisting of components with different rates of biodegradation.

Table 13.2 Some key modeling attributes.

	Characteristic			
Objective	**Easily understood**	**Easily measured parameters**	**Comprehensiveness**	**Addresses multiple biofiltration objectives**
Design	♦♦	♦♦	♦♦	♦♦ (?)
Operation	♦♦♦	♦♦♦♦	♦	♦♦ (?)
Research	♦	♦	♦♦♦♦	♦♦ (?)

Source: Huck and Sozański (2008). Reprinted with permission.

In their critical review of biological filtration, Urfer *et al.* (1997) summarized the practical usefulness of the most widely known models available at that time. Those authors note that although the models discussed provide important process insights for dinking water biofiltration, they are relatively complex and cannot be directly used by practitioners because they require parameters that cannot be readily obtained.

Table 13.3 is an updated version of a summary table provided by Urfer *et al.* (1997) showing the processes considered by each of the models, the major assumptions of each, the number of model parameters and the calculated outputs. The most important practical output, the substrate concentration, is an output of all of the models. The table has been updated by adding the models developed by Uhl (2000, 2001) and Hozalski and Bouwer (2001a, b). These newer models are also complex. Thus all of the models in Table 13.3 fall into the category of 'research models' in Table 13.2.

In terms of a simpler approach, a first-order empirical model was presented by Huck and Anderson (1992) and Huck *et al.* (1994) for the prediction of approximate performance. Those authors showed that the amount of BOM removed in a given biofilter was directly proportional to the influent concentration. Therefore removals can be approximated as a first-order process and a biofilter at apparent steady-state will essentially achieve a constant percentage removal, for a given EBCT and temperature. In addition to applying to the removal of AOC and BDOC, the relationship was shown to hold for trihalomethane formation potential (THMFP) and chlorine demand. Gagnon *et al.* (1997) demonstrated that it applied to the removal of carboxylic acids.

With the exception of the work by Hozalski and Bouwer (2001a, b) all of the models discussed above address steady-state or pseudo steady-state conditions. These are the most important in practice because, except for a few special circumstances such as process startup, biofilters can be generally considered to be operating at pseudo steady-state. Hozalski and Bouwer's non-steady-state model was developed to capture the impacts on performance of the startup phase of a biological filter and recurring events such as backwashing. Their modeling provided good qualitative and quantitative agreement with full-scale data (Hozalski and Bouwer 2001b). As mentioned previously, their simulations demonstrated that backwashing would have to remove a relatively high percentage of the attached biomass (60% or greater) to have an impact on performance. Thus as indicated by the results of the full/demonstration scale study (Huck *et al.* 2000) summarized earlier, for practically-oriented modeling a steady-state approach is normally adequate.

Although the book on biological filtration by Prévost *et al.* (2005) devotes several pages to a review of modeling, no overall modeling approach is proposed. Quantitative design-related criteria are not presented, except for the provision of some distribution system threshold BOM values. Also, Westerhoff *et al.* (2005) conducted a detailed investigation of ozone-biofiltration for geosmin and MIB removal. They conducted some first-order modeling of removal rates but did not provide an overall modeling approach. They did provide design criteria for new full-scale biofilters for geosmin/MIB removal. The process-related design recommendations were 48 inches (1220 mm) of GAC and an EBCT of 7.5 minutes. It is evident (even in very recent publications) that an overall quantitative yet practically-based approach to biofiltration design has not been developed.

Table 13.3 Summary of selected mechanistic models applicable to drinking water biofiltration.

Reference	Processes Considered	Major Assumptions	Number of Parameters			Calculated Output
			Measured or Calculated	Assumed	Total	
Rittmann and McCarty (1980a)	Mass transport of substrate to biofilm, simultaneous diffusion and bioreaction of substrate in biofilm, definition of S_{min}	Substrate must be growth-limiting, predation not directly considered	5	3	8	Biofilm thickness, substrate flux into biofilm, bulk substrate concentration
Wanner and Gujer (1986)	General model with steady-state or dynamic cases possible, mass transport of substrate to biofilm, diffusion and bioreaction of substrate in biofilm, biomass exchange with bulk liquid	Biofilm biomass treated as continuum, one spatial dimension considered	Dependent on model formulation			Biofilm thickness, concentration of suspended microbes, bulk substrate concentrations
Billen *et al.* (1992)	Uptake of rapidly and slowly biodegradable DOC, temperature dependence of bacterial activity, physicochemical adsorption of bacteria, biological attachment of	No resistance to mass transfer of BDOC	10	2	12	Attached and suspended biomass, rapidly and slowly biodegradable DOC concentration

	bacteria (biosynthesis of EPS), bacterial mortality and grazing by protozoa					
Wang and Summers (1995)	Mass transport of substrate to biofilm with substrate utilization on biofilm surface, definition of rapidly, slowly, and nonbiodegradable DOC fractions, biomass dependence on filter depth	Thin biofilm, expression of biomass dependence on filter depth must be known	7	3	10	Concentration of rapidly and slowly biodegradable DOC with filter depth
Zhang and Huck (1996)	Application of Rittmann and McCarty (1980a) approach to plug-flow bioreactors with AOC as substrate, definition of dimensionless EBCT	AOC is growth-limiting substrate and sole carbon and energy source, predation not directly considered	5	4	9	Substrate flux into biofilm, bulk effluent substrate concentration
Uhl (2001)	Substrate utilization, bacterial growth, maintenance and surface catalysis,	No substrate concentration too low for degradation,	12	3	15	Substrate concentration, removal of suspended

(*Continued*)

Table 13.3 Summary of selected mechanistic models applicable to drinking water biofiltration (*Continued*).

Reference	Processes Considered	Major Assumptions	Number of Parameters			Calculated Output
			Measured or Calculated	Assumed	Total	
	suspended bacteria, grazing, non-steady state formulation also possible	growth of suspended bacteria neglected, mass transfer not limiting				bacteria and concentration of attached biomass as a function of filter depth
Hozalski and Bouwer (2001a)	As a function of time: transport of substrate to, and utilization by, biofilm, deposition of suspended bacteria, biofilm growth, decay and detachment, loss of biomass due to backwashing, mixtures of BOM	Biofilm homogeneous within a given filter segment, single substrate both diffusion and reaction limiting, no substrate competition in mixtures	17	4	21	Biofilm thickness and substrate concentration as a function of time and filter depth, effect of mixture composition on effluent BOM concentration (Hozalski and Bouwer 2001b)

Source: Huck and Sozański (2008). Reprinted with permission. (Adapted from Urfer *et al.* (1997)).

Urfer *et al.* (1997) describe the introduction of the steady-state biofilm model by Rittmann and McCarty (1980a) (Table 13.3) as being "seminal" work. The model considers microbial (Monod) kinetics and mass transport (Fick's second law). The approach assumes that all nutrients are in excess except the rate-limiting substrate, S. An important contribution from this work was the introduction of the concept of S_{min}, which is defined as the bulk substrate concentration below which a steady-state biofilm cannot exist. Urfer *et al.* (1997) note that several studies (Zhang 1996; Zhang and Huck 1996; Rittmann and McCarty 1980b; Namkung and Rittmann 1987) have shown that the principles of the steady-state biofilm model are applicable to the low substrate (oligotrophic) conditions usually encountered in drinking water biofiltration.

The basic equations for the model, which are written in dimensionless form, are shown below:

$$S_b^* = S_s^* + J^* L^* / D^* \quad (1)$$

$$S_{min}^* = b/(Yk - b) \quad (2)$$

$$L_f^* = \frac{1 + S_{min}^*}{S_{min}^*} J^* = J^* kY/b \quad (3)$$

Where:

$$S_b^* = S_b/K_s \qquad S_s^* = S_s/K_s \quad S_{min}^* = S_{min}/K_s$$
$$J^* = J\tau/K_s D_f \quad L_f^* = L_f/\tau \quad L^* = L/\tau$$
$$D^* = D/D_f \qquad (\tau = [K_s D_f/kX_f]^{1/2}, \mathrm{L})$$

where S_b is the bulk liquid concentration of the substrate (M_sL^{-3}); D is the free liquid diffusivity (L^2T^{-1}); D_f (L^2T^{-1}) is the diffusivity in the biofilm; K_s is the half velocity constant in the Monod expression (M_sL^{-3}); k is the maximum specific rate of substrate utilization (T^{-1}); Y is the yield coefficient, M_x/M_s; b denotes the overall biofilm decay rate coefficient (T^{-1}); X_f is the biofilm density ($M_x\ L^{-3}$); L_f is the biofilm thickness (L) and L denotes the thickness of the effective diffusion layer (L).

Because a direct analytical solution for this model is not possible, Rittmann and coworkers developed pseudo-analytical solutions (Sáez and Rittmann 1988, 1992).

Zhang (1996) and Zhang and Huck (1996) obtained an analytical solution for this model for a plug flow reactor (i.e. a filter) after demonstrating that axial dispersion could be ignored. They then obtained the following equation for the depth of filter (X) required to achieve a specific concentration of substrate (at the biofilm outer surface):

$$X = \frac{\upsilon\tau}{\alpha D_f} \int_{S_{sx}^*}^{S_{so}^*} \frac{dS_s^*}{J^*} + \frac{\upsilon L}{\alpha D} \ln \frac{J_0^*}{J_X^*} \quad (4)$$

where v is the superficial flow velocity or hydraulic loading rate (LT^{-1}); α denotes the specific surface area which is the biofilm surface area in each unit volume of the bioreactor (L^{-1}); x is the longitudinal distance along the column (L); J_0 is the flux of substrate into the biofilm at the inlet end of the column ($M_sL^{-2}T^{-1}$); J_x is the flux at the outlet end of the column ($M_sL^{-2}T^{-1}$); $S_{so}^* = S_{so}/K_s$ with S_{so} as the substrate concentration on the biofilm surface at the inlet end, (M_sL^{-3}); and $S_{sx}^* = S_{sx}/K_s$ with S_{sx} as the substrate concentration on the biofilm surface at depth X, (M_sL^{-3}).

13.3 DEVELOPMENT OF A UNIFYING PARAMETER – DIMENSIONLESS CONTACT TIME (X*)

The most important contribution by Zhang (1996) and Zhang and Huck (1996) was the further transformation of Equation 4 to introduce the concept of an index for contact time, X*. This parameter, referred to as dimensionless EBCT, is given by Equation 5:

$$X^* = \theta\frac{\alpha D_f}{\tau} = \int_{S_{sx}^*}^{S_{so}^*} \frac{dS_s^*}{J^*} + \frac{L^*}{D^*}\ln\frac{J_0^*}{J_X^*} \tag{5}$$

where $\theta = X/v$, that is, the EBCT.

Although Equation 5 provides the definition for X*, the factors that it includes are shown more clearly in Equation 6, where τ is replaced by its constituent parameters.

$$X^* = \theta\alpha D_f^{1/2}(kX_f/K_s)^{1/2} \tag{6}$$

It is thus evident that X* includes, in addition to the actual contact time, the reactor specific surface area (surface area per unit filter volume), BOM diffusivity in the biofilm, biofilm density and biodegradation kinetic parameters.

Making use of pilot-scale data from an earlier study (Huck *et al.* 1991), Zhang (1996) and Zhang and Huck (1996) modelled the removal of AOC and provided extensive interpretations related to the X* parameter. They demonstrated a linear relationship between substrate removal and influent substrate concentration. The slope of this relationship (essentially the percentage removal) increases with X*, as shown in Figure 13.2. (The values for several other dimensionless parameters are shown in the figure, as is the relative importance of the external mass transfer resistance (L*/D*)). Although the curves show some small influence of an increase in the external mass transfer resistance, the major message of the figure is that the slope (essentially percentage removal) initially increases essentially linearly with increasing X* and then tapers off to approach a maximum value. Huck (1999) used this information to calculate approximate percentage removals, which are shown in Figure 13.3. It is thus evident that increasing the value of X*, for example by increasing the residence time in a biofilter, will provide improvement, as would be expected, but with diminishing returns. This is consistent with, for example, the results for BDOC shown in Figure 13.1.

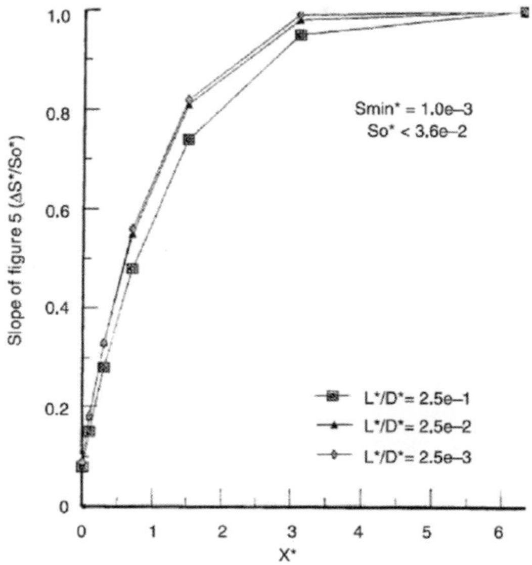

Figure 13.2 Impact of X* on change in substrate concentration.
Source: Zhang and Huck (1996). Reprinted with permission.

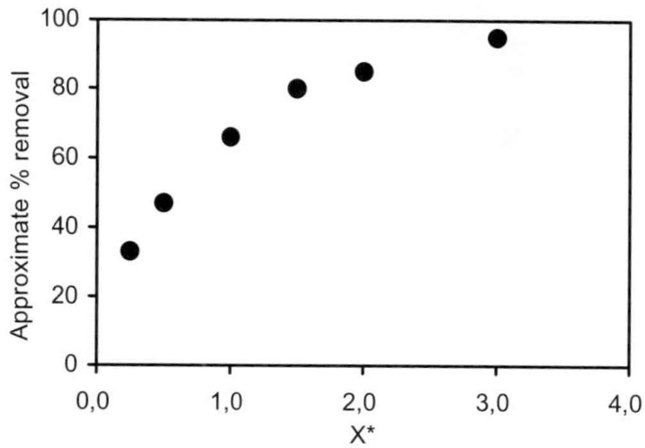

Figure 13.3 Impact of X* on percentage removal of substrate.
Source: Huck and Sozański (2008). Reprinted with permission.

An important aspect of X* is that even though the physical parameters of a reactor (media and flow rate) may remain constant (and therefore contact time remains constant), X* may vary, and therefore reactor efficiency may vary. This is illustrated conceptually in Figure 13.4a, where the removal profile on the right represents a coarser media, which of course has a lower value of α, leading to a reduction in X*, other factors such as flow rate remaining the same. A conceptually useful way to think of this is shown in Figure 13.4b. If, in general, we consider that reactor effectiveness increases with reactor size, the effect of, for example, using coarser media is to "shrink" the filter with respect to its effectiveness as a bioreactor.

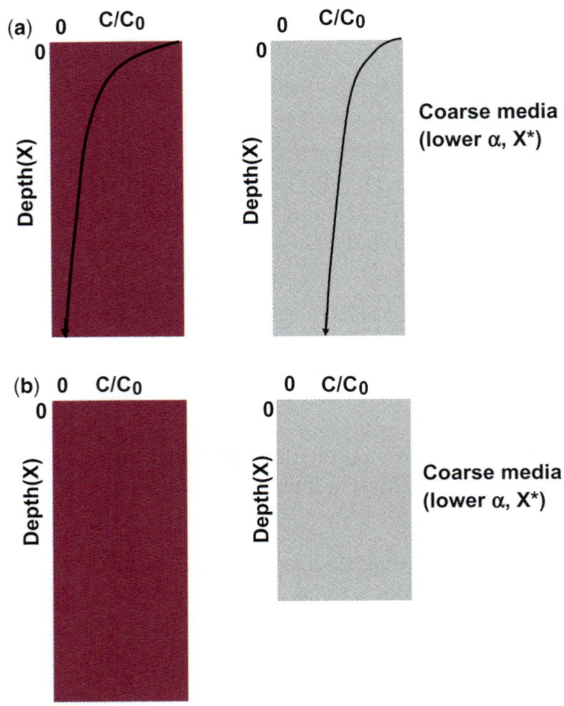

Figure 13.4 (a) Conceptual illustration of impact of X* on substrate removal in biofilter. (b) Illustration of concept that reduced X* "shrinks" biofilter (i.e. makes it less effective) (X* is the vertical dimension in the figure).
Source: Huck and Sozański (2008). Reprinted with permission.

Based on their extensive analysis of the pilot-scale AOC data from Huck *et al.* (1991), Zhang and Huck (1996) provided several important conclusions or inferences. These include the unimportance of external mass transport resistances and the fact that first-order kinetics would be appropriate. Both of these conclusions have been subsequently confirmed in other experimental investigations, as will be discussed later.

Huck (1999) used the X* concept to compare the expected bioreactor effectiveness of different typical filter designs. The hypothetical sand, dual-media and deep-bed monomedia filters considered all had an L/d ratio of 1100, and hydraulic loadings of 5, 10 and 15 m/h respectively. The EBCT of the deep-bed monomedia filter was one third less than that of the sand filter, however the product of α and EBCT (and therefore X*) decreased by a factor of three. In the same article he also used the X* concept to interpret results from different filter configurations that had been evaluated by Ødegaard (1996) in an investigation of ozone/biofiltration for the removal of humic substances.

The next section discusses further development of the X* parameter, including its usefulness in additional applications of biofiltration, and therefore its appropriateness as a unifying parameter. However, in order to make X* useful in practice, it needs to be directly linked to filter design and operating conditions. This is done later in this chapter through the introduction of a new directly-related parameter, the Biofiltration Factor.

13.4 FURTHER DEVELOPMENT OF THE X* PARAMETER

Work completed since the original development of X* indicates that it can be simplified, which would assist in its acceptance by practitioners. In addition, the X* concept can be extended, at least conceptually, to other applications of biological filtration (besides the classical ones of reduction in bacterial regrowth potential, chlorine demand and chlorination byproduct formation potential) such as trace contaminant removal and membrane pre-treatment.

13.4.1 Simplification of X*

Two opportunities for simplification of drinking water biofiltration modeling, and of the X* parameter in particular, are the ability, in most situations, to use first-order rather than Monod kinetics and to ignore mass transfer limitations. As indicated previously, the fact that external mass transfer limitations could be ignored and first-order kinetics could be used was identified by Zhang and Huck (1996). The appropriateness of first-order kinetics and the non-importance of both external and internal mass transfer resistances have been confirmed by further experimental work.

Booth *et al.* (2004) assessed the importance of both internal and external mass transfer resistances in an investigation of the removal of easily biodegradable compounds in pilot-scale biofilters. Those authors calculated both an external mass transfer modulus, as suggested by Karel *et al.* (1985) and an internal mass transfer modulus, as suggested by Weisz (1973). In both cases, if the modulus is less than unity, biodegradation kinetics control the rate of substrate removal. For their system, Booth *et al.* (2004) showed that both moduli were much less than unity, indicating that mass transfer limitations could be neglected. The authors note that the biomass values measured in their research (using the phospholipid method of Findlay *et al.* (1989) with minor modifications) were in the general range of those measured in full-scale biofilters, as reported by Huck *et al.* (2000).

They thus concluded that it would be reasonable to neglect, at least as a first approximation, mass transfer limitations for easily biodegradable substances in drinking water biofilters.

The conclusion drawn by Booth *et al.* (2004) is supported by the work of several others. Wang (1995) showed that the rate of removal of BDOC in biological filtration was controlled by reaction rate, and not mass transfer. Urfer-Frund (1998) demonstrated that both external and internal mass transfer were of only minor importance for the removal of easily biodegradable compounds in his pilot-scale biofilters. Gagnon and Huck (2001) showed that neither external nor internal mass transfer limited removals of easily biodegradable compounds in model drinking water distribution systems. Further, Elhadi (2004) demonstrated that mass transfer was not limiting for the removal of the odorous compounds geosmin and MIB in pilot-scale biofilters.

The appropriateness of first-order kinetics to represent the removal of easily biodegradable carbon in drinking water biofilters has been demonstrated in several instances. As described previously, Huck *et al.* (1994) demonstrated this empirically, and Zhang and Huck (1996) also showed that a linear relationship between amount removed and influent concentration (i.e. first-order behaviour) could be derived from the steady-state biofilm model. Although they regarded their estimates of kinetic parameters for AOC as being tentative, Zhang and Huck (1996) also determined that the K_s value for AOC was relatively high (in the mg/L range) indicating that removals would be first-order at concentrations normally experienced in drinking water. An exception to this might be heavily polluted raw waters with very high concentrations (several mg/L) of easily biodegradable carbon.

Trace contaminants are removed in biological processes through what is referred to as secondary utilization, where the biomass is sustained by biodegradation of the primary substrate. For the removal of geosmin and MIB at the ng/L level by biofiltration, Elhadi (2004) demonstrated that first-order kinetics were reasonable. Ho *et al.* (2007) also determined the biodegradation of these two compounds to be a pseudo-first-order reaction.

In terms of the simplification of X^*, if mass transfer is not limiting and kinetics can be considered to be first-order, Equation 6 can be simplified. Both the parameters D_f and K_s then become unimportant and can be grouped into a constant term, leading to Equation 7 below:

$$X^* = \theta\alpha K\,(kX_f)^{1/2} \tag{7}$$

where K is a constant.

The constant term is required to preserve the non-dimensionality of X^* and to maintain the relationship as an equation as opposed to a proportionality.

Since X^* was developed from a solution to the steady-state biofilm model, the elimination of two important aspects of that model (the substrate diffusivity and a parameter in the Monod relationship) make a strict theoretical interpretation of the simplified X* more difficult. However, from a practical point of view the simplified X* parameter can be very useful and provides important insights. Certainly θ and α can be relatively easily determined, and kX_f can be experimentally measured. K can be approximately determined from the relationship between percentage removal and X* previously presented (Figure 13.3).

Equation 7 clearly shows that for a given substrate at a given temperature (i.e. given k), performance is essentially directly proportional to both EBCT and media surface area. Thus this expression allows comparison among different process conditions. Equation 7 also shows that the biomass term is less important. This is supported by results reported in the literature, as reviewed earlier in this chapter.

The practically-oriented Biofiltration Factor introduced later in this paper does not require the simplified X*. However the Biofiltration Factor may be more accessible to practitioners with the simplified X* as an intermediate step.

13.4.2 Secondary utilization (trace contaminant removal)

Since biofiltration can provide removals of trace contaminants, it would be useful (for design purposes) to extend the X* concept to secondary utilization. Biofiltration can be successful for the removal of odorous compounds (e.g. Elhadi *et al.* 2006; Metz *et al.* 2006; Huck and Sozański 2011). Figure 13.5 is an example of such removals, and incidentally also shows that removals are not quantitatively related to biomass (as measured by the phospholipid method). In addition, recent work has shown removals of some pharmaceuticals at the ng/L level by biological filters (Hallé 2010; Zearley and Summers 2012) and by biologically active roughing filters (Guay *et al.* 2007).

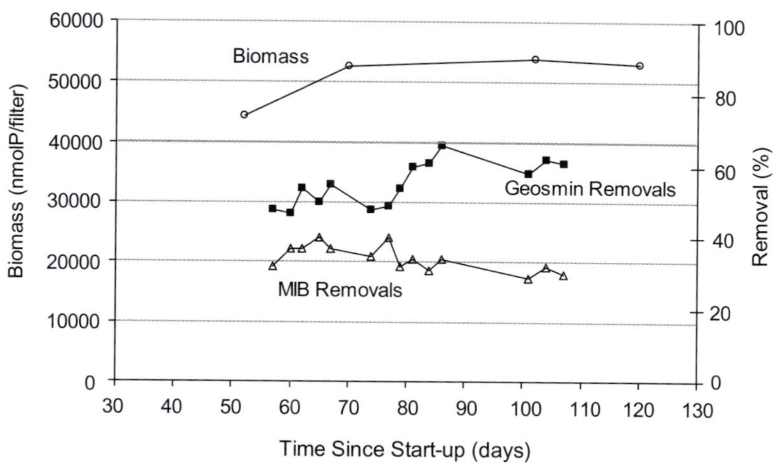

Figure 13.5 Biomass and geosmin/MIB removals as a function of time in a biofilter (exhausted GAC, 20°C).

Using geosmin as an example, mass transfer is not limiting and first-order kinetics can be considered reasonable, as was noted previously. The biomass in a biofilter used for the removal of trace contaminants is determined by primary utilization, so the parameter X* remains important. (From a practical perspective it is reasonable to consider the entire biomass as a surrogate for organisms capable of degrading geosmin, although Hoefel *et al.* (2006) have shown that a consortium of three specific bacteria is necessary for geosmin biodegradation.) Therefore, the only additional parameter that is required is the first-order rate constant for geosmin biodegradation. Although it is not necessary to use the simplified expression for X* to apply it to trace contaminant removal, it is conceptually simpler to do so.

Based on Equation 7, we can consider conceptually the percentage removal of a trace contaminant for two cases: in one case the first-order rate constant for secondary utilization (k_2) is greater than that for the removal of easily biodegradable carbon (k) and in the second case it is less. Figure 13.3 has shown that percentage removal is related to X*. If $k_2 > k$, the percentage removal of the trace contaminant will be greater than that of easily biodegradable carbon for a given value of X* if the other parameters in Equation 7 remain the same, that is, for a given filter and flow rate. The opposite will be true if $k_2 < k$. Thus as a first approximation, removal of a trace contaminant (probably following acclimation) could be estimated for a given biofilter by laboratory determination of the biodegradation rate constant in comparison to that for the removal of easily biodegradable carbon. This would give an indication of what θ and α (contact time and media diameter) would be required for the desired percentage removal of the trace contaminant. Based on Equations 6 and 7, the relative adjustments in θ and/or α for a given removal of a secondary contaminant will be proportional to $(k_2/k)^{1/2}$. This should be experimentally verified.

13.4.3 Biofiltration as membrane pre-treatment

Recent work has shown biofiltration to have promise as a membrane pre-treatment to reduce fouling (e.g. Hallé *et al.* 2009, Uhl *et al.* 2007). The extension of X* to this application requires considerable new development, which is discussed in this section.

In this section we first review the types of organic matter most responsible for organic and biofouling, two important types of membrane fouling. We then relate this information to biofiltration. In considering a biofilter as a way of reducing such fouling, it is important to consider the fact that biofilters not only remove organic matter but also produce both soluble and particulate organic material.

13.4.3.1 Organic matter and fouling

A number of earlier studies (e.g. Yuan and Zydney 2000; Mosqueda-Jimenez *et al.* 2004) determined that organic material with high molecular weight and high hydrophobic content contributes to the bulk of organic membrane fouling.

However, Carroll *et al.* (2000) suggested that a lower molecular weight, hydrophilic fraction is responsible. Investigating natural waters, Howe and Clark (2002) determined that very small colloids of 3–20 nm in diameter (primarily organic material) seemed to be most important. Lee *et al.* (2004) have demonstrated that the NOM fractions contributing most to fouling are colloidal biopolymers (polysaccharides, protein and amino sugars).

For charged ultrafiltration membranes, Cho *et al.* (2000) showed that the NOM foulants were the larger sized neutral and/or basic components. Investigating fouling during ultrafiltration of solutions of NOM fractions isolated from surface waters, Makdissy *et al.* (2004) found that the most significant flux decline (i.e. fouling) was due to the organic colloid fraction, a hydrophilic fraction consisting mostly of bacterial cell wall residues.

Jarusutthirak *et al.* (2002) investigated the fouling effects of wastewater effluent organic matter (EfOM) isolates on nanofiltration and ultrafiltration membranes. The colloidal EfOM fraction primarily consisted of polysaccharides, proteins and/or aminosugars and produced a high level of fouling. In particular, polysaccharides and/or aminosugars within the colloid fraction were found to be important in fouling. Haberkamp *et al.* (2007) also demonstrated the importance of EPS (primarily polysaccharides) for fouling of ultrafiltration membranes.

For RO membranes, Gabelich *et al.* (2004) attributed higher levels of fouling at pilot-scale (compared to full-scale) to production of low levels of natural polymers by biological communities upstream of the pilot-scale membrane.

More recent work (e.g. Hallé *et al.* 2009; Peiris *et al.* 2010; Haberkamp *et al.* 2011; Peldszus *et al.* 2011, 2012) has demonstrated the importance of biopolymers (polysaccharides and proteins or protein-like material) in the fouling of low-pressure (ultrafiltration) membranes. A key analytical tool in establishing this has been size exclusion chromatography with continuous organic carbon detection (LC-OCD) analysis, as described by Huber *et al.* (2011). This technique allows quantitation of major TOC components such as biopolymers and humic substances.

It is thus evident that specific fractions of organic matter are important for fouling of various types of membranes. The use of biofiltration to reduce this organic fouling has received systematic investigation only recently. Studies that have been conducted are reviewed briefly in a later section.

13.4.3.2 Biofouling

The traditional understanding of biofouling is that microorganisms in the feedwater become trapped on the membrane surface and proliferate because of the available nutrients (Uhl *et al.* 2003). In general, the limiting nutrient for bacterial growth in drinking water (not specifically related to membrane fouling) is considered to be biodegradable carbon. However it has been demonstrated that phosphorus may sometimes be limiting (Lehtola *et al.* 2001). In addition, recent work by

Lauderdale *et al.* (2012) has shown that the performance of biological filters can be improved by nutrient addition.

Biofouling has been investigated by a number of researchers (e.g. Vrouwenvelder *et al.* 2000, 2007). Flemming and Wingender (2001) discuss the role of extracellular polymeric substances (EPS) in biofouling. Flemming (1997) found that biofilm on an RO membrane contained more EPS under conditions of nutrient depletion. Of significance for the present chapter is the statement (e.g. Flemming 2002) that nutrients are the most important factor for biofouling. Hijnen *et al.* (2009) have demonstrated a relationship between AOC concentration and biofouling of high-pressure (spiral-wound) membranes. Flemming *et al.* (1997) proposed a biofilter as a membrane pre-treatment for the removal of easily biodegradable material, but provided no specifics.

Uhl *et al.* (2003) have presented calculations to show that microbial cell material transported to the membrane surface (as opposed to growth of microorganisms on the membrane) can be sufficient to account for observed accumulations of biological foulants in some situations.

Thus, although the literature indicates that reducing the concentration both of bacterial nutrients (with organic carbon usually being the limiting one) and also of microbial product material will be important for reducing biofouling, the use of biofiltration to do this has received virtually no systematic investigation for biofouling.

13.4.3.3 Microbial product material

Since it is evident that microbial product material can play a significant role in organic fouling and biofouling, the paragraphs below briefly review the different types of soluble and particulate material that may be produced. This discussion does not include organic material from other types of organisms (e.g. algae) that may be present in the raw water.

Laspidou and Rittmann (2002a) identify three groups of microbial products (EPS, soluble microbial products (SMP) and inert biomass) and provide a critical review of the relationships among them. SMP are defined as "soluble cellular components that are released during cell lysis, diffuse through the cell membrane, are lost during synthesis, or are excreted for some purpose". Those authors report that these substances have moderate formula weights and are biodegradable. They note that although early biofilm research often assumed that polysaccharides were the most abundant component of EPS, in fact proteins and nucleic acids also appear in significant amounts.

SMP consists of two components. The first, defined as utilization-associated products (UAP), are produced as a direct result of substrate utilization. The second component is biomass-associated products (BAP), which are formed from biomass, presumably as part of the decay process. Both of these components are relatively biodegradable, however they have distinct degradation kinetics (Laspidou and Rittmann 2002a).

Although SMP are completely soluble, EPS are mostly, although not completely, associated with the solid phase and are therefore mostly insoluble (Laspidou and Rittmann 2002a). Those authors note that some researchers divide EPS into "bound" and "soluble".

In a companion paper Laspidou and Rittmann (2002b) report on non-steady-state modeling of EPS, SMP and active and inert biomass. To conduct the modeling they used kinetic parameters for UAP and BAP obtained from other work. Of significance for the present chapter is the fact that these kinetic parameters reflect much lower biodegradation rates for UAP and BAP than for glucose. The maximum specific utilization rate for UAP is more than an order of magnitude lower than that for glucose, and the rate for BAP is in turn more than an order of magnitude lower than that for UAP. Although the authors note that the set of parameters used would not necessarily apply to all systems, and although easily biodegradable organic matter in drinking water would not necessarily be degraded as quickly as glucose, it is reasonable to expect that biodegradation rates for microbial products in drinking water biofilters would be substantially lower than for easily degradable organic matter.

Thus, microbial products (either entering the filter or generated within it) can be biodegraded and/or removed by physical filtration within the filter. The bed depths or contact times required are currently unknown but are likely to be greater than for the removal of easily biodegradable organic matter.

13.4.3.4 Effectiveness of biofiltration as a membrane pre-treatment

Before attempting to quantitatively assess the requirements for biofiltration as a membrane pre-treatment in subsequent sections, this section briefly reviews investigations where biofiltration has been used for this application.

Work related to bank filtration and aquifer recharge has demonstrated that the organic components of interest as membrane foulants have the potential to be removed by biofiltration processes (Dillon *et al.* 2005; Sattler *et al.* 2005; Grünheid and Jekel 2005). Although these results suggest that biofiltration could be promising for the removal of specific organic components leading to membrane fouling, the detention times in rapid filters are much less than in bank filtration and aquifer recharge, and therefore X^* will be much lower.

Results from investigations of rapid biofiltration have, however, also demonstrated its promise as a membrane pre-treatment. For low-pressure (UF) membranes, initial work feeding a biofilter (without prior coagulation or ozonation) with synthetic water containing easily biodegradable substances and humic acid was reported by Basu and Huck (2004). In subsequent experiments Mosqueda-Jiménez and Huck (2006a) confirmed these results and also investigated pre-treatment for nanofiltration (NF) membranes (Mosqueda-Jiménez and Huck 2006b). It was evident that the biofilter was capable of removing both low molecular weight (easily biodegradable) organic matter and particulate material retained on a 0.45 μm filter (colloidal humic material). In examining low flux NF installations for NOM removal from surface water, Uhl *et al.* (2007) found that biological activated carbon filtration at an EBCT of 34 minutes could provide significant mitigation of fouling. Dual-media filtration at an EBCT of 5.6 minutes had a limited effect. However, Xiong (2005) found that the effluent from a full-scale biofilter did not produce a lower rate of fouling with low-pressure membranes than did the filter influent.

In more recent work using LC-OCD analysis, Hallé *et al.* (2009) demonstrated that rapid biofiltration (without prior coagulation) was capable of reducing the concentration of biopolymers (polysaccharides and proteins), the major organic fouling components, and, as shown in Figure 13.6, also the rate of hydraulically reversible and hydraulically irreversible fouling of a UF membrane. Followup work at pilot-scale by Peldszus *et al.* (2012) over a two-year period confirmed these results for the same impacted river water, with temperatures varying between 1 and 25°C.

Thus, although biofiltration shows considerable promise as a membrane pre-treatment, to be effective it needs to achieve a net reduction in both the dissolved and particulate components most important for fouling, including the easily biodegradable organic matter that would facilitate actual bacterial growth on the membrane.

As indicated by Laspidou and Rittmann (2002a, b), the relationships among the original substrate and dissolved and particulate bacterial material within the filter will be complex. In the next sections we demonstrate the general conceptual relevance of X* with respect to dissolved and particulate bacterial material exiting a biofilter.

13.4.3.5 Net production of biomass

The net production of biomass by a biofilter is determined by the following factors:

- The amount of biomass entering the filter
- The amount of biomass generated within the filter
- The biomass removed or consumed within the filter.

The biomass concentration entering the filter will be a function of raw water levels and/or upstream processes. The generation of biomass within the filter will be controlled by the consumption (removal) of BOM, in large part the easily-degradable portion. Since the latter is related to X*, biomass generation will also be related to X* via a yield coefficient. The operating conditions of the filter may be important in determining the yield coefficient. For example Uhl (2000) determined a much lower yield coefficient (based on biodegraded TOC) than is typically used (e.g. Laspidou and Rittmann 2002b). This may have been because he was investigating second-stage biological filters, which would have lower influent concentrations of easily biodegradable carbon than many first-stage filters.

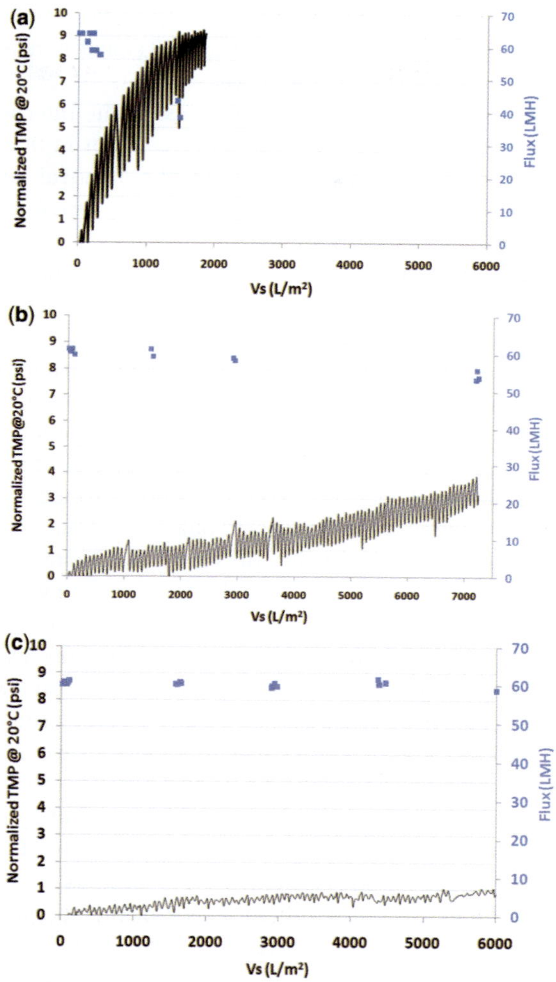

Figure 13.6 Results showing the effect of biofiltration pre-treatment on TMP of a UF membrane during the summer of 2007. The UF membrane was fed with (a) RF effluent, (b) B1 (5 min EBCT) effluent and (c), B2 (14 min EBCT) effluent. The right y-axis and the squares represent the permeate flux measured throughout the experiment. Different membrane modules were used for the RF-effluent and the biofilters effluent. The normalised TMPs were calculated by subtracting the clean water TMP of each membrane module measured prior to the experiment from the TMP obtained during the experiment.

Source: Reprinted (adapted) with permission from Cynthia Hallé, Peter M. Huck, Sigrid Peldszus, Jens Haberkamp, and Martin Jekel. Assessing the Performance of Biological Filtration As Pretreatment to Low Pressure Membranes for Drinking Water. Environmental Science & Technology, 2009, 43 (10), 3878–3884. Copyright 2009 American Chemical Society.

Biomass produced within the filter is of course subject to decay and detachment. Detachment is a complex phenomenon (Stewart 1993) and is related both to the amount of biomass present and the growth rate.

Filtration mechanisms can act to remove biomass in the filter influent. Hubele (1985) noted that filtration mechanisms can remove, lower in the filter, detached biomass from higher in the filter (facilitating the removal of more slowly biodegradable DOC lower in the filter), and that therefore the filtration characteristics of biofilter media are important. Filtration mechanisms will apply to both active and inert biomass as well as to insoluble or solid-associated EPS. Uhl (2000, 2001) included the attachment/detachment of bacteria in his modeling of biofilter performance.

For the minimization of membrane fouling, it is important to minimize the level of biomass exiting the filter. The generation of biomass will largely occur in the initial part of the filter where most of the easily biodegradable material is removed. The concentration and type of easily biodegradable material will be a function of the incoming water quality and may not be appreciably affected by any upstream processes, with the exception of ozonation and other oxidation processes.

From a design and operating point of view, the major opportunities for minimizing the amount of biomass exiting the filter lie in maximizing biomass removal within the filter and/or in minimizing its generation. In general, this would mean increasing the contact time to facilitate filtration (and at least some consumption) of both incoming biomass and biomass generated within the filter.

With respect to quantifying filtration, the factors affecting its performance are well known (e.g. MWH 2005). These include flow rate and media depth and diameter as well as characteristics of the media and particles affecting attachment. Characteristics affecting attachment and the conditions affecting detachment can be represented in filtration modeling through coefficients. Usually in rapid filtration upstream coagulation plays a crucial role and has a substantial impact on the coefficients. In the case of biofiltration as a membrane pre-treatment, it is also important to consider the case where such filters may be operated without coagulation. This is possible (and desirable from a "green" perspective) since the membrane is capable of providing particle removal. The filter is therefore functioning as a bioreactor and to reduce turbidity peaks in the case of varying influent water quality. The filtration performance in terms of particle removal therefore does not need to be as high as if the membrane were not also present.

In terms of the major controllable physical factors affecting filtration performance (media depth (L) and diameter (d)), the ratio L/d is often used to guide design (e.g. Kawamura 1999). For a given flowrate, L directly determines θ, the EBCT. Thus L/d is directly proportional to θ/d. Since the specific surface area of the media, α, is inversely proportional to d, θ/d is directly proportional to the product $\theta\alpha$, that is, the physical elements of X^* (Equations 6 and 7).

The L/d ratio is sometimes used as part of the exponent in a first-order relationship to approximately describe particulate removal by filtration. Such a relationship implies diminishing returns with increases in filter depth, a behaviour which parallels diminishing returns in terms of biofiltration performance for increasing values of X*.

It is thus evident that the parameter X* has relevance in terms of the net production of biomass by biofilters. Elements of this parameter are important both for the generation and the removal of biomass within a filter. Further work is required to quantitatively apply or adapt the X* parameter to the net production of biomass by biological filters. The modeling work of Uhl (2000, 2001) and Hozalski and Bouwer (2001a, b) that includes bacterial attachment and detachment may provide useful insights in this regard.

13.4.3.6 Net generation of dissolved microbial product material

Although the considerations regarding the net production of dissolved microbial product material (SMP and dissolved EPS) in filters bear some similarity to those for the net generation of biomass, there is an important difference in that these soluble materials will not be subject to removal by physical filtration.

As is the case for biomass, generation of dissolved microbial product material within the filter is controlled largely but not entirely by the removal of easily biodegradable organic matter. The relationship is more complicated than for biomass (Laspidou and Rittmann 2002a), but should generally be related to X*. The biofilter influent will likely contain some of this and/or similar material. Whether or not this is a significant amount in relation to that generated in the filter will be site-specific.

The removal of dissolved microbial product material within a biofilter is by biodegradation. Although the X* value determined for the primary limiting substrate entering the filter cannot be directly applied to this removal, in principle, a second X* value could be calculated for the second part of a biofilter (i.e. the part beginning where the original easily biodegradable substrate has been largely removed). The kinetics of biodegradation of dissolved microbial product material will undoubtedly be slower than for easily biodegradable material (Laspidou and Rittmann 2002b). Slowly biodegradable material in the filter influent will also be removed in this region of the filter. As noted earlier, Hubele (1985) discussed the importance of filtration mechanisms for retaining biomass lower in the filter for the removal of more slowly biodegradable DOC. Based on their simulations, Hozalski and Bouwer (2001b) note that the removal of slowly biodegradable material is significantly enhanced by the presence of easily biodegradable material, primarily because of the ability to maintain greater biomass levels in the filter. Biomass transport into the lower part of the filter will require modification to the X* concept. Uhl (2001) has noted that biodegradation can occur at concentrations below the S_{min} value, and has attributed this to biomass transport into the region of the filter where this biodegradation is occurring. First-order kinetics may not be applicable for dissolved microbial product material (and for influent slowly-biodegradable material) and diffusion limitations may not be able to be neglected. Whether or not it might be possible to use, as an approximation, the beginning of the filter bed as the starting point in terms of time or bed depth for the second X*, would depend on how low the degradation rates were.

Minimization of the concentration of dissolved microbial product material in a biofilter effluent will likely involve increasing filter contact time above that that might be used for easily removable BOM. As mentioned previously, Uhl *et al.* (2007) reported much better performance in reducing fouling for BAC pre-treatment compared to dual-media pre-treatment (which would also have been biologically active). Although other factors may have played a role, the much longer contact time provided in the BAC filter was likely an important contributor to this result. In general, whether the contact time required for the effective removal of dissolved microbial product material (or specific components that are most important for fouling) or for the minimization of particulate biomass exiting the filter will govern, will likely be site-specific. The overall process treatment train optimization will also depend on the extent to which it is desired to minimize bio- and organic fouling through pre-treatment. These issues represent fruitful ground for well-conducted experimental investigations and further theoretical development.

13.5 DEFINITION OF A NEW PERFORMANCE PARAMETER: THE BIOFILTRATION FACTOR (BF)

This chapter has demonstrated that the concept of an index for contact time, the dimensionless contact time or X*, is relevant for carbon removal for various objectives in drinking water biofiltration. These objectives include the original classic one of the removal of easily biodegradable organic matter, the removal of trace contaminants, and the emerging and potentially important application of biofiltration as a membrane pre-treatment. However, for this concept to be useful in practice, it needs to be concretely related to process conditions. We therefore propose a new parameter, the Biofiltration Factor (BF), which is a simple linear transformation of X*, but with a numerical reference point value tied to practice.

Examination of Figure 13.3 shows that approximately 50% AOC removal would be obtained at an X* value of 0.5. For convenience, we therefore assign a BF value of 0.5 to a filter capable of achieving 50% removal of AOC at approximately 20°C. This is a relatively easy definition to remember. The BF value required for other treatment objectives (or at other temperatures) can then be directly related to this.

It is likely that a BF of 0.5 would correspond to EBCTs in the range of 3 to 5 minutes, for the usual filtration media diameters. More accurate quantification of this from the literature is outside the scope of the present article and is a topic for further work.

The Biofiltration Factor is thus in a general sense analogous to the parameters G or GT for coagulation, L/d for filtration or CT for disinfection. That is to say, that although these parameters do not capture all the subtleties of the relevant processes, they are extremely useful as good engineering approximations for design. With the development of the appropriate database, it is expected that the Biofiltration Factor can also be used as a design and operating parameter. Although the BF is a parameter rather than a model, it is consistent with the desirable attributes proposed for design and operational models in Table 13.2.

Figure 13.7 shows conceptually the expected general sequence of BFs anticipated for various biofiltration applications. The exact value required, such as for reduction in membrane fouling in a particular situation will depend on site-specific water quality and operational conditions. Huck (1999) estimated X* values of more than 100 for the slow sand filters investigated by Ødegaard (1996), and noted that they were considerably over-designed with respect to biodegradable TOC removal. This means that processes such as bank filtration are also likely considerably over-designed in this regard.

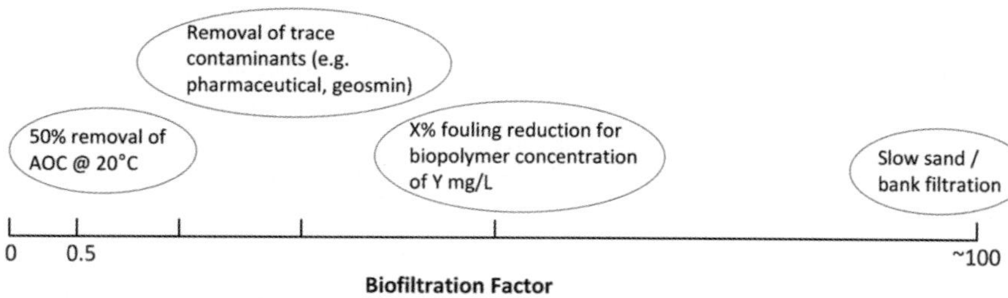

Figure 13.7 Expected general sequence of BF required for various treatment goals.

Considerable work remains to determine quantitative values of BF required for specific applications, so that a series of nomographs can be developed. For example, for a given parameter at a given temperature, a nomograph could show the required EBCT as a function of media size, for different desired percentage removals. Additional nomographs could be prepared for several representative temperatures, and potentially for ozonated and non-ozonated water.

Initial progress in developing nomographs can likely be made with data from the literature for classical applications of biofiltration and for the removal of at least some trace contaminants including geosmin and MIB, although additional experimental work will undoubtedly be required. The nature of these additional investigations can be better determined once information from the existing literature on biofiltration has been thoroughly analyzed for possible contributions to the quantification of X* and BF. For membrane pre-treatment, additional theoretical work is required as previously indicated and experimental work will also be necessary. Initially, the experiments will likely require use of a control filter for which AOC removals are measured, to provide a reference point for quantification of BF values for various applications and situations. Although it is likely that most of the work would be conducted at pilot-scale, ultimately confirmation at full or demonstration scale should be undertaken.

13.6 SUMMARY

This chapter presents the development of a unifying practically-oriented conceptual approach and newly defined performance parameter for biological filtration in drinking water treatment for various applications. These applications include the classical objectives of removing easily biodegradable carbon that leads to bacterial regrowth in distribution systems, and the usually related objectives of reducing chlorine demand and concentrations of chlorination by-product precursors in systems that use chlorine, by reducing the concentration of biodegradable TOC. The additional applications addressed in the chapter are the use of biological filtration for the removal of specific trace organic contaminants and as a pre-treatment for membrane filtration.

The major factors affecting biofiltration performance have been examined in a number of investigations. These factors include contact time, surface area of the media for biofilm growth, type of media, temperature, the nature and concentration of the influent BOM, and the potential disruption of the biofilm due to backwashing. With the exception of temperature and to some extent the nature of the influent BOM (depending on the presence or absence of upstream ozonation) these factors are subject to control through either design or operation. Of importance with respect to the development of a practically-oriented modeling approach, full-scale investigations have shown that BOM removals increase with increasing contact time (but in a less-than-proportional way), that backwashing likely does not have a measurable effect on BOM removals and that biomass (as measured by the often-used phospholipid method) is not quantitatively related to BOM removal.

Over the years a number of mathematical models have been developed for drinking water biofiltration, focusing largely on the classical and related process objectives mentioned above. Although some of these models have found localized use in practice, no generally accepted modeling approach has emerged. The major models, reviewed briefly in this chapter, are generally complex, incorporating a number of parameters. Thus often their use has been restricted to the research investigation in which they have been developed. A practical modeling approach or conceptual framework that could be used in the design and operation of full-scale biological filters should be relatively simple and incorporate the key controllable process parameters.

Arguably the seminal model for biofilm processes is the steady-state biofilm model developed by Rittmann and McCarty (1980a, b). Although not originally developed for drinking water biofilters, it has been used in this application. From the perspective of a practitioner, this model, which is presented in dimensionless form, is complex. However an important development for the understanding of drinking water biofiltration, the concept of dimensionless contact time, X* (Zhang 1996; Zhang and Huck 1996), is based on this model. The parameter X* is effectively an index for contact time and incorporates, in addition to contact time, other factors: the reactor specific surface area, substrate diffusivity and biodegradation kinetic parameters, and the biofilm density. The effect of temperature is incorporated through the substrate-related parameters. Biofiltration performance (i.e. removal) increases essentially linearly with X* for values of X* up to about 0.5 and thereafter increases in a less than proportional way.

Although EBCT is often used as a parameter for biological filtration, and in fact may be the single design criterion used, the use of X* is preferable as it allows for the meaningful comparison of biofilters incorporating different media sizes as well as having different contact times. It also provides insights into the removal, by the same biofilter, of substrates with different biodegradabilities or that are differently defined (e.g. AOC vs. BDOC).

In this chapter we demonstrate that the X* parameter is also relevant for additional biofiltration process objectives. However, we first show that X* can be simplified, because in a number of cases mass transfer limitations can be neglected and first-order biodegradation kinetics can be used. The first of the additional applications demonstrated is the use of X* to describe the removal of trace contaminants through secondary utilization. Such contaminants include the odorous compounds geosmin and MIB and biodegradable pharmaceutically active and endocrine disrupting compounds.

More recent investigations have shown that biological filtration can be effective as a membrane pre-treatment. It can act to reduce concentrations of both particulate matter and dissolved biodegradable material that have been shown to lead to membrane fouling. A section of this chapter is devoted to this use of biofiltration and to the extension of X* to this application.

The application of biofiltration as a membrane pre-treatment places several important additional requirements on the biofiltration process. In addition to reducing concentrations of easily biodegradable carbon that could lead to bacterial growth on the membrane, the biofilter should minimize the concentrations of both particulate and dissolved microbial product material and other biopolymers in the filter effluent.

Particulate biomass can be present in the biofilter influent and also be generated in the biofilter itself. Thus physical filtration mechanisms are important with regard to the level of particulate biomass exiting the filter. Important physical factors influencing filtration are bed depth and media diameter. Since filter specific surface area (α) is inversely proportional to media diameter, the two physical factors contained within X* (EBCT and α) are important for physical biomass removal by the filter. This physical filtration can be expected to show the usually observed pattern of decreasing returns with increasing filter depth, which parallels the relationship between biodegradation and X*.

Some dissolved microbial product material may enter the biofilter, and such material will certainly be generated within the filter. The generation will be related primarily to the removal of easily biodegradable material, and thus to X*. The removal of this dissolved microbial product material by biodegradation within the filter will also be related to X*, however it will likely be necessary to define a second X* for the filter relating to this material and also more slowly biodegradable carbon entering the filter. Because biomass transported from the upper part of the filter may be important for this biodegradation, some conceptual modification to the X* parameter will likely be required for this application. This area is a fruitful one for further experimental work and theoretical development.

In order to be useful for biofilter design and operation (regardless of the application), the index for contact time represented by X* must be transformed into a parameter directly related to practice. We therefore introduce a new parameter, the Biofiltration Factor (BF), which is directly linearly related to X*. For convenience, this parameter is assigned a value of 0.5 for a filter capable of achieving 50% AOC removal at approximately 20°C. BFs for the removal of many trace contaminants and for the use of biofiltration as a membrane pre-treatment will undoubtedly be larger than this value. Although the initial development of a series of nomographs for the BF for various applications can be undertaken based on data in the literature, additional experimental work will be required to complete this.

Although this chapter focuses on rapid biofiltration applications for organic carbon removal, it should be possible to extend the X* concept to other drinking water applications of biofiltration (e.g. nitrogen and iron/manganese removal) and to processes with much slower filtration rates such as bank filtration/ground passage and slow sand filtration. Since X* is based on a solution to the steady-state biofilm model in which axial dispersion is neglected, it will be important to determine whether axial dispersion can also be neglected for processes with much lower filtration rates.

Through the analysis presented in this chapter several areas for further work have been identified. These include:

- Experimental assessment of the predicted relationship $(k_2/k)^{1/2}$ for the relative removal of a trace contaminant with first-order rate constant k_2 and AOC with a rate constant of k,
- Further development of analytical methods for quantitation of the various organic fractions important for membrane fouling, and determination of biodegradation rate constants for these fractions,
- Additional investigations into the production of dissolved microbial product material by biofilm organisms and the production and detachment

of biomass from biofilms under conditions relevant for drinking water biofiltration,

- Further development to apply X* to the microbial product materials mentioned in the point just above, including the role of biomass transported to the lower regions of a filter,
- With respect to membrane fouling, determination of which is governing in terms of the necessary X*: particulate or dissolved microbial product material (or specific components thereof),
- Quantification, based on the literature and further experimental studies, of Biofiltration Factors for various biofiltration applications, leading to the development of nomographs for design and operational use, and
- Extension of the X* parameter and the Biofiltration Factor to other drinking water applications of biofiltration (e.g. removal of nitrogen and Fe/Mn), and evaluation of the applicability of these two parameters to biological processes other than rapid filtration (e.g. slow sand and bank filtration, and subsurface passage).

Acknowledgments

This chapter is based on an earlier publication by Huck and Sozański (2008). The authors acknowledge Shulin Zhang for developing the concept of dimensionless contact time (X*), and Graham Gagnon, Daniel Urfer and Jess Brown for helpful discussions. The first author acknowledges the support of the Alexander von Humboldt-Foundation in Germany and of Professor Martin Jekel, Technical University of Berlin during a sabbatical spent there several years ago. He also acknowledges the ongoing support of the partners in the NSERC Industrial Research Chair in Water Treatment at the University of Waterloo (www.civil.uwaterloo.ca/watertreatment). Financial support for the second author has been provided by NSERC, the Ontario Research Fund and the Deutsche Forschungsgemeinschaft (DFG).

13.7 REFERENCES

Amirtharajah A., McNelly N., Page G. and McLeod J. (1991). Optimum Backwash of Dual Media Filters and GAC Filter-Adsorbers with Air Scour. AWWA Research Foundation Report. Denver, Colorado.

Basu O. D. and Huck P. M. (2004). Integrated biofilter-immersed membrane system for the treatment of humic waters. *Wat. Res.*, **38**(3), 655–662.

Billen G., Servais P., Bouillot P. and Ventresque C. (1992). Functioning of biological filters used in drinking water treatment: the CHABROL Model. *J. Wat. Supply Res. Tech. – AQUA*, **41**(4), 231–241.

Booth S. D. J., Huck P. M., Slawson R. M., Butler B. J. and Ndiongue S. (2004). Removal of selected ozonation by-products in pilot scale drinking water biofilters: compound interactions and mass transfer considerations. *J. Wat. Supply Res. Tech. – AQUA*, **53**(4), 207–225.

Carroll T., King S., Gray S. R., Bolto B. A. and Booker N. A. (2000). The fouling of microfiltration membranes by NOM after coagulation treatment. *Wat. Res.*, **34**(11), 2861–2868.

Cho J., Amy G. and Pellegrino J. (2000). Membrane filtration of natural organic matter: factors and mechanisms affecting rejection and flux decline with charged ultrafiltration (UF) membrane. *J. Membrane Sci.*, **164**, 89–110.

Dillon P. J., Toze S., Pavelic P., Vanderzalm J. L., Barry K., Ying G.-G., Skjemstad J., Nicholson B. and Miller R. (2005). Water quality improvements during aquifer storage and recovery. *Abstract of presentation at 5th International Symposium on Management of Aquifer Recharge*, June 12–16, Berlin, Germany.

Elhadi S. L. N. (2004). Removal of earth/musty odour compounds from drinking water by biological filtration. Ph.D. Thesis, Department of Civil Engineering, University of Waterloo, Waterloo, Ontario, Canada.

Elhadi S. L. N., Huck P. M. and Slawson R. M. (2006). Factors affecting the removal of Geosmin and MIB in drinking water biofilters. *J. AWWA*, **98**(8), 108–119.

Farr A. J. and Stampone P. (2007). World's largest municipal ultrafiltration, biologically active carbon, and ozone plant prepares for startup. *J. AWWA*, **99**(5), 54–60.

Findlay R. H., King G. M. and Watling L. (1989). Efficacity of phospholipid analysis in determining microbial biomass in sediments. *Appl. Environ. Microbiol.*, **55**, 2888–2893.

Flemming H.-C. (2002). Biofouling in water systems – cases, causes and countermeasures. *Appl. Microbiol. Biotech.*, **59**, 629–640.

Flemming H.-C. and Wingender J. (2001). Relevance of microbial extracellular polymeric substances (EPSs) – Part II: technical aspects. *Wat. Sci. Tech.*, **43**(6), 9–16.

Flemming H.-C., Schaule G., Griebe T., Schmitt J. and Tamachkiarowa A. (1997). Biofouling – the Achilles heel of membrane processes. *Desalination*, **113**, 215–225.

Gabelich C. J., Yun T. I., Ishida K. P., Leddy M. B. and Safarik J. (2004). The effect of naturally occurring biopolymers on polyamide membrane fouling during surface water treatment. *Desalination*, **161**, 263–276.

Gagnon G. A. and Huck P. M. (2001). Removal of easily biodegradable organic compounds by drinking water biofilms: analysis of kinetics and mass transfer. *Wat. Res.*, **35**(10), 2554–2564.

Gagnon G. A., Booth S. D. J., Peldszus S., Mutti D., Smith F. and Huck P. M. (1997). Carboxylic acids: formation and removal in full-scale plants. *J. AWWA*, **89**(8), 88–97.

Gimbel R., Graham N. J. D. and Collins M. R. (2006). Recent progress in slow sand and alternative biofiltration processes. IWA Publ., London, ISBN: 9781843391203.

Grünheid S. and Jekel M. (2005). Fate of bulk organics during bank filtration and groundwater recharge of wastewater-impacted surface waters. *Abstract of presentation at 5th International Symposium on Management of Aquifer Recharge*, June 12–16, Berlin, Germany.

Guay C., Peldszus S., Huck P. M., McPhail B. and Mosqueda-Jimenez D. B. (2007). Removal of selected pharmaceutically active compounds and endocrine disrupting substances in surface water by membrane filtration with biological filtration pre-treatment. In: Chemical Water and Wastewater Treatment IX, (Proceedings, 12th Gothenburg Symposium of 2007, Ljubljana, Slovenia), pp. 297–306.

Haberkamp J., Haberkamp J., König A., Würtele A., Ernst M. and Jekel M. (2007). Extraction of bacterial extracellular polymeric substances and investigation of their impact on organic fouling of UF membranes. Poster presentation, 4th IWA International Conference on Membranes for Water and Wastewater Treatment, Harrogate, UK, May 15–17. International Water Association, London.

Haberkamp J., Ernst M., Paar H., Pallischeck D., Amy G. and Jekel M. (2011). Impact of organic fractions identified by SEC and fluorescence EEM on the hydraulic reversibility of ultrafiltration membrane fouling by secondary effluents. *Desalination & Water Treatment*, **29**, 73.

Hallé C. (2010). Biofiltration in drinking water treatment: Reduction of membrane fouling and biodegradation of organic trace contaminants. Doctoral Thesis, Department of Civil and Environmental Engineering, University of Waterloo.

Hallé C., Huck P. M., Peldszus S., Haberkamp J. and Jekel M. (2009). Assessing the performance of biological filtration as pretreatment to low pressure membranes for drinking water. *Environ. Sci. Technol.*, **43**(10), 3878–3884.

Hijnen W. A. M., Biraud D., Cornelissen E. R. and van der Kooij D. (2009). Threshold concentration of easily assimilable organic carbon in feedwater for biofouling of spiral-wound membranes. *Environ. Sci. Tech.*, **43**, 4890–4895.

Ho L., Hoefel D., Bock F., Saint C. P. and Newcombe G. (2007). Biodegradation rates of 2-methylisoborneol (MIB) and geosmin through sand filters and in bioreactors. *Chemosphere*, **66**, 2210–2218.

Hoefel D., Ho L., Aunkofer W., Monis P. T., Keegan A., Newcombe G. and Saint C. P. (2006). Cooperative biodegradation of geosmin by a consortium comprising three gram-negative bacteria isolated from the biofilm of a sand filter column. The Society for Applied Microbiology. In: Letters in Appl. Microbiol., **43**, 417–423, ISSN 0266-8254.

Howe K. J. and Clark M. M. (2002). Fouling of microfiltration and ultrafiltration membranes by natural waters. *Environ. Sci. Tech.*, **36**, 3571–3576.

Hozalski R. M. and Bouwer E. J. (2001a). Non-steady state simulation of BOM removal in drinking water biofilters: model development. *Wat. Res.*, **35**(1), 198–210.

Hozalski R. M. and Bouwer E. J. (2001b). Non-steady state simulation of BOM removal in drinking water biofilters: applications and full-scale validation. *Wat. Res.*, **35**(1), 211–223.

Hubele C. (1985). Adsorption und biologischer Abbau von Huminstoffen in Aktivkohlefiltern. Ph.D. dissertation (in German), Water Chemistry Division, Engler-Bunte Institute, University of Karlsruhe, Germany.

Huber S. A., Balz A., Abert M. and Pronk W. (2011). Characterisation of aquatic humic and non-humic matter with size-exclusion chromatography – Organic Carbon Detection – Organic Nitrogen Detection (LC-OCD-OND). *Wat. Res.*, **45**, 879–885.

Huck P. M. (1990). Measurement of biodegradable organic matter and bacterial growth potential in drinking water. *J. AWWA*, **82**(7), 78–86.

Huck P. M. (1999). Development of a framework for quantifying the removal of humic substances by biological filtration. *Wat. Sci. Tech.*, **40**(9), 149–156.

Huck P. M. and Anderson W. B. (1992). Quantitative relationships between the removal of NVOC, chlorine demand and AOX formation potential in biological water treatment. *Vom Wasser*, **78**, 281–303.

Huck P. M. and Sozański M. M. (2008). Biological filtration for membrane pre-treatment and other applications: Towards the development of a practically-oriented performance parameter. *J. Wat. Supply Res. Tech. – AQUA*, **57**(4), 203–224.

Huck P. M. and Sozański M. M. (2011). Chemical basis for water technology. In: Treatise on Water Science, P. Wilderer (ed.), Vol. 1, Oxford: Academic Press, pp. 429–469.

Huck P. M., Fedorak P. M. and Anderson W. B. (1991). Formation and removal of assimilable organic carbon during biological treatment. *J. AWWA*, **83**(12), 69–80.

Huck P. M., Zhang S. and Price M. L. (1994). BOM removal during biological treatment: a first-order model. *J. AWWA*, **86**(6), 61–71.

Huck P. M., Coffey B. M., Amirtharajah A. and Bouwer E. J. (2000). Optimizing Filtration in Biological Filters. AWWA Research Foundation and American Water Works Association, Report No. 90793. Denver, Colorado. 268 pp. ISBN 1-58321-065-2.

Jarusutthirak C., Amy G. and Croué J.-P. (2002). Fouling characteristics of wastewater effluent organic matter (EfOM) isolates on NF and UF membranes. *Desalination*, **145**, 247–255.

Karel S. F., Libicki S. B. and Robertson C. R. (1985). The immobilization of whole cells: engineering principles. *Chem. Eng. Sci.*, **40**, 1321–1354.

Kawamura S. (1999). Design and operation of high-rate filters. *J. AWWA*, **91**(12), 77–90.

Kijowska E., Leszczyńska M. and Sozański M. M. (2001). Metabolic activity test in investigation of biodegradation in biological filters. *Wat. Sci. & Tech.: Water Supply*, **1**(2), 151–158.

Laspidou C. S. and Rittmann B. E. (2002a). A unified theory for extracellular polymeric substances, soluble microbial products, and active and inert biomass. *Wat. Res.*, **36**(11), 2711–2720.

Laspidou C. S. and Rittmann B. E. (2002b). Non-steady state modeling of extracellular polymeric substances, soluble microbial products, and active and inert biomass. *Wat. Res.*, **36**(8), 1983–1992.

Lauderdale C., Chadik P., Kirisits M. J. and Brown J. (2012). Engineered biofiltration: Enhanced biofilter performance through nutrient and peroxide addition. *J. AWWA* (Online only), **104**, E298–E309.

Lee N., Amy G., Croué J. P. and Buisson H. (2004). Identification and understanding of fouling in low-pressure membrane (MF/UF) filtration by natural organic matter (NOM). *Wat. Res.*, **38**(20), 4511–4523.

Lehtola M. J., Miettinen I. T., Vartiainen T., Myllykangas T. and Martikainen P. J. (2001). Microbially available organic carbon, phosphorus, and microbial growth in ozonated drinking water. *Wat. Res.*, **35**(7), 1635–1640.

Makdissy G., Croué J.-P., Amy G. and Buisson H. (2004). Fouling of a polyethersulfone ultrafiltration membrane by natural organic matter. *Wat. Sci. and Tech.: Water Supply*, **4**(4), 205–212.

Metz D. H., Pohlman R. C., Vogt J. and Summers R. S. (2006). Removal of MIB and geosmin by full-scale biological sand filters. In: Recent Progress in Slow Sand and Alternative Biofiltration Processes, R. Gimbel, N. Graham and R. Collins (eds), IWA Publishing, London, UK, pp. 352–359.

Mosqueda-Jiménez D. B. and Huck P. M. (2006a). Characterization of membrane foulants in drinking water treatment. *Desalination*, **198**, 173–182.

Mosqueda-Jiménez D. B. and Huck P. M. (2006b). Fouling analysis of ultrafiltration and nanofiltration membranes. *Wat. Practice and Tech.*, **1**(4), 85–92.

Mosqueda-Jiménez D. B., Narbaitz R. M. and Matsuura T. (2004). Membrane fouling test: apparatus evaluation. *ASCE J. Environ. Eng.*, **130**, 90–99.

MWH (2005). Water Treatment: Principles and Design, J. C. Crittenden (ed.), 2nd Edition, Wiley.

Namkung E. and Rittmann B. E. (1987). Evaluation of bisubstrate secondary utilization kinetics by biofilm. *Biotechnol. Bioengr.*, **29**, 335.

Niquette P., Prévost M., Merlet N. and Lafrance P. (1999). Influence de facteurs contrôlant l'enlèvement de la demande en chlore et de précurseurs de sous-produits de chloration dans les filtres biologiques. *Wat. Res.*, **33**, 2329–2344.

Ødegaard H. (1996). The development of an ozonation/biofiltration process for the removal of humic substances. In: Advances in Slow Sand and Alternative Biological Filtration, N. Graham and R. Collins (eds), John Wiley & Sons, New York, pp. 39–49.

Peiris R. H., Hallé C., Budman H., Moresoli C., Peldszus S., Huck P. M. and Legge R. L. (2010). Identifying fouling events in a membrane-based drinking water treatment process using principal component analysis of fluorescence excitation-emission matrices. *Wat. Res.*, **44**(1), 185–194.

Peldszus S., Hallé C., Peiris R. H., Hamouda M. A., Jin X., Legge R. L., Budman H., Moresoli C. and Huck P. M. (2011). Reversible and irreversible low-pressure membrane foulants in drinking water treatment: Identification by principal component analysis of fluorescence EEM and mitigation by biofiltration pretreatment. *Wat. Res.*, **45**(16), 5161–5170.

Peldszus S., Benecke J., Jekel M. and Huck P. M. (2012). Direct biofiltration pretreatment for fouling control of ultrafiltration membranes. *J. AWWA* (Online only), **104**(7), E430–E445.

Prévost M., Desjardins R., Duchesne A. and Poirier C. (1991). Reduction de la demande en chlore dans une filière de traitement biologique en eaux froides. *Environ. Technol.*, **12**, 561–580.

Prévost M., Laurent P., Servais P. and Joret J.-C. (eds) (2005). Biodegradable organic matter in drinking water treatment and distribution. American Water Works Association, Denver, CO, USA, 300 pp.

Rice R. G. and Robson C. M. (1982). Biological Activated Carbon – Enhanced Aerobic Biological Activity in GAC Systems. Ann Arbor Science Publishers, Ann Arbor, Michigan, USA, 611 pp.

Rittmann B. E. and Huck P. M. (1989). Biological treatment of public water supplies. *CRC. Crit. Rev. Envir. Control.*, **19**, 119–184.

Rittmann B. E. and McCarty P. E. (1980a). Model of steady-state biofilm kinetics. *Biotechnol. Bioengr.*, **22**, 2343–2357.

Rittmann B. E. and McCarty P. E. (1980b). Evaluation of steady-state biofilm kinetics. *Biotechnol. Bioengr.*, **22**, 2359.

Sáez P. B. and Rittmann B. E. (1988). Improved pseudo-analytical solution for steady-state biofilm kinetics. *Biotechnol. Bioengr.*, **32**, 379–385.

Sáez P. B. and Rittmann B. E. (1992). Accurate pseudo-analytical solution for steady-state biofilm kinetics. *Biotechnol. Bioengr.*, **39**, 790–793.

Sattler A., Amy G. and Drewes J. E. (2005). Fate of wastewater effluent organic matter (EfOM) through soil aquifer treatment. *Abstract of presentation at 5th International Symposium on Management of Aquifer Recharge*, June 12–16, Berlin, Germany.

Servais P., Billen G. and Hascoët M. C. (1987). Determination of the biodegradable fraction of dissolved organic matter in waters. *Wat. Res.*, **21**, 445–450.

Sontheimer H., Heilker E., Jekel M., Nolte H. and Vollmer F.-H. (1978). The Műlheim process. *J. AWWA*, **70**(7), 393–396.

Sozański M. M. (1995). The effect of increased water salinity of the removal of micropollutants by biological carbon filters. *Environ. Technol.*, **16**(11), 1061–1071.

Sozański M. M. and Olańczuk–Neyman K. (2002). Stan wiedzy i perspektywy rozwojowe technologii uzdatniania wody jako współczesnej dyscypliny nauki, Monografie Komitetu Inżynierii Środowiska P.A.N., Vol. 9 – Inżynieria Środowiska Stan Obecny i Perspektywy Rozwoju, Lublin.

Stewart P. S. (1993). A model of biofilm detachment. *Biotechnol. Bioengr.*, **41**, 111–117.

Uhl W. (2000). Einfluss von Schüttungsmaterial und Prozessparametern auf die Leistung von Bioreaktoren bei der Trinkwasseraufbereitung. Ph.D. thesis (in German), University of Duisburg, Germany.

Uhl W. (2001). Modeling biological DOM-removal in drinking water treatment filters – a new approach. Proceedings AWWA Water Quality Technology Conference, Nashville, USA, CD-ROM.

Uhl W. (2007). Impact of biofiltration pretreatment on performance and fouling of NF membranes. Proceedings AWWA Membrane Technology Conference and Exhibition, Tampa, Florida. CD-ROM, paper Wed4-3, 7 pp.

Uhl W., Hagmeyer G. and Huck P. M. (2003). Biofouling of NF and RO membranes in water production: a theoretical study on prerequisites and possible prevention strategies. Proceedings AWWA Membrane Technology Conference and Exhibition, Atlanta, Georgia, USA, CD–ROM.

Urfer D., Huck P. M., Booth S. D. J. and Coffey B. M. (1997). Biological filtration for BOM and particle removal: a critical review. *J. AWWA*, **89**(12), 83–98.

Urfer-Frund D. (1998). Effects of oxidants on drinking water biofilters. Ph.D. Thesis, Department of Civil Engineering, University of Waterloo, Waterloo, Ontario, Canada.

van der Kooij D., Visser A. and Hijnen W. A. M. (1982). Determining the concentration of easily assimilable organic carbon in drinking water. *J. AWWA*, **74**(10), 540–545.

Vrouwenvelder J. S., Manolarakis S. A., Veenendaal H. R. and van der Kooij D. (2000). Biofouling potential of chemicals used for scale control in RO and NF membranes. *Desalination*, **132**, 1–10.

Vrouwenvelder J. S., Bakker S. M., Cauchard M., Le Grand R., Apacandié M., Idrissi M., Lagrave S., Wessels L. P., van Paassen J. A. M., Kruithof J. C. and van Loosdrecht M. C. M. (2007). The membrane fouling simulator: a suitable tool for prediction and characterisation of membrane fouling. *Wat. Sci. & Technol.*, **55**(8–9), 197–205.

Wang J. Z. (1995). Assessment of biodegradation and biodegradation kinetics of natural organic matter in drinking water biofilters. Ph.D. Thesis, Department of Civil and Environmental Engineering, University of Cincinnati, Ohio.

Wang J. and Summers R. S. (1995). A heterogeneous biofiltration model for natural organic matter utilization. Proceedings AWWA Annual Conference, Anaheim, CA.

Wang J., Summers R. S. and Miltner R. J. (1995). Biofiltration performance: part 1, relationship to biomass. *J. AWWA*, **87**(12), 55–63.

Wanner O. and Gujer W. (1986). A multispecies biofilm model. *Biotechnol. Bioengr.*, **27**, 314.

Weisz P. B. (1973). Diffusion and chemical transformation. *Science*, **179**, 433–440.

Westerhoff P., Summers R. S., Chowdhury Z. and Kommineni S. (2005). Ozone-Enhanced Biofiltration for Geosmin and MIB Removal. AWWA Research Foundation and American Water Works Association, Report No. 91075. Denver, Colorado. 188 pp. ISBN 1-58321-396-1.

Xiong S. (2005). Investigation of low-pressure membranes at the Brantford water treatment plant. Master's Thesis, Department of Civil Engineering, University of Waterloo, Waterloo, Ontario, Canada.

Yuan W. and Zydney A. (2000). Humic acid fouling during ultrafiltration. *Environ. Sci. Tech.*, **34**(23), 5043–5050.

Zearley T. L. and Summers R. S. (2012). Removal of trace organic micropollutants by drinking water biological filters. *Environ. Sci. Technol.*, **46**(17), 9412–9419.

Zhang S. (1996). Modeling biological drinking water treatment processes. Ph.D. Thesis, Department of Civil Engineering, University of Alberta, Edmonton, Alberta, Canada.

Zhang S. and Huck P. M. (1996). Removal of AOC in biological water treatment processes: a kinetic modeling approach. *Wat. Res.*, **30**(5), 1195–1207.

Chapter 14

Optimization of design and operation of distribution systems to preserve water quality and maintain customer satisfaction

J. H. G. Vreeburg

14.1 INTRODUCTION

Water authorities and water companies put a lot of effort into making safe and good drinking water that meets all the necessary and legal requirements to be called drinking water. Within the realm of drinking water supply there is a long tradition of applying scientific research leading to a large amount of knowledge about water quality and production methods (van Dijk and van der Kooij 2005).

Once the water is finished and collected in the clear water reservoir, it enters the distribution system. Traditionally, the drinking water distribution system has merely been the working field of technicians. The design and construction of the network is considered to be more the domain of skilled labour and less that of engineers and scientists. The explanation for this might be the perception that the network is a relatively simple construction with the first priority to provide enough transport capacity and integrity to fulfil the basic goal of a water company: 'supply sufficient and safe water with a high quality as is in the interest of public health'. The network was long considered to play a minor role in the total scheme of water quality management. The focus on capacity led to a situation, especially for the distribution network, of over-dimensioning, resulting in long residence times and even in the stagnation of flows. A special issue in this is the fire flow, the amount of water required for fire fighting. Seemingly a minor topic as it is only one of the many requirements, in fact this is a very dominating design boundary (Snyder and Deb 2003).

The priorities in the design and operation of the traditional network engineer are first the quantity and pressure, second the continuity or reliability of delivery and third the quality of the water. In the beginning of public water supply the emphasis was mostly on the first two items in network design. The safeguarding of the microbiological water quality aiming at public health protection was assured by maintaining a disinfectant residual in the water during distribution. The objectives of the measure are preventing the contamination of the water with faecal pathogens, prevention of microbial growth and to have a sentinel in case the water gets faecal contaminated. This approach led to water quality control being mainly focused on monitoring the residual disinfectant levels.

In the Netherlands in the early seventies of the last century the policy of water companies distributing drinking water prepared from surface water was changed to the distribution of drinking water without residual disinfectant. An important driver for that was the discovery of disinfection by-products with toxic properties (Rook 1974). Also the taste and odour objections of customers gained in weight, which led to banning the chlorine residual from the water. Nowadays, in almost all networks in the Netherlands drinking water is distributed without a residual disinfectant.

Research over recent years into the changes in water quality in the network has shown that the traditional approach to the network of skilled labour and common sense has to be extended with a more scientific approach. Discolouration is not so much the result of corrosion, but has much to do with particles in the finished water in combination with hydraulic circumstances that cause these particles to accumulate in the network. It is hypothesized that this physical process will have its impact on biofilm growth. Biofilms will develop on any surface in contact with the water, including the surface offered by the loose particles.

This hypothesis, if confirmed, adds some new possibilities for biological water quality management. Proper management of particles for reducing the discolouration potential may also improve the control of biological processes. In other words: an operational key to managing biological water quality might be managing the physical aspects of water quality. This scientific approach to the network and the practical implications are described in more detail in this chapter.

14.2 ROLE OF THE RESIDUAL DISINFECTANT

The primary function of chlorine or another disinfectant residual is to neutralize the biological and public health impact of faecal contaminated water that has penetrated into the piped system. Secondly, the multiplication of omnipresent bacteria in the water should be prevented. This goes for bacteria in general, but of course for human or animal (opportunistic) pathogenic bacteria in particular.

If a disinfectant residual is left out then prevention of contamination or regrowth should be achieved by alternative measures. The most effective method of achieving this is the prevention of these processes. Within the concept of the Unified Biofilm Approach (van der Kooij *et al.* 2003), the regrowth of bacteria and consequential threats to public health is prevented alternatively. In popular terms, instead of attacking the growth potential of bacteria by disinfection it is aimed at starving the bacteria. In the Netherlands this philosophy behind treatment has been applied for many years now, resulting in biologically stable drinking water (Smeets *et al.* 2009).

Regrowth of bacteria in the network is, however, complex and depends on more factors than only the concentration of growth-promoting compounds in the treated water entering the network. Presence of sediments, biofilms, residence time, materials and temperature also play major roles in the process of regrowth.

The approach to prevent contamination through ingress is basically preventing the ingress in the first place. This can be achieved through good asset management and sufficient monitoring of pressure and quality.

14.3 PROCESSES IN THE NETWORK

14.3.1 Introduction

Water quality changes during transport from the production location to the customer's tap. The processes in the network are complex and interrelated. The initial quality and stability of the water is an important first condition, but the residence time, temperature, pipe material and hydraulic circumstances are almost equally important for the water quality.

The most important distribution problems related to water quality are discoloured water, regrowth and contamination due to ingress. Discoloured water is the main reason for customers to complain to their water companies about water quality, whereas non-compliance with microbiological water quality criteria may threaten public health. Traditionally, the discolouration is blamed on the corrosion of cast-iron pipes and biological problems on the presence of a biofilm. However, also in networks that have no or very limited amounts of cast iron, discolouration may occur and even to levels that are well comparable to cast-iron networks (Vreeburg 2007). In addition, non-compliance with biological water quality criteria also occurs in networks with a sufficient disinfectant residual.

The research into the nature and causes of discoloured water resulted in new insights in the design, operation and maintenance of distribution networks and is briefly described in the following sections.

14.3.2 Cause and nature of discolouration

Though referred to as discoloured water, complaints mostly concern turbidity caused by particles. Typically, if a sample of discoloured water is left to stand, particles sediment, resulting in a layer of particles at the bottom of the container. Hence, it can be concluded that particles play a key role in the generation of discoloured water. Therefore, research into the nature and causes of discolouration was focused on particles. Particle accumulations are also known to have a relation with biological activity (Gauthier *et al.* 1999; Vreeburg *et al.* 2009a). Of the organic matter 1 to 12% in the particulate accumulations may consist of bacterial biomass, making the deposits an important factor in affecting the microbiological quality of drinking water during distribution.

Based on several experiments (Vreeburg *et al.* 2008) a conceptual model of particle related processes has been derived. The relevant processes in this model are presented graphically in Figure 14.1.

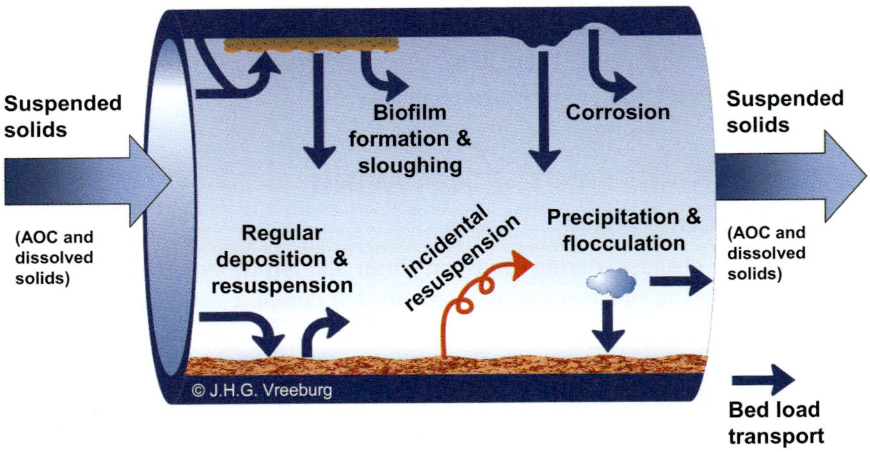

Figure 14.1 Particle related processes in a distribution network; AOC: assimilable organic carbon.

The main mechanism of the discolouration problem is the incidental resuspension of particles present in sediments, which leads to the actual discolouration event (Vreeburg and Boxall 2007). That means that there has to be a hydraulic disturbance causing the velocity to increase above a certain threshold that resuspends the particles from the sediment layer.

The particles have accumulated at the bottom of the pipe wall under the hydraulic circumstances that are common in that particular part of the network. In transport mains with relatively high velocities and shear stress other types of particles will accumulate than in distribution areas with low velocity (Blokker *et al.* 2009; Vreeburg *et al.* 2009b). This process is represented in Figure 14.1 as "Regular deposition and resuspension" and results in a layer of particles at the bottom of the pipe.

The main source for particles in non-corroding networks is the particle content in the finished water that leaves the treatment plant (Vreeburg *et al.* 2008). The contribution of the corrosion process to the total particle load is probably limited (Richardt and Korth 2009) and isn't always the cause of discolouration problems as is always assumed in literature. This makes the "corrosion" process as a particle-producing process in the Netherlands less relevant as a unique source of particles.

The process of biofilm formation might have several contributions to water quality problems. It can generate particles in the form of sloughed biomass, but it can also entrap particles in the biofilm. Within the concept of the Unified Biofilm Approach (van der Kooij *et al.* 2003) the Biofilm Formation Rate is an indication for the biological stability of the water. However, the exact role of biofilm in the particle accumulation is not clear.

14.3.3 Biological regrowth and sediments

According to the Unified Biofilm Approach, regrowth is dependent on the biofilm formation potential of the water and the enhanced microbial growth characteristics of the materials in contact with water (van der Kooij *et al.* 2003). Within the biofilm concept the actual biofilm is generally envisioned as a layer of biomass on the pipe wall. The model is that the biofilm is stuck to the wall occasionally letting particles escape from it (Flemming and Wingender 2001). However, the layer of loose deposits, responsible for the phenomenon of discoloured water, also renders a surface area that can be utilized by biofilm to grow on. As an example a representative amount of deposits of 1 g/m in a 110 mm polyvinyl chloride (PVC) pipe can be considered (Vreeburg *et al.* 2008). Assuming that these deposits consist of sphere-shaped particles with an average diameter of 15 μm and a specific weight of 1050 kg/m^3 gives 0.38 m^2 surface area of the sediment particles. This area is slightly greater than the surface area of the pipe wall itself (0.32 m^2). This simplified estimation, in which the amount and shape of the sediment is very conservative, suggests that the surface area of the sediment is larger than that of the surface area of the pipe itself.

The consequence of this ratio in surface areas is that the biofilm will probably not only grow on the pipe wall, but also on the sediment. In several trials the composition of the sediment is analyzed, showing a considerable amount of it to be organics (LeChevallier *et al.* 1987). In the presence of sediment, more organic material can be flushed out than in a comparable network without sediment. (Vreeburg *et al.* 2008). This observation is consistent with earlier reports which show that microorganisms not only grow on the pipe wall, but also on the sediment (Zacheus *et al.* 2001; Lipponen *et al.* 2002).

14.3.4 Experimental verification of biofilm on sediments

The hypothesis of microorganisms growing on sediments was tested during investigations in 2010 and 2011 in which the sediment built up in the supply area of three different treatment plants was studied. The treatment plants are identified as T1, T2 and T3. In each supply area two subareas were isolated by closing valves in such a way that a single feed was left. In each area, identified as A1 and A2 respectively, the pipes were flushed with a volume flow sufficient to cause a velocity of 1.5 m/s and during several flush actions samples were taken during the first refreshment of the pipe content. In these samples Total Suspended Solids

(TSS), turbidity and adenosine triphosphate (ATP) were determined. The first sample series were collected directly after the isolation of the 6 areas and repeated after a year. The first sample series is identified as T# A#-0 and the second series as T# A#-1. In the first series the sediment that had been built up in previous years was sampled and in the second series the sediment accumulated during one year was sampled. The relationship between the turbidity and the suspended solids is shown in Figure 14.2. For each sampling series the linear relation is determined between these parameters and the coefficients are presented in Table 14.1. Only the relationships with $R^2 > 0.6$ are presented.

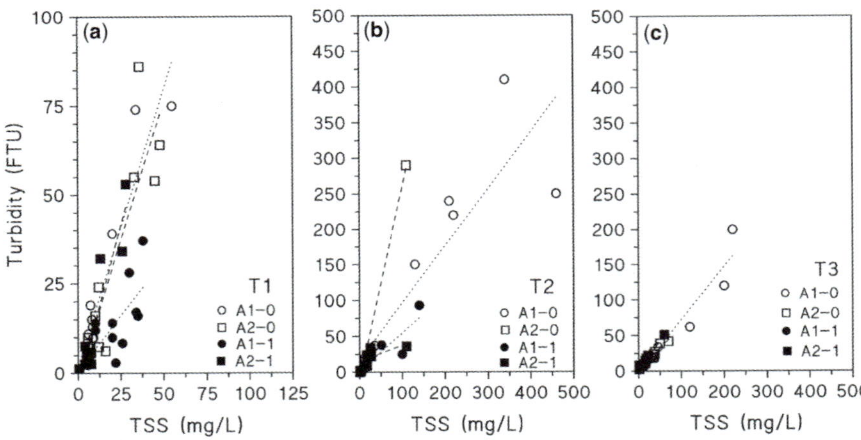

Figure 14.2 Relation between turbidity (FTU, Formazin turbidity unit) and suspended solids (mg/L) in samples of water flushed from mains in two isolated areas (A1 and A2) of three groundwater supplies (T1, T2 and T3).

Table 14.1 Relationship between Turbidity (FTU) and Total Suspended Solids (TSS).

Treatment/ Area-series	Turbidity (FTU) = a × TSS (mg/litre) + b			
	a (±sd)	b (±sd)	R^2	n
T1 A1-0	1.55 ± 0.17	1.8 ± 3.7	0.90	12
T1 A2-0	1.55 ± 0.77	–1.7 ± 7	0.78	12
T1 A2-1	1.8 ± 0.22	–3.7 ± 2.7	0.87	12
T2 A1-0	0.8 ± 0.13	16.4 ± 26	0.81	11
T2 A1-1	0.53 ± 0.1	1.8 ± 5.9	0.80	8
T2 A2-0	27 ± 0.1	–11 ± 3.7	0.99	9
T3 A1-0	0.75 ± 0.08	–3.3 ± 7.9	0.92	9

(*Continued*)

Table 14.1 Relationship between Turbidity (FTU) and Total Suspended Solids (TSS) (*Continued*).

Treatment/ Area-series	Turbidity (FTU) = a × TSS (mg/litre) + b			
	a (±sd)	b (±sd)	R^2	n
T3 A1–1	0.53 ± 0.13	2.6 ± 1.2	0.68	9
T3 A2–0	0.58 ± 0.06	3.4 ± 2.5	0.91	10
T3 A2–1	0.81 ± 0.05	2.1 ± 1.2	0.97	10

Table 14.1 shows that turbidity and TSS in flushed water at most locations are significantly related with each other. However, the slope of the relationship (FTU/TSS) differs between water types. The relationships between the suspended solids and ATP levels are resented in Figure 14.3. At 4 locations, a statistically-significant linear relationship was observed between the ATP concentration of the flushed water and the concentration of suspended solids. These relations are shown in Table 14.2.

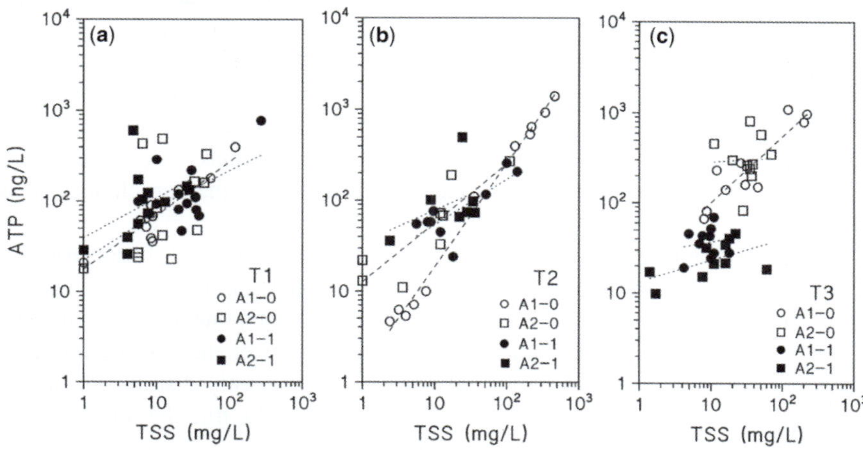

Figure 14.3 Relation between ATP and suspended solids in samples of water flushed from mains in two isolated distribution areas (A1, A2) of three groundwater supplies (T1, T2, T3).

The relations between the turbidity and the suspended solids are stronger than those between the suspended solids and ATP as is obvious from the figures. The significant relations are all in the first sample series, being the old sediment. These are also the series with the highest turbidity and amounts of sediments.

Table 14.2 Statistically significant relations between ATP and Total Suspended Solids (TSS).

Treatment / Area-series	ATP [ng/l] = a × TSS (mg/litre) + b			
	a (±sd) (FTU/TSS)	b (±sd	R^2	*p*
T1 A1–0	3.0 ± 0.2	+34 ± 0.1	0.96	<0.001
T2 A1–0	2.9 ± 0.1	–8.7 ± 14	0.99	<0.001
T2 A2–0	2.3 ± 0.5	29.1 ± 17	0.75	0.001
T3 A1–0	4.3 ± 0.8	102 ± 82	0.78	0.001

14.3.6 Discussion

Sediments found in distributions systems have an organic component as is reported by several authors (Besner *et al.* 2001). In this experiment the sampling of the sediments is standardized and restricted to sediments that can be resuspended by a velocity of 1.5 m/s and are taken in the first volume flushed from the pipe.

At 4 of the 12 sampling series there is a strong relation ($R^2 > 0.75$; $p < 0.001$) between the amount of ATP and the amount of TSS; in the other series the relation is weak or non-existent. The samples of flush water were taken from relatively clean networks. The absolute value of turbidity seldom exceeds 100 FTU. These samples were taken during unidirectional flushing and represented the highest values during the flushing (excluding the first draw from the hydrant itself). Every sample with a significant amount of solids (and turbidity) contains an elevated amount of ATP and especially at the higher values the relationship is stronger. These observations indicate that more active biomass is present in the sediment, either because there is a surface area available or the particle itself is a source for growth.

The observation that available surface area is an important factor for the development of biofilms should be considered seriously. If this surface area is provided by loose sediments this adds a hydraulic component to the already complex processes. The distribution of loose deposits over the distribution area seems to be erratic, and also biological activity is not evenly distributed. Given the ratio of particle surface and pipe wall surface it may be true that the majority of the biofilm grows on particles instead of on pipe walls.

14.4 MANAGING PARTICLES IN THE DISTRIBUTION NETWORK: A 3-STAGE APPROACH

14.4.1 Introduction

Regardless of their origin, particles play an important role in the water quality changes in the distribution system. The accumulation and resuspension of particles is the key process in the discolouration events in the distribution network. Particles may contain a certain amount of biofilm, which might contribute to the biological instability of the water. Control of the particle processes gives control of the discolouration problem and possibly also of the network component of the Unified Biofilm Approach. As sediments are the core of the problem, control of this layer is the key to control of the discolouration problem. A major component of the sediment in the distribution system is particles in the finished water of the treatment plant. Thus, preventing these particles from entering the network can reduce the problem effectively. A second measure, if the presence of particles in treated water is unavoidable, is to control the deposition and resuspension process preventing accumulation of the particles to layers. The last measure, if the two earlier measures are not applicable, is the control of the sediments through regular removal by cleaning of the network. These three measures are called the three-stage-approach and handle the three main aspects of particle accumulation in the network and the associated water quality problems.

The first stage is controlling the input of particles. The main input originates from the treated water and is not a constant value; discontinuities in the production process, for example changes in volume flow contribute significantly to the total particle load. This first stage is the most sustainable solution because it intervenes at the commencement of the problem.

The second stage is managing the hydraulics to limit the accumulation process. In the Netherlands the concept for self-cleaning networks has been developed for distribution networks. The basic principle is that distribution networks are designed following branched type of valve sections instead of the traditional looped type of sections. This approach can only be applied in new or redesigned networks. Conventional distribution networks are dimensioned on the fire flow demand and therefore experience very low flow velocities during normal operation.

The third stage is aimed at the operation of the network: timely and effectively cleaning of the network. Though this is an operational measure that does not prevent sediment accumulation, it can serve well as an effective and efficient method of management. In fact this is the aspect of network operation where the traditional skilled labour and experience is confronted with theoretical and scientific knowledge. All three stages will be described in more detail in the following sections.

14.4.2 First stage: control of input

Following the hypothesis on particles and the relation with water quality, in most supplies in the Netherlands, the particle input is the most important parameter to control. An experiment has been described in which two similar networks are compared on sediment built up under the condition that one is fed with normal drinking water and the other with the same water post treated with membrane ultrafiltration resulting is particle free water (Vreeburg *et al.* 2008) The effect on the discolouration potential, measured with the Resuspension Potential Method (RPM)(Vreeburg *et al.* 2004), is shown in Figure 14.4.

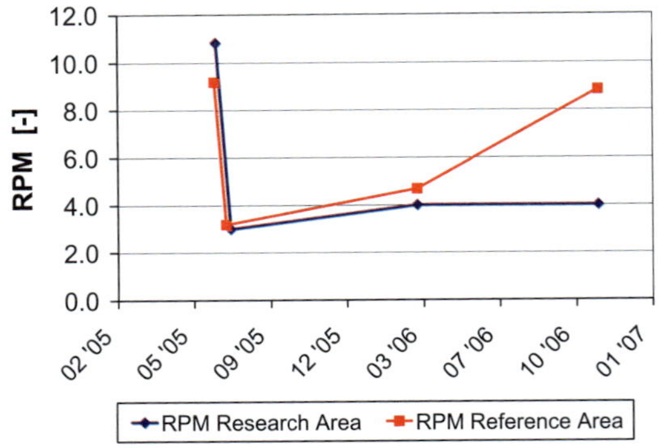

Figure 14.4 Development of the dimensionless RPM value in two similar networks. The feeding water supplied to the network in the research area is post treated with membrane ultrafiltration and does not contain particles.

The results clearly show the effect of the particles on the discolouration potential in the pipes. Practically, however, it is not feasible to provide each treatment facility with ultrafiltration to remove the particles.

Generally, the turbidity of treated water leaving the treatment facility is low. However fluctuations in turbidity at the entry point will load the network with particles. Two dominant discontinuities in water treatment affecting turbidity are (i) backwashing of filters and (ii) ripening period following the backwashing and changes in filtration rate. Especially accelerations of the filtration rate give peaks in turbidity. Suggestions for controlling are to minimize the number of switches in filtration rate by good use of balancing reservoir. The second suggestion is to optimize the backwash programme of the filter beds concentrating on the procedure for ripening of the filter (Eikebrokk 2000). During the ripening period the concentration of particles is very high.

14.4.3 Second stage: control of hydraulics

The second stage may be characterised as "keep the particles moving". The velocity is one of the main driving parameters that govern the settling and resuspension process (Thienen *et al.* 2011). The main goal of the so-called self-cleaning networks is to achieve a critical velocity in the distribution network at least once a day to prevent the sediment from accumulating. Discolouration is typically a problem of the distribution network, because the lowest velocities occur and the customer connections are located in the distribution network.

Design of drinking water distribution systems (DWDS) has historically been a matter of experience rather than science. The design criteria for household demand dominated DWDS are (1) the minimal pressure, (2) the continuity of supply, (3) the fire flow demand and (4) the actual drinking water demand. Based on these criteria, conventionally a design is made for a looped layout of the network with pipe diameters that are sufficient to supply the dominant fire flow demands and are usually 100 mm or larger. In conventional distribution networks the actual flow velocities are low because the design is based on the fire flow demands that surpass the drinking water demand several times (Snyder and Deb 2003). Maximum velocities in this type of network are in the range of a few centimetres per second (Blokker *et al.* 2006, 2011).

Without explicit knowledge about the adverse effects of low velocity and long residence times, the importance of quantity and pressure surpasses the importance of water quality. However, with growing insight into the effect of velocity on the formation of sediments (Boxall and Saul 2005) together with the awareness that discolouration problems are not exclusively related to cast iron but a combination of particle load and flow velocity, more attention to the detailed design of the DWDS became necessary (Vreeburg *et al.* 2008). For the design of self-cleaning networks an extra criterion is added: the daily maximum flow velocity. This flow velocity, which should occur daily for at least a few minutes, is capable of resuspending particles that were allowed to settle during low flow periods. The daily maximum velocity should surpass the threshold value.

Adding the daily maximum velocity criterion changes the layout of the DWDS from a completely looped structure and oversized pipes radically, to a branch-type network with uni-directional high velocities in smaller pipes as is shown in Figure 14.5. Seemingly, the extra criterion surpasses other criteria such as supply continuity and fire flows. However, the new branched layout does not compromise the continuity of supply or the necessary fire flows. In fact it enhances the continuity of supply, though only slightly, because of the shorter time needed to isolate and re-install the section. The fire flows have been downsized to realistic values for modern residential areas.

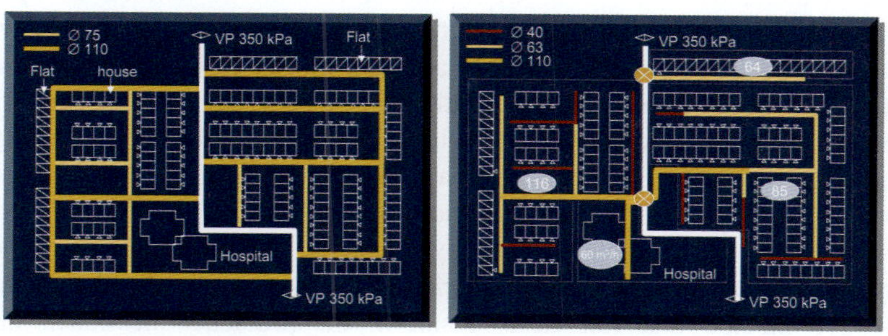

Figure 14.5 Left a conventional distribution network, typically looped and diameters of 75 and 110 mm (PVC). Right a self-cleaning high velocity network with three branches and declining pipe diameter.

The new layout proved to reduce costs by around 20% compared to the traditional network. This is due to the shorter length of the network because the loops are not closed and to the smaller pipes, which are also less expensive.

Applying a self-cleaning network results in a distribution network layout that differs from a transport network layout. The looped transport system ensures the reliability of the supply to larger areas. Failure of a transport pipe doesn't affect individual connections directly, but it affects areas that are known as nodes or intersection joints in a network model with a concentration of connections. The branched cut-off sections are the smallest unit that affects the actual and direct supply to a number of customers. In the self-cleaning network this cut-off section is connected with only one valve, whereas in a looped network at least two valves are needed and mostly more. The single valve makes the actual cut-off more reliable and easier to perform (Vreeburg *et al.* 2009b).

14.4.4 Third stage: control of sediment layer by cleaning

14.4.4.1 Introduction

Control of the sediment layer basically requires timely and effective cleaning of the pipes to prevent the accumulation of sediment from growing to levels that can cause problems like discolouration.

Water flushing is the most common applied method for cleaning networks (Antoun *et al.* 1999; Friedman *et al.* 2003). The principle of the method is that an increased water flow causes an increased velocity in the pipe, which leads to an additional shear stress on the loose deposits in the pipes. These loose deposits are whirled up and removed by the flushed water. The extra flow is most commonly induced by opening a hydrant and blowing off the extra water. Despite the long history, there is little known about the operational requirements for effective flushing.

Conventional flushing is the approach used by most utilities in the UK and USA (Friedman *et al.* 2003). This was also the method most frequently applied in the Netherlands until 1990. Conventional flushing is defined as opening hydrants in a specific area of the distribution system until pre-selected water quality criteria like turbidity or colour are met, mostly by a visual assessment. Though this seems to be a clear definition, in practice there is little uniformity in the application of conventional flushing. This means that it can hardly be regarded as a standard method. There are even examples that the end point of a flushing, expressed as a certain turbidity of the water, is reached by slowly decreasing the velocity of flushing by gradually closing the hydrant, even compromising the few operational criteria. This makes conventional flushing an ineffective method that can easily result in the problems mentioned in the previous sections.

In the 1990s at several locations conventional flushing was refined to unidirectional flushing (Oberoi 1994; Slaats *et al.* 2002). The advantage of the unidirectional flushing is that a clearer operational guideline is applied that leaves less room for ambiguity. The basic characteristic of unidirectional flushing is that the flush flow is directed in one direction through manipulation of valves aimed at reaching a maximal velocity. Sometimes dedicated flushing points are applied instead of hydrants to reduce outflow resistance in the hydrants resulting in higher flows in the pipes. The benefit of the unidirectional flushing is that it requires careful planning to meet the operational parameter of unidirectional flow. The careful planning also ensures that all pipes are actually cleaned.

The intended effect of flushing is the increase in flow resulting in increased shear stress to resuspend the sediments. For effective flushing requirements must be set for the velocity and the amount of water flushed. Next to that requirements should also be defined regarding planning of the order of flushing to prevent recontamination of cleaned pipes. In other words: water used for flushing should originate from clean (ed) pipes. These requirements will be described in the following paragraphs and illustrated with experiments.

14.4.4.2 Minimum velocity

The value set as minimum velocity in several studies is 1.5–1.8 m/s (Brashear 1998; Slaats *et al.* 2002). The reasoning behind this is partly the theoretical approach based on Shields' and Stokes' equations, though this leads to lower velocities than 1.5 m/s for example (Boxall *et al.* 2003). For the other part, the value of 1.5 m/s is based on practical experience that these velocities are significantly above the normal daily maximum velocities. With 1.5 m/s all the sediment that could cause discolouration would then be removed. Finally the velocity of 1.5 m/s is well attainable, with a number of valve manipulations.

The flow resistance of a hydrant-standpipe-hose combination often limits the maximum capacity to 60 to 90 m^3/h. At this flow the resistance approaches the available head in the network. Typically, the hydrants are installed on 100 mm pipes and a 60 m^3/h flow pushes 2.1 m/s through those pipes.

Apart from the theoretical approach looking at the possibilities of obtaining a velocity of 1.5 m/s, a practical approach is to see what happens if the velocity is not reached. The adverse effects of low velocity are best demonstrated by the negative effects of conventional flushing (Antoun *et al.* 1997): viz. the increase of complaints after flushing. This effect can be explained with the results of turbidity measurements during an experiment with two flushing velocities (Figure 14.6). In this experiment in a 3-inch cast-iron pipe first a flushing/disturbance was applied with a velocity of 0.4 m/s. The pipe was isolated by closing a valve near the hydrant that was located in such a way that a unidirectional flushing was caused. Turbidity was monitored with the Sigrist KT65 with a measuring frequency of 10 minutes.

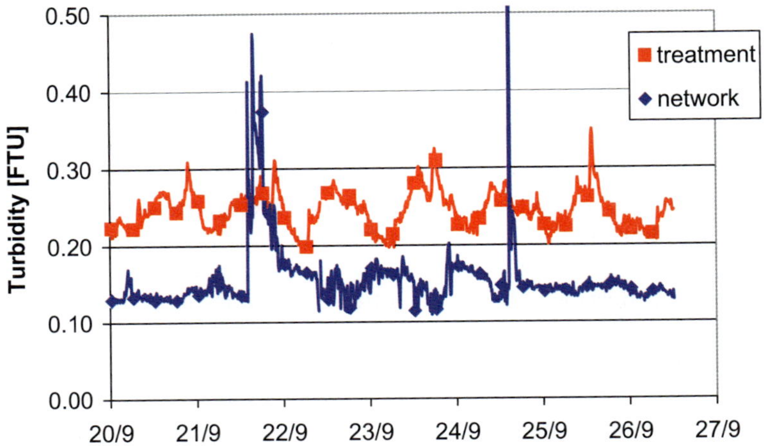

Figure 14.6 Influence of flushing velocity on sediment mobility; first flushing on 21/9 was with velocity less than 1.5 m/s (at least 0.4 m/s), second flushing with at least 1.5 m/s.

Monitoring started on 20 September and there was a distinct difference between the turbidity at the treatment plant and at the research location in the network. This difference indicated that the network was being loaded with particles. The turbidity at the monitoring location was stable until the disturbance/flushing on 21 September. The stability in turbidity indicated that the layer of sediment was not mobile under normal flow circumstances. In the early afternoon of 21 September, a disturbance/flushing took place with a flow of 6.5 m^3/h. Because the pipe was an old 3-inch unprotected cast-iron pipe the velocity is as least 0.4 m/s, but presumably higher because the effective flow area can be increased by corrosion products.

The turbidity following the disturbance is higher than the turbidity prior to the disturbance. It took almost 24 hours before the turbidity decreased, though not to the starting level. After that, the turbidity pattern had more variation than the pattern before the disturbance. In the afternoon of 24 September the hydrant was opened again, but now intentionally causing a unidirectional flushing with a velocity of at least 1.5 m/s (flow is 27 m^3/h) under the same unidirectional conditions as the original flushing. The total amount flushed during the second flushing was three times the pipe content and the water was conveyed through clean pipes (a clear water front was used). The turbidity during the opening of the hydrant was high, but only for a short period. Directly after closing the hydrant, the turbidity dropped again to the initial level and stayed stable.

These observations indicate that the flushing/disturbance with a lower velocity mobilized the sediment without actually removing it completely. During the disturbance, though, a part of the sediment was removed. Flushing with a high velocity removed the sediment instantaneously and limited the customer inconvenience resulting from the flushing to the actual flushing time.

These results lead to the conclusion that a velocity of 1.5 m/s is sufficient to remove the sediments in this case, while the lower velocity had an adverse effect. Though this is only one experiment, the results make the adverse effects of conventional flushing understandable: viz. a too low velocity disturbs and mobilizes the sediment without actually removing it.

The operational requirement for velocity in water flushing is set at 1.5 m/s amongst others based on this experiment. It must be noted that lower velocities could also effectively remove sediment, but the experience gained over the years reveal that it is universally applicable for many combinations of network and treatment. Moreover, the awareness that a too low flow velocity can have adverse effects because of the mobilization of the sediment, contributed to the acceptance of this requirement for minimum velocity for flushing.

14.4.4.3 Flushed volume

The materials that must be removed from the pipe are the loose particles. With a sufficient increase in velocity, the relevant particles will resuspend and be flushed out with the water. In principle all these particles will be resuspended immediately after the shear stress from the increased velocity is exercised. The depth of the particle layers is in the order of some micrometres or millimetres (Vreeburg 2007). Therefore, a scouring effect is only necessary during a short period. The velocity profile over the cross-section of a pipe with a turbulent flow is not evenly divided. These two effects mean that if just a bit more than the volume of the pipe is flushed no additional sediment removal will occur and that the minimal flushed volume should be more than one times the volume of the pipe to be cleaned.

The result of one of many experiments in pipes in the Netherlands is presented in Figure 14.7. It shows that the nature of the sediment removed was actually loose deposits that resuspended immediately and completely when the shear stress was increased. Figure 14.7 shows the measurements of the turbidity of the flushed water during flushing of a 300 mm AC pipe with a velocity of 1.5 m/s. The distance between the flushing point and the clear water front is 3600 metres and the flushing is unidirectional from the clear water front to the flushing point. Turbidity is measured with a Dr Lange Ultraturb at a frequency of once per minute. The theoretical first turnover was around 40 minutes, which was confirmed during the experiment with the sharp decrease of turbidity. The further exponential decrease can be explained by the non-linear velocity distribution: the lower turbidity is due to the mixing of the fluid layers closer to the wall that move slower than the bulk of the fluid.

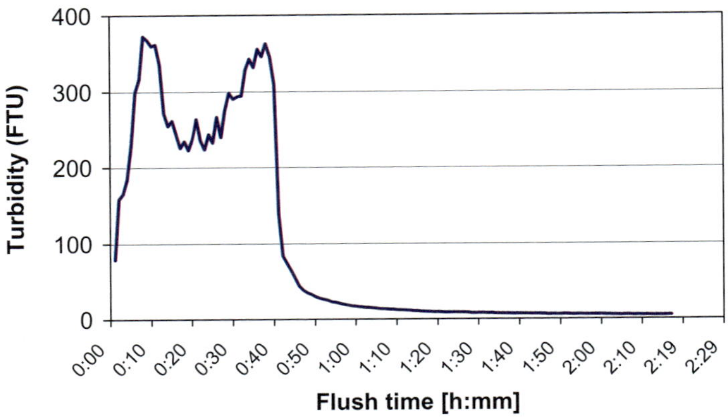

Figure 14.7 Turbidity measurements at the flushing point of a 300 mm asbestos cement (AC) pipe, length 3600 m, volume flow 680 m^3/h (1.5 m/s). Horizontal axis is the time elapsed after the start of the experiment.

The drop in turbidity after ten minutes of flushing and the gradual increase after 25 minutes of flushing are probably caused by the irregular nature of the deposits in the pipe. The layer of sediment is not distributed evenly over the complete pipe which may be caused by several factors such as the horizontal level profile or irregularities in the surface area of the pipe. A sharp drop in turbidity after one turnover of the pipe has been observed in many experiments. It leads to the second operational condition for water flushing to effectively remove the sediment, viz. the flushed volume should be at least two times the content of the pipe to overcome the velocity dispersion. In the Netherlands the first set of operational requirements stated a refreshment rate of three, but in practice it has been proven with measurements as presented in Figure 14.7 that two turnovers are enough as the water clears fairly well after two refreshments.

14.4.4.4 Clear-water front

The turbidity pattern in Figure 14.7 shows the effect when the water used for flushing itself is not clear enough. First, this clouds the effective end point of a flushing when this is defined in the form of a turbidity threshold. Carriere (2005) explicitly reports a stop criterion for flushing of 1 FTU, but at several locations, values below 5 FTU could not be reached. Probably the effective removal of sediment from the pipe has stopped, but the water used for flushing has a turbidity of 5 FTU. The second effect is that with the flushing 'up-stream-sediment' is carried to the flushing location, already partly loading the pipe with sediment. A third effect is that the 'upstream-sediment' is resuspended with velocities lower than the threshold velocity of 1.5 m/s, with effects as shown in Figure 14.7.

This sets the third operational requirement to effective water flushing: the water used to flush the pipes should come from pipes with no resuspendable sediment. This concept is called "Working from a clear water front."

14.5 MANAGING NETWORK INTEGRITY/TOTAL QUALITY AWARENESS

14.5.1 Introduction

With respect to water quality, the main features that indicate the quality of the network are the leakage rate and the failure rate. Both show the integrity of the network and thus the risk of ingress. Next to these figures also the pressure in the network is important; as long as pressure is sufficiently high water will not penetrate into the pipes.

For the Netherlands the leakage rate is about 5% (Geudens 2010) and the failure rate is 0.07 failure/km/year (Vloerbergh and Blokker 2010), respectively. Internationally, these figures can be put in perspective. Especially with regard to the leakage figures, some comparative research has been performed (Beuken *et al.* 2006). The main outcome of this research is that there is not one particular reason for the excellent performance figures. It is rather a combination of factors that contribute to the performance. These factors have in common that they are all quality driven, both in effect and in performance.

14.5.2 Total quality awareness

The geographic location of the Netherlands, in the delta of two large rivers in a highly industrialized area, has always forced water authorities to be very aware of the threats posed by the presence of water and the challenges of using it beneficially. One of the keys to the successful application of chlorine free water (Smeets *et al.* 2009) is the awareness of quality in all stages of the drinking water supply. From selecting the source for the drinking water to the material for the pipes and the restrictions set on the inner installations. Next to that, the operation of the network, during repair or installation is of importance. The first so-called "Code for hygienic work in the distribution network" originates from the early 1980s. The latest update/revision dates from 2010 (Meerkerk and Kroesbergen 2010). This code of practice has developed from an industry standard to a more general legislative standard which is incorporated into the latest version of the Drinking Water Decree in the Netherlands (Ministerie van Infrastructuur en Milieu [I&M] 2011).

Examples of this total quality awareness are numerous; for instance the policy of most of the water companies in the Netherlands to keep the pipes under so-called "open coverage". That means that pipes are located in sidewalks or streets that are covered with a tiled pavement. If the open cover of the road is 'closed', usually by putting a layer of concrete of tarmac on the road, most companies will relocate the pipe to another open location. The main reason for this being so that the pipe remains easy to reach and easy to monitor for leakage.

Figure 14.8 shows some examples of the open coverage and the sandy soil that is the predominant soil condition in the Netherlands.

Figure 14.8 Pipes under open coverage and in sandy soft soils.

14.6 DISCUSSION

Research on discolouration of the water has led to new insights in the design and operation of distribution systems, contributing to the preservation of water quality and maintaining customer confidence. A new element was the introduction of the dynamics of water velocity as a significant factor in the presence of sediment in a network and the consequences for the aesthetic and possibly also microbiological water quality aspects.

Based on these results, classical patterns towards reliability of networks have been reconsidered. In general, looping of networks is considered to enhance the reliability of supply. For transport networks this is undoubtedly true, provided that valves are present and sufficiently operational. At the distribution level, however, the smallest unit that can be isolated, the valve section, is normative for the continuity of supply. Looping at this level is even counterproductive as it takes longer to isolate a part of a loop than a single branch: it takes closing of at least two valves as opposed to one valve for the branch. Many advantages can be attributed to the high velocity branched distribution networks that resulted from the research, such as shorter residence times and no accumulation of sediment. Lower cost investments in new networks are, however, the most powerful arguments. With the power of these arguments, it is shown that the paradigm of "looped networks = more reliable networks" is only valid for transport networks and not for distribution networks.

Drinking water is distributed without a disinfectant residual in a number of European countries (e.g. the Netherlands, Denmark, Germany and Switzerland). To some people in other countries this approach seems to feel like walking on a tightrope without a safety net. However, when well considered, adding a residual disinfectant may mask important information on the quality of the distribution process.

The formation of disinfection by-products is a reason to minimize the level of chlorine dosage as many water companies try to do. Maintaining chlorine residual however may mask the biological stability of the water. If the water is sufficiently stable, the chlorine level may not decrease much because there is probably not much to oxidize. If the water is not sufficiently stable, the chlorine will be consumed, which is a time and temperature dependent process. Both these parameters cannot be controlled fully in the network. Moreover, the residence times in networks may be underestimated, especially in the smaller District Meter Area (DMA) level networks. Residence time in a normal 500 connections large DMA in the Netherlands may be 24 hours and longer at some places (Blokker *et al.* 2009). Furthermore, the residual level of chlorine may not always be sufficiently effective to neutralize contamination by ingress as a result of low pressure in combination with apparent leakage, especially when the residual is minimal to limit formation of disinfection by-products or taste and odour problems.

As mentioned in this chapter, loose deposits in the network may also have significance in the microbial safety of the network. However, with the three-stage approach towards water quality, the possible microbiological significance of loose deposits can be neutralized. The essential point in chlorine-free distribution is "being in control". Most of what is described above is about 'being in control', based on the approach of total quality awareness and unbiased consideration of the processes in the network.

14.7 REFERENCES

Antoun E. N., Dyksen J. E. and Hiltebrand D. (1999). Unidirectional flushing: a powerful tool. *J. Am. Water Works Assoc.*, **91**(7), 62–71.

Antoun E. N., Tyson T. and Hiltebrand D. (1997). Unidirectional flushing: a remedy to water quality problems such as biologically mediated corrosion. AWWA Annual Conference 1997, Denver, CO.

Besner M.-C., Gauthier V., Barbeau B., Millette R., Chapleau R. and Prévost M. (2001). Understanding distribution system water quality. *J. Am. Water Works Assoc.*, **93**(7), 101–114.

Beuken R., Lavooij K., Bosch A. and Schaap P. (2006). Low leakage in the Netherlands confirmed. Water Distribution System Analysis #8. Cincinnati, Ohio, USA, American Society of Civil Engineers.

Blokker E. J. M., Beverloo H., Vogelaar A. J., Vreeburg J. H. G. and van Dijk J. C. (2011). A bottom-up approach of stochastic demand allocation in a hydraulic network model; a sensitivity study of model parameters. *J. Hydroinform.*, **13**(4), 714–728.

Blokker E. J. M., Schaap P. G. and Vreeburg J. H. G. (2009). Self-cleaning networks put to the test. Strategic Asset Management of Water Supply and Wastewater Infrastructures; Invited papers from the IWA Leading edge conference on strategic asset management (LESAM), Lisbon, October 2007. H. Alegre and M. d. C. Almeida: 400–417.

Blokker E. J. M., Vreeburg J. H. G., Vogelaar A. J. *et al.* (2006). Combining the probabilistic demand model SIMDEUM with a network model. Water Distribution System Analysis #8. Cincinnati, Ohio, USA, American Society of Civil Engineers.

Boxall J. B. and Saul A. J. (2005). Modeling discoloration in potable water distribution systems. *J. Environ. Eng.*, **131**(5), 716–725.

Boxall J. B., Skipworth P. J. and Paul A. J. (2003). Aggressive flushing for discolouration event mitigation in water distribution networks. *Water Sci. Technol.: Water Supply*, **3** (1–2), 179–186.

Brashear K. (1998). Distribution water quality problems created by upgraded water treatment plants. 1998 AWWA WQTC, San Diego, USA.

Carriere A., Gauthier V., Desjardins R. and Barbeau B. (2005). Evaluation of loose deposits in distribution systems through unidirectional flushing. *J. Am. Water Works Assoc.*, **97**(9), 82–92.

Eikebrokk B. (2000). Optimization of coagulation-filtration processes. (In Norwegian). Drinking Water Research towards the year 2000. Trondheim.

Flemming H. C. and Wingender J. (2001). Relevance of microbial extracellular polymeric substances (EPSs), Part I: structure and ecological aspects. *Water Sci. Tech.*, **6**(43), 1–8.

Friedman M. J., Martle K., Hill A., Holt D., Smith S., Ta T., Sherwin C., Hiltebrand D., Pommenrenk P., Hinedi Z. and Camper A. (2003). Establishing site-specific flushing velocities. AwwaRF & Kiwa WR, 6666 West Quincy Avenue, Denver, CO.

Gauthier V., Gérard B., Portal J.-M., Block J. C. and Gatel D. (1999). Organic matter as loose deposits in a drinking water distribution system. *Water Research*, **33**(4), 1014–1026.

Geudens P. J. J. G. (2010). Water Supply Statistics 2008. VEWIN. Rijswijk, The Netherlands.

I&M (2011). Drinking Water Decree. Staatsblad 293: 1–90. Ministry of Infrastructure and Environment. The Hague, The Netherlands.

LeChevallier M., Babcock T. M. and Lee R. J. (1987). Examination and characterization of distribution system biofilms. *Appl. Environ. Microbiol.*, **53**(12), 4014–4018.

Lipponen M. T. T., Suutari M. H. and Martikainen P. J. (2002). Occurrence of nitrifying bacteria and nitrification in Finnish drinking water distribution systems. *Water Res.*, **36**, 4319–4329.

Meerkerk M. A. and Kroesbergen J. (2010). Hygiënecode drinkwater; opslag transport en distributie. Nieuwegein, KWR.

Oberoi K. (1994). Distribution flushing program: Benefits and results. AWWA Annual Conference, New York, AwwaRF.

Richardt S. and Korth A. (2009). Concept for optimized flushing of potable water networks. Integrating Water Systems. J. M. Boxall and C. Boca Raton, CRC Press-Taylor & Francis Group, 411–414.

Rook J. J. (1974). Formation of haloforms during chlorination of natural waters. *Water Treat. Exam.*, **23**(2), 234–243.

Slaats N., Rosenthal L. P. M., Siegers W. G., van den Boomen M., Beuken R. H. S. and Vreeburg J. H. G. (2002). Processes Involved in Generation of Discolored Water, American Water Works Association Research Foundation/Kiwa, The Netherlands.

Smeets P. W. M. H., Medema G. J. and van Dijk J. C. (2009). The Dutch Secret: how to provide safe drinking water without chlorine in the Netherlands. *Drinking Water Eng. Sci.*, **2**(2), 1–14.

Snyder J. K. and Deb A. K. (2003). Impacts of Fire Flow on Distribution System Water Quality, Design, and Operation. AWWARF report 2554, AWWARF

van der Kooij D., Vrouwenvelder J. S. and Veenendaal H. R. (2003). Elucidation and control of biofilm formation processes in water treatment and distribution using the unified biofilm approach. *Water Sci. Tech.*, **47**(5), 83–90.

van Dijk J. C. and van der Kooij D. (2005). Water Quality 21 research programme for water supplies in The Netherlands. *Water Sci. Technol: Water Supply*, **4**(5–6), 181–188.

van Thienen P., Vreeburg J. H. G. and Blokker E. J. M. (2011). Radial transport processes as a precursor to particle deposition in drinking water distribution systems. *Water Res.*, **45** (4), 1807–1817.

Vloerbergh I. and Blokker M. (2010). Sharing failure data to gain insight into network deterioration. *Water Asset Manage. Int.*, **6**(2), 9–14.

Vreeburg J. H. G. (2007). Discolouration in Drinking Water Systems: A Particular Approach. PhD Thesis, TU Delft.

Vreeburg J. H. G. and Boxall J. B. (2007). Discolouration in potable water distribution systems: a review. *Water Res.*, **41**, 519–529.

Vreeburg J. H. G., Schaap P. and van Dijk J. C. (2004). Measuring discolouration risk: resuspension potential method. IWA Leading Edge Conference, Prague, IWA.

Vreeburg J., Schippers D., Verberk J. Q. and van Dijk J. C. (2008). Impact of particles on sediment accumulation in a drinking water distribution system. *Water Res.*, **42**(16), 4233–4242.

Vreeburg J. H. G., Arsénio A. and Leijssen H. (2009*a*). Origin and behaviour of particles in drinking water transport networks. CCWI 2009 'Integrating water systems', Sheffield, CRC Press.

Vreeburg J. H. G., Blokker E. J. M., Horst P. and van Dijk J. C. (2009*b*). Velocity based self cleaning residential drinking water distribution systems. *Water Sci. Technol*, **9**(6), 635–641.

Zacheus O. M., Lehtola M. J., Korhonen L. K. and Martikainen (2001). Soft deposits, the key site for microbial growth in drinking water distribution networks. *Water Res.*, **35**(7), 1757–1765.

Chapter 15

Research needs

D. van der Kooij and P. W. J. J. van der Wielen

15.1 INTRODUCTION

Multiplication of microorganisms in drinking-water distribution systems and in potable-water installations has been observed and investigated for more than a century, but is still a concern and a challenge to microbiologists and engineers involved with water supply. The problems caused by regrowth are recognized worldwide for systems distributing drinking water with or without a disinfectant residual. As described in previous chapters, much progress has been made in the past decades, in identifying the nature and causes of regrowth, as well as in developing control measures in treatment and distribution. The large amount of collected knowledge includes a variety of methods developed for the detection of microorganisms involved in regrowth, for assessing the microbial-growth potential of treated water and for studying biofilms. The diversity of methods for assessing specific parameters illustrates that many efforts have been made to clarify the nature and causes of the complex regrowth phenomena and also reflects regional differences in practices in water treatment and distribution. Scientific developments in the 20th century, for example principles of microbial growth kinetics and genetics, discovery of infectious microorganisms, chemical and physical methods for water quality analysis, and so on. also affected the progress made. Future scientific developments, for example related to molecular methods, most likely will also impact the nature of the research needed and conducted for the assessment and control of regrowth. Last, but not least, incorporation of water quality standards in regulations also is a major driving force for research aiming at defining measures to achieve compliance.

The presentations at the Joint Research Programme (Bedrijfstakonderzoek [BTO]) conference held at Santpoort, the Netherlands, on 11 and 12 May 2011 and the plenary session on research needs served as a basis for summarizing the issues related with microbial growth in drinking-water supplies that require further investigation. These issues depend on type of source water and water treatment and differ between water supplies within countries, and also between countries, due to effects of climate, traditions and regulation. To facilitate the classification of the research needs, a differentiation is made between the various aspects of regrowth, viz. (i) the impact of regrowth on drinking-water quality, (ii) the conditions affecting the nature and extent of microbial growth in distribution systems and in potable-water installations and (iii) control measures.

15.1.1 Ranking of regrowth problems

Microbial growth in distribution systems and potable-water installations can affect the quality of drinking water in several manners: (i) non-compliance with water quality criteria included in legislation; (ii) presence of opportunistic pathogens, (iii) aesthetic problems, for example, discoloured water, off-flavour, presence of invertebrates visible to the naked eye and (iv) technical problems, for example, corrosion of pipe material, pressure loss or leakages. These issues are summarized in Chapter 1 and have been addressed in more detail in the other chapters.

The primary task of water supply companies is to distribute safe drinking water, and for this purpose water quality standards have been defined in regulations. Noncompliance with these standards can result in consumers becoming ill, can have a negative impact on the reputation of the water supply company and might diminish consumer's confidence. However, some parameters, that is, heterotrophic plate counts (HPCs), coliform bacteria (not *E. coli*) and also the aeromonads (included in regulations in the Netherlands), provide information about regrowth, and are not related to safety. Hence, non-compliance with the standards for HPCs, coliforms or *Aeromonas* generally is not considered a problem directly related to safety, but indicates that biological processes in the distribution system are insufficiently under control. However, as described in previous chapters, HPCs, coliforms and *Aeromonas* are not ideal parameters for regrowth assessment.

Opportunistic pathogenic microorganisms that are capable of growing in the drinking water distribution system have been observed in a wide variety of distribution systems and potable-water installations (see Chapters 6, 7, 8 and 11). The potential impact of these organisms on public health gives this issue the highest priority of the regrowth concerns. However, for most of these organisms specific criteria are not included in drinking water legislation. There are several reasons for the absence of criteria:

- the role of drinking water strains of opportunistic pathogens in disease has not been proven for all opportunistic pathogens that are detected in drinking water.

- there is a lack of standardized methods for the quantitative detection and identification of these microorganisms in water and in clinical material.
- in temperate regions, opportunistic pathogens have mainly been detected in potable-water installations, which are beyond the direct responsibility of the water supply companies.
- defining quantitative criteria for each opportunistic pathogen that has been observed in drinking water is not desirable. The use of one or more indicator parameters that correlate with these opportunistic pathogens in drinking water distribution systems may be more efficient.

Furthermore, the public-health significance of target organisms in a viable but nonculturable state (VBNC) is unclear.

Aesthetic problems related to regrowth, in particular discoloured water, the presence of animal invertebrates visible to the naked eye and off-flavours, do not directly affect the well-being of the consumer, but can reduce consumer satisfaction and confidence. This in turn might lead to media attention and reputation damage for the drinking water companies.

Microorganisms can enhance corrosion of certain metallic materials, and corroded iron pipes enhance regrowth in the distribution system (Chapter 4). Clogging of water meters by invertebrate animals is another problem related to growth of invertebrates. Research needs for invertebrates in the drinking water ecosystem will be discussed in relation to the aesthetic problems.

15.2 OPPORTUNISTIC PATHOGENS

15.2.1 Identity and priority

A variety of opportunistic pathogens, including bacteria, fungi and free-living protozoa, are capable of multiplication in the drinking water distribution system and/or in potable-water installations. The pathogens of concern in the Netherlands have been identified and prioritized based on a literature survey (Chapter 7). Differences between climate regions and countries regarding water temperature, water treatment and water composition affect the nature and numbers of the opportunistic pathogens in drinking water. The results of the survey described in Chapter 7 can serve as a starting point for the identification and prioritization of opportunistic pathogens in different regions on the globe.

15.2.2 Detection

Detection methods are needed to specifically quantify the selected organisms in drinking water, biofilms and sediments. For some of these opportunistic pathogens methods are available (Chapters 6 and 7), but for other organisms (e.g. *Burkholderia cepacia* complex, *Afipia* spp., *Bosea* spp., *Simkania negevensis*) methods for detection and quantification need to be developed.

Methods currently available for the specific detection of opportunistic pathogens can be assigned into two categories, that is, cultivation-based methods and molecular methods. Cultivation-based methods use (semi)selective (agar) media, incubated at a specific temperature, followed by confirmation tests for a certain organism or group of organisms. For *L. pneumophila*, nontuberculous mycobacteria (NTM) and *Pseudomonas aeruginosa* such (semi)selective agar media in combination with confirmation tests have been published and validated with water samples (ISO 16266; Neumann *et al.* 1997). The advantages of using selective agar media are (i) quantification of the opportunistic pathogen is based on cells that are alive and metabolically active and (ii) organisms can be isolated to determine the physiological and genetic characteristics of the strains. These characteristics can be used to compare (drinking) water strains with clinical strains. Disadvantages of using selective agar media include (i) the incubation time can be quite long (1 to 12 weeks), (ii) growth of disturbing microflora on the medium can yield false-negative results and (iii) organisms that are alive, but (transiently) unable to grow on an agar medium, are not detected.

Molecular methods are based on the detection of DNA or RNA and a variety of methods have been developed. However, the quantitative polymerase chain reaction (qPCR) can be considered the standard molecular method for the detection of opportunistic pathogens in water. qPCR methods are available for *L. pneumophila*, NTM, *Mycobacterium avium*, fungi, *Aspergillus fumigatus*, *P. aeruginosa*, *Stenotrophomonas maltophilia* and *Acanthamoeba* (Qvarnstrom *et al.* 2006; Valster *et al.* 2011; Wullings *et al.* 2011; Chapter 7). The major advantages of using qPCR assays compared to cultivation-based methods are the short analysis time (<8 hours) and the specificity of qPCR assays. The main disadvantage of qPCR assays is that the detected DNA is present in both dead and viable cells. As a result, qPCR data cannot be used to conclude that viable opportunistic pathogens, capable of infecting humans, are present in drinking water. Still, the results can be used to decide whether further investigations or measures are needed. Also fluorescent in situ hybridization (FISH) methods can be used for the detection of the target organisms. However, their application is limited because the method has the same disadvantage as qPCR (it detects both dead and viable cells), has in general a higher detection limit and is more time-consuming than qPCR. The large genetic diversity within species of opportunistic pathogens also complicates the interpretation of a positive test result. Multilocus sequencing of *L. pneumophila* has resulted in the detection of more than 1000 sequence types (ST). A number of these STs have been associated with Legionnaires' disease, some of which are rarely observed in the environment. STs which are more frequently observed in environmental samples, have not, or only rarely, been isolated from patients (Harrison *et al.* 2009; Kozak *et al.* 2009; Chapter 6). Variations in virulence are also observed within species of other genera, for example *P. aeruginosa* (Campbell *et al.*

2000; Potvin *et al.* 2003). Hence, detection of microorganisms belonging to a species identified as an opportunistic pathogen does not prove the presence of the virulent genotype(s), but is (at least) a warning signal for environmental conditions promoting the growth of an undesired species. Further refinement of qPCR methods is needed to detect the (most) virulent genotypes of the involved species.

Differentiation between viable and dead cells of opportunistic pathogens is important for the interpretation of the test result, for example to assess the health risk or the effect of a control measure. Viability stains in combination with cell counting or qPCR and detection of mRNA (Heijnen and Medema 2005; Berney *et al.* 2008; Nocker and Camper 2009) might overcome the problem with viable and dead cells (see also Chapter 8). However, further testing and validation of these methods with water sampled from drinking-water supplies is needed. In addition, preliminary results of these techniques indicate that the 'holy grail' for discriminating viable from dead cells has not been developed yet, but is the subject of intense research.

Recently, next generation sequencing has been proposed as a tool to identify opportunistic pathogens of concern in drinking water (Hong *et al.* 2010). However, the resolution of next generation sequencing methods is currently too low to identify microorganisms to the species level, although this will probably improve in the near future. Currently, a variety of methods is available for the quantitative detection of a number of opportunistic pathogens. However, differences between methods used for the detection of a specific organism hamper comparison and interpretation of the results.

Future research on methods for the detection of opportunistic pathogens should, therefore, focus on:

- designing of qPCR methods for the specific detection of the (most) virulent genotypes;
- further testing of current concepts to discriminate viable and dead cells of opportunistic pathogens in drinking water;
- development of new technologies and/or concepts to determine the viability of opportunistic pathogens present in drinking water;
- harmonization, validation and standardization of methods to improve (inter) national comparison and interpretation of the observations.

15.2.3 Surveys

A large variety of opportunistic pathogenic organisms has been observed in drinking water in several countries (Chapters 6, 7 and 8). However, quantitative information about the presence of these organisms in drinking-water supplies and potable-water installations in different types of drinking water, and in different seasons and climates, is only scarcely available for some organisms (e.g. *Legionella,*

Mycobacterium spp.) and lacking for others. This lack of information hampers the assessment of the public-health significance of the presence of these organisms in water supplies as well as their control. Additional research in this field is needed and should focus on the following aspects:

- obtaining quantitative information about the occurrence of selected opportunistic pathogenic organisms in different types of drinking-water supplies in different seasons and climates;
- determining possible microbial health risks associated with drinking water, based on quantitative information.

Surveys for opportunistic pathogens in drinking water distribution and in plumbing systems should be deployed at locations with an elevated risk for regrowth. The results from such surveys can also be used to:

- identify conditions (e.g. temperature, microbial-growth potential, pipe material, and plumbing systems) that support growth of these organisms in the drinking water ecosystem;
- identify correlations with other (microbiological) water quality parameters, for example HPC, adenosine triphosphate (ATP), total direct cell count (TDC), *Aeromonas*, dissolved organic carbon (DOC);
- investigate if water quality parameters can be used as indicator parameters for opportunistic pathogens in drinking water distribution systems.

15.2.4 Public-health significance

The public health significance of opportunistic pathogens observed in drinking water distribution and premise plumbing systems is still largely unclear. Further research should therefore focus on:

- Demonstrating an epidemiological link between drinking-water strains of opportunistic pathogens and disease cases by:
  - development of selective cultivation methods that can be used to cultivate the organism of interest from drinking water and patients;
  - development of integrated genotyping methods for opportunistic pathogens using the latest techniques (e.g. multi-locus sequence typing (MLST), genome sequencing);
  - genotypic characterization of drinking water strains and comparison with genotypes of clinical strains.
- The public health significance of opportunistic pathogens present in a viable but nonculturable (VBNC) state (Chapter 8).
- Quantitative microbial risk assessment (QMRA).

The quantitative data about opportunistic pathogens in drinking water distribution and plumbing systems as obtained from surveys at different drinking water locations, with or without relation to cases of disease, can be used as a starting point for QMRA. However, additional data are required to perform QMRA:

- quantitative data on exposure routes from drinking water (e.g. showering, drinking, skin contact) to patients for each opportunistic pathogen should be available.
- dose response data for each opportunistic pathogen must be available. Currently, dose response data are either (i) lacking for many of the opportunistic pathogens or (ii) have been obtained from studies with animals or healthy humans and/or (iii) are not based on the use of the most virulent genotype(s). How available dose-response data relate to immunocompromised humans is at present unknown.

The results from epidemiological surveys and QMRA can subsequently be used to define guideline criteria for opportunistic pathogenic organisms that are related to public health and to develop a monitoring programme for the drinking water distribution system and premise plumbing installations.

15.2.5 Control measures

Regrowth-control strategies include (i) maintenance of a disinfectant residual and/or (ii) reducing the microbial-growth potential of treated water and/or (iii) replacement of materials and/or (iv) cleaning of the distribution system (see Sections 14.3–14.7). Prevention or limiting the growth of opportunistic pathogens in drinking-water supplies and in potable-water installations therefore requires information about the conditions promoting their growth. Effects of certain conditions (easily assimilable organic carbon (AOC) level, pipe material, sediments, temperature) on the growth of some opportunistic pathogens (*L. pneumophila*, *M. avium*) have been demonstrated (Falkinham *et al.* 2001; Torvinen *et al.* 2004; van der Kooij *et al.* 2005; Valster *et al.* 2011; Wullings *et al.* 2011), but are not known for other opportunistic pathogens. Research in this field should focus on:

- determining the quantitative and qualitative role of biofilm and/or sediment on growth of opportunistic pathogens. In these studies, the presence and behaviour of free-living protozoa, that serve as host for these pathogens, should be included.
- assessment of the effects of environmental conditions, for example, temperature, AOC, disinfectant residual, pipe material, plumbing system, on the presence of opportunistic pathogens in biofilms and sediments.

15.3 OTHER REGROWTH PROBLEMS

15.3.1 Coliforms and aeromonads

Non-compliance with standards for coliforms (and/or for aeromonads in the Netherlands) is a problem in a number of water supplies. Elevated AOC concentrations, corrosion, sediments and a temperature above 15°C have been identified as conditions promoting the growth of coliforms in distribution systems (Chapters 2 and 4). The presence of coliforms therefore indicates elevated levels of biofilms and sediments, possibly with anaerobic niches, for example, in corroding pipes (Chapter 4). Growth of aeromonads is also associated with biofilm formation (Chapter 11), most likely at (much) lower concentrations than needed for growth of coliforms. The value of coliforms and aeromonads as indicators for regrowth can be increased by:

- establishment of quantitative relationships between the numbers of coliforms and/or aeromonads in drinking water and the concentration and composition of biofilm and sediment;
- establishment of quantitative relationships between coliforms or aeromonads and parameters that directly relate to regrowth problems (e.g. opportunistic pathogens, off-flavour compounds or microorganisms associated with off-flavour and invertebrate animals).

15.3.2 Organisms involved in taste and odour problems

Taste and odour problems of drinking water related to microbial growth have gained much attention in the past (Chapter 1). Major causes for regrowth-associated off-flavour problems are (i) formation of trichloroanisoles by actinomycetes and fungi from chlorinated phenols in chlorinated supplies, (ii) formation of (in) organic sulphides in anaerobic niches in distribution systems and (iii) formation of geosmin or 2-methylisoborneol by actinomycetes. The microorganisms involved in these problems are difficult to identify or cannot be detected with culture methods.

Assessment of the significance of microbially-induced off-flavour problems requires information about (i) the identity and presence of the organisms producing the off-flavour-causing compounds, (ii) identification of the environmental conditions promoting the multiplication of these organisms and (iii) effects of control measures. The research needs include:

- identification of the microorganisms involved with microbially-induced off-flavour problems, based on literature surveys and analysis of the microbial community at locations with off-flavour problems;
- development of qPCR methods for the quantitative detection of the organisms of concern (e.g. fungi, actinomycetes, and other microorganisms);
- surveys on the occurrence of these organisms in a variety of drinking water supplies, with different levels of regrowth, to identify their role in odour and taste problems in drinking water;
- tracing of hot spots of odour-causing organisms, as they are usually not uniformly distributed in the biofilms of the system;

- assessment of the quantitative and qualitative role of biofilms, sediments, temperature, disinfectant and/or oxygen on growth of the microorganisms producing flavour-affecting compounds, so that control measures can be defined.

15.3.3 Invertebrates

A large variety of invertebrates can proliferate and/or accumulate in distribution systems (Chapter 9). These organisms do not have direct health significance, but consumers associate observations of these organisms with unhygienic conditions. As this association most likely is false in the majority of water supplies complying with the microbiological water quality standards, communication about the presence of (micro)organisms in drinking water supplies should be improved, for example 'drinking water is a natural product and is not sterile'. As soon as invertebrates become a nuisance, however, there is a need for control measures. Invertebrates are part of the food web in distribution systems, feeding on microbial biofilms and sediments. Reducing the formation of biofilms and sediments therefore will limit the abundance of invertebrates.

Methods for collecting and counting invertebrates, as well as monitoring schemes, have been developed and correlations between numbers of invertebrates and water quality parameters have been established (Chapter 9). Still, the conditions favouring the local accumulation of (certain) invertebrates are not sufficiently understood to define efficient preventive and/or control measures. Studies of the ecology are needed to identify the causes of the variability in abundance and diversities within and between water supplies. Furthermore, determination of the numbers and types of organisms using microscopy is laborious, requires a skilled person and yields only estimations of the amount of biomass represented by these organisms. Also, it is not clear to what level invertebrate abundance should be limited and to what extent identification is needed. Consequently, research efforts should be directed at the quantification of the biomass amount of invertebrates. For this purpose, gauze filters with different mesh sizes can be used to collect the organisms from tap water and from water flushed from the mains. Subsequently, the amount of biomass collected can be determined as ATP, Total Organic Carbon, carbohydrates or proteins (see Section 15.4). Finally, molecular methods might also be useful to detect environmental DNA of invertebrates in water, as has been shown for endangered vertebrates in surface water (Thomsen *et al.* 2011). The following research activities are proposed:

- optimization and standardization of methods for quantification of invertebrate biomass collected by gauze filters of different mesh sizes. Such method(s) should enable differentiation between large animals visible to the naked eye, and the total amount of invertebrate biomass.
- evaluation of the use of molecular methods for detection of specific invertebrates (e.g. Asellidae and Oligochaeta) in drinking water supplies.

- surveys in different types of drinking water supplies to establish relationships between the invertebrate biomass and (i) microscopic observations, (ii) concentrations of biofilms and sediments, (iii) other environmental conditions, primarily in distribution systems and (iv) to obtain a database of the amount of invertebrate biomass present in drinking water supplies.
- developing an ecological model, including hydraulics, of invertebrate variability in distribution systems, focusing on invertebrates that may be a nuisance, that is Asellidae and Oligochaeta.
- definition of guideline values for the evaluation of results, development of efficient monitoring strategies and definition of control measures based on the relationship between invertebrates and the concentration of biofilms and sediments.
- optimization of methods for the removal of invertebrates, for example ice pigging or rinsing of affected parts of the system with carbonated water in problem situations.

15.3.4 Technical problems

The impact of corrosion on microbial growth has been shown in many studies (Chapters 1 and 4). However, the mechanism and importance of microbially-influenced corrosion (MIC), also called biologically-enhanced corrosion, in drinking-water distribution systems (cast iron) and potable-water installations (copper) are still not resolved, due to a lack of information about the identities and properties of the responsible microorganisms. With the use of molecular methods, the organisms involved in MIC may be identified. The investigations should aim at:

- characterization of the microbial community in biofilms on corroding surfaces;
- identification of the microorganisms of interest;
- design and application of qPCR methods to quantify the presence of these organisms in relation to MIC;
- determination of the presence of these organisms in biofilms and sediments observed in corroding pipes;
- elucidation of the interactions between these microorganisms and corrosion;
- assessment of the quantitative and qualitative role of biofilm and sediment on MIC;
- development of strategies to suppress the growth of MIC-involved bacteria and to prevent, mitigate or sanitize MIC.

15.3.5 Schematic overview

Figure 15.1 gives a schematic overview of the research needs for the described regrowth problems. This figure shows that the structure of the research needed

Research phase	Type of research	1	2	3	4	5
A	Identification and prioritization of the organism(s) of concern by: (i) survey of literature and (ii) microbial community analysis	■	□	■	□	■
B	Development of quantitative detection methods for defined species (and genotypes of opportunistic pathogens); standardization of methods	■	□	■	■	■
C	Surveys in distribution systems and connected potable-water installations of a variety of water supplies with different levels of regrowth. Also collection of data about additional water quality parameters (HPC, TDC, ATP, DOC)	■	□	■	■	■
D	The concentration of the organism(s) in relation to the concentration and composition of biofilms and sediments in water supplies, model systems and pilot plants Effects of environmental conditions (e.g. temperature, disinfectant, oxygen, material) on the growth of the organism(s) in biofilms and sediments	■	■	■	■	■
E	Develop strategies for implementation of research results into practice to solve problems	■	■	■	■	■

Figure 15.1 Schematic of main research needs for regrowth of (micro)organisms creating water quality problems; 1, opportunistic pathogens; 2, coliforms and aeromonads; 3, microorganisms affecting taste and odour; 4, invertebrates and 5, microbially-induced corrosion. Grey box: research needed; open box, no research needs identified.

for elucidation of the nature, impact and causes of a specific problem are similar for all problems. A key point is the quantitative and qualitative information about biofilm and sediments in relation with the concentration of the organisms that cause problems. The information obtained with this research strategy shows the ecological niches of the organisms that cause problems and might result in guideline values for biofilm and sediment concentration.

15.4 ASSESSMENT OF REGROWTH

Several microbiological parameters included in drinking-water quality regulations (HPCs, coliforms, aeromonads) provide information about whether or not regrowth occurs in the distribution system, but not about the total amount of bacterial biomass (Chapters 1, 10, 11). Therefore, regrowth assessment requires quantitative information about the total concentration of biomass in water and deposits in the distribution system. Biomass suspended in drinking water, attached biomass (biofilms) and biomass in sediments can be both the result of regrowth and a cause of regrowth. For the quantification and characterization of biomass, appropriate methods are needed.

15.4.1 Biomass

Assessment and monitoring of the level of microbial regrowth requires a method for quantification of the biomass of all active bacteria. The following research activities are proposed to achieve this goal:

- selection of (an) appropriate method(s), for example, ATP, total cell counts, concentration of DNA;
- harmonization and standardization of the selected method(s);
- surveys of a variety of drinking water types with the selected method(s), and comparison with traditional and other generally accepted HPC methods (e.g. spread plate counts on R2A medium);
- development of schemes for routine monitoring of drinking-water supplies in different climate zones based on the selected method(s).

As has been described in Chapter 11, the use of ATP as a parameter for active biomass in water, biofilms and sediments can lead to the construction of a framework of quantitative data that facilitates the interpretation of the observations.

Information about the total concentration of biomass (active and dead) in water, biofilms and sediments is needed for the assessment of (i) the biofouling potential of treated water and (ii) the level of biofouling of the distribution system and the potable-water installations. Total biomass includes (components of) live and dead bacteria, archaea and eukaryotes (algae, protozoa, fungi and invertebrates). Quantification of iron and manganese associated with biomass in biofilms and sediments should be included, because these compounds may contribute to discoloured water (Chapter 14). The following research activities are proposed:

- selection, optimization and validation of methods to isolate and concentrate biomass present in water (e.g. membrane filtration, centrifugation);
- selection, optimization and validation of methods for determination of the amount of biomass (carbohydrates, protein, DNA) in water, biofilms and sediments;
- surveys in a variety of drinking-water supplies (treated water and water in distribution systems) and connected potable-water installations to identify the concentration ranges and relationships with water quality problems (regrowth, turbidity, discolouration).

15.4.2 Detection of microorganisms with specific physiological properties

The involvement of certain (groups of) compounds in regrowth is still largely unknown. Quantitative detection of microorganisms utilizing specific compounds as a source of energy by culture is either not possible or complicated and incomplete (e.g. nitrifying microorganisms, sulphate-reducing bacteria). The use of molecular methods enables the detection of microorganisms with specific physiological characteristics in water treatment processes and in distribution systems. Research should focus on:

- detection of methane-oxidizing bacteria, iron- and manganese-oxidizing bacteria, sulphate-reducing bacteria and archaea, ammonia-oxidizing bacteria and archaea, nitrite-oxidizing bacteria, actinomycetes and bacteria with specific nutritional requirements or biodegradation capabilities (e.g. utilization of certain polysaccharides);
- detection of specific free-living amoeba and ciliates, which may be used as an indicator for (protozoan activity in) biofilms and sediments.

15.5 MICROBIAL-GROWTH POTENTIAL

15.5.1 Driving force for regrowth

The microbial-growth potential of treated water reflects the amount of energy available in the present biodegradable compounds, both organic and inorganic, which is the main driving force for microbial growth in distribution systems and in potable-water installations. However, the extent of utilization of the available energy source(s) in the distribution system depends on (i) the nature and the concentration of the biodegradable compounds, (ii) the residence (contact) time, (iii) material in contact with drinking water and (iv) other environmental conditions (e.g. temperature, disinfectant). Assessment of regrowth is based on measurements in water, but most microorganisms are present in biofilms and sediments. Defining regrowth control strategies therefore requires information about the relationship between the microbial-growth potential of treated water and the formation and accumulation of biomass in distribution systems.

15.5.2 Assessment of the microbial-growth potential of water

A variety of methods is available for determining the microbial-growth potential of treated water (Chapters 2, 10, 11). Two main approaches can be distinguished, that is, measuring the production of biomass (AOC-related tests) and measuring the decrease of the DOC concentration (biodegradable dissolved organic carbon (BDOC)-related tests). Furthermore, the methods include static (batch) tests and dynamic (flow-through) tests and a combination may be needed for a proper assessment of the microbial-growth potential of treated water. Evaluation, harmonization and standardization of such methods are needed. Research should focus on:

- detection limits, reproducibility and costs as critical performance aspects for methods to determine the microbial-growth potential, including selection, standardization and validation. This research should include a comparison of AOC-related and BDOC-related methods in a number of water supplies in different countries.
- evaluation of test methods for determining the concentration of growth-limiting inorganic nutrients (P, N), with emphasis on the required detection limit and reproducibility.
- application of methods for natural organic matter (NOM) characterization in combination with the method(s) selected for assessing the microbial-growth potential in water.

15.5.3 Biofilms and sediments

The interpretation of studies on relationships between the microbial-growth potential and regrowth is hampered by differences between methods used for measuring the level of regrowth, as well as the microbial-growth potential. Regrowth observed in drinking water can arise from growth in the water and/or detachment or re-entrainment of microorganisms from biofilms and sediments in the distribution system. Information about the microbial-growth potential and the concentration and composition of these deposits is lacking. Studies with improved methods should focus on:

- establishment of the relationship(s) between the microbial-growth potential of treated water, the level of regrowth in drinking water and the accumulation of biofilms and sediments;

- the impact of environmental conditions (e.g. hydraulics, retention time, (corrosion of) pipe material, disinfectant residual) on the (local) accumulation of biofilms and sediments in distribution systems.

Establishing relationships between the microbial-growth potential of treated water and the accumulation of biofilms and sediments in water supplies and potable-water installations in full-scale systems is complicated. Therefore, model systems, for example annular reactor, biofilm monitor, and/or pilot plants may be used in the laboratory, and connected with selected supplies to establish these relationships under well-defined conditions.

15.5.4 Enhancement of microbial growth by materials in contact with drinking water

Corroding iron can enhance biofilm formation and sediment accumulation (Chapter 4). Furthermore, release of growth-promoting compounds by synthetic materials in contact with drinking water can also enhance microbial growth. Several standardized methods for assessing the enhancement of microbial growth by materials and chemicals in contact with drinking water are available (Chapter 12), but differences between these methods complicate the definition of internationally-acceptable pass-fail criteria. Further research should, therefore, focus on the development of one internationally-acceptable test method to determine enhancement of microbial growth by materials and the definition of pass-fail criteria. Such pass-fail criteria might be based on the enhancement of growth of opportunistic pathogens by the material (see Section 15.2).

15.5.5 Guideline values

To support the evaluation of the effects of water treatment and distribution on the involved regrowth parameters, guideline values are needed for:

- the concentration of active bacterial biomass in finished and distributed drinking water;
- the total concentration of biomass in finished drinking water, and in biofilms and sediments in the distribution system;
- the microbial-growth potential of finished drinking water.

Guideline values may differ for different climate regions and/or for water with or without disinfectant residual.

15.5.6 Overview

Figure 15.2 gives a schematic overview of the research issues related to the assessment of regrowth and the microbial-growth potential of water in the distribution system.

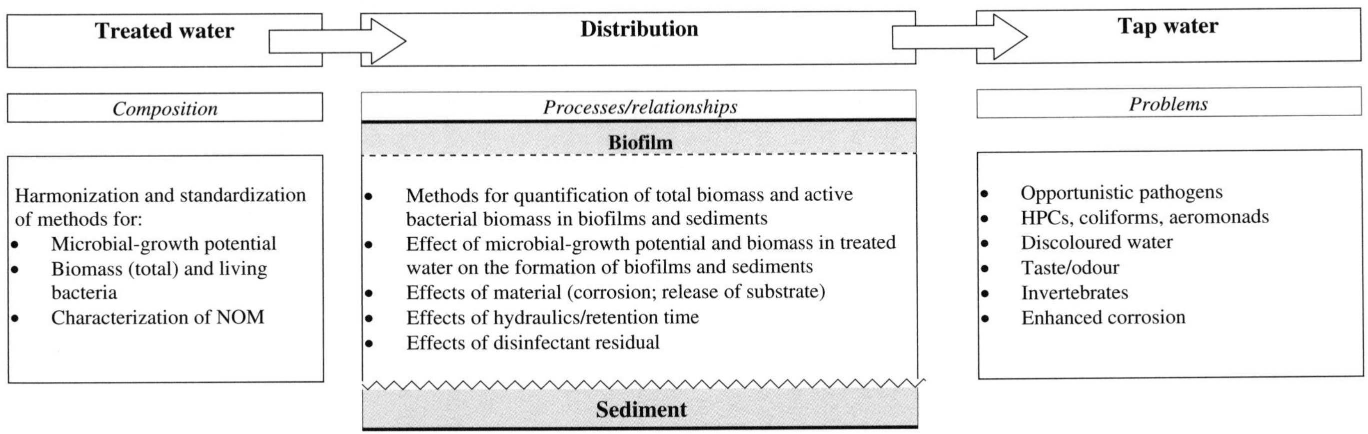

Figure 15.2 Schematic overview of research issues related to the assessment of regrowth and microbial-growth potential in distribution systems and potable-water installations.

15.6 EFFECTS OF WATER TREATMENT ON THE MICROBIAL-GROWTH POTENTIAL

15.6.1 Developments in drinking water treatment

Water treatment includes barriers for the removal of chemical contaminants, for example, pesticides and pharmaceuticals, compounds affecting taste and odour, pathogenic microorganisms, inorganic contaminants and constituents of natural origin, that is ammonia, iron, manganese and calcium and natural organic matter (NOM). Physico-chemical treatment processes either enhance the removal of growth-promoting substances or result in the formation of easily biodegradable compounds and/or the stimulation of biological processes in water treatment. Biological filtration processes (sand, dual media, granular activated carbon (GAC)) play a central role in many water treatment systems in Europe and their importance for reducing the microbial-growth potential of water is also recognized in Northern America (Chapters 2 and 13). The continuing developments in water treatment technologies and the need to control regrowth require a thorough analysis of the effects of water treatment processes on biostability.

15.6.2 Removal of biodegradable compounds and biomass

The effect of water treatment processes on the concentration of suspended biomass and the effects of operational changes and changes in water composition on the performance of processes related to the microbial-growth potential require further attention. The following research issues are identified:

- the effect of biofiltration characteristics (empty bed contact time, filtration rate, size and nature of filter bed material, intensity and frequency of backwashing, and temperature) on the total concentration of particles, biomass and microbial-growth potential in the filtrate;
- the effects of physico-chemical processes in pre-treatment and post treatment, for example, following biofiltration, or filtration without intended microbial activity, on the total concentration of particles, biomass and biodegradable compounds;
- the impact of changing operational conditions and/or raw water quality on the performance of treatment processes. Assessment of these effects requires monitoring of the concentration of microbial biomass and the microbial-growth potential, chemical analysis and in-line monitoring of a number of water quality parameters, for example, turbidity, particle counts, and also the rate of biomass accumulation in the continuous biofouling monitor.

A framework of data about effects of unit processes on the microbial-growth potential and the biomass concentration would facilitate (i) the design or optimization of water treatment and (ii) the evaluation of a water treatment system without detailed monitoring of the microbial-growth potential, (iii) the optimization of the process in terms of efficacy and economy.

15.7 EFFECT OF DISTRIBUTION ON BIOMASS ACCUMULATION

15.7.1 Developments in water distribution

Delivering drinking water that complies with all quality criteria in sufficient quantity to the consumers is the main obligation of the water supply companies. Preventing growth of opportunistic pathogens and a high aesthetic quality are important objectives to protect public health and maintain consumers' confidence. Apart from water supplies in a number of countries in Northern and Central Europe, the application of a disinfectant residual is common practice across the globe. However, in distribution systems with a disinfectant residual in the water additional measures are also needed to ensure safety (Chapter 5) and to control regrowth (see Chapters 2 and 4).

15.7.2 Impact of distribution system conditions

15.7.2.1 Hydraulics and residence time

Changing existing systems, for example, by replacement of materials, reduction of residence time, increase of flow rates, may lead to a reduction of regrowth (Chapter 14), but such measures usually are expensive. Furthermore, cleaning is labour intensive and may not have the desired effect, particularly when water quality does not meet criteria for regrowth. Selection of the most cost-effective approach in distribution for limiting regrowth therefore requires knowledge about the anticipated effects of the potential measures. The following research activities are proposed:

- identify knowledge gaps about the effects of distribution system conditions (e.g. residence time, hydraulics, pipe materials) and operational measures (type and frequency of cleaning) on the formation of biofilms and sediments, based on a survey of the literature. The development of a model that predicts regrowth in the distribution system may be helpful both for identification of the most critical conditions and for prioritizing further research efforts.
- elucidate the effects of the distribution system conditions and operational measures on regrowth to enable timely intervention. Tools and strategies (parameters, locations, duration) to monitor water-quality related regrowth phenomena have to be developed.
- determine the effects of measures within distribution systems (e.g. self-cleaning network (see Chapter 14), flushing) on the level of regrowth observed in drinking water and the amount and composition (microbiological and chemical) of biofilms and sediments.

15.7.2.2 Disinfectant residual

Maintaining a disinfectant residual is used to ensure the microbiological safety of drinking water, including aspects related to faecal contamination (Chapters 2, 4, 5). Maintaining a residual for safety purposes (Chapters 2 and 4) may require precursor (NOM) reduction in water treatment. In many supplies in the USA and in a few European supplies, monochloramine is applied. This disinfectant is less reactive than free chlorine and enables maintaining a residual in the peripheral parts of the distribution system (see Chapter 6). However, monochloramine decay is accelerated in climates with warm temperatures, causing in the development of a nitrifying microbial community, which results in formation of biomass from CO_2. Information is needed about:

- the effect of disinfectant usage on the accumulation and the composition of deposits in parts of the distribution system where the residual is depleted;
- the effect of maintaining a disinfectant residual to prevent microbial growth in potable-water systems, on the quality (e.g. disinfectant by-products, corrosion, taste and odour) of drinking water;
- water quality parameters and materials promoting the growth of disinfectant-resistant microorganisms, including nontuberculous mycobacteria, in the presence of a disinfectant residual.

15.7.3 Potable-water installations

Potable-water installations are a critical part of the pathway of finished water from the treatment plant to the consumer's tap. Characteristics of these potable-water installations include: (i) the surface to volume ratio is (more than) ten times higher than in the distribution system, (ii) water temperature may increase, in particular when cold water pipes are close to warm water pipes, (iii) water may stagnate during weekends and holidays, (iv) a wide range of materials is used in newly-constructed systems, and (v) limited monitoring in comparison to public networks. Consequently, quality changes may occur in potable-water installations in buildings and family homes, for example growth of *Legionella* and other opportunistic pathogens. The composition of the supplied drinking water plays a role in microbial quality changes in plumbing systems and water supply companies should have information about (the most critical aspects of) these effects. A number of research issues have been identified (see also Section 15.3):

- the effect of drinking water quality (including the microbial-growth potential) on the formation of biomass and sediments in potable-water installations;
- the disinfectant residual in delivered drinking water required to prevent growth of *L. pneumophila* or other opportunistic pathogens in potable-water systems.

Assessment and control of conditions in existing installations is complicated, as is collecting representative samples. The use of a generally accepted and standardized test system would be very helpful to study the effects of water quality, temperature, hydraulic conditions and materials on microbial growth.

These studies should lead to guideline values for water quality and eventual revision of criteria for design and maintenance of potable-water installations. Furthermore, the effect of drinking water quality (disinfectant residual, microbial-growth potential) on microbial growth in potable-water systems and the responsibility of the building owner/contractors for the quality of the water coming from this installation should lead to establishing a cooperative approach to prevent water quality deterioration.

15.8 COSTS AND BENEFITS

Research on the effects of regrowth on water quality and on the effects of water treatment and distribution on regrowth should lead to measures that are beneficial for the consumers and/or the involved water supply. The costs of preventive and/or curative measures will be much higher than the costs of research, in particular when research is based on a structured approach and conducted as a cooperative effort. Elucidation of the balance between the costs and benefits of potential control measures should be part of the research efforts. As explained above, public health aspects are the most important issue in relation to regrowth, followed by quality aspects affecting the appreciation of the water quality by the consumer. Therefore, the costs of preventive and curative measures should be evaluated in terms of public-health protection and maintenance of consumers' confidence in distributed drinking water.

Acknowledgement

The chapter is based on the presentations at the BTO Conference and the contributions of the participants to the discussions in the Research Needs session of this conference. The authors thank Frans L. Schulting for chairing this session and Anne K. Camper, Hans–Curt Flemming and members of the BTO Programme Committee Microbiology for their comments made on the draft.

15.9 REFERENCES

Berney M., Vital M., Hulshoff I., Weilenmann H. U., Egli T. and Hammes F. (2008). Rapid, cultivation-independent assessment of microbial viability in drinking water. *Wat. Res.*, **42**, 4010–4018.

Campbell M., Mahentiralingam E. and Speert D. (2000). Evaluation of random amplified polymorphic DNA typing of *Pseudomonas aeruginosa*. *J. Clin. Microbiol.*, **38**(12), 4614–4615.

Falkinham J. O. III, Norton C. D. and LeChevallier M. W. (2001). Factors influencing numbers of *Mycobacterium avium*, *Mycobacterium intracellulare*, and other mycobacteria in drinking water distribution systems. *Appl. Environ. Microbiol.*, **67**, 1225–1231.

Harrison T. G., Afshar B., Doshi N., Fry N. K. and Lee J. V. (2009). Distribution of *Legionella pneumophila* antibody subgroups and DNA sequence types in recent clinical and environmental isolates from England and Wales. *Eur. J. Microbiol. Infect. Dis.*, **28**, 781–791.

Heijnen L. and Medema G. J. (2005). Quantitative detection of *E. coli*/*E. coli* O157 and other shiga toxin producing *E. coli* in water samples using a culture method combined with real-time PCR. *J. Wat. Health*, **4**, 487–498.

Hong P. Y., Hwang C., Ling F., Andersen G. L., LeChevallier M. W. and Liu W. T. (2010). Pyrosequencing analysis of bacterial biofilm communities in water meters of a drinking water distribution system. *Appl. Environ. Microbiol.*, **76**, 5631–5635.

Kozak N. A., Benson R. F., Brown E., Alexander N. T., Taylor T. H., Shelton B. G. and Fields B. S. (2009). Distribution of *lag-1* alleles and sequence-based types among *Legionella pneumophila* serogroup 1 clinical and environmental isolates in the United States. *J. Clin. Microbiol.*, **47**(8), 2525–2535.

Neumann M., Schulze-Robbecke R., Hagenau C. and Behringer K. (1997). Comparison of methods for isolation of mycobacteria from water. *Appl. Environ. Microbiol.*, **63**, 547–552.

Nocker A. and Camper A. K. (2009). Novel approaches toward preferential detection of viable cells using nucleic acid amplification techniques. *FEMS Microbiol. Lett.*, **291**, 137–142.

Potvin E., Lehoux D. E., Kukavica-Ibrulj I., Richard K. L., Sanschagrin F., Lau G. W. and Levesque R. C. (2003). In vivo functional genomics of *Pseudomonas aeruginosa* for high-throughput screening of new virulence factors and antibacterial targets. *Environ. Microbiol.*, **5**, 1294–1308.

Qvarnstrom Y., Visvesvara G. S., Sriram R. and da Silva A. J. (2006). Multiplex real-time PCR assay for simultaneous detection of *Acanthamoeba* spp., *Balamuthia mandrillaris*, and *Naegleria fowleri*. *J. Clin. Microbiol.*, **44**, 3589–3595.

Thomsen P. F., Kielgast J., Iversen L. L., Wiuf C., Rasmussen M., Gilbert M. T., Orlando L. and Willerslev E. (2011). Monitoring endangered freshwater biodiversity using environmental DNA. *Mol. Ecol.*, **21**, 2565–2573.

Torvinen E., Suomalainen S., Lehtola M. J., Miettinen I. T., Zacheus O., Paulin L., Katila M. L. and Martikainen P. J. (2004). Mycobacteria in water and loose deposits of drinking water distribution systems in Finland. *Appl. Environ. Microbiol.*, **70**(4), 1973–1981.

Valster R. M., Wullings B. A., van den Berg R. and van der Kooij D. (2011). Relationships between free-living protozoa, cultivable *Legionella* spp., and water quality characteristics in three drinking water supplies in the Caribbean. *Appl. Environ. Microbiol.*, **77**, 7321–7328.

van der Kooij D., Veenendaal H. R. and Scheffer W. J. (2005). Biofilm formation and multiplication of *Legionella* in a model warm water system with pipes of copper, stainless steel and cross-linked polyethylene. *Wat. Res.*, **39**, 2789–2798.

Wullings B. A., Bakker G. and van der Kooij D. (2011). Concentration and diversity of uncultured *Legionella* spp. in two unchlorinated drinking water supplies with different concentrations of natural organic matter. *Appl. Environ. Microbiol.*, **77**(3), 634–641.

Index

£ 129.00 CI 305 8-10-2013